ELECTRICALLY INDUCED VORTICAL FLOWS

MECHANICS OF FLUIDS AND TRANSPORT PROCESSES
Editors: R. J. Moreau and G. Æ. Oravas

V. BOJAREVIČS, J. A. FREIBERGS, E. I. SHILOVA,
and E. V. SHCHERBININ

*Institute of Physics, Latvian S.S.R. Academy of Sciences,
Riga, Salaspils, U.S.S.R.*

Electrically Induced Vortical Flows

Kluwer Academic Publishers

Dordrecht / Boston / London

Library of Congress Cataloging in Publication Data

```
Ėlektrovikhrevye techenii͡a. English.
   Electrically induced vortical flows / by V. Bojarevičs ... [et
al.] ; edited by E.V. Shcherbinin ; translated and edited by
V. Bojarevičs.
      p.  cm. -- (Mechanics of fluids and transport processes ; 9)
   Translation of: Ėlektrovikhrevye techenii͡a.
   Bibliography: p.
   Includes index.
   ISBN-13: 978-94-010-7017-1       e-ISBN-13: 978-94-009-1163-5
   DOI: 10.1007/978-94-009-1163-5
   1. Vortex-motion.  2. Magnetohydrodynamics.  3. Eddy currents
(Electric)   I. Boi͡arevich, V. V. (Valdis Vladislavovich)
II. Shcherbinin, Ė. V. (Ėduard Vasil'evich)   III. Title.
IV. Series.  V. Series: Mechanics of fluids and transport processes
; v. 9.
QC159.E4413 1988
532'.0595--dc19                                            88-8823
                                                              CIP
```

Originally published under the title 'Elektro-vikhrevye techenija'
Translated by V. Bojarevičs

Published by Kluwer Academic Publishers,
P.O. Box 17, 3300 AA Dordrecht, The Netherlands

Kluwer Academic Publishers incorporates the publishing programmes
of D. Reidel, Martinus Nijhoff, Dr W. Junk, and MTP Press.

Sold and distributed in the U.S.A. and Canada
by Kluwer Academic Publishers,
101 Philip Drive, Norwell, MA 02061, U.S.A.

In all other countries, sold and distributed
by Kluwer Academic Publishers Group,
P.O. Box 322, 3300 AH Dordrecht, The Netherlands

All Rights Reserved
© 1989 by Kluwer Academic Publishers
Softcover reprint of the hardcover 1st edition 1989

No part of the material protected by this copyright notice may be reproduced or
utilized in any form or by any means, electronic or mechanical,
including photocopying, recording or by any information storage and
retrieval system, without written permission from the copyright owner

Contents

Preface		ix
Introduction		1
Chapter 1 / Basic Properties of Axially Symmetric Motions in Magnetohydrodynamics		18
1.1.	Magnetohydrodynamic Equations	18
1.2.	Some Facts about Orthogonal Curvilinear Coordinates	21
1.3.	Differential Operators in Orthogonal Curvilinear Coordinates	24
1.4.	The Most Commonly Used Rotational Coordinate Systems	25
1.5.	Axisymmetric Motions	29
1.6.	Relation between Stokes Stream Function and Self-Magnetic Field of an Electric Current in Problems with Axial Symmetry	32
1.7.	Feasible Schemes for Axially Symmetric Electric Current Distributions	35
1.8.	Magnetic Field in Axisymmetric Flow	41
1.9.	Electric Field in Axisymmetric Flow	55
1.10.	Full Set of Equations for Axisymmetric Motion	57
Chapter 2 / Solutions in Spherical Coordinates		62
2.1.	Definition of the Class of Exact Solutions	63
2.2.	Low Magnetic Reynolds Number Approximation. Electric Current and External Magnetic Fields	67
2.3.	Integral Flow Characteristics and Dimensionless Criteria	72
2.4.	Review of the Class of Exact Solutions in Spherical Coordinates	74
2.5.	Electrically Induced Vortex Flow in a Cone	84
2.6.	Gas Flow in an Electrical Arc	91
2.7.	Problems of the Nonlinear Solution	95
2.8.	Landau-Squire Flows in the Presence of a Radially Diverging Electric Current	99

2.9. Effect of the Induced Electric Current on the Flow at a Point Current Source — 104
2.10. Electrically Induced Flows at Finite Size Electrodes — 107

Chapter 3 / Electrically Induced Vortex Flow at a Point Electrode and Azimuthal Rotation — 120

3.1. Integral Properties of the Flows Driven by Rotational Electromagnetic Forces — 121
3.2. A Model Demonstrating the Effect of Viscosity — 123
3.3. Flow at an Immersed Electrode — 125
3.4. Asymptotic Solution for High S — 128
3.5. Electrically Induced Flow with Differential Rotation — 133
3.6. Growth of Azimuthal Disturbance in the Electrically Induced Flow at a Point Electrode — 136
3.7. Intensification of Rotation in a Closed Volume — 143
3.8. Mechanism of Rotation Intensification in an Axisymmetric Vortex — 148

Chapter 4 / Flows with Cylindrical Symmetry — 154

4.1. External Electric Current and Magnetic Field in Cylindrical Coordinates — 154
4.2. Similarity Solutions — 155
4.3. Electrically Induced Flow between Two Parallel Walls — 162
4.4. Flow with Line Source in a Circular Cone — 169
4.5. Magnetohydrodynamic Model of Tornado — 174
4.6. EVF in a Cylindrical Container — 184
4.7. Effect of Electric Current Configuration on Flow in a Cylindrical Container — 190

Chapter 5 / Periodic Electrically Induced Flows — 196

5.1. Periodic Distributions of Current and Magnetic Field in Cylindrical Coordinates — 196
5.2. Integral Action of Electromagnetic Force — 198
5.3. A Method Used to Construct Linear Solution of Periodic EVF in Tubes — 203
5.4. EVF in a Tube with Radial Current Supply — 209
5.5. EVF in a Tube with Longitudinal Electric Current — 216
5.6. EVF in an Annular Tube — 224
5.7. Periodic EVF in a Longitudinal Magnetic Field — 228
5.8. Longitudinal Magnetic Field Effect on Integral Features of EVF — 233
5.9. Nonlinear Interaction of Periodic EVF with Through-Flow — 236
5.10. Electrically Induced Flow in a Loosely Coiled Tube — 240

Contents   vii

Chapter 6 / Bodies in a Current-Carrying Fluid   245

6.1. Effect of Potential Forces on a Body in a Current-Carrying Fluid   245
6.2. Effect of the Rotational Electromagnetic Forces on Axisymmetric Bodies 249
6.3. Flow at a Stationary Sphere   257
6.4. Drag of a Sphere in the Flow of Current-Carrying Fluid   259
6.5. Flows at Spheroids   266
6.6. Discharge between Electrodes of Hyperboloidal Form   269
6.7. Flow at a Cone with an Electric Current Source in the Apex   274
6.8. Motion of a Sphere with a Current-Source   277

Chapter 7 / Heat and Mass Transfer in Electrically Induced Vortical Flows 282

7.1. Equations of Heat and Mass Transfer, and the Nondimensional Numbers   282
7.2. Mass Transfer from a Stationary Spherical Particle in Current-Carrying Fluid 287
7.3. Mass Transfer from a Translating Spherical Particle in a Current-Carrying Fluid 291
7.4. Mass Transfer from a Stationary Sphere in a Longitudinal Magnetic Field 294
7.5. Heat and Mass Transfer in a Cylindrical Container   296
7.6. Thermal Convection in Electrically Induced Flows   307

Chapter 8 / Experimental Investigations of EVF and Applications   311

8.1. Electroslag Welding   311
8.2. Electroslag Remelting   324
8.3. Electric-Arc Furnaces   333
8.4. Hydrodynamics of Furnaces with Multiple Electrodes   335
8.5. Electrical Jet Thrusters   341
8.6. Induction Channel Furnaces   346
8.7. Electrically Induced Flows in a Flat Layer between Ferromagnetic Masses 352
8.8. Electrolytic Aluminium Production   358

References   365

Index   379

Preface

Every scientific subject probably conceals unexplored or little investigated strata, which may show up at the proper time when favourable conditions coincide (practical demands, a circle of scientists prepared to recognize the novelty and capable of giving impetus to the development of a new theory, etc.). Something like this occurred in early seventies for magnetohydrodynamics, which at the time was considered to be a relatively complete branch of hydrodynamics with no apparent broad, unexplored areas.

It was unexpectedly realized that, in addition to the traditional methods of affecting an electrically conducting medium, there is yet another way, one which subsequently lead to a new direction in magnetohydrodynamics. In the Soviet scientific literature this direction has been termed 'electrically induced vortex flows', the essence of which are hydrodynamic effects due to the interaction of an electric current passing through the fluid with its own magnetic field.

It cannot be said that this direction was created *ex nihilo*: individual studies related to the flows driven in a current-carrying medium in the absence of external magnetic fields appeared in the sixties; in the thirties the flows themselves were known to take place within electrical arcs; and yet the first observations on the behaviour of liquid current-carrying conductors were made at the beginning of this century.

The growing number of publications in the seventies, of course, could not be explained by professional interest alone. The subject was 'lucky' (and this significantly stimulated the development of the investigations), because it was obvious from the very beginning that the studies were closely related to high-current technologies. Moreover, the theory of electrically induced flows had to be developed, since it was necessary for electro-technology, in particular, for electrometallurgy.

Electrically induced flows have evolved in a separate direction, which is not only due to the specific electromagnetic force generation; the distinguishing feature of these flows is the rotational character of the driving electromagnetic force, i.e. the flows can be induced even with homogeneous (zero) boundary conditions for the velocity field.

A situation in which an electric current density is nonuniform in space (if we restrict ourselves to uniform material properties) is typical of the rotational electromagnetic force. Hence, the electrically induced flows should be investigated, first of all, as spatial flows.

It is well known in general hydrodynamics that the solution of the equations for spatial nonlinear viscous flows is a difficult problem. Even the contemporary achievements in numerical techniques and computer design have not resolved the problem of reliable solutions for spatial flows. The study of electrically induced spatial flows therefore necessarily involve certain simplifications, the most significant of which is the assumption of axial symmetry for all the quantities governing and characterizing the fluid flow.

The assumption of axial symmetry for physical fields, on the one hand, retains a spatial distribution while, on the other these quantities are described with reference to only two spatial coordinates. Of course, the transformation from the three-dimensional description of electrically induced flows to the two-dimensional one significantly increases the amount of information which can be extracted by analytical and numerical solutions of this specific class of flow equations.

The assumption of axial symmetry made it necessary to consider the general properties of axisymmetric velocity, electric current, magnetic and electric fields. One of the properties is the possibility of representing the solenoidal vector fields by their meridional and azimuthal components. The meridional components can be expressed by introducing a common concept of stream function (hydrodynamic, electric, magnetic), which permits us to describe each of the vector fields by the use of two scalar quantities: the relevant stream function and the azimuthal component of the vector field. This reduces the number of functions to be determined.

In general hydrodynamics different methods are used to simplify the full Navier-Stokes equations. The same simplifications may be used in the study of electrically induced flows. Thus, many qualitative features of the electrically induced flows can be explained on the basis of the Stokes approximation, linearizing the equations of motion. However, the nonlinear terms in the equation of motion, which complicate the solution substantially, could lead to fundamentally new results that are unattainable in the linear statement of the problem. As an example of the nonlinear effect we may mention the nonlinear interaction of meridional electrically induced flows with an imposed azimuthal rotation leading to an increase of the rotation energy and also to a significant change in the meridional flow configuration.

From among the approximations that are totally unacceptable for the characterisation of electrically induced flows we shall mention only the following two. First, the approximation of quasi-rigid motion widely applied in engineering MHD is absolutely meaningless. The electrically induced flow problems are primarily hydrodynamic problems. Moreover, in spite of the fact that the electromagnetic force drives the flow and determines the flow configuration, the integral flow characteristics may be in conflict with the integral electromagnetic force. As an example, consider the periodic electrically induced flows in tubes,

Preface

where the electrically induced flow rate is directed oppositely to the integral electromagnetic force.

The second doubtful approximation is the assumption of inviscid fluid flow. Such an approximation is unacceptable since the rotational electromagnetic force constantly generates a vorticity in the fluid which, to avoid an infinite growth, must be balanced by viscous or turbulent dissipation. In the absence of dissipation, as was shown by Shercliff in a particular problem, the velocity field inevitably contains a singularity.

As far as purely magnetohydrodynamic approximations are concerned, an electrodynamic approximation is widely used in the book. This approximation neglects not only the magnetic field deformation by the fluid motion (which is the basis of the widely adopted noninduction approximation in MHD), but also that electric current induced by the fluid motion. In other words, the electromagnetic force is constructed by taking into account only the electric current supplied by an external source of e.m.f. and its self-magnetic field. Yet if an electrically induced flow takes place in an additional external magnetic field, then the additional electromagnetic force is assumed to result from the interaction of this field with the unperturbed electric current. Such approximation is motivated by a significant reduction of computational efforts without the loss of generality.

The authors were primarily interested in the scope of physical effects occurring within a current-carrying fluid. Therefore we have tried to give a physical interpretation to each result obtained, which, of course, do not exclude other interpretations. We are also sure that the theory of electrically induced flows contains more unsolved problems than solved ones. Several of these are mentioned in this book.

The book has been written on the basis of the monograph *Electro-vortex Flows* (in Russian) by the same authors, published in 1985, publishing house 'Zinātne', Riga (USSR). A significant part of the contents (and practically all the experimental material) is composed from the studies by the authors and their colleagues in the Laboratory of Magnetic Hydraulics, the Institute of Physics, Latvian SSR Academy of Sciences. Equally important contribution to the development of electrically induced flow theory have been made by a great number of scientists from different countries (mostly from England and the USA). The English edition of the book was prepared by E. V. Shcherbinin and V. Bojarevičs; the latter, in addition, translated the book in English (and he is totally responsible for the imperfect English in this book).

As far as it is known to the authors, this book is the first monographic publication in English to present the fundamentals of electrically induced flow theory and to classify the material concerning electrically induced flows.

We are aware that there will be reasons enough for criticism, and we would greatly appreciate any constructive comments concerning both the contents of the book and the exposition of the material.

THE AUTHORS

Introduction

Physical principles of electrically induced vortical flows

The magnetic hydrodynamics of an incompressible fluid, in addition to the usual hydrodynamic forces (inertial, viscous, and pressure gradient), take into account the electromagnetic force, which, in fact, sets magnetohydrodynamics apart as an individual branch of general hydrodynamics. The inclusion of additional force is not only an efficient tool to control the flow, but also leads to a variety of purely hydrodynamic phenomena which are not realizable by the usual methods.

The electromagnetic force $\mathbf{f}_e = \mathbf{j} \times \mathbf{B}$, as follows from its definition, is generated in the fluid when the electric current $\mathbf{j}$ interacts with the magnetic field $\mathbf{B}$ in the same fluid volume. By the methods of force generation we may discern, first, the conductive method, in which the electric current passes into the fluid from an external source of e.m.f., and the fluid volume is placed in the external magnetic field associated with the electric currents in an external circuit. Various conductive MHD-devices work on this principle. The second method is based on the electric current being induced in the fluid by an external alternating magnetic field. The principle of the interaction of alternating electric current with the alternating magnetic field is the object of investigation by another branch of magnetohydrodynamics; induction MHD-devices operate on this principle. Finally, a large part of magnetohydrodynamic studies is related to an electrically conducting fluid flow in an external magnetic field. The electric current in the fluid is induced, according to Ohm's law $\mathbf{j} = \sigma(\mathbf{E} + \mathbf{v} \times \mathbf{B})$, by the interaction of an externally driven fluid flow with the magnetic field.

The physical basis of electrically induced flows is the electromagnetic force generated due to the interaction of the electric current, supplied from an external source of e.m.f., with the self-magnetic field. In this case the source of electromagnetic force is the electric current passing into the fluid. Therefore, the first step of the evaluation of the force field is a solution or specification of the distribution of current density $\mathbf{j}$ in the fluid volume and the supplying circuit. Generally speaking, a boundary value problem for the potential Φ of the electric field should be solved to find the distribution of $\mathbf{j}$. If the current induced by the fluid flow may be neglected, the problem of the steady current field determination is reduced to the solution of Laplace's equation $\nabla^2 \Phi = 0$ with boundary conditions corresponding to a specific situation.

In the case of plane or axisymmetric current distribution the problem is substantially simplified and reduced to the solution of a two-dimensional Laplace equation:

$$\nabla^2 \psi_1 = \frac{\partial^2 \psi_1}{\partial x^2} + \frac{\partial^2 \psi_1}{\partial y^2} = 0$$

in the plane case and

$$E^2 \psi_1 = r \left(\frac{\partial}{\partial z} \frac{1}{r} \frac{\partial \psi_1}{\partial z} + \frac{\partial}{\partial r} \frac{1}{r} \frac{\partial \psi_1}{\partial r} \right) = 0$$

in the axisymmetric case for the electric current stream function ψ_1. In these cases the conditions at the boundaries of the current-carrying region are sufficient to determine the current distribution uniquely. Moreover, it appears that the same equations describe irrotational fluid motions, and hence the description of electric current density distribution can be reduced to the well-established theory of irrotational flows (see, for instance, the monographs [12, 15]).

The next step in the construction of electromagnetic force is the evaluation of magnetic field induction (flux density) by the known current density distribution. The magnetic field of a continuously-distributed steady electric current $\mathbf{j}$ is usually evaluated according to the Biot-Savart law:

$$\mathbf{B} = \frac{\mu_0}{4\pi} \int_V \frac{\mathbf{j} \times \mathbf{R}}{R^3} \, d\tau,$$

where $\mathbf{R}$ is the radius-vector of the observation point (i.e. the point where the field is evaluated) drawn from the current element $\mathbf{j}\, d\tau$, and the integration is performed over the whole current-carrying volume. The Biot-Savart formula determines the magnetic field both outside the current-carrying volume and within it, yet a numerical computation within the volume requires a certain caution in dealing with the singularity when $R \to 0$ in the integral.

Sometimes, particularly for an axisymmetric current distribution, the integral form of Maxwell's equation is useful:

$$\frac{1}{\mu_0} \oint_L \mathbf{B} \, d\mathbf{l} = I,$$

according to which the circulation of the magnetic field vector around a closed contour is equal to the total current passing through the surface resting on the contour.

Introduction

In the general case the differential form of Maxwell's equation should be used:

$$\mathbf{j} = \frac{1}{\mu_0} \operatorname{curl} \mathbf{B}, \tag{I.1}$$

which, in general, requires a knowledge of conditions at the boundary of the current-carrying region. After the field $\mathbf{B}$ is found, the electromagnetic force $\mathbf{f}_e = \mathbf{j} \times \mathbf{B}$ can be evaluated at any point at which the current density $\mathbf{j}$ is specified.

The following important question concerns the nature of the electromagnetic force; specifically, whether it is potential (curl $\mathbf{f}_e = 0$) or rotational (curl $\mathbf{f}_e \neq 0$). Suppose, the medium is at rest in the absence of the electric current. Then the question of the potential or rotational nature of the electromagnetic force is equivalent to the question of whether the fluid will remain motionless or will it be set in motion if the electric current passes through it, for homogeneous boundary conditions of velocity field. Generally, the electromagnetic force can be represented as the superposition of a purely potential part and the remaining part, which may include the potential and rotational parts of the force:

$$\mathbf{f}_e = \mathbf{j} \times \mathbf{B} = \frac{1}{\mu_0} \operatorname{curl} \mathbf{B} \times \mathbf{B} = -\nabla \left(\frac{|\mathbf{B}|^2}{2\mu_0} \right) + \frac{1}{\mu_0} (\mathbf{B} \cdot \nabla)\mathbf{B}. \tag{I.2}$$

When the fluid motion is slow, the vorticity may be assumed to grow at a rate proportional to the curl of the driving force ($\nu\rho$ curl curl curl $\mathbf{v}$ = curl $\mathbf{f}_e$). If the force is potential, and the fluid was motionless prior to the force's application, i.e. if the vorticity were absent, then it would remain motionless during the whole time period of the force's action and zero boundary conditions for the velocity field. Formally the potential part of electromagnetic force can be included in the pressure gradient term. This means that this part of the force merely causes a pressure redistribution in the current-carrying volume.

Let us consider the conditions which make the electromagnetic force rotational, thereby initiating a motion in the current-carrying region. Let the electric current $j_x(y, z)$ pass along a cylindrical liquid conductor of arbitrary cross section (Fig. I.1). In this simplest situation the electric current induces a magnetic field (inside the conductor as well as outside it), the induction vector of which lies in the yz plane. In other words, the magnetic field has B_y and B_z components, and, in the light of (I.1), the magnetic field in an arbitrary point in space within the conductor is related to the current density in the same space point by the relation

$$j_x = \frac{1}{\mu_0} \left(\frac{\partial B_z}{\partial y} - \frac{\partial B_y}{\partial z} \right),$$

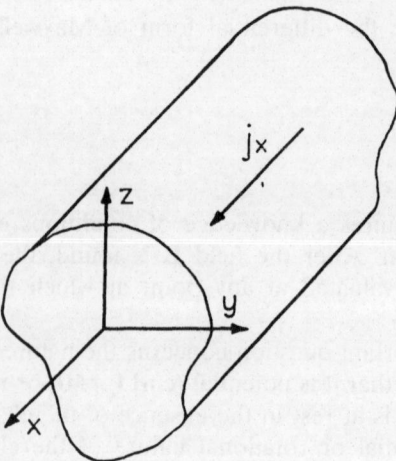

Fig. I.1. Electric current passing along a cylindrical conductor.

and beyond the conductor, where the current is absent, the magnetic field is governed by the equation

$$0 = \frac{\partial B_z}{\partial y} - \frac{\partial B_y}{\partial z}.$$

The electromagnetic force $\mathbf{f}_e = \mathbf{j} \times \mathbf{B}$, similarly to the magnetic field, has the components $f_y = -j_x B_z$ and $f_z = j_x B_y$. Now let us express $\operatorname{curl} \mathbf{f}_e$. Since all quantities are invariant along the length of the conductor, then among all the components of curl, only the x-component differs from zero:

$$(\operatorname{curl} \mathbf{f}_e)_x = \frac{\partial f_z}{\partial y} - \frac{\partial f_y}{\partial z} = B_z \frac{\partial j_x}{\partial z} + B_y \frac{\partial j_x}{\partial y} \tag{I.3}$$

(the magnetic field derivatives $\partial B_z/\partial z + \partial B_y/\partial y$ vanish due to the solenoidal condition div $\mathbf{B} = 0$). On the basis of expression (I.3) we can conclude that, if the electric current density is constant in the cross-section of the conductor, the right-hand side of (I.3) is zero, i.e. curl $\mathbf{f}_e = 0$, and the effect of electromagnetic force is reduced to a pressure redistribution over the conductor's cross section, and the fluid motion is absent. An example of this situation is the z-pinch effect.

If the electrical conductivity is nonuniform, due, for instance, to a nonuniform heating of the medium, the electromagnetic force may also become rotational for a unidirectional current distribution [6]. Henceforth we shall deal only with uniform conductivity, and the flows due to the nonuniformity will not be considered.

Now let the component of electric current along the conductor's length be absent. We assume that electrodes are embedded in the walls bounding the conductor (Fig. I.2). When the electrodes are set at different potentials, a

Introduction

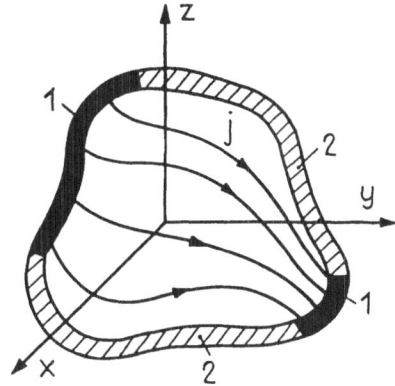

Fig. I.2. Scheme of a plane current flow: 1 — electrodes, 2 — insulating walls.

current with the components j_y and j_z passes in the conductor, setting up a longitudinal magnetic field B_x:

$$j_y = \frac{1}{\mu_0} \left(\frac{\partial B_x}{\partial z} - \frac{\partial B_z}{\partial x} \right) = \frac{1}{\mu_0} \frac{\partial B_x}{\partial z},$$

$$j_z = \frac{1}{\mu_0} \left(\frac{\partial B_y}{\partial x} - \frac{\partial B_x}{\partial y} \right) = -\frac{1}{\mu_0} \frac{\partial B_x}{\partial y}.$$

(I.4)

Here the conductor is assumed to be sufficiently long to make all the quantities invariant along the x coordinate. On evaluating $\mathbf{f}_e = \mathbf{j} \times \mathbf{B} = j_z B_x \mathbf{i}_y - j_y B_x \mathbf{i}_z$ we see that

$$(\operatorname{curl} \mathbf{f}_e)_x = \frac{\partial f_z}{\partial y} - \frac{\partial f_y}{\partial z} = 0$$

in view of the fact that div $\mathbf{j} = 0$ and (I.4).

Consequently, a plane current flowing across the long liquid conductor cannot drive a fluid flow. Yet if the conductor is of a finite length and the current passes, as previously stated, in the yz plane, then all three components of curl $\mathbf{f}_e$ may be different from zero. The final conclusion depends on the specific shape of the conductor and the current leads closing the current beyond the fluid region.

Consider now the most interesting case in which the electric current is nonuniform in space. In a simple situation, which, nevertheless, is quite instructive in the explanation of several effects, the current is supplied to a point electrode O situated at the centre of a nonconducting plane S, and then passes in the upper half-space along the spherical radii, i.e. the second electrode is a hemisphere of sufficiently large radius (Fig. I.3).

On assuming the radial current density to be equal for all points of a

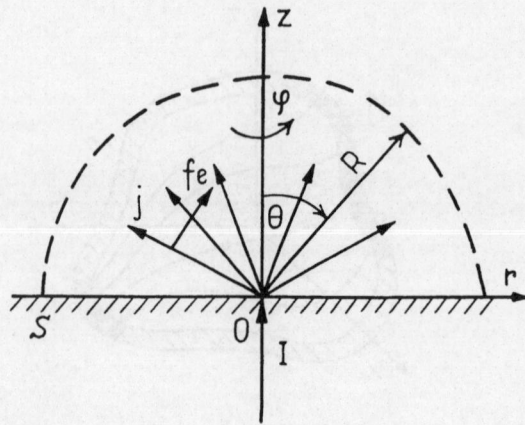

Fig. I.3. Scheme of the spatial electric current diverging from a point electrode.

hemisphere of radius R, we obtain a relation between the total current, the density j_R and the surface area of hemisphere $2\pi R^2$: $I = 2\pi j_R R^2$. Whence the current density $j_R = I/2\pi R^2$ decreases at a rate inverse to the square of distance from the point electrode.

From equation (I.1), written in spherical coordinates for the axisymmetric situation

$$j_R = \frac{1}{\mu_0}(\text{curl } \mathbf{B})_R = \frac{1}{\mu_0}\frac{1}{R^2 \sin\theta}\frac{\partial R \sin\theta \, B_\varphi}{\partial \theta},$$

it follows that the radial current induces an azimuthal magnetic field $B_\varphi = \mu_0 I(1 - \cos\theta)/2\pi R \sin\theta$.

The relevant electromagnetic force $\mathbf{f}_e$, normal to the vectors $\mathbf{j}$ and $\mathbf{B}$, is oriented in the direction of decreasing meridional angle θ (see Fig. I.3), and it is equal to

$$f_\theta = -\frac{\mu_0 I^2}{4\pi^2}\frac{1}{R^3}\frac{1 - \cos\theta}{\sin\theta}. \tag{I.5}$$

On evaluating the curl of force, we make sure that the azimuthal component of the curl is non-zero

$$(\text{curl } \mathbf{f}_e)_\varphi = \frac{\mu_0 I^2}{2\pi^2}\frac{1}{R^4}\frac{1 - \cos\theta}{\sin\theta} \neq 0, \tag{I.6}$$

i.e. the electromagnetic force is rotational, and a fluid motion should be induced by the passage of electric current.

The fluid flow character may be suggested *a priori*. According to (I.5), the force decreases as R^{-3} along the radius $\theta = $ const (Fig. I.4). Consider a fluid line element coinciding with the radius. If the fluid element is free to move in space, then under the action of the force it will move, first, as a whole along the principal vector of the parallel set of forces; second, it will rotate in such a way that the elements of the line closer to the origin O will move along the

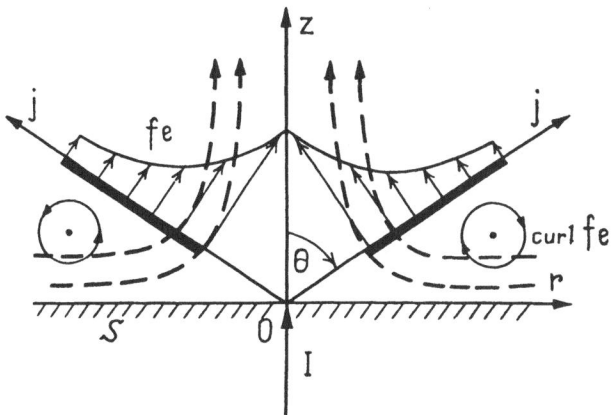

Fig. I.4. The electromagnetic force distribution along the radial current element and the expected steady flow stream lines (dashed lines).

symmetry axis, while the more distant ones will move to the plane S, and then along the plane to the symmetry axis. The motions of fluid induced by the passage of electric current in it we shall call 'electrically induced vortical flows' (EVF). The origin of the term is clear: these flows are induced, first, by an electric current in the absence of external magnetic fields and, second, they are initiated by the vortical nature of the electromagnetic force when the electric current interacts with the self-magnetic field.

Historical review

Some months ago, my friend, Carl Hering described to me a surprising and apparently new phenomenon which he had observed. He found, in passing a relatively large alternating current through a non-electrolytic, liquid conductor contained in a trough, that the liquid contracted in cross section and flowed uphill lengthwise of the trough, climbing up upon the electrode. With a further increase of current, he found that this contraction of cross section became so great at one point that a deep depression was formed in the liquid with steeply-inclined sides like the letter V. This depression extended in the case of a liquid metal as deep as six inches. With a current of constant value, the condition was a stable one, but the liquid on the inclined surfaces showed great agitation. With a still greater increase of current, the depression reaches the bottom of the trough, thereby rupturing the circuit; this, of course, resulted in the liquid flowing together again, and again breaking .... Mr. Herring suggested the idea that this contraction was probably due to the elastic action of the lines of magnetic force which encircle the conductor, which lines, he said, acted on the conductor like stretched rubber bands, tending to compress it, especially at its weakest point. As the action of the forces on the conductor is to squeeze or pinch it, he jocosely called it the 'pinch phenomenon'.

Thus Edwin F. Northrup began his paper 'Some Newly Observed Manifestations of Forces in the Interior of an Electric Conductor' [18] published in 1907 (this year might fix the date of the beginning of studies in current-carrying fluid hydrodynamics, though observations of the interaction of current-carrying mercury with floating wires in it could be traced to Ampère, Faraday, and their contemporaries). Northrup conducted the first systematic experimental investigations of the electromagnetic force effect on liquid conductors. In one of his experiments he filled a box of variable cross section with a liquid alloy of potassium and sodium in (Fig. I.5) and observed a drop of the liquid metal level in the narrow section of the conductor. The surface of the alloy was covered by an oil layer; therefore, probably, the organized motion of metal was not observed.

In the second experiment the current passed through a central circular tube 1 (Fig. I.6) of diameter $2R_2 = 2.54$ cm filled with liquid mercury. An insulating ring 2 of internal diameter $2R_1 = 1.27$ cm was inserted in the tube, and the openings 3 in the ring communicated the mercury in the central tube to an annular tube 4. One of the electrodes 5 has an opening 6. When the electric current was passed, the mercury flowed in a continuous stream from the opening 6 and circulated along the contour 6—4—3—1—6. With a current of 1800 A the mercury column above 6 was approximately 1.5 cm above the general level of the mercury in the annular tube.

Using the equations of magnetostatics, Northrup evaluated the pressure drop along the axis of conductor carrying a total current I:

$$p_{R_1} - p_{R_2} = \frac{\mu_0 I^2}{\pi^2}\left(\frac{1}{R_1^2} - \frac{1}{R_2^2}\right), \tag{I.7}$$

and explained the necessity of the sudden constriction, where the electromagnetic compression was higher than in the wide section, hence the mercury was expelled from the narrow section. Northrup also suggested a practical application of the 'pinch effect'; specifically, in devices built on the principle of summing up the relatively small pressure drops in the single elements similar to that in Fig. I.6.

The next step in the investigation of the motion of current-carrying media is associated with the advent and development of electric-arc welding. The first measurements of the momentum exerted by an electric arc on the electrodes

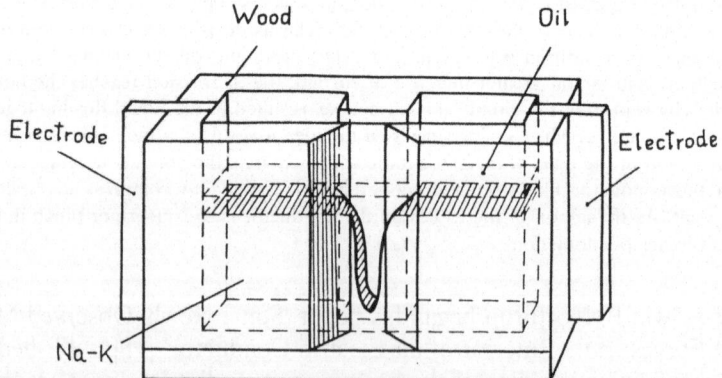

Fig. I.5. Scheme of the first Northrup experiment. (During the passage of current the level of liquid metal decreases in the narrow section.)

Introduction

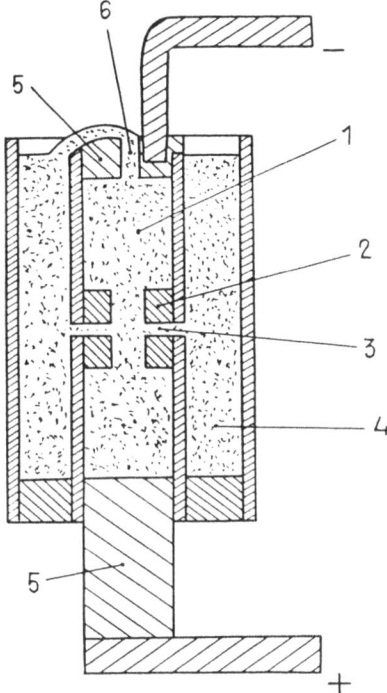

Fig. I.6. Scheme of the second Northrup experiment (See explanation in the text).

were made by Tauberg and Kobel in 1930, yet the greatest interest in the gas dynamics of the electric arc came in the mid-fifties (a review of these works may be found in [5]). From the investigations of this period we would note the experimental study by Wienecke [29] where, with the help of high-speed cinematography of the decay of a carbon arc, the velocity field was measured within the arc. It was found that the plasma jet had a high velocity: the maximum velocity close to the cathode reached up to 300 m/s for a current of 200 A.

Different hypotheses were proposed to explain such high velocities. The assumption of jet development due to a fast volatilization of the electrode's material was not confirmed experimentally. Up to the present time it has been accepted that the main driving force of plasma flow in an electric arc is the electromagnetic interaction of the arc current with the self-magnetic field [5, 10, 21]. In explaining the mechanism of flow development, Maecker's theory [14] gained ground; he applied Northrup's result (I.7) and proposed an expression for the estimation of the axial velocity v: $p_{max} = \mu_0 Ij/\pi = \rho v^2/2$, where j is the current density at the cathode. Unfortunately, this approach, based on the concept of different electromagnetic compression, is limited to the above expression. At the same time a growing interest was developing in the magnetostatic effects of current-carrying media in relation to the plasma confinement problem.

Zhigulev [31] was probably the first to note that the electromagnetic force in a spatially divergent electrical discharge could not be compensated by pressure forces and, hence, the discharge should drive the gas ejection. A year later Grozdovskij and co-workers [8] constructed an approximate solution for a conical discharge in a nonviscous fluid and, in 1962, Uberoi [28] emphasized that under certain conditions the electromagnetic force due to the interaction of electric current and its self-magnetic field was rotational and had to drive the fluid.

In the sixties the papers by Chow and co-workers appeared (see references to Chapter 6), on the electrically-induced flows at bodies, and also those by Smyslov and Reznikov (flow at front critical point), Pao (rotating disc), which could be related to electrically induced vortical flows (see references to Chapter 4). Note that these investigations were not in the main stream of magnetohydrodynamics of the time and were not widely known. This might explain why different investigators later independently came to the same conclusion as had been expressed by Zhigulev and Uberoi.

In 1969 Lundquist [13] published a Stokes solution of the problem represented schematically in Fig. I.3; he said, when visiting the Institute of Physics in Riga, that the problem was suggested by a situation where lightning was striking a metallic surface. A little later the same problem was studied by Shercliff [22], Sozou [26], Shilova and Shcherbinin [25], but on the basis of full equations of motion. The increase of the number of publications on the electrically induced flows (the term emerged in 1977) from this time on, could be explained, first, by the close relation of the problem considered of a point electric current source to a well-known class of exact solutions in hydrodynamics, and, second, by the unexpected problem of the nonlinear solutions, which led to a breakdown of the nonlinear viscous solution for the meridional electrically induced flow, even for a relatively small magnitude of the current passing through the fluid.

At the present time the theory of electrically induced flows is developing in several directions. Historically one of the first directions is associated with the EVF in tubes [28]. However, from the moment when the first studies were published by Uberoi, containing the description of toroidal vortices driven by a periodically disturbed electric current in a straight tube, 15 years had passed until the conditions were determined for which the periodic EVF gained a new quality, leading, in principle, to a practical application. These investigations were performed in the Institute of Physics, Latvian SSR Academy of Sciences, at the end of the seventies (see the review in Chapter 5).

The second adequately developed direction — electrically induced flows at bodies — is based on Chow's works [1], and has been extended by Oreper (see Chapter 7) who investigated the mass transfer at bodies in a current-carrying fluid.

A great number of studies are devoted to the flows at a small electrode, relative to the flow region. It is of interest to note that the investigations in this direction have led to a new insight in the theory of rotating vortices.

A separate line of investigation may be termed the magnetic control of electrically induced flows, which is of both practical importance for applications, and also involves new physical effects in a moving fluid.

Of course, the sphere of application of the electrically induced flow theory is not limited to the directions listed. We shall discuss this in greater detail in the following section.

Applications in science and technology

The relationship of current-carrying fluid motion and the hydrodynamic effects in it could be found to quite useful either in the solution of technological problems, or the development of new kinds of technological equipment. At the present time there are no publications that adequately cover the basic application aspects of the electrically induced flow theory although many authors, of course, have considered applications of their particular investigations. With the aim of filling this gap to a certain extent, we shall sketch several examples of practical applications (existing or proposed) of the theoretical results, and we express a hope that the electrically induced flows are steadily reaching the threshold at which the theory can be realized as an engineering science.

Electrometallurgy

The electrically induced flows are driven when an electric current passes through the fluid even in the absence of external magnetic fields. This explains the interest in this direction of magnetohydrodynamics — the high-current technological processes are in the first line of its 'users'.

The technology of using high electric current is widely employed in various industries, a typical example being electrometallurgy. There are several reasons for bringing electrotechnology into commercial practice. First, some metals are produced by an electrolytic process on the basis of their available natural resources. Second, smelting of metals in electrical furnaces permits us to obtain a high-quality metal and to integrate the smelting process with the necessary shaping of the ingots. Finally, considerations related to the protection of the environment are playing a not unimportant role.

At the present time there is a steady trend towards increasing the power of electrometallurgical equipment. For instance, the electrolysis cells for aluminium production employing currents of magnitude 50—100 kA are being replaced, or will be replaced, by cells of 180—280 kA; the power of induction channel furnaces is increasing from 300—500 kW up to 1000—5000 kW (depending on the specific metal). This tendency will, most likely, be maintained in the future. New factors in the processes are thereby becoming important, such as the increase in electromagnetic forces. Indeed, an increase of the electric current by an order of magnitude leads to a similar growth of the self-magnetic field. The resulting electromagnetic force, growing by two orders in magnitude, may significantly influence the character of the process in comparison with the previous moderate current. Hence, it is necessary to consider accurately the role of the electromagnetic forces and to operate these in order to damp any disadvantageous effects or to increase the efficiency of the technological process. Consider now two instructive examples.

The first example is electroslag welding. In this process a melting electrode is continuously fed ino the slag melt zone. If the rate of feed is high, the end of the electrode approaches the interface between the slag and the liquid metal, and the electrically induced jet in the slag causes a deep deformation of the interface. Thereby the ohmic heat generated in the slag is transported deep into the metallic pool, which leads to a radial growth of crystallites in the solidifying metal and gives rise to an axial weakness in the weld. This example illustrates a disadvantageous effect of electrically induced flow.

The example of metal smelting in a two-phase induction channel furnace will show how the specific features of EVF can be employed to increase the efficiency of technological processes. The active part of the furnace consists of two magnetic circuits, the primary windings of which are connected to a power source. The secondary windings are the short-circuited channels within the liquid metal, and the main part of the ohmic heat is generated within the channels. The problem facing metallurgists may be formulated in the following way: how can an intense transit flow of metal be generated in the channels in order to ensure an effective heat exchange between the heated metal in the channels and the cold metal in the bath of the furnace. In the case of traditional in-phase connection of the inductors, vortical flows are induced by the electric current in the zones of the side and central channel openings, and the pump-effects of these are mutually compensated. A proposal to connect the inductors in opposite phases has led to the elimination of electrically induced vortices in the central channel opening which, in turn, have ensured a steady transit flow due to the EVF in the side channel openings (see Section 8.5).

Note, electric current is an inseparable part of the technological processes discussed above, hence it is reasonable in such situations to seek an efficient application of the electrically induced flows to improve the technology.

Electric-arc technology

We have already mentioned electrically induced flows in electric arcs. We could add to what was previously said, that the electric arcs are widely applied in different technological operations (from manufacturing ultra-disperse powders to building up thin metallized coatings), and the hydrodynamics of the electric arc can crucially affect the course of technological processes.

MHD-technology

Production of composite materials by electromagnetic means is a rather new technological operation. A uniform composition of two metals of different density and electrical conductivity can be produced under terrestrial conditions if the crystallization takes place in crossed electric (E) and magnetic (B) fields. The idea of obtaining such materials is based on the expression of force acting

Introduction

in the crossed fields on a particle of density ρ_1, electrical conductivity σ_1, and volume V, surrounded by a fluid with the respective parameters ρ, σ:

$$F = g \left[\rho_1 - \rho \pm (\sigma_1 - \sigma) \frac{EB}{g} \right] V. \tag{I.8}$$

From this it follows that, with an appropriate variation of magnitude and direction of the fields, the particle or a drop of different metal can be made of a greater 'weight' or 'weightlessness'. Equating the effective 'densities' of the two metals during the crystallization process, one can obtain a uniform distribution of a dopant metal in the base metal. The same principle also allows us to separate particles of different density or electrical conductivity. The expression (I.8) is derived under the assumption that the particle does not disturb the uniformity of the electric current. Generally, it is possible merely for $\sigma_1 = \sigma$. Yet if $\sigma_1 \neq \sigma$, the current density varies in the vicinity of the particle, which immediately makes the electromagnetic force rotational and drives a vortical motion. The flow is partially electrically induced; the other flow component is generated by the interaction of the current perturbation with the external magnetic field [6].

The effect of vortical motion on the transport of particles significantly complicates the separation process. In addition, it must be taken into account that the vortical motion in the vicinity of the particles intensifies both heat- and mass-transfer. This factor should be kept in mind not only in relation to the separation process and production of composite materials, but also during alloy production in electrical furnaces.

The passage of electric current through the fluid changes the pressure distribution in it. Under the action of the pressure gradient electrically nonconducting particles travel to the lateral boundary of the current-carrying region. The particles of higher electrical conductivity than the fluid are transported in the opposite direction. If a group of particles is placed in a stream of current-carrying fluid, their individual hydrodynamic drags depend essentially on the electrical conductivities. These principles could be used to separate particles according to their conductivities; the same effects explain the deposition of impurities on the walls of current-carrying channels in electrical furnaces.

The electrically induced flows are usually not able to develop a high pressure drop, since even for currents of order 10 kA the magnitude of the representative magnetic field induction is of the order of 10^{-2} T for a typical size of the current-carrying region of 10^{-1} m. Hence, equipment employing the interaction of current and its self-magnetic field cannot compete with pumps using a strong external magnetic field if the purpose is to obtain a significant pressure drop. However, if the need arises for an intense motion in a limited fluid volume, where the hydraulic resistance is minimum, EVF is fully competitive (e.g. for mixing of melts). The electrically induced flows can be organized by a variety of techniques (different surface of electrodes, varying cross-sections of the current-carrying volume, etc.). Yet if, for instance, due to the hostility of the melt, the electrodes cannot come into contact with it, the current may be generated by induction. The quadratic dependence of electromagnetic force on the current ($f_e \sim I^2$) guarantees unaltered orientation of the force in any point of the volume.

Space technology. Accelerators

Astronautics is paradoxical. Anywhere it imposes, at the first glance, conflicting requirements. On the one hand, boosters developing a thrust force of hundreds and thousands of tons are needed to put spacecrafts in orbit. On the other hand, after spacecrafts are placed in orbit, as a rule, engines of low thrust, measured in milligrams or grams — in kilograms at most — are sufficient.

The most promising engines of low thrust — electrical jet thrusters (EJT). The main advantage of these engines — high outlet velocities — are unattainable by usual thermochemical engines.

These words, opening Morozov's book [17], could characterize the role and purpose of EJT.

The presence of an electrical arc is a common feature of all types of EJT. Its main purpose is heating for the first type of EJT, while for others it is the ionization of the working material. Finally, in the electromagnetic EJT (plasma-end-thrusters) the arc also generates the thrust force due to the arc current's interaction with the self-magnetic field. The electrically induced flow theory could be useful in the analysis of the performance of EJT, the more so in that the physical understanding of processes, even in the most intensively investigated types of plasma EJT, remains unclear in many respects.

The principle of generating a force by the interaction of an electric current with its self-magnetic field is also the basis for accelerators of small objects with masses of order 1 g—1 kg. It is remarkable that the methods used to accelerate such macroscopic projectiles are still based on chemical combustion, which has remained essentially unchanged since the time of the invention of gunpowder in ancient China. The estimates in [4] show that the projectiles can be accelerated by the self-magnetic field up to velocities of the order of 100 km/s. However, at the present time, velocities up to 10 km/s have been obtained for projectiles of masses of several grams, which is close to the velocity necessary to launch the objects into an Earth-orbital position [4].

The simplest electromagnetic accelerator — the railgun — consists of two rigid electrically conducting parallel rails, which are connected in an electrical circuit by a moving bridge — the projectile. An impulse current ~ 1 MA passes through one rail, the bridge, and the other rail; the bridge is thrust by the electromagnetic force over a millisecond time interval with an acceleration of $\sim 10^6$ g. Apart from the ultra-high current generation problem, another problem is the projectile's material disintegration due to the acceleration and (to an even higher degree) due to ohmic heating. The problem is resolved by replacing the bridge with a plasma discharge between the rails. Then the plasma pressure pushes the projectile, i.e. due to the electrically induced flow.

Another type of accelerator [27] is based on the z-pinch and plasma focus ideas: in a long tube ring electrodes are embedded, between each subsequent pair of which the arc discharge is initiated just at the moment when the rigid projectile reaches them. Passing the long row of discharges, the projectile is accelerated to a high velocity, and the pinch force holds it stably on the tube's axis. The electrically induced flows between the periodic electrodes here play the basic role, both accelerating the projectile and also stabilizing it. However, the solutions in a similar situation (see Chapter 5) show that the hydrodynamic effects can sometimes lead to unexpected results, viz., the projectile under certain conditions would move in the opposite direction to that expected on the basis of electrodynamic considerations.

Biology

Strong electric fields (of intensity 10^5 V/m) are employed in biology, for instance, in the separation of dead and living cells. A minute volume of water suspension of the cells having a conductivity σ of order $10^{-2}-10^{-3}$ $(\Omega\,\text{m})^{-1}$ is placed in a nonuniform electric field between two electrodes of different surface areas. The separation process is based on different permittivities (and also conductivities) of the particles and fluid. According to the dielectrophoresis theory, the particle within the fluid moves to one of the electrodes under the action of nonuniform electric field. However, several unpredicted effects were observed in the experiments [23]: the fluid was driven in two ring vortices at the needle electrode, thereby entraining the particles; the velocity of particles was an order of magnitude higher than that predicted; the particles were not deposited at the needle electrode, instead, after approaching it, they are repelled to the centre of the fluid volume. This kind of motion at the needle electrode was explained by the action of electrostatic forces in weakly conducting fluids ($\sigma = 10^{-8}-10^{-11}$ $(\Omega\,\text{m})^{-1}$) [30]. Yet for conductivities of the order $10^{-2}-10^{-3}$ the electrostatic forces should be lower than the electromagnetic forces, due to the respective currents. The velocities estimated for the electrically induced flows [24] are in both a qualitative and quantitative agreement with the experiment.

Geophysics and astrophysics

High electric currents exist not only in man-made conditions. A well-known example of natural electrical discharge is lightning, in the channel of which the impulse current reaches a magnitude of the order of 10^5 A. The role of electromagnetic forces is not sufficiently clear in this phenomenon, yet with a certain definiteness we could suggest the existence of electrically induced flows in the zones of electric current convergence to the narrow channel of discharge.

Plasma flows most likely play an essential role in certain kinds of natural discharges. Thus, the long-term stability of a globular lightning is probably associated with the formation of a stable spherical vortex structure, the energy of which is accumulated at the moment of initial discharge. This possibility is suggested by the analogy to the green coloured fire-ball that is sometimes generated when the electrical power circuit of a submarine engine is disrupted [3].

An intense electrical activity is observed in the funnel of destructive atmospheric vortices — tornadoes. According to the measurements of terrestrial magnetic field distortion in the vicinity of a passing tornado, it has been estimated that the electrical current coming down the tornado funnel is of the order of several hundreds of ampères. The experiments [20] have confirmed the possibility of sustaining slender vortices by the discharge between two electrodes. The channel of electrical discharge is stabilized in the presence of swirl, and a stable vortical structure is formed (a tornado vortex model based on EVF is discussed in Section 4.7, and the development of a rotating vortex in Chapter 3).

Electric currents in near-Earth space play a substantial role in a variety of processes in the magnetosphere. The shape of the magnetosphere itself depends on a large-scale magnetospheric current system. It is now accepted that the currents are associated with different changes in magnetospheric plasma flows. The system of currents consists of the following main components [19]: ionospheric currents aligned with the Earth's magnetic field (or Birkeland currents), ring currents, and currents at the boundary and tail of the magnetosphere, which all are mutually connected in a complex three-dimensional circuit. The system of currents near our planet is subject to significant changes during an increase of Solar activity, and it is closely associated with the Solar wind. Long term electric currents have not been detected within the Solar wind, although, in view of its high electrical conductivity, the possibility cannot be ruled out.

The Solar atmosphere is a good conductor. Electric currents there are associated with such phenomena as flares and prominences [9]. The electric currents in the Solar atmosphere are of two kinds: the current coming from sources beneath the photosphere, and the currents located above the photosphere. The first kind of currents are associated with the formations on the Solar surface of the Sun-spot type. At the present time we have no generally accepted physical model of the spot's structure, not of the motion of material in it [7, 11]. It is known from observations [11] that the motion within the spot is of quite a complex nature. Obviously, the electric currents in the Solar atmosphere are responsible not only for distribution of magnetic field in the spot, but also for the motion of material within it.

The structures of currents on the Solar surface, of course, are related to the electric currents in the interior, which we can detect by the global Solar magnetic field. The magnetic field of stars and planets is sustained by the action of a dynamo mechanism within the liquid core [16]. The dynamo mechanism could be represented in the following way: in the existing magnetic field a convective motion of liquid material induces electric currents, whose magnetic field is close to the initial value. Hence, the magnetic field in the liquid core is the self-magnetic field of the currents passing in it, and the respective electromagnetic force drives a specific motion of the liquid material, which can affect the initial motion. The inclusion of the back effect of the magnetic field on the motion is a composite part of any dynamically closed theory of the magnetic field generation; however, such models have been little developed at the present time.

Another possible source of the electric currents in the liquid core is the thermoelectric effect due to a nonuniform heating of different fluid layers. In recent times thermocurrents have been considered as a possible source of a seeding magnetic field within large-scale liquid metal power generation installations, which is able to activate the dynamo mechanism.

1

Basic properties of axially symmetric motions in magnetohydrodynamics

The theory of electrically induced vortical flows should be constructed primarily as a theory of spatial flows since these are developed only as the result of a spatially inhomogeneous electric current density. The difficulties associated with the mathematical description of such flows in hydrodynamics can be greatly reduced by restricting our attention to axisymmetric situations. To maintain laminar axisymmetric magnetohydrodynamic flow, it is essential to operate with only such electromagnetic fields as do not disturb the axial symmetry of motion. Axisymmetric electromagnetic fields meet this condition. The main part of this chapter is devoted to a study of the properties of such fields, which may also be of interest in other areas of MHD. It is necessary to derive the full system of magnetohydrodynamic equations in the form which can be used in any curvilinear coordinate system having axial symmetry, and which, to our knowledge, is hard to find in the monographic literature. The last section of the chapter contains this material.

1.1. Magnetohydrodynamic equations

Derivation of the magnetohydrodynamic equations for the media most often used in the laboratory or in technology (liquid metals, fused salts), has been given in many well-known monographs (see, for example, [11, 3, 5, 9]). We shall therefore repeat only the basic assumptions used to obtain the equations: the medium is nonmagnetic (its magnetic permeability differs little from the permeability of free space); the medium is electrically conducting (its conductivity is high enough), and, due to the relatively small dielectric permittivity, polarization can be neglected; for processes that do not vary too rapidly (low-frequency approximation) and for velocities much smaller than the velocity of light, the displacement current and the convection current can be neglected in comparison with the conduction current. The physical properties of the incompressible medium are assumed to be homogeneous, isotropic, and, with the exception of the density, they are independent of temperature and other conditions.

Basic properties of axially symmetric motions in magnetohydrodynamics

With these assumptions the set of magnetohydrodynamic equations consists of the following equations:

momentum transport

$$\rho\left[\frac{\partial \mathbf{v}}{\partial t} + (\mathbf{v}\nabla)\mathbf{v}\right] = -\nabla p + \nu\rho\nabla^2\mathbf{v} + \rho\mathbf{g} + \mathbf{j}\times\mathbf{B}, \qquad (1.1.1)$$

heat transfer

$$\rho c\left(\frac{\partial T}{\partial t} + \mathbf{v}\cdot\nabla T\right) = \kappa\nabla^2 T + \frac{|\mathbf{j}|^2}{\sigma}, \qquad (1.1.2)$$

concentration transport

$$\frac{\partial C}{\partial t} + \mathbf{v}\cdot\nabla C = D\nabla^2 C, \qquad (1.1.3)$$

continuity

$$\nabla\cdot\mathbf{v} = 0; \qquad (1.1.4)$$

Maxwell's equations

$$\nabla\cdot\mathbf{B} = 0, \qquad (1.1.5)$$

$$\frac{1}{\mu_0}\nabla\times\mathbf{B} = \mathbf{j}, \qquad (1.1.6)$$

$$-\nabla\times\mathbf{E} = \frac{\partial \mathbf{B}}{\partial t}, \qquad (1.1.7)$$

$$\nabla\cdot\mathbf{E} = \frac{\rho_e}{\varepsilon_0} \qquad (1.1.8)$$

Ohm's law for a moving medium

$$\mathbf{j} = \sigma(\mathbf{E} + \mathbf{v}\times\mathbf{B}), \qquad (1.1.9)$$

and charge conservation

$$\nabla\cdot\mathbf{j} = 0, \qquad (1.1.10)$$

following from (1.1.6). The equation for density must be added to these equations:

$$\rho = \rho(T) \qquad (1.1.11)$$

which is necessary when a nonhomogeneous current density, and the associated Joule-heating, causes convection in the fluid.

We neglected viscous dissipation in equation (1.1.2), assuming that heating due to the friction is unimportant compared to the Joule-heating, and in equation (1.1.3) we neglected the change of concentration resulting in chemical reactions. Note, also, that equation (1.1.8) is used more often in electrostatics; in magnetohydrodynamics it is used to determine the electric charge density ρ_e after the electric field **E** is found.

The set of equations (1.1.1)–(1.1.11) simplifies for a stationary situation. Moreover, in this case the electric field is irrotational (from (1.1.7) it follows that curl **E** = 0) and the electric field potential Φ can be defined:

$$\mathbf{E} = -\text{grad } \Phi. \tag{1.1.12}$$

The equation for the potential follows from Ohm's law substituting there (1.1.12), taking div, and using (1.1.10):

$$\nabla^2 \Phi = \text{div}(\mathbf{v} \times \mathbf{B}). \tag{1.1.13}$$

From (1.1.6), (1.1.8), (1.1.5), and curl **E** = 0 we may obtain the equation for the magnetic field:

$$\nabla^2 \mathbf{B} = -\mu_0 \sigma \, \text{curl}(\mathbf{v} \times \mathbf{B}). \tag{1.1.14}$$

The equation (1.1.1) for stationary flow with ρ = const is frequently used in the form eliminating the pressure p after curl operator is applied to (1.1.1):

$$\text{curl}(\text{curl } \mathbf{v} \times \mathbf{v}) + \nu \, \text{curl curl curl } \mathbf{v} = \frac{1}{\mu_0 \rho} \text{curl}(\text{curl } \mathbf{B} \times \mathbf{B}). \tag{1.1.15}$$

Another form of the equation (1.1.1) was introduced by Landau [6]:

$$\rho \frac{\partial \mathbf{v}}{\partial t} + \text{div } \Pi = \rho \mathbf{g}. \tag{1.1.16}$$

Here Π is the second order symmetric tensor of momentum density flux. The components of the tensor, including Maxwell stress, are

$$\Pi_{ik} = \left(p + \frac{|\mathbf{B}|^2}{2\mu_0} \right) \delta_{ik} + \rho v_i v_k - 2\rho \nu \varepsilon_{ik} - \frac{B_i B_k}{\mu_0}, \tag{1.1.17}$$

where δ_{ik} is Kronecker symbol, and ε_{ik} are the components of the strain tensor ε:

$$\varepsilon = \tfrac{1}{2}[\nabla \mathbf{v} + (\nabla \mathbf{v})^*] \tag{1.1.18}$$

($\nabla \mathbf{v}$ is defined by (1.3.8) and * indicates a transposed tensor). The momentum density flux tensor may be divided into two parts: $\Pi_{ik} = \rho v_i v_k - \Pi^0_{ik}$, where the part $\rho v_i v_k$ is responsible for mechanical momentum transport, and the remaining part

$$\Pi^0_{ik} = -\left(p + \frac{|\mathbf{B}|^2}{2\mu_0} \right) \delta_{ik} + 2\rho \nu \varepsilon_{ik} + \frac{B_i B_k}{\mu_0} \tag{1.1.19}$$

is called the modified stress tensor; it is responsible for the forces acting on a fluid volume element, and is not directly related to its motion.

1.2. Some facts about orthogonal curvilinear coordinates

Successful solution of problems in hydro- and magnetohydrodynamics greatly depends on the choice of coordinate system for the particular problem. A good choice of coordinate system simplifies the representation and the mathematical description of the fields investigated, permitting us, for instance, to transform partial differential equations into ordinary differential equations, i.e. to solve the problem in a self-similar formulation; to separate variables in the basic equations; and to satisfy boundary conditions without great difficulties. The choice of system is usually determined by the flow domain boundary, the topology of the electric current, or the external magnetic field. Sometimes the analysis of physical phenomena in different coordinate systems opens a possibility for solving new problems of practical significance. Let us recall the basic formulae for the orthogonal curvilinear coordinates, and also some information for later use concerning the basic coordinate systems (see also [1, 2]).

Vector and scalar fields can always be expressed in Cartesian coordinates x, y, z. A generalised system of curvilinear coordinates q_1, q_2, q_3 is defined by the three groups of surfaces, expressed in Cartesian coordinates by the equations:

$$q_k = q_k(x, y, z), \qquad k = 1, 2, 3. \tag{1.2.1}$$

Any three surfaces q_k = const with different k should intersect in only one point of the space region considered. Then the equations (1.2.1) can be solved for x, y, z:

$$x = x(q_1, q_2, q_3); \qquad y = y(q_1, q_2, q_3); \qquad z = z(q_1, q_2, q_3) \tag{1.2.2}$$

and for any point of the space there is the single correspondence between the three numbers (x, y, z) and (q_1, q_2, q_3). Then the values q_k uniquely define a point in space and can be used as the new coordinate system.

For a constant value of one coordinate q_k, varying other two, we get a coordinate surface q_k defined by the equation (1.2.1) or (in parametric form) by (1.2.2). If two coordinates are fixed, then varying the third q_k we obtain a coordinate line q_k — a curve which is the intersection line of the two coordinate surfaces. Three surfaces and three coordinate lines intersect in any point of the space.

A vector field in the space is described by three projections of the vector on a given basis of three vectors. The basis in Cartesian coordinates is $\mathbf{i}_x$, $\mathbf{i}_y$, $\mathbf{i}_z$. For instance, the radius vector of the point (x, y, z) may be expressed in the form:

$$\mathbf{R} = x\mathbf{i}_x + y\mathbf{i}_y + z\mathbf{i}_z. \tag{1.2.3}$$

For curvilinear coordinates the local basis $\mathbf{i}_k(q_1, q_2, q_3)$, $k = 1, 2, 3$ is defined for any space point, and it consists of the unit vectors $\mathbf{i}_k$ tangent to the coordinate lines at the point. The vectors $\mathbf{i}_k$ are oriented in the increasing direction of the corresponding coordinate q_k, they are of unit length $|\mathbf{i}_k| = 1$, but, in general, change their orientation from point to point.

By definition the vectors $\mathbf{i}_k$ can be represented in the form $\mathbf{i}_k = \partial \mathbf{R}/\partial \ell_k$ (where $\partial \ell_k$ is the arc length differential along the line q_k) or

$$\mathbf{i}_k = \frac{\partial \mathbf{R}}{\partial q_k} \bigg/ \left| \frac{\partial \mathbf{R}}{\partial q_k} \right| = \frac{\partial \mathbf{R}}{\partial q_k} \left| \frac{\partial q_k}{\partial \ell_k} \right| = \frac{1}{H_k} \frac{\partial \mathbf{R}}{\partial q_k}. \tag{1.2.4}$$

The quantities $H_k(q_1, q_2, q_3) = |\partial \mathbf{R}/\partial q_k| = |\partial q_k/\partial \ell_k|^{-1}$ are called Lame coefficients;* the variation of coordinate dq_k corresponds to the arc length $d\ell_k = H_k\, dq_k$ along the coordinate line. The coefficients H_k define thereby the variation of the coordinate line's arc length in space. All the differential expressions for vector and scalar fields, which will be derived and then repeatedly used, are defined using Lame coefficients.

According to (1.2.3) the expression for $\partial \mathbf{R}/\partial q_k$ in Cartesian coordinates is

$$\frac{\partial \mathbf{R}}{\partial q_k} = \frac{\partial x}{\partial q_k} \mathbf{i}_x + \frac{\partial y}{\partial q_k} \mathbf{i}_y + \frac{\partial z}{\partial q_k} \mathbf{i}_z. \tag{1.2.5}$$

Knowing the relation (1.2.2) between Cartesian and curvilinear coordinates, the Lame coefficients are evaluated by the formulae

$$H_k^2 = \left(\frac{\partial x}{\partial q_k} \right)^2 + \left(\frac{\partial y}{\partial q_k} \right)^2 + \left(\frac{\partial z}{\partial q_k} \right)^2. \tag{1.2.6}$$

Systems of nonorthogonal coordinates are not usually used in hydrodynamic problems. Therefore we shall consider only orthogonal coordinates. The necessary conditions for orthogonality of the curvilinear coordinates are $\mathbf{i}_j \cdot \mathbf{i}_k = 0$ or

$$\frac{\partial \mathbf{R}}{\partial q_j} \cdot \frac{\partial \mathbf{R}}{\partial q_k} = 0 \quad (j \neq k, \quad j \text{ and } k = 1, 2, 3).$$

The sequence of the unit vectors is usually chosen to form a right-handed coordinate system.

Let us emphasize that the system of unit vectors, remaining orthogonal, generally changes its orientation in space from point to point. This explains why derivatives of the unit vectors with respect to the coordinates are not zero (as is the case for Cartesian systems). As shown in [2], the derivatives of the unit vectors can be expressed by the formulae:

$$\frac{\partial \mathbf{i}_k}{\partial q_j} = \mathbf{i}_j \frac{1}{H_k} \frac{\partial H_j}{\partial q_k}, \quad j \neq k; \tag{1.2.7}$$

$$\frac{\partial \mathbf{i}_k}{\partial q_k} = -\mathbf{i}_j \frac{1}{H_j} \frac{\partial H_k}{\partial q_j} - \mathbf{i}_\ell \frac{1}{H_\ell} \frac{\partial H_k}{\partial q_\ell}, \quad j \neq k \neq \ell. \tag{1.2.8}$$

---

* Some authors, e.g. [2], use the inverse quantities $h_k = H_k^{-1}$.

Basic properties of axially symmetric motions in magnetohydrodynamics 23

The theory of electrically induced flows is based on the use of orthogonal curvilinear coordinates associated with surfaces of rotation. In the general case a coordinate system of rotation may be obtained in the following way. An orthogonal coordinate system (q_1, q_2) with Lame coefficients $H_1(q_1, q_2)$, $H_2(q_1, q_2)$ is introduced in a plane. A straight line lying in the plane is chosen as the symmetry axis. One half of the plane is rotated relative to the axis. Every plane containing the symmetry axis is called a meridional plane. The position of a point in it is defined by the coordinates q_1, q_2. The third coordinate, defining the position of the meridional plane, is the angle by which the plane is rotated ($q_3 = \varphi$), and it is called the azimuthal angle.

The axisymmetric coordinate system (q_1, q_2, φ) so obtained is orthogonal in view of the orthogonality of the system (q_1, q_2), and of the azimuthal displacement $H_3 \, d\varphi$ relative to the meridional plane φ = const. In the simplest case the introduced system in the plane is Cartesian (z, r). Then $H_1 = H_2 = 1$ and the z axis is chosen as the symmetry axis. The coordinate system is then termed cylindrical.

Rotational coordinate systems are commonly related to the rotational cylindrical coordinates z, r, φ.* Then instead of (1.2.2) we have the relations:

$$z = z(q_1, q_2), \qquad r = r(q_1, q_2), \qquad \varphi = q_3. \tag{1.2.9}$$

Let us express the Lame coefficients (1.2.6) with the use of relations (1.2.9). Cartesian and cylindrical coordinates are related to each other by

$$x = r \cos \varphi, \qquad y = r \sin \varphi, \qquad z = z. \tag{1.2.10}$$

Substituting (1.2.10) in (1.2.6) we find the expressions for Lame coefficients with the known relationships (1.2.9):

$$H_1^2 = \left(\frac{\partial r}{\partial q_1}\right)^2 + \left(\frac{\partial z}{\partial q_1}\right)^2,$$

$$H_2^2 = \left(\frac{\partial r}{\partial q_2}\right)^2 + \left(\frac{\partial z}{\partial q_2}\right)^2, \tag{1.2.11}$$

$$H_3 = r(q_1, q_2);$$

orthogonality of the system is ensured if

$$\frac{\partial r}{\partial q_1} \frac{\partial r}{\partial q_2} + \frac{\partial z}{\partial q_1} \frac{\partial z}{\partial q_2} = 0.$$

It is evident from (1.2.11) that the coefficient H_3 has a clear geometrical meaning — it expresses the radial distance from the symmetry axis to the point where the coefficient is evaluated.

---

* The coordinate sequence is chosen to agree with the common sequence of other rotational coordinate systems.

1.3. Differential operators in orthogonal curvilinear coordinates

Let the operation grad a be defined for a scalar function a in Cartesian coordinates:

$$\text{grad } a = \nabla a = \frac{\partial a}{\partial x}\mathbf{i}_x + \frac{\partial a}{\partial y}\mathbf{i}_y + \frac{\partial a}{\partial z}\mathbf{i}_z. \tag{1.3.1}$$

Multiplying (1.3.1) by the unit vectors (1.2.4) and using (1.2.5), we obtain expressions for the components of grad a in the orthogonal curvilinear coordinates:

$$(\text{grad } a) \cdot \mathbf{i}_k = \nabla a \cdot \mathbf{i}_k = \frac{1}{H_k} \nabla a \cdot \frac{\partial \mathbf{R}}{\partial q_k}$$

$$= \frac{1}{H_k}\left(\frac{\partial a}{\partial x}\frac{\partial x}{\partial q_k} + \frac{\partial a}{\partial y}\frac{\partial y}{\partial q_k} + \frac{\partial a}{\partial z}\frac{\partial z}{\partial q_k}\right)$$

$$= \frac{1}{H_k}\frac{\partial a}{\partial q_k}.$$

Thence the operator ∇ in the curvilinear coordinates has the form

$$\nabla = \mathbf{i}_1 \frac{1}{H_1}\frac{\partial}{\partial q_1} + \mathbf{i}_2 \frac{1}{H_2}\frac{\partial}{\partial q_2} + \mathbf{i}_3 \frac{1}{H_3}\frac{\partial}{\partial q_3}. \tag{1.3.2}$$

Applying the operator ∇ to a vector function

$$\mathbf{u} = u_1 \mathbf{i}_1 + u_2 \mathbf{i}_2 + u_3 \mathbf{i}_3$$

and using the formulae (1.2.7), (1.2.8) we obtain

$$\text{div } \mathbf{u} = \nabla \cdot \mathbf{u} = \frac{1}{H_1 H_2 H_3}\left[\frac{\partial}{\partial q_1}(H_2 H_3 u_1) + \frac{\partial}{\partial q_2}(H_3 H_1 u_2) + \frac{\partial}{\partial q_3}(H_1 H_2 u_3)\right]; \tag{1.3.3}$$

$$\text{curl } \mathbf{u} = \nabla \times \mathbf{u} = \frac{1}{H_1 H_2 H_3}\begin{vmatrix} H_1 \mathbf{i}_1 & H_2 \mathbf{i}_2 & H_3 \mathbf{i}_3 \\ \dfrac{\partial}{\partial q_1} & \dfrac{\partial}{\partial q_2} & \dfrac{\partial}{\partial q_3} \\ H_1 u_1 & H_2 u_2 & H_3 u_3 \end{vmatrix}. \tag{1.3.4}$$

To derive the equations of motion we shall need expressions in the curvilinear coordinates for

$$[(\mathbf{v} \cdot \nabla)\mathbf{u}]_1 = [(\mathbf{v} \cdot \text{grad})\mathbf{u}]_1$$

$$= \frac{v_1}{H_1}\frac{\partial u_1}{\partial q_1} + \frac{v_2}{H_2}\frac{\partial u_1}{\partial q_2} + \frac{v_3}{H_3}\frac{\partial u_1}{\partial q_3} +$$

$$+ \frac{u_2}{H_1 H_2}\left(v_1 \frac{\partial H_1}{\partial q_2} - v_2 \frac{\partial H_2}{\partial q_1}\right) +$$

$$+ \frac{u_3}{H_1 H_3}\left(v_1 \frac{\partial H_1}{\partial q_3} - v_3 \frac{\partial H_3}{\partial q_1}\right). \quad (1.3.5)$$

Other components of (1.3.5) may be obtained by cyclic permutation of the indexes. Substituting $\mathbf{u} = \nabla$ in (1.3.3), Laplace's operator for a scalar function is expressed as

$$\nabla \cdot \nabla = \nabla^2 = \frac{1}{H_1 H_2 H_3}\left[\frac{\partial}{\partial q_1}\left(\frac{H_2 H_3}{H_1}\frac{\partial}{\partial q_1}\right) + \right.$$

$$\left. + \frac{\partial}{\partial q_2}\left(\frac{H_3 H_1}{H_2}\frac{\partial}{\partial q_2}\right) + \frac{\partial}{\partial q_3}\left(\frac{H_1 H_2}{H_3}\frac{\partial}{\partial q_3}\right)\right]. \quad (1.3.6)$$

The Laplacian for a vector function $\nabla^2 \mathbf{u}$ can be obtained using the vector identity:

$$\nabla^2 \mathbf{u} = \nabla(\nabla \cdot \mathbf{u}) - \nabla \times (\nabla \times \mathbf{u}), \quad (1.3.7)$$

where, in order to express the first term, it is necessary to apply (1.3.2) to (1.3.3), and to get the second term we use (1.3.4) twice. From (1.3.5) we can obtain the expression for $[(\mathbf{i}_k \cdot \text{grad})\mathbf{u}]_\ell$, which is adopted as the definition of the tensor components

$$[\text{grad } \mathbf{u}]_{k\ell} = [(\mathbf{i}_k \cdot \text{grad})\mathbf{u}]_\ell. \quad (1.3.8)$$

1.4. The most commonly used rotational coordinate systems

All the rotational coordinate systems have the symmetry axis (z-axis below) and the coordinate φ, which is the azimuthal angle measured from a certain half plane, containing the z-axis, counterclockwise if viewing from the positive direction of the z-axis. The coordinate systems considered below are represented by the schematic drawings (1—7), by the relations to rotational cylindrical system (cylindrical coordinates are related to Cartesian), and by the expressions for Lame coefficients. To obtain Lame coefficients the formulae (1.2.11) and the relations (1.2.9) are used. Knowing the Lame coefficients, the expressions for differential operators (1.3.2)—(1.3.7) may be obtained.

1. *Rotational cylindrical coordinates*

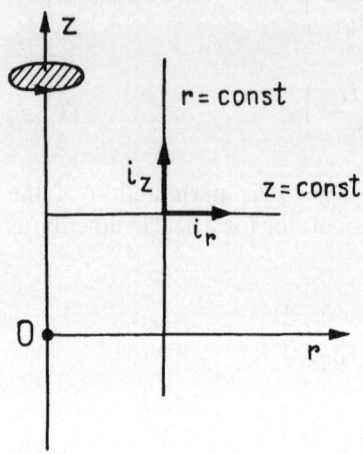

$x = r \cos \varphi, \quad y = r \sin \varphi, \quad z = z;$
$q_1 = z, \quad q_2 = r, \quad q_3 = \varphi;$
$H_1 = 1, \quad H_2 = 1, \quad H_3 = r.$

2. *Spherical coordinates*

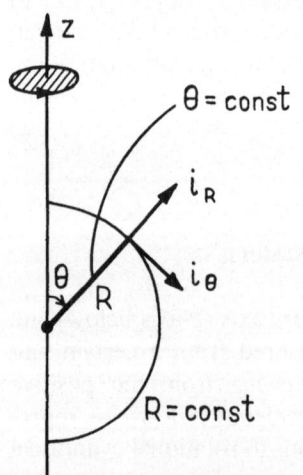

$z = R \cos \theta, \quad r = R \sin \theta, \quad \varphi = \varphi;$
$q_1 = R, \quad q_2 = \theta, \quad q_3 = \varphi;$
$H_1 = 1, \quad H_2 = R, \quad H_3 = R \sin \theta.$

3. Coordinates of prolate spheroid

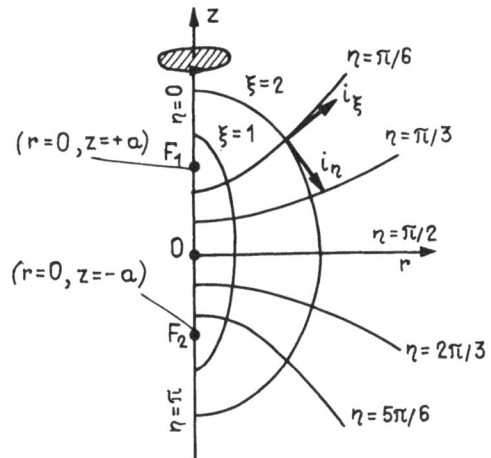

$$z = a \cosh \xi \cos \eta, \quad r = a \sinh \xi \sin \eta;$$
$$q_1 = \xi, \quad q_2 = \eta, \quad q_3 = \varphi;$$
$$H_1 = H_2 = a(\sinh^2 \xi + \sin^2 \eta)^{1/2},$$
$$H_3 = a \sinh \xi \sin \eta.$$

4. Coordinates of oblate spheroid

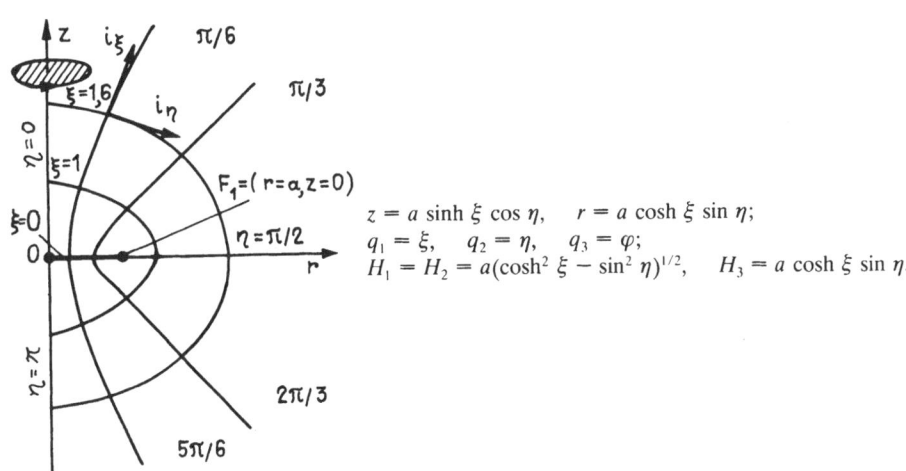

$$z = a \sinh \xi \cos \eta, \quad r = a \cosh \xi \sin \eta;$$
$$q_1 = \xi, \quad q_2 = \eta, \quad q_3 = \varphi;$$
$$H_1 = H_2 = a(\cosh^2 \xi - \sin^2 \eta)^{1/2}, \quad H_3 = a \cosh \xi \sin \eta.$$

5. Paraboloidal coordinates

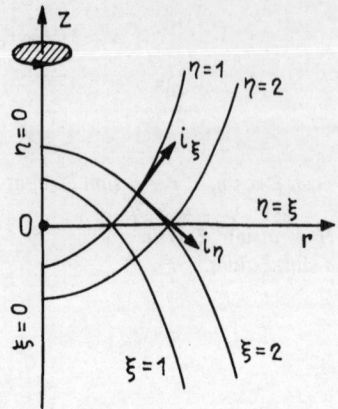

$$z = a(\xi^2 - \eta^2), \quad r = 2a\xi\eta;$$
$$q_1 = \xi, \quad q_2 = \eta, \quad q_3 = \varphi;$$
$$H_1 = H_2 = 2a(\xi^2 + \eta^2)^{1/2},$$
$$H_3 = 2a\xi\eta.$$

6. Toroidal coordinates

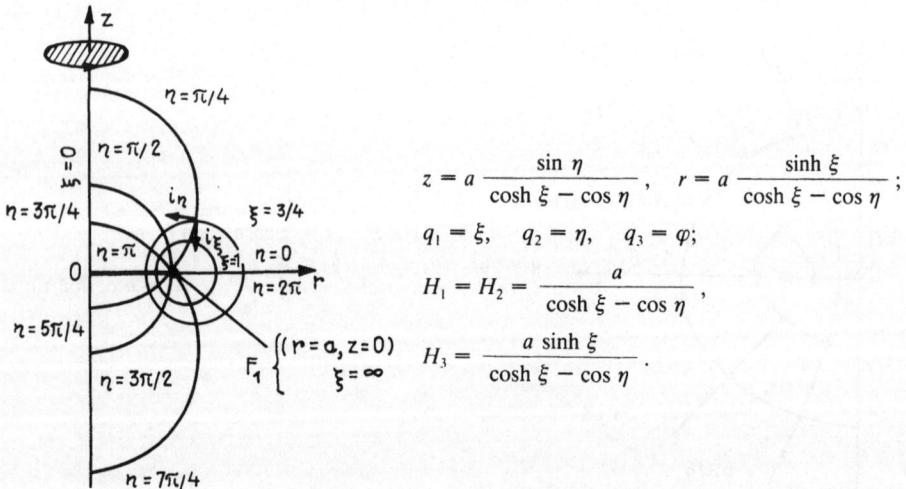

$$z = a\,\frac{\sin\eta}{\cosh\xi - \cos\eta}, \quad r = a\,\frac{\sinh\xi}{\cosh\xi - \cos\eta};$$
$$q_1 = \xi, \quad q_2 = \eta, \quad q_3 = \varphi;$$
$$H_1 = H_2 = \frac{a}{\cosh\xi - \cos\eta},$$
$$H_3 = \frac{a\sinh\xi}{\cosh\xi - \cos\eta}.$$

7. Bipolar coordinates

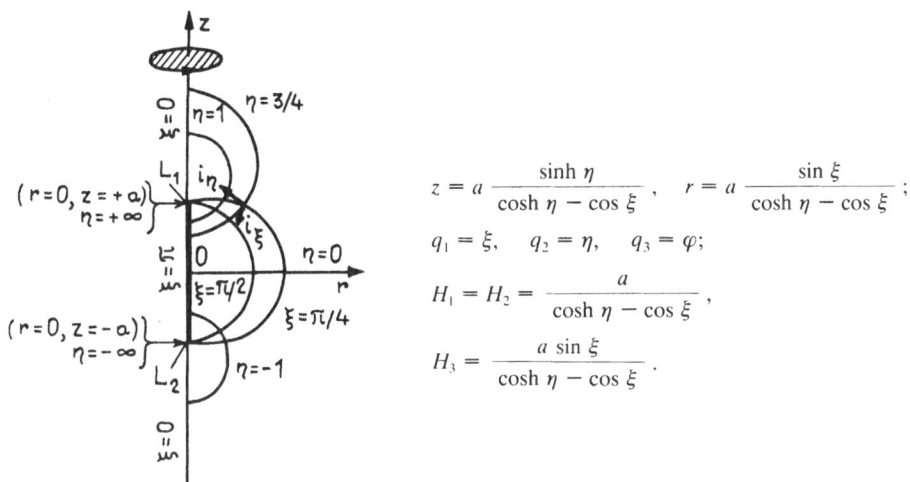

$$z = a \frac{\sinh \eta}{\cosh \eta - \cos \xi}, \quad r = a \frac{\sin \xi}{\cosh \eta - \cos \xi};$$

$$q_1 = \xi, \quad q_2 = \eta, \quad q_3 = \varphi;$$

$$H_1 = H_2 = \frac{a}{\cosh \eta - \cos \xi},$$

$$H_3 = \frac{a \sin \xi}{\cosh \eta - \cos \xi}.$$

1.5. Axisymmetric motions

We shall consider the axisymmetric motion for which all the factors affecting the motion remain invariable after an arbitrary rotation relative to the symmetry axis. The motion is also invariable in all the meridional planes; in other words, it is independent of the position of the meridional plane $\varphi = $ const. To describe such a motion the rotational coordinate system (q_1, q_2, φ) is used. In the axially symmetric case the motion is spatial, although its dynamic characteristics depend on the two curvilinear coordinates q_1 and q_2 defined in the meridional plane, i.e. the problem becomes two-dimensional.

The vector fields, which we shall use (**v**, **j**, **B**), are solenoidal, i.e. their divergence is zero. In particular, the velocity field **v** is solenoidal if the fluid is incompressible. For such a field the equation (1.1.4), by use of (1.3.3) and with axial symmetry, may be expressed in the form

$$\frac{\partial}{\partial q_1}(H_2 H_3 v_{q_1}) + \frac{\partial}{\partial q_2}(H_3 H_1 v_{q_2}) = 0. \tag{1.5.1}$$

From this it follows that a scalar function ψ can be introduced identically satisfying the equation (1.5.1), by use of which the meridional flow field is defined by

$$v_{q_1} = \frac{1}{H_2 H_3} \frac{\partial \psi}{\partial q_2}, \quad v_{q_2} = -\frac{1}{H_3 H_1} \frac{\partial \psi}{\partial q_1}. \tag{1.5.2}$$

The azimuthal velocity v_φ does not appear in (1.5.1) because of the axial symmetry ($\partial v_\varphi/\partial \varphi = 0$), and therefore v_φ is the second of the two scalars completely defining the solenoid velocity field. Then the spatial velocity field of axisymmetric motion can be represented by the vectorial sum of the meridional and the azimuthal motions, i.e. the sum of the meridional vector $\mathbf{v}_m$ lying in the meridional plane and the azimuthal vector normal to it:

$$\mathbf{v} = \mathbf{v}_m(q_1, q_2) + v_\varphi(q_1, q_2)\mathbf{i}_\varphi.$$

The meridional velocity $\mathbf{v}_m$ may be expressed by the scalar function ψ in a different way. From div $\mathbf{v} = 0$ it follows that $\mathbf{v}$ may be represented by two scalar functions if we apply the well-known relation div $\mathbf{A} \times \mathbf{B} = \mathbf{B} \cdot \text{curl } \mathbf{A} - \mathbf{A} \cdot \text{curl } \mathbf{B}$:

$$\mathbf{v}_m = \nabla \psi \times \nabla \chi.$$

Inserting instead of function χ the coordinate $q_3 = \varphi$, for which $\nabla q_3 = H_3^{-1} \mathbf{i}_\varphi$, we obtain:

$$\mathbf{v}_m = H_3^{-1} \nabla \psi \times \mathbf{i}_\varphi. \tag{1.5.3}$$

It is easy to see that the expressions (1.5.3) and (1.5.2) define the same function ψ.

Let us introduce the concept of the meridional flow stream line: this is a line whose tangent at any point coincides with the velocity vector of the flow. The stream line L in the meridional plane P (Fig. 1.1) is specified by the equation $L(q_1, q_2) = $ const. The tangent is defined by the relation $\tan \alpha = d\ell_1/d\ell_2 = $

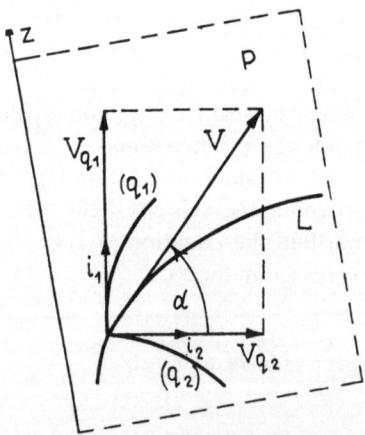

Fig. 1.1. To determine the meaning of stream function.

$H_1 dq_1/H_2 dq_2$. On the other hand, the definition of stream line implies $\tan \alpha = v_{q_1}/v_{q_2}$. From these we obtain the equation for the stream line

$$\frac{H_1 dq_1}{v_{q_1}} = \frac{H_2 dq_2}{v_{q_2}}. \tag{1.5.4}$$

Substituting (1.5.2) in (1.5.4), we conclude that the equation (1.5.4) is of the total differential type:

$$d\psi = \frac{\partial \psi}{\partial q_1} dq_1 + \frac{\partial \psi}{\partial q_2} dq_2 = 0, \tag{1.5.5}$$

from the solution of which

$$\psi(q_1, q_2) = \text{const.} \tag{1.5.6}$$

it follows that the value of ψ is constant along the stream line. This explains why the name of the function ψ is the stream function or the Stokes stream function. In what follows the hydrodynamic stream function (stream line) will be called simply the stream function (stream line) in contrast to the stream function (line) of electric current (Section 1.6).

The stream function is closely related to the volume flow rate through a surface, which forms a rotating curve, given in the meridional plane, relative to the symmetry axis. Consider the curve connecting the points A and B (Fig. 1.2) in this plane. The curve, rotated through the angle $\varphi = 2\pi$, will form a surface of rotation. Construct the unit normal $\mathbf{n} = \mathbf{i}_1 \cos \alpha + \mathbf{i}_2 \cos \beta$ to the curve's element $d\mathbf{l}$, directed from A to B, in such a way that the triangle $\mathbf{n}, d\mathbf{l}, \mathbf{i}_\varphi$ is right-angled. Assuming the flow rate to be positive in the direction of the normal $\mathbf{n}$, we have

$$Q = \int_S \mathbf{v} \cdot \mathbf{n} \, dS = 2\pi \int_A^B \mathbf{v} \cdot \mathbf{n} r \, d\ell$$

where r is the distance from the element of length $d\ell$ to the symmetry axis. Since $d\ell \cos \alpha = d\ell_2$, $d\ell \cos \beta = -d\ell_1$, and $r = H_3$, from (1.5.2) we obtain

$$Q = 2\pi \int_A^B H_3(v_{q_1} d\ell_2 - v_{q_2} d\ell_1)$$

$$= 2\pi \int_A^B \left(\frac{\partial \psi}{\partial q_1} dq_1 + \frac{\partial \psi}{\partial q_2} dq_2 \right)$$

$$= 2\pi \int_A^B d\psi = 2\pi[\psi(B) - \psi(A)]. \tag{1.5.7}$$

Thus the difference between the values of the function in the two points, apart from the multiple 2π, determines the volume flow rate through the surface, which is made rotating any curve connecting the points A and B. Particularly, if

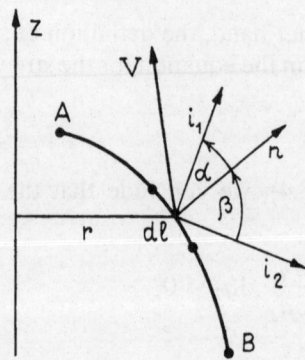

Fig. 1.2. To determine the relation between stream function and the flow rate across a rotational surface.

the fluid sources are absent on the symmetry axis, then, placing the point A on the symmetry axis, we conclude that the flow rate is independent of the position of this point on the axis. Therefore it may be adopted $\psi(A) = 0$ on the axis, and the axis is regarded as the zero stream line. Then the constant in (1.5.6) and the value of $\psi(B)$ in (1.5.7) are determined by the flow rate, depending merely on the position of the point B, and the stream line $\psi = $ const separates the domain where the flow rate remains constant. From (1.5.7) there follows the dimension of the stream function: $[\psi] = L^3/T$, where L and T are the dimensions of length and time respectively.

1.6. Relation between Stokes stream function and self-magnetic field of an electric current in problems with axial symmetry

To begin the computation of electrically induced flows it is advantageous to start with the construction of expressions for the electromagnetic force. At this stage we assume an electric current density distribution, find its self-magnetic field, and finally the electromagnetic force, neglecting the effect of fluid motion on the electric current and on the magnetic field as smaller perturbations. To determine the current density and the magnetic field from (1.1.10), (1.1.6), and (1.1.14) we have in this case:

$$\text{div}\,\mathbf{j} = 0, \tag{1.6.1}$$

$$\mathbf{j} = \frac{1}{\mu_0}\,\text{curl}\,\mathbf{B}, \tag{1.6.2}$$

$$\nabla^2 \mathbf{B} = 0. \tag{1.6.3}$$

The last equation is equivalent with the previous assumption to curl $\mathbf{j} = 0$.

A solution of the equations (1.6.1)–(1.6.3) can be found more easily after considering the analogy between the electromagnetic quantities and the hydrodynamic quantities of irrotational flow in an axisymmetric situation. Indeed, since for the irrotational incompressible flow

$$\text{div}\,\mathbf{v} = 0, \qquad \text{curl}\,\mathbf{v} = 0, \tag{1.6.4}$$

Basic properties of axially symmetric motions in magnetohydrodynamics 33

the vector potential **A** can be defined:

$$\mathbf{v} = \text{curl } \mathbf{A} \tag{1.6.5}$$

and it is determined by the equation

$$\text{curl curl } \mathbf{A} = \text{grad div } \mathbf{A} - \nabla^2 \mathbf{A} = 0$$

Assuming that div **A** = 0, we have

$$\nabla^2 \mathbf{A} = 0. \tag{1.6.6}$$

In some cases the vector potential may be related to the usual hydrodynamic quantities [7]. If $\mathbf{A} = (0, 0, A_3)$ and $H_1 H_2 A_3$ does not depend on q_3, then from (1.3.3) div $\mathbf{A} \equiv 0$ and the components of (1.6.5) are

$$v_{q_1} = \frac{1}{H_2 H_3} \frac{\partial H_3 A_3}{\partial q_2}, \qquad v_{q_2} = -\frac{1}{H_3 H_1} \frac{\partial H_3 A_3}{\partial q_1}.$$

Comparing these with (1.5.2), we see that the vector potential is related to the stream function ψ by

$$\psi = H_3 A_3. \tag{1.6.7}$$

The curl of the velocity and the Laplacian of the vector potential $\mathbf{A} = (0, 0, A_3)$ are related by the use of (1.6.7):

$$\text{curl } \mathbf{v} = -\nabla^2 \mathbf{A} = -\frac{1}{H_3} E^2 (H_3 A_3) \mathbf{i}_\varphi = -\frac{1}{H_3} E^2 \psi \mathbf{i}_\varphi,$$

where the operator E^2 is

$$E^2 = \frac{H_3}{H_1 H_2} \left(\frac{\partial}{\partial q_1} \frac{H_2}{H_1 H_3} \frac{\partial}{\partial q_1} + \frac{\partial}{\partial q_2} \frac{H_1}{H_2 H_3} \frac{\partial}{\partial q_2} \right). \tag{1.6.8}$$

The Stokes stream function for the axisymmetric irrotational flow is thus determined, due to (1.6.6), by the equation

$2C_1, j_r = 0$, i.e. we have an electric current of uniform density, directed along

The operator E^2 is closely related to the Laplacian of the scalar function (1.3.6):

$$E^2 = \nabla^2 + H_3^2 \nabla \left(\frac{1}{H_3^2} \right) \cdot \nabla$$

$$= \nabla^2 + \frac{H_3^2}{H_1^2} \frac{\partial}{\partial q_1} \left(\frac{1}{H_3^2} \right) \frac{\partial}{\partial q_1} + \frac{H_3^2}{H_2^2} \frac{\partial}{\partial q_2} \left(\frac{1}{H_3^2} \right) \cdot \frac{\partial}{\partial q_2},$$

therefore the solutions of equation (1.6.9) are closely related to the solutions of Laplace equation $\nabla^2 a = 0$.

Comparison of (1.6.1), (1.6.2), and (1.6.3) with the corresponding (1.6.4), (1.6.5), and (1.6.6) gives evidence for the analogy between **v** and **j**, **A** and **B** (if boundary conditions for the comparable quantities are identical), i.e. between the velocity field of irrotational flow and the electric current field, the Stokes stream function and the self-magnetic field of electric current. Thus, if in the axisymmetric case the electric current density were to have only the meridional components j_{q_1} and j_{q_2}, the magnetic field of this current has only a single, azimuthal component $B_\varphi = B_3$, hence the electric current stream function may be introduced

$$\psi_1 = \frac{1}{\mu_0} H_3 B_\varphi \tag{1.6.10}$$

defining the current density $\mathbf{j} = H_3^{-1} \nabla \psi_1 \times \mathbf{i}_\varphi$ (by analogy with (1.5.3)) by the components

$$j_{q_1} = \frac{1}{H_2 H_3} \frac{\partial \psi_1}{\partial q_2}, \quad j_{q_2} = -\frac{1}{H_3 H_1} \frac{\partial \psi_1}{\partial q_1}, \tag{1.6.11}$$

and the stream lines of the electric current by

$$\psi_1 = \frac{1}{\mu_0} H_3 B_\varphi = \text{const}; \tag{1.6.12}$$

in order to determine the magnetic field B_φ it is necessary to solve the equation

$$E^2 \psi_1 = \frac{1}{\mu_0} E^2 H_3 B_\varphi = 0. \tag{1.6.13}$$

The electromagnetic force and its curl may be expressed in this case by the function ψ_1 in the following way:

$$\mathbf{j} \times \mathbf{B} = -\mu_0 \psi_1 H_3^{-2} \nabla \psi_1,$$

$$\text{curl } \mathbf{j} \times \mathbf{B} = -\mu_0 \nabla \left(\frac{\psi_1}{H_3^2} \right) \times \nabla \psi_1 \tag{1.6.14}$$

$$= \mathbf{i}_\varphi \mu_0 \frac{\psi_1}{H_1 H_2} \left[-\frac{\partial}{\partial q_1} (H_3^{-2}) \frac{\partial \psi_1}{\partial q_2} + \frac{\partial}{\partial q_2} (H_3^{-2}) \frac{\partial \psi_1}{\partial q_1} \right];$$

the curl of the force has only a single component coinciding in direction with that of the self-magnetic field, i.e. parallel to the azimuthal unit vector $\mathbf{i}_\varphi$.

If curl of the force differs from zero, then the fluid flow is driven in the meridional planes; i.e. the flow direction corresponds to the sign of curl of the force.

It is of interest to analyse the circumstances for which patterns of the axisymmetric electric current curl of the electromagnetic force is zero, and the

Basic properties of axially symmetric motions in magnetohydrodynamics 35

flow of fluid, being initially at rest, is not generated. Without loss of generality we constrain the system to cylindrical coordinates. In this system the condition

$$\text{curl } \mathbf{j} \times \mathbf{B} = \mathbf{i}_\varphi \mu_0 \psi_1 \frac{\partial}{\partial r}\left(\frac{1}{r^2}\right) \frac{\partial \psi_1}{\partial z} = 0 \qquad (1.6.15)$$

means that $\psi_1 = \psi_1(r)$. Taking this into account, we see that the equation (1.6.13) determines ψ_1 in the form

$$\frac{\partial}{\partial r} \frac{1}{r} \frac{\partial \psi_1}{\partial r} = 0, \qquad (1.6.16)$$

which has the solution

$$\psi_1 = C_1 r^2 + C_2 \qquad (1.6.17)$$

in the whole fluid-containing domain.

Assigning $C_1 = 0$ in (1.6.17), and substituting the solution $\psi_1 = C_2$ in (1.6.11) and (1.6.12), we may see that no currents are present in the fluid; the magnetic field $B_\varphi = C_2/r$ can be interpreted as the field of a thin current conductor on the symmetry axis. For $C_2 = 0$ the same substitution gives $j_z = 2C_1$, $j_r = 0$, i.e. we have an electric current of uniform density, directed along the symmetry axis z, with the corresponding self-magnetic field $B_\varphi = C_1 r$. Superposition of the two cases, of course, leads to the same result. Consequently, the fluid flow is not generated only for the two cases of the current distribution considered. In all the other cases the axially symmetric electric current must generate a flow.

The condition (1.6.15) actually means that the curl of the electromagnetic force is equal to zero only in the absence of the radial component of electric current ($j_r = -r^{-1}\partial\psi_1/\partial z = 0$). The second component j_z, at the same time, may depend arbitrarily on the radius (i.e. curl $\mathbf{f}_e = 0$ is not restricted merely to the distributions of j_z considered above). The dependence of j_z on radius with $j_r = 0$ may appear, for instance, if the conductivity is varying in the radial direction.

In conclusion let us determine dimensions of the electric current stream function ψ_1 and the self-magnetic field B_φ, which follow from (1.6.11) and (1.6.10):

$$[\psi_1] = j_0 L^2 = I, \qquad [B_\varphi] = \frac{\mu_0 I}{L}.$$

Here I is a typical total current in the fluid, and L a typical length scale. The electric current stream function and the current distribution, consequently, is independent of magnetic permeability of the fluid.

1.7. Feasible schemes for axially symmetric electric current distributions

The analogy between the velocity field and the electric current field permits us to apply the known solutions for the axisymmetric irrotational flow stream function [8] in order to construct the electric current field and the electromagnetic force for the analysis of the electrically induced vortex flows. Moreover, it becomes possible to use solutions which are not considered in hydrodynamics due to their small practical significance, but which do make sense for the problems of electric current distribution.

Following the usual practice in the theory of irrotational flow, we may choose the electrically induced flow boundaries as the surfaces of natural coordinates for the given current distribution. One of the possible variants is to choose the surface determined by constant values of the current stream function; the second variant is to choose the surfaces of equal potentials, which are orthogonal to the surfaces $\psi_1 = $ const. Indeed, from $\mathbf{j} = \sigma \mathbf{E} = -\sigma \nabla \Phi$ it follows

$$\mathbf{j} = -\sigma \left(\mathbf{i}_1 H_1^{-1} \frac{\partial \Phi}{\partial q_1} + \mathbf{i}_2 H_2^{-1} \frac{\partial \Phi}{\partial q_2} \right) = -\sigma \nabla \Phi.$$

On the other hand, by (1.6.11) and (1.5.3)

$$\mathbf{j} = \mathbf{i}_1 \frac{1}{H_2 H_3} \frac{\partial \psi_1}{\partial q_2} - \mathbf{i}_2 \frac{1}{H_1 H_3} \frac{\partial \psi_1}{\partial q_1} = H_3^{-1} \nabla \psi_1 \times \mathbf{i}_\varphi.$$

Comparing the two expressions for current density, we conclude that $\nabla \psi_1$ and $\nabla \Phi$, and, consequently, the surfaces

$$\psi_1(q_1, q_2) = \text{const}, \qquad \Phi(q_1, q_2) = \text{const} \tag{1.7.1}$$

are orthogonal. The set of equations for ψ_1 and Φ is solvable, since the Jacobian $D(\Phi, \psi_1)/D(q_1, q_2) \neq 0$ due to the fact that $j^2 \neq 0$. Finally, from

$$-\frac{\sigma}{H_1} \frac{\partial \Phi}{\partial q_1} = \frac{1}{H_2 H_3} \frac{\partial \psi_1}{\partial q_2}, \qquad \frac{\sigma}{H_2} \frac{\partial \Phi}{\partial q_2} = \frac{1}{H_1 H_3} \frac{\partial \psi_1}{\partial q_1}$$

it follows that, if the solution of $E^2 \psi_1 = 0$ is found, the surfaces of equal potential can be determined by the integral

$$\sigma \Phi = \int_{q_{10}, q_{20}}^{q_1, q_2} \frac{H_2}{H_1 H_3} \frac{\partial \psi_1}{\partial q_1} dq_2 - \frac{H_1}{H_2 H_3} \frac{\partial \psi_1}{\partial q_2} dq_1,$$

where q_{10}, q_{20} is an initial point on the surface. If the potential $\Phi(q_1, q_2)$ is found by solving the equation $\nabla^2 \Phi = 0$, the corresponding stream function can be determined by

$$\psi_1(q_1, q_2) = \sigma \int_{q_{10}, q_{20}}^{q_1, q_2} \left(\frac{H_1 H_3}{H_2} \frac{\partial \Phi}{\partial q_2} dq_1 - \frac{H_2 H_3}{H_1} \frac{\partial \Phi}{\partial q_1} dq_2 \right).$$

Thus the functions ψ_1 and Φ are mutually connected, their level lines (1.7.1) are orthogonal in every point, and they form a system of natural coordinates $f_1(\psi_1), f_2(\Phi)$. Solutions of the equations $E^2 \psi_1 = 0$ and $\nabla^2 \Phi = 0$ determine the functions ψ_1 and Φ apart from constant terms. Choice of these constants specifies the position of the initial point of the coordinates f_1, f_2 relative to the coordinates q_1, q_2.

Basic properties of axially symmetric motions in magnetohydrodynamics 37

If the flow domain is bounded by the coordinate surfaces (1.7.1) of this system, the former ones describe the surfaces of electrical insulators and the latter are the surfaces of ideally conducting electrodes.

Let us consider some feasible schemes of the axisymmetric electric current flow which result from solving equation (1.6.13) in spherical coordinates:

$$\frac{\partial^2 \psi_1}{\partial R^2} + \frac{\sin \theta}{R^2} \frac{\partial}{\partial \theta} \frac{1}{\sin \theta} \frac{\partial \psi_1}{\partial \theta} = 0. \quad (1.7.2)$$

Separating the variables in (1.7.2), $\psi_1 = W(R)M(\theta)$, we get the following equations to determine $W(R)$ and $M(\theta)$:

$$R^2 W'' - n(n-1)W = 0, \quad (1.7.3)$$

$$M'' - \cot \theta M' + n(n-1)M = 0. \quad (1.7.4)$$

The solutions of these equations are well known [4]. For positive integers n the solution of (1.7.3) is

$$W(R) = A_n R^n + B_n R^{1-n}. \quad (1.7.5)$$

The substitution $\mu = \cos \theta$ is made in the second equation; it then transforms to the Gegenbauer equation of order $-1/2$:

$$(1 - \mu^2)M'' + n(n-1)M = 0, \quad (1.7.6)$$

the particular solutions of which are Gegenbauer functions of the two kinds of order $-1/2$ and degree n. In the degenerate cases ($n = 0$ and $n = 1$) these functions are defined as

$$J_0 = 1, \qquad H_0 = -\mu;$$
$$J_1 = -\mu, \qquad H_1 = 1;$$

for $n \geq 2$

$$J_2 = \frac{1}{2}(1-\mu^2), \qquad H_2 = \frac{1}{2} J_2 \ln \frac{1+\mu}{1-\mu} + \frac{\mu}{2};$$

$$J_3 = \frac{1}{2}\mu(1-\mu^2), \qquad H_3 = \frac{1}{2} J_3 \ln \frac{1+\mu}{1-\mu} + \frac{1}{6}(3\mu^2 - 2);$$

$$J_4 = \frac{1}{8}(1-\mu^2)(5\mu^2 - 1), \qquad H_4 = \frac{1}{2} J_4 \frac{1+\mu}{1-\mu} + \frac{1}{24}\mu(15\mu^2 - 13);$$

$$J_5 = \frac{1}{8}\mu(1-\mu^2)(7\mu^2 - 3), \qquad H_5 = \frac{1}{2} J_5 \ln \frac{1+\mu}{1-\mu} +$$

$$+ \frac{1}{120}(105\mu^4 - 115\mu^2 + 16)$$

etc.

The solution of (1.7.6) is

$$M(\mu) = C_n J_n(\mu) + D_n H_n(\mu). \quad (1.7.7)$$

Gegenbauer functions for $n \geq 2$ are closely related to the Legendre polynomials P_n, Q_n:

$$(1 - \mu^2) \frac{d}{d\mu} P_{n-1} = n(n-1)J_n,$$

$$(1 - \mu^2) \frac{d}{d\mu} Q_{n-1} = n(n-1)H_n. \tag{1.7.8}$$

Another relation is [2]:

$$J_n(\mu) = \frac{P_{n-2}(\mu) - P_n(\mu)}{2n - 1},$$

$$H_n(\mu) = \frac{Q_{n-2}(\mu) - Q_n(\mu)}{2n - 1}, \qquad n \geq 2. \tag{1.7.9}$$

Thus the solution of (1.7.2) is

$$\psi_1 = \sum_{n=0}^{\infty} (A_n R^n + B_n R^{1-n}) \genfrac{}{}{0pt}{}{J_n(\mu)}{H_n(\mu)}. \tag{1.7.10}$$

Let us derive solutions of the equation (1.6.13) for other rotational coordinate systems. For the coordinate system of a prolate spheroid (see Section 3.1.4) the new variables $\mu = \cos \eta$, $\lambda = \cosh \xi$ are introduced, then $z = a\lambda\mu$, $r = a\sqrt{(\lambda^2 - 1)(1 - \mu^2)}$,

$$H_1 = H_2 = a(\lambda^2 - \mu^2)^{1/2}, \qquad H_3 = a\sqrt{(1 - \mu^2)(\lambda^2 - 1)},$$

and the equation (1.6.13) takes the form

$$E^2\psi_1 = \frac{1}{a^2(\lambda^2 - \mu^2)} \left[(\lambda^2 - 1) \frac{\partial^2}{\partial \lambda^2} + (1 - \mu^2) \frac{\partial^2}{\partial \mu^2} \right] \psi_1 = 0. \tag{1.7.11}$$

Separating variables, $\psi = M(\mu)L(\lambda)$, we obtain the known equations

$$(1 - \mu^2)M'' + n(n-1)M = 0,$$

$$(1 - \lambda^2)L'' + n(n-1)L = 0,$$

the solutions of which are Gegenbauer functions. Thence the solution of the equation (1.7.11) can be represented as

$$\psi_1 = \sum_{n=0}^{\infty} [A_n J_n(\mu) + B_n H_n(\mu)] \genfrac{}{}{0pt}{}{J_n(\lambda)}{H_n(\lambda)}. \tag{1.7.12}$$

For the coordinates of oblate spheroid (Section 4.1.4), after the substitution $\mu = \cos \eta$, $\lambda = \sinh \xi$ $(r = a\sqrt{(1 - \mu^2)(1 + \lambda^2)}, z = a\lambda\mu$,

$$H_1 = H_2 = a\sqrt{\lambda^2 + \mu^2}, \qquad H_3 = a\sqrt{(1 - \mu^2)(1 + \lambda^2)},$$

Basic properties of axially symmetric motions in magnetohydrodynamics   39

the equation (1.6.13) is of the form

$$E^2\psi_1 = \frac{1}{a^2(\lambda^2+\mu^2)}\left[(1+\lambda^2)\frac{\partial^2}{\partial\lambda^2}+(1-\mu^2)\frac{\partial^2}{\partial\mu^2}\right]\psi_1 = 0. \quad (1.7.13)$$

Separating variables, $\psi_1 = M(\mu)L(\lambda)$, we obtain the two equations:

$$(1-\mu^2)M'' + n(n-1)M = 0,$$
$$(1+\lambda^2)L'' - n(n-1)L = 0,$$

the solution of the first equation is the aforementioned Gegenbauer functions, and the second has as solution the functions:

for $n = 0, n = 1$,

$$L_0 = 1, \quad G_0 = -\lambda; \quad L_1 = -\lambda, \quad G_1 = 1;$$

for $n \geq 2$

$$L_2 = \tfrac{1}{2}(1+\lambda^2), \quad\quad G_2 = L_2 \arctan \lambda + \frac{\lambda}{2};$$

$$L_3 = \tfrac{1}{2}\lambda(1+\lambda^2), \quad\quad G_3 = L_3 \arctan \lambda + \frac{1}{6}(3\lambda^2+2);$$

$$L_4 = \tfrac{1}{8}(1+\lambda^2)(5\lambda^2-1), \quad G_4 = L_4 \arctan \lambda + \frac{1}{24}\lambda(15\lambda^2+13); \quad (1.7.14)$$

$$L_5 = \tfrac{1}{8}\lambda(1+\lambda^2)(7\lambda^2+3), \quad G_5 = L^2 \arctan \lambda + \frac{1}{120}(105\lambda^4 + 115\lambda^2+16)\ldots$$

Thus the solution of the equation (1.7.13) may be written

$$\psi_1 = \sum_{n=0}^{\infty}[A_n J_n(\mu)+B_n H_n(\mu)]\substack{L_n(\lambda)\\G_n(\lambda)}.$$

For the paraboloidal coordinates (Section 5.1.4) solution of the equation

$$E^2\psi_1 = \frac{\xi\eta}{4a^2(\xi^2+\eta^2)}\left[\frac{1}{\eta}\frac{\partial}{\partial\xi}\left(\frac{1}{\xi}\frac{\partial}{\partial\xi}\right)+\frac{1}{\xi}\frac{\partial}{\partial\eta}\left(\frac{1}{\eta}\frac{\partial}{\partial\eta}\right)\right]\psi_1 = 0 \quad (1.7.15)$$

can be sought in the form $\psi_1 = \xi \eta L(\xi) M(\eta)$. Then to determine $L(\xi)$ and $M(\eta)$ we have Bessel equations:

$$\xi^2 L'' + \xi L' - (n^2 \xi^2 + 1) L = 0,$$

$$\eta^2 M'' + \eta M' + (n^2 \eta^2 - 1) M = 0,$$

whence the solution of (1.7.15) is (for $n \neq 0$)

$$\psi_1 = \eta \xi \sum_{n=1}^{\infty} [A_n I_1(n\xi) + B_n K_1(n\xi)]_{Y_1(n\eta)}^{J_1(n\eta)}, \qquad (1.7.16)$$

where J_1, Y_1 are Bessel functions of the first order, respectively, the first and the second kind of real argument; I_1, K_1 are functions of imaginary argument. For $n = 0$

$$\psi_1 = (A_0 \xi^2 + B_0)(C_0 \eta^2 + D_0). \qquad (1.7.17)$$

A generalized method to separate variables in the equation (1.6.13) can be suggested. Making the substitution $\psi_1 = \sqrt{H_3} F$ in (1.6.13), we obtain

$$\frac{\sqrt{H_3}}{H_1 H_2} \left(\frac{\partial^2 F}{\partial q_1^2} + \frac{\partial^2 F}{\partial q_2^2} \right) + FE^2 \sqrt{H_3} = 0. \qquad (1.7.18)$$

Recall that Lame coefficient H_3 has the meaning of cylindrical radius ($H_3 = r$). Expressing the last term of (1.7.18) in cylindrical coordinates we obtain

$$E^2 \sqrt{H_3} = E^2 \sqrt{r} = -\tfrac{3}{4} r^{-3/2} = -\tfrac{3}{4} H_3^{-3/2},$$

and the equation (1.7.18) in the form

$$\frac{\partial^2 F}{\partial \xi^2} + \frac{\partial^2 F}{\partial \eta^2} - \frac{3}{4} \frac{H_1 H_2}{H_3^2} F = 0, \qquad (1.7.19)$$

permitting us to separate variables straightforwardly for each coordinate system from Section 1.4, with the exception of the spherical ones. So, for toroidal coordinates the equation (1.7.19) is

$$\frac{\partial^2 F}{\partial \xi^2} + \frac{\partial^2 F}{\partial \eta^2} - \frac{3}{4} \frac{1}{\sinh^2 \xi} F = 0.$$

Separating variables, $F = L(\xi) M(\eta)$, we obtain the equations

$$\sinh^2 \xi L'' - \{\tfrac{3}{4} + [\tfrac{1}{4} + n(n-1)] \sinh^2 \xi \} L = 0,$$

$$M'' + [\tfrac{1}{4} + n(n-1)] M = 0.$$

The first of them with the substitutions $\lambda = \cosh \xi$ and $L = (\lambda^2 - 1)^{-1/4} P$ transforms to

$$(1 - \lambda^2) P'' + n(n-1) P = 0,$$

i.e. to the equation (1.7.6); the second is solved in elementary functions:

$$M = A \sin \sqrt{\tfrac{1}{4} + n(n-1)} \, \eta + B \cos \sqrt{\tfrac{1}{4} + n(n-1)} \, \eta.$$

For bipolar coordinates the substitution $F = L(\xi)M(\eta)$ in the equation

$$\frac{\partial^2 F}{\partial \xi^2} + \frac{\partial^2 F}{\partial \eta^2} - \frac{3}{4}\frac{1}{\sin^2 \xi} F = 0$$

yields:

$$\sin^2 \xi L'' - \{\tfrac{3}{4} - [\tfrac{1}{4} + n(n-1)]\sin^2 \xi\} L = 0,$$
$$M'' - [\tfrac{1}{4} + n(n-1)]M = 0.$$

The first of the equations with the substitutions $\lambda = \cos \xi$ and $L = (1-\lambda^2)^{-1/4}P$ transforms to the known equation (1.7.6)

$$(1-\lambda^2)P'' + n(n-1)P = 0$$

solution of the second is

$$M = A \exp\sqrt{\tfrac{1}{4} + n(n-1)}\,\eta + B \exp(-\sqrt{\tfrac{1}{4} + n(n-1)}\,\eta).$$

Consider now some of the particular solutions drawn from the above variety, together with the corresponding topology of the electric current field and the equipotential surfaces. All information is presented in Table 1.1, which contains the expressions for the electric current stream function and the electric potential. With these functions, using (1.6.11), (1.6.10), and (1.6.14), we may write down expressions for the electric current components, the azimuthal magnetic field, and its curl. The expression can be transformed to cylindrical coordinates with the use of the relations in Section 1.4, hence all the solutions in Table 1.1 may be regarded as particular solutions of (1.6.13) in cylindrical coordinates.

The particular solutions in the table correspond to the given values of n in (1.7.10), (1.7.12), and (1.7.16). Apart from these, the linear combinations of the solutions with various n may be considered. The number of such combinations, of course, is infinite, and therefore we restrict our listing of them to the simplest and more interesting combinations. The superpositions are indicated in the first column by parentheses containing the numbers of the elemental solutions.

1.8. Magnetic field in axisymmetric flow

One way to operate the electrically induced flows is to apply an external magnetic field to the flow domain. In this context the question arises concerning which configurations of the magnetic field do not destroy the axisymmetric situation, and thereby permit retention of the axisymmetric model. Another aspect of the flow of electrically conducting fluid is distortion of the magnetic field, which also ought to be considered if we wish to retain this effect.

Consider first the origin of meridional and azimuthal magnetic fields. From

Table 1.1. Axisymmetric fields of electric current.

No.	Scheme	The form of current field	ψ_1; Φ
1	2	3	4
		Spherical coordinates	
1	$n=0$ (linear current with B_φ, I, z)	Linear current at the symmetry axis	$\psi_1 = $ const; $\Phi = $ const
2	$n=0$ (semi-infinite source/sink diagram with I, z, r)	Semi-infinite line source and sink	R; $\ln \tan \theta/2$
3	$n=1$ (infinite line source diagram with I, z, r)	Infinite line source or sink	$-R \cos \theta$; $-\ln R \sin \theta$

Table 1.1. (continued)

1	2	3	4
4	$n=1$	Point source or sink with the current supply along the symmetry axis to the origin	$-\cos\theta$; $\dfrac{1}{R}$
5	$n=2$	Uniform current parallel to the symmetry axis	$R^2 \sin^2\theta$; $-2R\cos\theta$
6	$n=2$	Current dipole	$\dfrac{\sin^2\theta}{R}$; $\dfrac{\cos\theta}{R^2}$

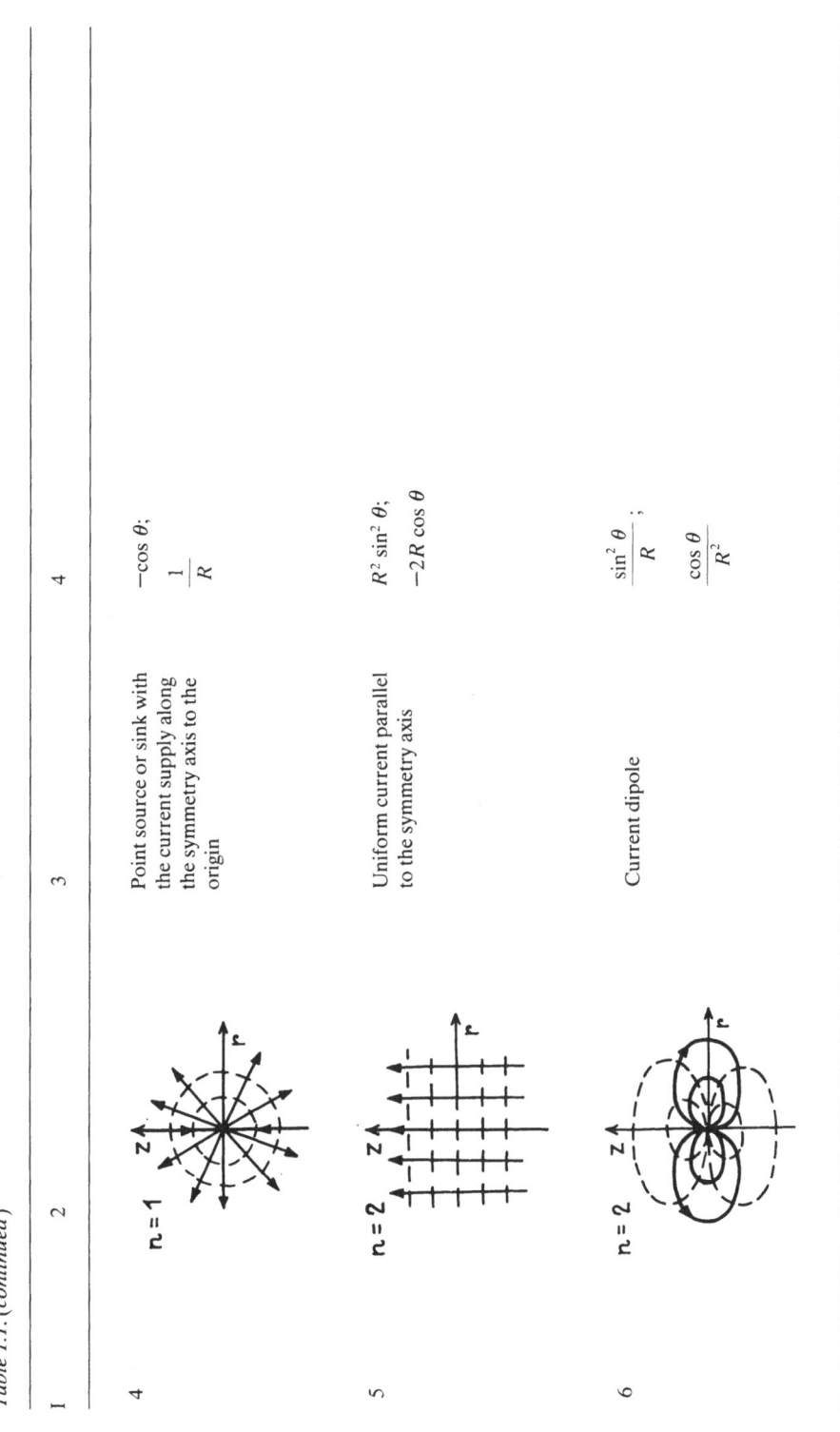

Table 1.1. (continued)

1	2	3	4
7		Current at critical point on nonconducting plane	$-R^3 \cos\theta \sin^2\theta$; $R^2(\cos^2\theta - \tfrac{1}{2}\sin^2\theta)$
8		Quadrupole	$\dfrac{\cos\theta \sin^2\theta}{R^2}$; $\dfrac{2\cos^2\theta - \sin^2\theta}{3R^3}$
9		Point source or sink in the point $z = a$	$\dfrac{R\cos\theta - a}{\sqrt{R^2 - 2aR\cos\theta + a^2}}$; $-\dfrac{1}{\sqrt{R^2 - 2aR\cos\theta + a^2}}$

Table 1.1. (continued)

1	2	3	4
10 (2–1)		Flow around a sphere in the field of semi-infinite line source and sink	$R - a$; $\ln \tan \theta/2 - a$
11 (1–3)		Line source between two planes with current supply from below	$a - R\cos\theta$; $a - \ln R \sin\theta$

Table 1.1. (continued)

1	2	3	4
12 (5–6)	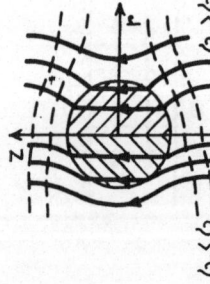	Flow around a sphere of radius a and conductivity σ_2 placed in fluid (σ_1) with uniform current; $\sigma = \dfrac{2(\sigma_1 - \sigma_2)}{2\sigma_1 + \sigma_2}$	$R > a$: $\left[\left(\dfrac{R}{a}\right)^2 - \dfrac{a}{R}\sigma\right]\sin^2\theta;$ 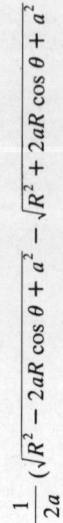 $\dfrac{-a\sigma\cos\theta}{R^2} - \dfrac{2R\cos\theta}{a^2}\sigma$ $R < a$: $-(\sigma - 1)\left(\dfrac{R}{a}\right)^2 \sin^2\theta;$ $\dfrac{(\sigma - 1)2R\cos\theta}{a^2}$
13	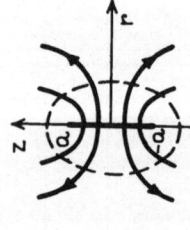	Line source of finite length $2a$	$\dfrac{1}{2a}\left(\sqrt{R^2 - 2aR\cos\theta + a^2} - \sqrt{R^2 + 2aR\cos\theta + a^2}\right);$ $\ln\dfrac{R\cos\theta + a + \sqrt{R^2 + 2aR\cos\theta + a^2}}{R\cos\theta - a + \sqrt{R^2 - 2aR\cos\theta + a^2}}$

Table 1.1. (continued)

1	2	3	4
14	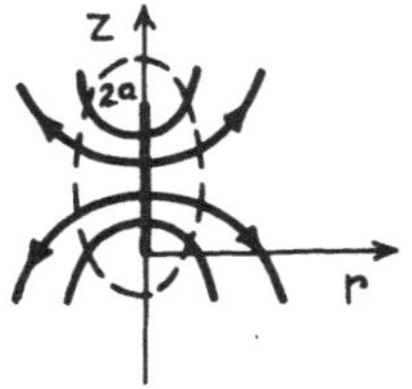	Displaced finite line source	$\dfrac{1}{a}(R - \sqrt{R^2 - 2aR\cos\theta + a^2}\,)$; expression for Φ see No. 13
15	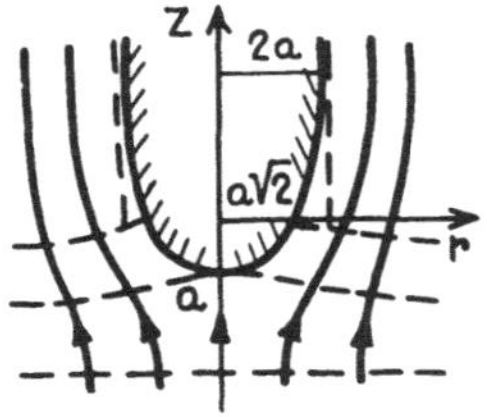	Flow around a cylindrical body of radius $2a$ with blunt end	$R^2 \sin^2\theta - 2a^2(1+\cos\theta)$; $-2R\cos\theta + \dfrac{2a^2}{R}$
16	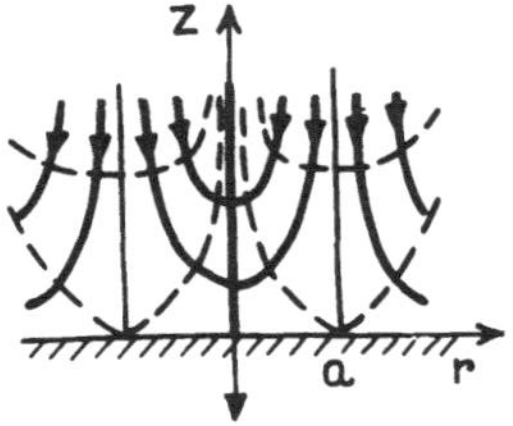	Flow external to a cylinder on the plane	$a^2 R \cos\theta - R^3 \cos\theta \sin^2\theta$; $a^2 \ln R \sin\theta + R^2(\cos^2\theta - \tfrac{1}{2}\sin^2\theta)$

Table 1.1. (continued)

1	2	3	4
17	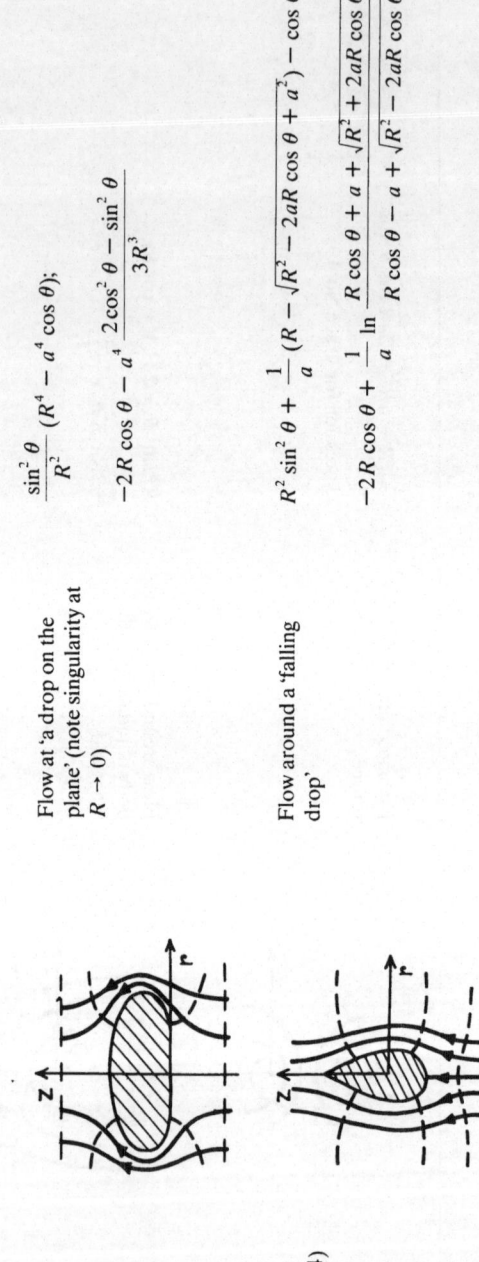	Flow at a nonconducting sphere of radius a	$\sin^2\theta \left[\dfrac{R^2}{a^2} - \dfrac{a}{R} \right];$ $-\dfrac{a}{R^2}\cos\theta - \dfrac{2R}{a^2}\cos\theta$
18 (5–8)		Flow at 'a drop on the plane' (note singularity at $R \to 0$)	$\dfrac{\sin^2\theta}{R^2}(R^4 - a^4\cos\theta);$ $-2R\cos\theta - a^4\,\dfrac{2\cos^2\theta - \sin^2\theta}{3R^3}$
19 (4+5+14)		Flow around a 'falling drop'	$R^2\sin^2\theta + \dfrac{1}{a}(R - \sqrt{R^2 - 2aR\cos\theta + a^2}) - \cos\theta;$ $-2R\cos\theta + \dfrac{1}{a}\ln\dfrac{R\cos\theta + a + \sqrt{R^2 + 2aR\cos\theta + a^2}}{R\cos\theta - a + \sqrt{R^2 - 2aR\cos\theta + a^2}}$

Table 1.1. (continued)

1	2	3	4
20 (2+3)	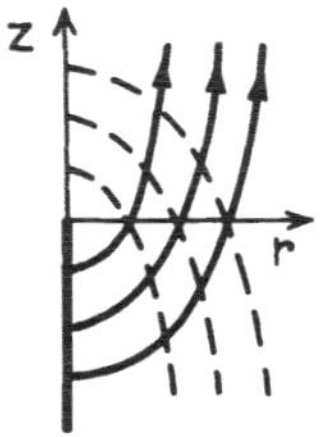 $n = 0; 1$ See scheme 13	Semi-infinite line source (sink)	$R(1 - \cos\theta);$ $-\ln R(1 + \cos\theta)$
21		*Coordinates of prolate spheroid*	
			$\mu;$ $\frac{1}{2}\ln\left|\frac{\lambda-1}{\lambda+1}\right|$
22	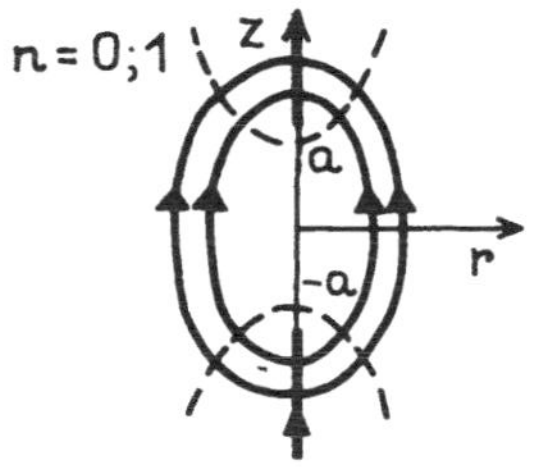 $n = 0; 1$	Semi-infinite line source and sink separated by distance $2a$ from each other	$\lambda;$ $\frac{1}{2}\ln\left|\frac{1-\mu}{1+\mu}\right|$

Table 1.1. (continued)

1	2	3	4
23	See scheme 3		$\lambda\mu$; $\ln\sqrt{(\lambda^2-1)(1-\mu^2)}$
24	See scheme 5		$(1-\mu^2)(\lambda^2-1)$; $-2\mu\lambda$
25	$n=2$ 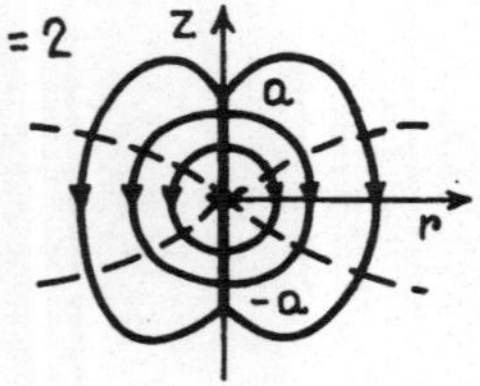	Line source and sink of finite length a	$(1-\mu^2)\left[\frac{1}{2}(1-\lambda^2)\times\ln\frac{\lambda+1}{\lambda-1}+\lambda\right]$; $\mu\left(\lambda\ln\frac{\lambda+1}{\lambda-1}-2\right)$
26	$n=3$ See scheme 7		$\mu(1-\mu^2)\lambda(\lambda^2-1)$; $\frac{1}{2}(\lambda^2-3\mu^2\lambda^2+\mu^2)$
27 (24 + 25)	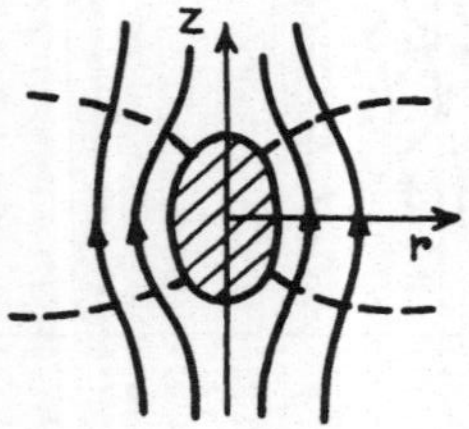	Flow around a nonconducting prolate spheroid placed in uniform current	$(1-\mu^2)\left[\frac{1}{2}(1-\lambda^2)\ln\frac{\lambda+1}{\lambda-1}+\lambda-c(\lambda^2-1)\right]$, $c=\frac{1}{2}\ln\frac{\lambda_0+1}{\lambda_0-1}+\frac{\lambda_0}{\lambda_0^2-1}$; $\mu\left[\lambda\ln\frac{\lambda+1}{\lambda-1}-2+2c\lambda\right]$

Table 1.1. (continued)

1	2	3	4
		Coordinates of oblate spheroid	
28	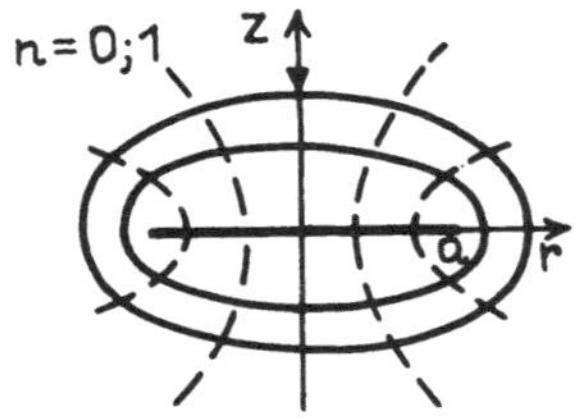 $n = 0; 1$	Semi-infinite line source and sink separated by a nonconducting disc $z = 0, r \leqslant a$	$\lambda;$ $\dfrac{1}{2a} \ln \dfrac{1-\mu}{1+\mu}$
29	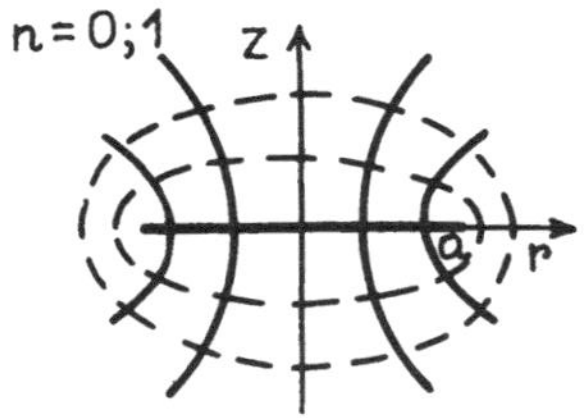 $n = 0; 1$	Current through a hole in plane $z = 0$ (Venturi)	$\mu;$ $\dfrac{1}{a} \arctan \lambda$
30	$n = 0; 1$ See scheme 3		
31	$n = 2$ See scheme 5		$\lambda\mu;$ $\dfrac{1}{2a} \ln[(1+\lambda^2)(1-\mu^2)]$ $(1+\lambda^2)(1-\mu^2);$ $-\dfrac{2\mu\lambda}{a}$

Table 1.1. (continued)

1	2	3	4
32	$n = 2$ 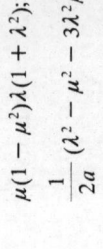	Flow around a nonconducting disc in uniform current	$(1-\mu^2)[(1+\lambda^2)\arctan\lambda + \lambda]$; $-\dfrac{2\mu}{a}(\lambda\arctan\lambda + 1)$
33	$n = 3$ See scheme 7		$\mu(1-\mu^2)\lambda(1+\lambda^2)$; $\dfrac{1}{2a}(\lambda^2 - \mu^2 - 3\lambda^2\mu^2)$
34 $(31+32)$	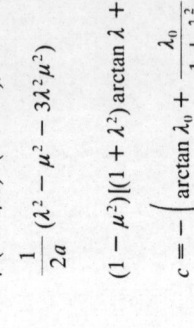	Flow around an oblate spheroid in uniform current	$(1-\mu^2)[(1+\lambda^2)\arctan\lambda + \lambda + c(1+\lambda^2)]$, $c = -\left(\arctan\lambda_0 + \dfrac{\lambda_0}{1+\lambda_0^2}\right)$; $-\dfrac{2\mu}{a}(\lambda\arctan\lambda + 1 + c\lambda)$
35 $(30+33)$	See scheme 16		$\mu(1-\mu^2)\lambda(1+\lambda^2) - \lambda\mu$; $\dfrac{1}{2a}[\lambda^2 - \mu^2 - 3\lambda^2\mu^2 - \ln(1+\lambda^2)(1-\mu^2)]$

Table 1.1. (continued)

1	2	3	4
			Paraboloidal coordinates
36	See scheme 20		η^2
37	See scheme 20		ξ^2
38	See scheme 5		$\xi^2\eta^2$
39	See scheme 3		$\frac{1}{2}(\eta^2 - \xi^2)$

Maxwell's equation $\mathbf{j} = 1/\mu_0 \text{ curl } \mathbf{B}$ expressed for the axisymmetric case in the form

$$\mathbf{j} = \frac{1}{\mu_0} \frac{1}{H_1 H_2 H_3} \begin{vmatrix} H_1 \mathbf{i}_1 & H_2 \mathbf{i}_2 & H_3 \mathbf{i}_3 \\ \dfrac{\partial}{\partial q_1} & \dfrac{\partial}{\partial q_2} & 0 \\ H_1 B_1 & H_2 B_2 & H_3 B_\varphi \end{vmatrix}$$

it follows that

$$j_1 = \frac{1}{\mu_0} \frac{1}{H_2 H_3} \frac{\partial}{\partial q_2} H_3 B_\varphi,$$

$$j_2 = -\frac{1}{\mu_0} \frac{1}{H_1 H_3} \frac{\partial}{\partial q_1} H_3 B_\varphi,$$

$$j_3 = j_\varphi = \frac{1}{\mu_0} \frac{1}{H_1 H_2} \left(\frac{\partial}{\partial q_1} H_2 B_2 - \frac{\partial}{\partial q_2} H_1 B_1 \right).$$

This means that the meridional currents j_1 and j_2 are the source of the azimuthal magnetic field; meridional magnetic fields in principle may originate from azimuthal currents in the fluid. Consider the electric field which can sustain the azimuthal currents in the fluid. Since $\mathbf{j} = \sigma \mathbf{E}$, the azimuthal current is related to the azimuthal electric field, $j_\varphi = \sigma E_\varphi$, i.e. $\mathbf{E} = (0, 0, E_\varphi)$. For the stationary fields curl $\mathbf{E} = 0$, and in the axisymmetric case the equation

$$\text{curl } \mathbf{E} = \frac{1}{H_1 H_2 H_3} \left(H_1 \frac{\partial H_3 E_\varphi}{\partial q_2} \mathbf{i}_1 - H_2 \frac{\partial H_3 E_\varphi}{\partial q_1} \mathbf{i}_2 \right) = 0 \qquad (1.8.1)$$

means that $H_3 E_\varphi = r E_\varphi = \text{const}$. Excluding the singularity of $E_\varphi = \text{const}/r$ at $r = 0$, we obtain $E_\varphi \equiv 0$, i.e. the stationary azimuthal current from an external source in the axisymmetric situation is impossible ($j_\varphi \equiv 0$). From this it follows that currents in homogeneous fluids cannot sustain meridional magnetic fields, which then can originate only from the sources external to the fluid domain. In other words, meridional magnetic fields $\mathbf{B}_m$ are determined in the fluid domain by the equations div $\mathbf{B}_m = 0$ and curl $\mathbf{B}_m = 0$.

Since the meridional magnetic field has two components B_1, B_2, then div $\mathbf{B}_m = 0$ would be satisfied, again introducing the scalar function ψ_2:

$$B_1 = \frac{1}{H_2 H_3} \frac{\partial \psi_2}{\partial q_2}, \qquad B_2 = -\frac{1}{H_3 H_1} \frac{\partial \psi_2}{\partial q_1}. \qquad (1.8.2)$$

Substitution of (1.8.2) in curl $\mathbf{B}_m = 0$ leads to the familiar equation

$$E^2 \psi_2 = 0, \qquad (1.8.3)$$

solutions to which are discussed in Section 1.7. The meaning of the function ψ_2, which is called the magnetic stream function, is elucidated by analogy with the hydrodynamic stream function ψ. Since ψ gives the instantaneous flow rate through the surface of rotation based on the circle of fixed ψ, then ψ_2 gives the magnetic field flux through the same surface, and the lines $\psi_2 = $ const are lines of magnetic flux.

The dimension of the magnetic stream function ψ_2 follows from (1.8.2):

$$[\psi_2] = B_0 L^2 = \mu_0 I L,$$

where B_0 is a characteristic magnitude of the external magnetic field. In contrast to the electric current stream function ψ_1 the dimension of ψ_2 contains the magnetic permeability of the medium μ_0 (which in this case is assumed to be equal to the permeability of free space).

The configuration of the magnetic field flux lines is the same as for the electric current lines represented in Section 1.7, thus problems with both kinds of the magnetic fields can be considered.

If it is necessary to take into account the electric current induced by the fluid flow, and the associated magnetic field, calculation of the electromagnetic force $\mathbf{j} \times \mathbf{B} = 1/\mu_0 \, \text{curl} \, \mathbf{B} \times \mathbf{B}$ must be performed using the equation (1.1.14), which means solving the mutually connected equations of motion (1.1.1) and of magnetic field (1.1.14). In such a formulation there is no need for any direct information concerning the electric field.

If the velocity and magnetic fields are decomposed into their meridional and azimuthal components, then the equation (1.1.14) for axisymmetric fields takes the form

$$\nabla^2 \mathbf{B}_\varphi = -\mu_0 \sigma \, \text{curl}(\mathbf{v}_\varphi \times \mathbf{B}_m + \mathbf{v}_m \times \mathbf{B}_\varphi),$$
$$\nabla^2 \mathbf{B}_m = -\mu_0 \sigma \, \text{curl}(\mathbf{v}_m \times \mathbf{B}_m). \tag{1.8.4}$$

From the equations (1.8.4) it follows that the azimuthal motion cannot directly disturb the meridional magnetic field, but in the presence of the external meridional magnetic field it is able to induce the azimuthal magnetic field. Yet the azimuthal magnetic field itself is deformed by the meridional flow.

Concerning the meridional magnetic field, its deformation occurs due to the meridional motion inducing the azimuthal current $\sigma \mathbf{v}_m \times \mathbf{B}_m$. Thus, the azimuthal current in the axisymmetric situation arises only as the result of the interaction of the meridional motion and the meridional magnetic field, but it cannot exist as a result of external sources of current or electric field.

1.9. Electric field in axisymmetric flow

In the previous section we mentioned the possibility of representing the electromagnetic force in the equation of motion using the magnetic field only. In this way the equation (1.1.14) is solved with the equation of motion to take into account the effect of the fluid flow on the magnetic field. An approximation of low magnetic Reynolds number is often used in magnetohydrodynamics supposing the external magnetic field, or the magnetic field of the electric current from external source, to be unperturbed by the fluid flow. Nevertheless, this does not mean that we neglect the induced electric current and its contribution to the electromagnetic force — it is merely supposed that its effect on the initial magnetic field is negligible.

In Ohm's law

$$\mathbf{j} = \sigma(\mathbf{E} + \mathbf{v} \times \mathbf{B}) \tag{1.9.1}$$

the term $\sigma \mathbf{v} \times \mathbf{B}$ is responsible for the induced electric current, and the electromagnetic force in (1.1.1) takes the form

$$\sigma(\mathbf{E} + \mathbf{v} \times \mathbf{B}) \times \mathbf{B},$$

where $\mathbf{B}$ is the unperturbed magnetic field determined by solutions of (1.6.13), and (1.8.3), (1.8.2).

The electric field in a moving liquid originates, firstly, due to the potential difference ensuring the current flow from the external source, and secondly, due to the fluid flow's interaction with the magnetic field. The latter case is referred to as the fluid flow induced electric field. (It is associated with the charges on the conductor's boundary.)

The relation between the electric current density $\mathbf{j}_0$, supplied to the fluid at rest, and its corresponding electric field is simply

$$\mathbf{j}_0 = \sigma \mathbf{E}_0. \tag{1.9.2}$$

If the distribution of current density $\mathbf{j}_0$ is determined by the solution of (1.6.13), i.e. with the help of its self-magnetic field, then $\mathbf{E}_0$ can be determined from (1.9.2) without difficulty.

Determination of the induced electric field is a more complicated problem. Decomposing the fields $\mathbf{v}$ and $\mathbf{B}$ as $\mathbf{v} = \mathbf{v}_m + \mathbf{v}_\varphi$, $\mathbf{B} = \mathbf{B}_m + \mathbf{B}_\varphi$ in equation (1.1.13), we find that the potential of the induced electric field has to be determined by solving the equation

$$\nabla^2 \Phi = \operatorname{div}(\mathbf{v}_m \times \mathbf{B}_\varphi + \mathbf{v}_\varphi \times \mathbf{B}_m). \tag{1.9.3}$$

In the right side of (1.9.3) there are absent the cross products $\mathbf{v}_\varphi \times \mathbf{B}_\varphi$ and $\mathbf{v}_m \times \mathbf{B}_m$. The first is identically zero, which may be interpreted as follows: the azimuthal fluid motion does not induce the electric field in the magnetic fluid of the meridional current. The second product is absent since the vector $\mathbf{v}_m \times \mathbf{B}_m$ has only the φ-component, but in the axisymmetric situation $\operatorname{div} \mathbf{v}_m \times \mathbf{B}_m = 0$. Taking account of the results of Section 1.8, this means that in the axisymmetric situation the E_φ-component is not generated either by external sources, or by the fluid motion, i.e. $\mathbf{E} = (E_1, E_2, 0)$. Yet note that the azimuthal current $\sigma \mathbf{v}_m \times \mathbf{B}_m$ will exist due to the interaction of the moving medium with the magnetic field, and, thereby, will take part in the electromagnetic force. Thus the electro-

Basic properties of axially symmetric motions in magnetohydrodynamics 57

magnetic force in the approximation of low magnetic Reynolds number can be written as

$$\mathbf{f}_e = \mathbf{j}_0 \times \mathbf{B} + \sigma(\mathbf{E}_i + \mathbf{v} \times \mathbf{B}) \times \mathbf{B}$$
$$= \frac{1}{\mu_0} \text{curl } \mathbf{B}_\varphi \times (\mathbf{B}_m + \mathbf{B}_\varphi) +$$
$$+ \sigma(\mathbf{E}_i + \mathbf{v}_m \times \mathbf{B}_m + \mathbf{v}_m \times \mathbf{B}_\varphi + \mathbf{v}_\varphi \times \mathbf{B}_m) \times (\mathbf{B}_m + \mathbf{B}_\varphi), \quad (1.9.4)$$

where $\mathbf{E}_i$ means, in contrast with (1.9.2), the induced electric field determined by (1.9.3); $\mathbf{B}_m$ and $\mathbf{B}_\varphi$ are the unperturbed fields.

1.10. Full set of equations for axisymmetric motion

Before writing down the set of equations governing the fields of the sought MHD-quantities in the axisymmetric case, let us list the expressions for the differential operators and for the vector fields in the axisymmetric case. If $a(q_1, q_2)$ is a scalar function and $\mathbf{A}(q_1, q_2)$ is a vector function, then

$$\nabla a = \mathbf{i}_1 \frac{1}{H_1} \frac{\partial a}{\partial q_1} + \mathbf{i}_2 \frac{1}{H_2} \frac{\partial a}{\partial q_2}, \quad (1.10.1)$$

$$\nabla \cdot \mathbf{A} = \frac{1}{H_1 H_2 H_3} \left[\frac{\partial}{\partial q_1} (H_2 H_3 A_1) + \frac{\partial}{\partial q_2} (H_3 H_1 A_2) \right], \quad (1.10.2)$$

$$\nabla \times \mathbf{A} = \frac{1}{H_1 H_2 H_3} \begin{vmatrix} H_1 \mathbf{i}_1 & H_2 \mathbf{i}_2 & H_3 \mathbf{i}_\varphi \\ \frac{\partial}{\partial q_1} & \frac{\partial}{\partial q_2} & 0 \\ H_1 A_1 & H_2 A_2 & H_3 A_3 \end{vmatrix}, \quad (1.10.3)$$

$$\nabla^2 a = \frac{1}{H_1 H_2 H_3} \left[\frac{\partial}{\partial q_1} \left(\frac{H_2 H_3}{H_1} \frac{\partial a}{\partial q_1} \right) + \right.$$
$$\left. + \frac{\partial}{\partial q_2} \left(\frac{H_3 H_1}{H_2} \frac{\partial a}{\partial q_2} \right) \right], \quad (1.10.4)$$

$$E^2 = \frac{H_3}{H_1 H_2} \left[\frac{\partial}{\partial q_1} \left(\frac{H_2}{H_1 H_3} \frac{\partial}{\partial q_1} \right) + \right.$$
$$\left. + \frac{\partial}{\partial q_2} \left(\frac{H_1}{H_2 H_3} \frac{\partial}{\partial q_2} \right) \right]. \quad (1.10.5)$$

The vector fields in the axisymmetric case can be expressed in the form:

$$\mathbf{v} = \left(\frac{1}{H_2 H_3} \frac{\partial \psi}{\partial q_2}, -\frac{1}{H_1 H_3} \frac{\partial \psi}{\partial q_1}, v_\varphi \right); \qquad (1.10.6)$$

$$\operatorname{curl} \mathbf{v} = \mathbf{w} = \left(\frac{1}{H_2 H_3} \frac{\partial H_3 v_\varphi}{\partial q_2}, \frac{1}{H_1 H_3} \frac{\partial H_3 v_\varphi}{\partial q_1}, -\frac{E^2 \psi}{H_3} \right); \qquad (1.10.7)$$

$$\mathbf{j} = \left(\frac{1}{H_2 H_3} \frac{\partial \psi_1}{\partial q_2}, -\frac{1}{H_1 H_3} \frac{\partial \psi_1}{\partial q_1}, j_\varphi \right), \quad \psi_1 = \frac{1}{\mu_0} H_3 B_\varphi; \qquad (1.10.8)$$

$$\mathbf{B} = \left(\frac{1}{H_2 H_3} \frac{\partial \psi_2}{\partial q_2}, -\frac{1}{H_1 H_3} \frac{\partial \psi_2}{\partial q_1}, B_\varphi \right); \qquad (1.10.9)$$

$$\mathbf{E} = \left(-\frac{1}{H_1} \frac{\partial \Phi}{\partial q_1}, -\frac{1}{H_2} \frac{\partial \Phi}{\partial q_2}, 0 \right). \qquad (1.10.10)$$

As shown in (1.10.6)—(1.10.10) the assumption of axial symmetry permits us to reduce the number of scalar functions to be determined to the five functions: ψ, ψ_1, ψ_2, v_φ, Φ, instead of the seven functions: v_1, v_2, v_3, B_1, B_2, B_3, Φ in the three dimensional case.

The electromagnetic force in (1.1.1) can be expressed, as noted earlier, in two ways: either using the magnetic field only, or using the magnetic field and the electric field (or the potential). The reason for this lies in the double expressions for the current density: (1.1.6) and (1.1.9). When using (1.1.6), the expression for the electromagnetic force is $(1/\mu_0)$ curl $\mathbf{B} \times \mathbf{B}$, where $\mathbf{B}$ is determined by solving the equation (1.1.14).

The equation of motion is given below in the form of (1.1.15) where pressure is eliminated using the curl operation on the equation (1.1.1). It may also appear to be useful to represent the equations in vector form (with differential operators and vector operations). In practice the set of equations is often used in various approximations. Thus, in the low magnetic Reynolds number approximation the magnetic field is supposed to be unperturbed by the fluid motion (more precisely, by the motion induced electric current), but the induced current is taken into account in the expression for the electromagnetic

Basic properties of axially symmetric motions in magnetohydrodynamics 59

force. Even the lower order approximation is one that also neglects the induced current. In this approximation, which we shall call the electrodynamic approximation, the electromagnetic force is the result of the interaction of the externally supplied electric current with the self- and external magnetic fields. The equations in this approximations are also given below.

The set of governing equations is obtained by substituting (1.10.6)–(1.10.10) in (1.1.15) and (1.1.14) (the last equation can be replaced by (1.8.4)).

Equations of motion

$$\frac{H_3}{H_1 H_2} \left[\frac{\partial \psi}{\partial q_1} \frac{\partial}{\partial q_2} \left(\frac{E^2 \psi}{H_3^2} \right) - \frac{\partial \psi}{\partial q_2} \frac{\partial}{\partial q_1} \left(\frac{E^2 \psi}{H_3^2} \right) \right] +$$

$$+ \frac{H_3}{H_1 H_2} H_3 v_\varphi \left[\frac{\partial}{\partial q_1} (H_3 v_\varphi) \frac{\partial}{\partial q_2} \left(\frac{1}{H_3^2} \right) - \right.$$

$$\left. - \frac{\partial}{\partial q_2} (H_3 v_\varphi) \frac{\partial}{\partial q_1} \left(\frac{1}{H_3^2} \right) \right] + \nu E^4 \psi$$

$$= \frac{\mu_0}{\rho} \frac{H_3}{H_1 H_2} \psi_1 \left[\frac{\partial \psi_1}{\partial q_1} \frac{\partial}{\partial q_2} \left(\frac{1}{H_3^2} \right) - \right.$$

$$\left. - \frac{\partial \psi_1}{\partial q_2} \frac{\partial}{\partial q_1} \left(\frac{1}{H_3^2} \right) \right] +$$

$$+ \frac{1}{\rho \mu_0} \frac{H_3}{H_1 H_2} \left[\frac{\partial \psi_2}{\partial q_1} \frac{\partial}{\partial q_2} \left(\frac{E^2 \psi_2}{H_3^2} \right) - \right.$$

$$\left. - \frac{\partial \psi_2}{\partial q_2} \frac{\partial}{\partial q_1} \left(\frac{E^2 \psi_2}{H_3^2} \right) \right]; \qquad (1.10.11)$$

$$\frac{1}{H_1 H_2 H_3} \left[\frac{\partial \psi}{\partial q_1} \frac{\partial}{\partial q_2} (H_3 v_\varphi) - \frac{\partial \psi}{\partial q_2} \frac{\partial}{\partial q_1} (H_3 v_\varphi) \right] + \nu E^2 (H_3 v_\varphi)$$

$$= \frac{1}{\rho} \frac{1}{H_1 H_2 H_3} \left(\frac{\partial \psi_2}{\partial q_1} \frac{\partial \psi_1}{\partial q_2} - \frac{\partial \psi_2}{\partial q_2} \frac{\partial \psi_1}{\partial q_1} \right). \qquad (1.10.12)$$

The first of these equations is the φ-component of the curl of the Navier-Stokes equations, and it describes the meridional motion; the second is the azimuthal component of the Navier-Stokes equations, and it describes azimuthal rotation.

Magnetic field equations

$$E^2\psi_1 = \sigma \frac{H_3}{H_1 H_2}\left[\frac{\partial \psi_2}{\partial q_1}\frac{\partial}{\partial q_2}\left(\frac{v_\varphi}{H_3}\right) - \frac{\partial \psi_2}{\partial q_2}\frac{\partial}{\partial q_1}\left(\frac{v_\varphi}{H_3}\right) + \right.$$

$$\left. + \mu_0\left(\frac{\partial \psi}{\partial q_2}\frac{\partial}{\partial q_1}\frac{\psi_1}{H_3^2} - \frac{\partial \psi}{\partial q_1}\frac{\partial}{\partial q_2}\frac{\psi_1}{H_3^2}\right)\right]; \qquad (1.10.13)$$

$$E^2\psi_2 = \mu_0 \sigma \frac{1}{H_1 H_2 H_3}\left(\frac{\partial \psi_2}{\partial q_1}\frac{\partial \psi}{\partial q_2} - \frac{\partial \psi_2}{\partial q_2}\frac{\partial \psi}{\partial q_1}\right). \qquad (1.10.14)$$

These equations in vector form are

$$\left[H_3 \nabla\psi \times \nabla\left(\frac{E^2\psi}{H_3^2}\right) + H_3 v_\varphi \nabla(H_3 v_\varphi) \times \nabla\left(\frac{1}{H_3^2}\right)\right]\cdot \mathbf{i}_\varphi + \nu E^4 \psi$$

$$= \frac{H_3}{\rho}\left[\mu_0 \psi_1 \nabla \psi_1 \times \nabla\left(\frac{1}{H_3^2}\right) + \right.$$

$$\left. + \frac{1}{\mu_0}\nabla\psi_2 \times \nabla\left(\frac{E^2\psi_2}{H_3^2}\right)\right]\cdot \mathbf{i}_\varphi; \qquad (1.10.15)$$

$$\frac{1}{H_3}\nabla\psi \times \nabla(H_3 v_\varphi)\cdot \mathbf{i}_\varphi + \nu E^2(H_3 v_\varphi) = \frac{1}{\rho H_3}\nabla\psi_2 \times \nabla\psi_1 \cdot \mathbf{i}_\varphi; \qquad (1.10.16)$$

$$E^2\psi_1 = \sigma H_3\left[\nabla\psi_2 \times \nabla\left(\frac{v_\varphi}{H_3}\right) + \mu_0 \nabla\left(\frac{\psi_1}{H_3^2}\right)\times \nabla\psi\right]\cdot \mathbf{i}_\varphi; \qquad (1.10.17)$$

$$E^2\psi_2 = \mu_0 \sigma \frac{1}{H_3}\nabla\psi_2 \times \nabla\psi \cdot \mathbf{i}_\varphi. \qquad (1.10.18)$$

In the electrodynamic approximation the magnetic field, determined by the functions ψ_1 and ψ_2, is found from simpler equations than (1.10.13), (1.10.14) i.e.:

$$E^2\psi_1 = 0, \qquad (1.10.19)$$

$$E^2\psi_2 = 0. \qquad (1.10.20)$$

Basic properties of axially symmetric motions in magnetohydrodynamics 61

Consequently, the first equation of motion (1.10.11) simplifies as ((1.10.12) remaining unchanged):

$$\frac{H_3}{H_1 H_2}\left[\frac{\partial \psi}{\partial q_1}\frac{\partial}{\partial q_2}\left(\frac{E^2\psi}{H_3^2}\right) - \frac{\partial \psi}{\partial q_2}\frac{\partial}{\partial q_1}\left(\frac{E^2\psi}{H_3^2}\right)\right] +$$

$$+ \frac{H_3}{H_1 H_2} H_3 v_\varphi \left[\frac{\partial}{\partial q_1}(H_3 v_\varphi)\frac{\partial}{\partial q_2}\left(\frac{1}{H_3^2}\right) - \right.$$

$$\left. - \frac{\partial}{\partial q_2}(H_3 v_\varphi)\frac{\partial}{\partial q_1}\left(\frac{1}{H_3^2}\right)\right] + vE^4\psi$$

$$= \frac{\mu_0}{\rho}\frac{H_3}{H_1 H_2}\psi_1\left[\frac{\partial \psi_1}{\partial q_1}\frac{\partial}{\partial q_2}\left(\frac{1}{H_3^2}\right) - \right.$$

$$\left. - \frac{\partial \psi_1}{\partial q_2}\frac{\partial}{\partial q_1}\left(\frac{1}{H_3^2}\right)\right]. \tag{1.10.21}$$

In the following the equation for the pure electrically induced flow will be often used with $v_\varphi = 0$:

$$\frac{H_3}{H_1 H_2}\left[\frac{\partial \psi}{\partial q_1}\frac{\partial}{\partial q_2}\left(\frac{E^2\psi}{H_3^2}\right) - \frac{\partial \psi}{\partial q_2}\frac{\partial}{\partial q_1}\left(\frac{E^2\psi}{H_3^2}\right)\right] + vE^4\psi$$

$$= \frac{\mu_0}{\rho}\frac{H_3}{H_1 H_2}\psi_1\left[\frac{\partial \psi_1}{\partial q_1}\frac{\partial}{\partial q_2}\left(\frac{1}{H_3^2}\right) - \right.$$

$$\left. - \frac{\partial \psi_1}{\partial q_2}\frac{\partial}{\partial q_1}\left(\frac{1}{H_3^2}\right)\right], \tag{1.10.22}$$

in which ψ_1 is determined by (1.10.19).

In the low magnetic Reynolds number approximation the equations (1.10.19), (1.10.20) are to be used instead of (1.10.13), (1.10.14).

2

Solutions in spherical coordinates

With this chapter we begin a systematic study of axisymmetric electrically induced vortex flows, the governing equations of which have been presented in the previous chapter. The study will begin with so-called exact solutions, by which we mean those solutions obtained from the full equations of motion without *a priori* estimates and omission of any terms, say, after the order-of-magnitude analysis (this does not apply to the electrodynamic quantities, which will conform to assumptions simplifying the electrodynamic part of problem).

The nonlinearity of the Navier-Stokes equations leads to a very limited number of exact solutions. In general, such solutions are obtained

(a) for the flows invariant along a certain direction in space;
(b) for flows permitting us to transform the governing initial partial differential equations into ordinary differential equations (self-similar flows).

The first kind of flows are those with parallel stream lines when the nonlinear terms are identically zero (e.g., Couette flow, one-dimensional flow in tubes, transient flow at an infinite plane, steady flow at an infinite plane with a uniform suction or with an homogeneous transverse magnetic field, flow along an infinite cylinder in a transverse field, and so on). The nonlinear terms are not zero for the second kind of flows (e.g., plane and axisymmetric flows at a critical point, flow at an infinite rotating disc, Landau-Squire jet flows, flow in a plane diffuser, Hamel flow, and so on).

Solutions related to the exact ones are those obtained under the assumption that separate nonlinear inertial terms are not zero, yet they mutually cancel out. The velocity field, found from this condition, serves to determine the magnetic field exactly. Such flow models are often used when solving the problems in astrophysics [15].

The exact solutions are valuable, not merely due to the direct information which they supply, but also because they give us the possibility to check various assumptions made when constructing approximate solutions, or as test problems for computational magnetohydrodynamics.

In this and the following chapters we shall consider some similarity solutions in spherical and cylindrical coordinates, which can be applied to electrically

Solutions in spherical coordinates

induced vortex flows. The analysis of exact solutions in other coordinate systems is left for the future.

2.1. Definition of the class of exact solutions

Similarity solutions to the equations (1.10.11)–(1.10.14), expressed in the spherical coordinates (R, θ, φ),

$$\sin\theta \left(\frac{\partial \psi}{\partial R} \frac{\partial}{\partial \theta} \frac{E^2 \psi}{R^2 \sin^2 \theta} - \frac{\partial \psi}{\partial \theta} \frac{\partial}{\partial R} \frac{E^2 \psi}{R^2 \sin^2 \theta} \right) +$$

$$+ \frac{\sin\theta}{R} \frac{\partial v_\varphi^2}{\partial \theta} - \cos\theta \frac{\partial v_\varphi^2}{\partial R} + \nu E^4 \psi$$

$$= \frac{\mu_0}{\rho} \left(\frac{1}{R^3 \sin\theta} \frac{\partial \psi_1^2}{\partial \theta} - \frac{\cos\theta}{R^2 \sin^2\theta} \frac{\partial \psi_1^2}{\partial R} \right) +$$

$$+ \frac{1}{\rho\mu_0} \sin\theta \left(\frac{\partial \psi_2}{\partial R} \frac{\partial}{\partial \theta} \frac{E^2 \psi_2}{R^2 \sin^2 \theta} - \right.$$

$$\left. - \frac{\partial \psi_2}{\partial \theta} \frac{\partial}{\partial R} \frac{E^2 \psi_2}{R^2 \sin^2 \theta} \right); \qquad (2.1.1)$$

$$\frac{\partial \psi}{\partial R} \frac{\partial}{\partial \theta} (R \sin\theta v_\varphi) - \frac{\partial \psi}{\partial \theta} \frac{\partial}{\partial R} (R \sin\theta v_\varphi) +$$

$$+ \nu R^2 \sin\theta E^2 (R \sin\theta v_\varphi)$$

$$= \frac{1}{\rho} \left(\frac{\partial \psi_2}{\partial R} \frac{\partial \psi_1}{\partial \theta} - \frac{\partial \psi_2}{\partial \theta} \frac{\partial \psi_1}{\partial R} \right); \qquad (2.1.2)$$

$$E^2 \psi_1 = \sigma \sin\theta \left[\frac{\partial \psi_2}{\partial R} \frac{\partial}{\partial \theta} \frac{v_\varphi}{R \sin\theta} - \frac{\partial \psi_2}{\partial \theta} \frac{\partial}{\partial R} \frac{v_\varphi}{R \sin\theta} + \right.$$

$$+ \mu_0 \left(\frac{\partial \psi}{\partial \theta} \frac{\partial}{\partial R} \frac{\psi_1}{R^2 \sin^2\theta} - \right.$$

$$\left.\left. - \frac{\partial \psi}{\partial R} \frac{\partial}{\partial \theta} \frac{\psi_1}{R^2 \sin^2\theta} \right) \right]; \qquad (2.1.3)$$

$$E^2\psi_2 = \frac{\mu_0 \sigma}{R^2 \sin\theta} \left(\frac{\partial \psi_2}{\partial R} \frac{\partial \psi}{\partial \theta} - \frac{\partial \psi_2}{\partial \theta} \frac{\partial \psi}{\partial R} \right); \qquad (2.1.4)$$

$$E^2 = \frac{\partial^2}{\partial R^2} + \sin\theta \frac{\partial}{\partial \theta} \frac{1}{R^2 \sin\theta} \frac{\partial}{\partial \theta},$$

can be obtained either by separating variables in the equations (2.1.1)–(2.1.4), or by introducing a new variable as a combination of coordinates. The second approach is advocated when setting up problems in Cartesian and cylindrical coordinates (see Chapter 4), and the first when dealing with cylindrical and spherical coordinates (we shall use it below).

Let us initially restrict our analysis to the hydrodynamic part of the problem without the electromagnetic terms and let $v_\varphi = 0$. Separating variables in an expression for the stream function $\psi = f(R)g(\theta)$ and substituting ψ in the left-hand side of (2.1.1), we obtain

$$f'g \sin\theta \left[\frac{f''}{R^2} \left(\frac{g}{\sin^2\theta} \right)' + \frac{f}{R^4} \left(\frac{1}{\sin\theta} \left(\frac{g'}{\sin\theta} \right) \right)' \right] -$$

$$- fg' \left[\frac{g}{\sin\theta} \left(\frac{f''}{R^2} \right)' + \left(\frac{g'}{\sin\theta} \right)' \left(\frac{f}{R^4} \right)' \right] +$$

$$+ \nu \sin\theta \left\{ \frac{g}{\sin\theta} f^{IV} + \left(\frac{g'}{\sin\theta} \right)' \left(\frac{f}{R^2} \right)'' + \left(\frac{g'}{\sin\theta} \right)' \frac{f''}{R^2} + \right.$$

$$\left. + \left(\frac{1}{\sin\theta} \left[\sin\theta \left(\frac{g'}{\sin\theta} \right)' \right]' \right)' \frac{f}{R^4} \right\} = 0, \qquad (2.1.5)$$

where differentiation is with respect to R or θ, depending on which argument enters the function. To separate the variables in the viscosity term, it is essential that the factors depending on R are directly proportional to each other: This is possible in the case when f is a power function of R ($f \sim R^\alpha$). Substituting this function in (2.1.5) we find that the inertial term is of the order of $R^{2\alpha-5}$ and the viscous term is $R^{\alpha-4}$; their orders are equal if $\alpha = 1$. Then $f = CR$. Further, dividing (2.1.5) by R^{-3}, we obtain an ordinary differential equation with the argument θ.

Since the dimension of the stream function is L^3/T, the dimension of factor C must be L^2/T. Setting the kinematic viscosity coefficient (its dimension is also

Solutions in spherical coordinates 65

L^2/T) instead of the constant C, we obtain a form of meridional flow stream function for the similarity solution:

$$\psi = vRg(\theta). \tag{2.1.6}$$

The corresponding to (2.1.6) velocity field of meridional flow is

$$v_R = \frac{1}{R^2 \sin\theta} \frac{\partial \psi}{\partial \theta} = \frac{v}{R} \frac{g'(\theta)}{\sin\theta},$$

$$v_\theta = -\frac{1}{R \sin\theta} \frac{\partial \psi}{\partial R} = -\frac{v}{R} \frac{g(\theta)}{\sin\theta}. \tag{2.1.7}$$

As follows from (2.1.7), the velocity field of the class of similarity solutions under consideration has meridional flow velocity components that are inversely proportional to the spherical radius. It is easy to see that the similarity of the equations (2.1.1), (2.1.2) would remain if we also choose the azimuthal velocity in the form

$$v_\varphi = \frac{v}{R} \frac{\ell(\theta)}{\sin\theta}. \tag{2.1.8}$$

Let the reader note that the structure of the electromagnetic term in (2.1.1) is analogous to the inertial term (e.g. the terms containing ψ_2 and ψ), hence the variables can also be separated in the electromagnetic term if the magnetic field is in a form similar to (2.1.7), (2.1.8):

$$B_R = \frac{\mu_0 I_1}{2\pi R} \frac{G'(\theta)}{\sin\theta}, \quad B_\theta = -\frac{\mu_0 I_1}{2\pi R} \frac{G(\theta)}{\sin\theta},$$

$$B_\varphi = \frac{\mu_0 I}{2\pi R} \frac{L(\theta)}{\sin\theta}. \tag{2.1.9}$$

The form of the dimensional factor in (2.1.9) is also determined by dimensional considerations: the dimension of the magnetic field $[B] = T = $ kg A^{-1}s^{-2} can be constructed from the dimensions of length $[L] = m$, magnetic permeability $[\mu_0] = H/m = $ m^{2}kg A^{-2}s^{-2}, and of a quantity with the dimension of electric current $[I] = [I_1] = A$. In (2.1.9) the meridional field (B_R, B_θ) and azimuthal B_φ have different sources, as was noted in Chapter 1. Hence these have different characteristic current scales I_1 and I.

The magnetic stream function ψ_2 is related to the function $G(\theta)$ in (2.1.9) by an expression similar to (2.1.6):

$$\psi_2 = \frac{\mu_0 I_1}{2\pi} RG(\theta), \tag{2.1.10}$$

the electric stream function ψ differs from the function $L(\theta)$ in (2.1.9) by a constant factor only:

$$\psi_1 = \frac{I}{2\pi} L(\theta). \tag{2.1.11}$$

Substituting (2.1.6), (2.1.8), (2.1.10), and (2.1.11) in (2.1.1)–(2.1.4) we obtain

$$g\left[\frac{1}{\sin\theta}\left(\frac{g'}{\sin\theta}\right)'\right]' + \frac{3g'}{\sin\theta}\left(\frac{g'}{\sin\theta}\right)' + 2\left(\frac{g'}{\sin\theta}\right)' +$$

$$+ \left\{\frac{1}{\sin\theta}\left[\sin\theta\left(\frac{g'}{\sin\theta}\right)'\right]'\right\}' + \frac{2\ell\ell'}{\sin^2\theta}$$

$$= S\frac{2LL'}{\sin^2\theta} + k^2 S\left\{G\left[\frac{1}{\sin\theta}\left(\frac{G'}{\sin\theta}\right)'\right]' + \right.$$

$$\left. + \frac{3G'}{\sin\theta}\left(\frac{G'}{\sin\theta}\right)'\right\}; \qquad (2.1.12)$$

$$\sin^2\left(\frac{\ell'}{\sin\theta}\right)' + g\ell' = kSGL'; \qquad (2.1.13)$$

$$\left(\frac{L'}{\sin\theta}\right)' = k\beta\left[G\left(\frac{\ell}{\sin\theta}\right)' + 2G'\frac{\ell}{\sin\theta}\right] +$$

$$+ \beta\left(\frac{2g'L}{\sin^2\theta} - \frac{2\cos\theta}{\sin^3\theta}gL + \frac{L'g}{\sin^2\theta}\right); \qquad (2.1.14)$$

$$\left(\frac{G'}{\sin\theta}\right)' = \frac{\beta}{\sin^2\theta}(Gg' - G'g). \qquad (2.1.15)$$

Here the meaning of the dimensionless parameters

$$S = \frac{\mu_0}{4\pi^2}\frac{I^2}{\rho v^2}, \qquad \beta = \mu_0 \sigma v, \qquad k = I_1/I,$$

will be discussed later in Section 2.3.

Solutions in spherical coordinates

The equation (2.1.12) can be integrated three times:

$$g' \sin\theta - 2g \cos\theta + \frac{1}{2} g^2 + (1 + \cos^2\theta) \int \frac{\cos\theta}{\sin^3\theta} \ell^2 \, d\theta -$$

$$- \cos\theta \int \frac{1 + \cos^2\theta}{\sin^3\theta} \ell^2 \, d\theta$$

$$= -a \cos^2\theta + b \cos\theta - c + k^2 S \frac{G^2}{2} +$$

$$+ S \left[(1 + \cos^2\theta) \int \frac{\cos\theta}{\sin^3\theta} L^2 \, d\theta - \right.$$

$$\left. - \cos\theta \int \frac{1 + \cos^2\theta}{\sin^3\theta} L^2 \, d\theta \right]. \qquad (2.1.16)$$

Thus the full set of equations, determining the whole class of exact solutions under consideration, contains the equations (2.1.13)–(2.1.16). If needed, the pressure p can be determined using the full pressure function $p_m = p + |\mathbf{B}|^2/2\mu_0$ from the expression [50]:

$$p_m = \frac{v^2 \rho}{R^2 \sin^2\theta} \left[2g' \sin\theta - 2g \cos\theta - b \cos\theta + a + c - \right.$$

$$- \cos\theta \int \frac{1 + \cos^2\theta}{\sin^3\theta} \ell^2 \, d\theta + 2 \int \frac{\cos\theta}{\sin^3\theta} \ell^2 \, d\theta +$$

$$\left. + S \left(\cos\theta \int \frac{1 + \cos^2\theta}{\sin^3\theta} L^2 \, d\theta - 2 \int \frac{\cos\theta}{\sin^3\theta} L^2 \, d\theta \right) \right]. \qquad (2.1.17)$$

2.2. Low magnetic Reynolds number approximation. Electric current and external magnetic fields

In Section 1.10 the set of equations for the low magnetic Reynolds number approximation were obtained by simply neglecting the right-hand sides in the equations (1.10.13) and (1.10.14). The approximation can be derived more formally by expanding the magnetic field functions in power series of the small parameter β, the Batchelor number (e.g. for mercury $\beta = 10^{-7}$): $L = L_0 +$

$\beta L_1 + \ldots$, $G = G_0 + \beta G_1 + \ldots$, and retaining the terms to the first power of β. Then the set of equations (2.1.16), (2.1.13)–(2.1.15) takes the form:

$$g' \sin \theta - 2g \cos \theta + \frac{1}{2} g^2 +$$

$$+ (1 + \cos^2 \theta) \int \frac{\cos \theta}{\sin^3 \theta} \ell^2 \, d\theta - \int \frac{1 + \cos^2 \theta}{\sin^3 \theta} \ell^2 \, d\theta$$

$$= -a \cos^2 \theta + b \cos \theta - c + S \left[(1 + \cos^2 \theta) \int \frac{\cos \theta}{\sin^3 \theta} L_0^2 \, d\theta - \right.$$

$$- \cos \theta \int \frac{1 + \cos^2 \theta}{\sin^3 \theta} L_0^2 \, d\theta \Bigg] + k^2 S \frac{1}{2} G_0^2 +$$

$$+ 2S\beta \left[(1 + \cos^2 \theta) \int \frac{\cos \theta}{\sin^3 \theta} L_0 L_1 \, d\theta - \right.$$

$$- \cos \theta \int \frac{1 + \cos^2 \theta}{\sin^3 \theta} L_0 L_1 \, d\theta \Bigg] + k^2 S \beta G_0 G_1; \qquad (2.2.1)$$

$$\sin^2 \theta \left(\frac{\ell'}{\sin \theta} \right)' + g\ell' = kSG_0 L_0' + kS\beta (G_0 L_1' + G_1 L_0'); \qquad (2.2.2)$$

$$\left(\frac{L_0'}{\sin \theta} \right)' = 0; \qquad (2.2.3)$$

$$\left(\frac{G_0'}{\sin \theta} \right)' = 0; \qquad (2.2.4)$$

$$\left(\frac{L_1'}{\sin \theta} \right)' = k \left[G_0 \left(\frac{\ell}{\sin \theta} \right)' + 2G_0' \frac{\ell}{\sin \theta} \right] +$$

$$+ \frac{2g' L_0}{\sin^2 \theta} - \frac{2 \cos \theta}{\sin^3 \theta} gL_0 + \frac{L_0' g}{\sin^2 \theta}; \qquad (2.2.5)$$

$$\left(\frac{G_1'}{\sin \theta} \right)' = \frac{1}{\sin \theta} (G_0 g' - G_0' g). \qquad (2.2.6)$$

Equation (2.2.3) describes the distribution of the meridional current supplied externally to the fluid (and the associated azimuthal magnetic field), and the

Solutions in spherical coordinates

equation (2.2.4) describes the external meridional magnetic field; both fields are not perturbed by the fluid motion. The electric currents induced by the fluid motion in these fields and the magnetic field perturbations are expressed by the functions L_1 and G_1 (equations (2.2.5), (2.2.6)).

Let us now seek the forms of current distributions and external magnetic fields determined by the equations (2.2.3), (2.2.4), which may be considered as belonging to the class of exact solutions. According to the expressions for ψ_1 (2.1.11) and ψ_2 (2.1.10), these functions correspond in Table 1.1 to numbers 1 and 4 for the electric current, and 2, 3, and 20 for the magnetic field. Consider these distributions in greater detail. From (2.1.11), (1.10.8) it follows that

$$j_R = \frac{I}{2\pi R^2 \sin\theta} \frac{\partial L(\theta)}{\partial \theta}; \quad j_\theta = -\frac{I}{2\pi R \sin\theta} \frac{\partial L(\theta)}{\partial R} = 0, \quad (2.2.7)$$

i.e. it is possible to consider only the radially diverging (converging) electric current in this class of exact solutions. Moreover, a general solution of (2.2.3) is

$$L_0 = B - A\cos\theta, \quad (2.2.8)$$

corresponding to the azimuthal magnetic field

$$B_\varphi = \frac{\mu_0 I}{2\pi} \frac{B - A\cos\theta}{R\sin\theta}. \quad (2.2.9)$$

For different relations between the constants A and B in (2.2.3) there may be various kinds of the radial electric current flow. Consider, for instance, a situation where a liquid conductor fills a cone with angle $2\theta_1$ at the apex (Figure 1 in Table 2.1). (The plane is a particular case of the cone when $\theta_1 = \pi/2$.) An electric current is supplied to the apex by a thin wire along $\theta = \pi$. We shall find the distribution of B_φ only inside the fluid conductor. On using the integral form of the equation $(1/\mu_0)\text{curl }\mathbf{B} = \mathbf{j}$, applied to a circle of radius $r = R\sin\theta$, and $\mathbf{j} = \mathbf{i}_R j_R$ (2.2.7), (2.2.8), we obtain

$$\frac{1}{\mu_0} \oint B_\varphi r\, d\varphi = \frac{2\pi}{\mu_0} rB_\varphi = 2\pi \int_0^\theta j_R R^2 \sin\theta\, d\theta$$

$$= IA(1 - \cos\theta). \quad (2.2.10)$$

and

$$B_\varphi = \mu_0 I \frac{A}{2\pi R} \frac{1 - \cos\theta}{\sin\theta},$$

i.e. in (2.2.8) and (2.2.9) $B = A$. At the cone's boundary $\theta = \theta_1$ and the right-hand side of (2.2.10) must be equal to the total current I passing through the cone's normal section $z = \text{const}$. Therefore $A = B = (1 - \cos\theta_1)^{-1}$ and

$$B_\varphi = \frac{\mu_0 \psi_1}{R\sin\theta} = \frac{\mu_0 I}{2\pi R\sin\theta} \frac{1 - \cos\theta}{1 - \cos\theta_1}. \quad (2.2.11)$$

The same result may be obtained by determining B_φ (2.2.9) from the conditions: $B_\varphi|_{\theta=0} = 0$ and $B_\varphi|_{\theta=\theta_1} = \mu_0 I/2\pi R \sin \theta$. Below, in Table 2.1, are

$\theta \leqslant \theta_1$	$\theta \leqslant \theta_1$	$\theta_1 \leqslant \theta \leqslant \theta_2$	$\theta_1 \leqslant \theta \leqslant \theta_2$	$\theta_1 \leqslant \theta \leqslant \pi - \theta_1$
1	2	3	4	5
$\dfrac{\mu_0 I (1-\cos\theta)}{2\pi R \sin\theta (1-\cos\theta_1)}$	$\dfrac{\mu_0 I (\cos\theta_1-\cos\theta)}{2\pi R \sin\theta (1-\cos\theta_1)}$	$\dfrac{\mu_0 I (\cos\theta_1-\cos\theta)}{2\pi R \sin\theta (\cos\theta_1-\cos\theta_2)}$	$\dfrac{\mu_0 I (\cos\theta_2-\cos\theta)}{2\pi R \sin\theta (\cos\theta_1-\cos\theta_2)}$	$\dfrac{\mu_0 I \cos\theta}{2\pi R \sin\theta \cos\theta_1}$

listed the expressions for B_φ obtained in a similar way for different current flow schemes. Heavy arrows show the current supply. The upper row in the table indicates a domain occupied by the liquid conductor, the drawings show the schemes of current flow and supply, the lower row contains the expressions for B_φ in the liquid conductor domains. These will be considered later, together with the vortical flows arising. The list of situations, of course, may be expanded.

Note that the expression for radial current component $j_R = (IA)/(2\pi R^2)$ does not contain any information concerning the current supply to the fluid. This information is contained in the expression for magnetic field (2.2.9) with both constants A and B.

Consider now which external magnetic fields are feasible in the given class of solutions. The solution to the equation (2.2.4) for $G_0(\theta)$, or to the equation $E^2 \psi_2 = 0$ for ψ_2, is

$$G_0 = C \cos \theta + D, \qquad \psi_2 = \frac{\mu_0 I_1 R}{2\pi} (C \cos \theta + D),$$

i.e. the magnetic lines of force lie on the surfaces

$$R(C \cos \theta + D) = C_1 = \text{const.} \tag{2.2.12}$$

For $D^2 > C^2$ the surfaces (2.2.12) are spheroidal in a coordinate system with the origin $r = 0$, $z = -(CC_1)/(D^2 - C^2)$, then semiminor axis $C_1/\sqrt{D^2 - C^2}$ is directed along the r-axis and semimajor axis $(DC_1)/(D^2 - C^2)$ along z-axis (Fig. 2.1a). For $D < C$ we have rotational hyperboloids in the coordinate system with the same origin, the real semiaxis is $(DC_1)/(D^2 - C^2)$ and the imaginary semiaxis $C_1/\sqrt{C^2 - D^2}$ (Fig. 2.1b). In the particular case $D = 0$ we have lines $R \cos \theta = z = \text{const}$, i.e. the cylindrical radial field (Fig. 2.1c); the case $C = 0$ gives us the lines $R = \text{const}$, i.e. a spherical field due to, for instance, conical pole ends of a magnet (Fig. 2.1d). The case $C = \pm D$ corresponds to parabolic field (the vertices of parabolas are in the points $z = \mp C/2$) due to, for instance, a magnet with one pole of paraboloidal form and the second pole at infinity

Solutions in spherical coordinates

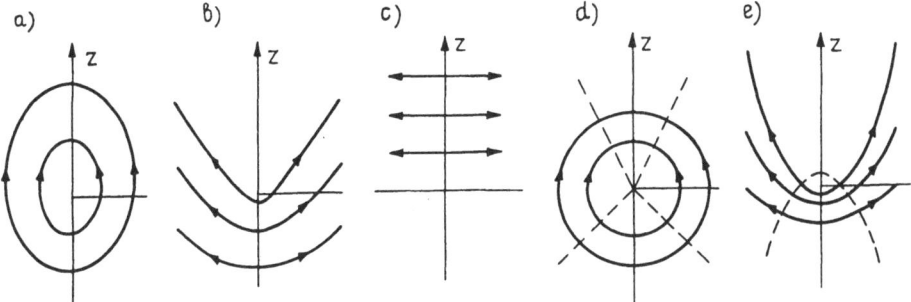

Fig. 2.1. External magnetic fields of the solution class (2.1.9).

(Fig. 2.1e). Of course, the magnetic fields shown in Fig. 2.1 are merely the limiting cases of some real fields. These all have singularities on the symmetry axis, this property being in general pertinent to the class of solutions considered.

Substitute the expressions for L_0 and G_0 in (2.2.1)–(2.2.2). The term G_0^2 is of the same form as $(-a\cos^2\theta + b\cos\theta - c)$ in the equation (2.2.1); therefore, redefining the constants a, b, and c in (2.2.1), we can neglect the term $k^2 SG_0^2/2$. Formally, this means that the electromagnetic force, corresponding to the external meridional magnetic field (curl $\mathbf{B}_0 \times \mathbf{B}_0$), is potential; moreover curl $\mathbf{B}_0 \equiv 0$ for G_0.

Considering only the interaction of an externally supplied electric current with its self-magnetic field, and with an external magnetic field, i.e. neglecting the induced electric currents (the electrodynamic approximation: $L_1 = G_1 = 0$), the set of equations is

$$g' \sin\theta - 2g\cos\theta + \tfrac{1}{2}g^2 +$$

$$+ (1 + \cos^2\theta) \int \frac{\cos\theta}{\sin^3\theta} \ell^2 \, d\theta - \cos\theta \int \frac{1 + \cos^2\theta}{\sin^3\theta} \ell^2 \, d\theta$$

$$= -a\cos^2\theta + b\cos\theta - c +$$
$$+ \tfrac{1}{2} S[A(A+B)(1+\cos\theta)^2 \ln(1+\cos\theta) +$$
$$+ A(A-B)(1-\cos\theta)^2 \ln(1-\cos\theta)]; \qquad (2.2.13)$$

$$\sin^2\theta \left(\frac{\ell'}{\sin\theta}\right)' + g\ell' = kSA(C\cos\theta + D)\sin\theta. \qquad (2.2.14)$$

If additionally, for some reason, rotation is also absent ($\ell = 0$), then the equation

$$g'\sin\theta - 2g\cos\theta + \tfrac{1}{2}g^2 = -a\cos^2\theta + b\cos\theta - c +$$
$$+ \tfrac{1}{2}S[A(A+B)(1+\cos\theta)^2 \ln(1+\cos\theta) +$$
$$+ A(A-B)(1-\cos\theta)^2 \ln(1-\cos\theta)] \qquad (2.2.15)$$

describes a pure electrically induced vortex flow and is related to the Ricatti type equation. The term on the right, proportional to S, is different in form than $-a\cos^2\theta + b\cos\theta - c$. This condition is a formal guarantee that a fluid motion will be generated due to the interaction of the electric current and its self-magnetic field (if $A \neq 0$).

2.3. Integral flow characteristics and dimensionless criteria

An electric current passing the fluid drives the considered vortex flows, therefore a total electric current, appearing in the expression (2.1.9) for B_φ,* is a characteristic of the flow. Apart from this, other quantities may also be ascribed to characterize a source of motion and its intensity. However, these characteristics have to obey certain conditions. First of all, the source of motion should not disturb the axial symmetry; second, since the expressions for velocity field (2.1.7), (2.1.8) contain the parameter ν — the kinematic viscosity, then any quantity of the same dimension as ν may be used instead. On satisfying these conditions, a flow may be described by the class of solutions (2.1.7), (2.1.8) (these are, of course, only the necessary conditions).

Since the velocity field in axisymmetric case is determined by the two 'independent' scalars ψ and v_φ, then each of them may be related to its own characteristic quantity. Consider further some characteristic quantities.

Denote by $Q = \int_\Omega v_n \, d\Omega$ a volume flux of fluid, by $\mathbf{F} = \int_\Omega \mathbf{v} v_n \, d\Omega$ a momentum flux through a closed surface Ω surrounding the coordinate origin, and by $\Gamma = \oint_\ell \mathbf{v} \cdot d\mathbf{l} = \int_{\Omega_1} \text{curl } \mathbf{v} \cdot d\mathbf{\Omega}$ a circulation along a closed curve ℓ lying on the surface Ω (Ω_1 is a part of Ω separated by ℓ and crossed by the symmetry axis). Dimensions of $Q, \mathbf{F}, \Gamma$, and ν are

$$[Q] = L^3 T^{-1}, \qquad [\mathbf{F}] = L^4 T^{-2}, \qquad [\Gamma] = L^2 T^{-1}, \qquad [\nu] = L^2 T^{-1},$$

where L = length scale and T = time scale.

Suppose the surface Ω is finite (e.g. a sphere of radius L). Since the dimensions of L and ν are independent, the momentum flux $\mathbf{F}$ of dimension ν^2 can be used as the flow characteristic parameter, yet the length scale L should not enter as an independent parameter in the formulation of the problem. This may be achieved drawing the surface Ω to a point, i.e. by defining the finite momentum flux at the point. During this the dimension of the circulation remains ν. Thus, the point momentum flux and the intensity of a point vortex may be taken as the flow characteristic parameter. The volume flux at this point should be zero or infinite.

Assume now that the surface Ω is in the form of a cylinder with a fixed generatrix of length L moving along the guiding curve ℓ, which circles the symmetry axis ($d\Omega = L \, d\ell$). Drawing the contour ℓ to a point on the axis, i.e.

---

* Note that I_1 in (2.1.9) for the meridional field cannot be used on its own to characterize the flow, but it can be used in combination with another quantity, e.g. with I.

Solutions in spherical coordinates

eliminating the length scale associated with ℓ, the length scale L remains and a new parameter Q/L (volume flux per unit length) has the dimension of v, the dimension of Γ remains v, although $[F/L]$ is independent of $[v]$. Consequently, the flow may be characterized by the volume flux for a unit length of line source and by the intensity of a vortex line on the symmetry axis.

Let us consider further the meaning of the parameters in the set of equations (2.1.12)–(2.1.15) and (2.2.1)–(2.2.6). The parameter $S = \mu_0 I^2 / 4\pi^2 \rho v^2$, which is called the parameter of electrically induced vortex flow, contains a square of the electric current penetrating the fluid. It characterizes the ratio of the electromagnetic force due to the interaction of current and self-magnetic field to the viscous forces (or inertial forces, since for the velocity field (2.1.7), (2.1.8) both the forces have equal orders of typical magnitudes), and it determines the intensity of the fluid flow generated by the electromagnetic force.

The parameter β is called the Batchelor number, and is equal to the ratio v/v_m of kinematic viscosity to magnetic viscosity $v_m = (\mu_0 \sigma)^{-1}$. The parameter $k^2 S = \mu_0 I_1^2 / 4\pi^2 \rho v^2$ has no special meaning (as noted before, the term $k^2 S G_0^2 / 2$ in (2.2.1) can be absorbed in the right-hand side term $a \cos^2 \theta - b \cos \theta + c$), but in conjunction with the parameter β, i.e. the parameter $k^2 S \beta$, it characterizes the interaction of the external magnetic field with the flow induced electric current. A similar meaning is attached to the parameter $S\beta$, but for the self-magnetic field of the supplied electric current.

The parameter $kS = \mu_0 II_1 / 4\pi^2 \rho v^2$ characterizes the electromagnetic force arising from the interaction of external magnetic field with the electric current supplied to the fluid. This force, as follows from (2.2.2), causes the fluid to rotate. The effect of the force from the interaction of meridional currents (supplied and induced) with meridional magnetic fields (external and induced) on rotation is accounted for by the parameter $kS\beta$. As a matter of fact, $S\beta$, $k^2 S\beta$, and $kS\beta$ are squares of Hartman numbers expressed by the squares of the self-magnetic field, the external magnetic field, and by the product of those fields, respectively, if the representative magnetic field scales are used instead of currents, e.g. $\mu_0 I_1 = B_1 L$. Finally, the parameter $k = I_1/I$ has the meaning of the ratio of a typical meridional magnetic field to the azimuthal self-magnetic field of the supplied current.

We have merely elucidated the meaning of the electromagnetic parameters, since there are no other parameters in the set (2.1.12)–(2.1.15). If, apart from the electromagnetic, there are other motive sources — those mentioned earlier in this paragraph, for instance — then, in general, instead of the factor v in (2.1.7) and (2.1.8), other quantities of the same dimension may be used. For these cases the equations of motion and of the magnetic field would contain parameters relevant to the motive sources. We proceed in a different way, however. The velocity field is retained in the form (2.1.7), (2.1.8), and then the set of equations (2.1.12)–(2.1.15) or (2.2.1)–(2.2.6) is considered. If, however, additional motive sources are considered, they will appear as additional constraints, e.g. boundary conditions. Indeed, in the absence of additional motive sources the boundary conditions are homogeneous and are thus on the symmetry axis $g(0) = 0$. If, for instance, there is a line mass source on the symmetry axis, the last condition is replaced by $g(0) = Q/\nu L = Re$ (see Section 4.4).

2.4. Review of the class of exact solutions in spherical coordinates

The class of exact solutions considered here may be traced historically to the work of Landau [27], in which there is described one of the few cases when the nonlinear Navier-Stokes equations can be integrated. Up to the present time, quite a number of papers have appeared, considerably extending the domain of application of the class of solutions. However, a thorough review of these works is lacking, therefore we hope that the one presented below will fill the gap to a certain extent.

One of the first studies in which an attempt was made to separate the variables in equation (2.1.1) is the work by Slezkin [45], wherein the author attempted to present a stream function for flow in a gap of two coaxial cones in the form of decreasing powers of the distance R from the common apex of the cones:

$$\psi = \sum_{i=0}^{\infty} \frac{g_i(\theta)}{R^i}, \qquad v_\varphi = 0. \tag{2.4.1}$$

Substitution of (2.4.1) in (2.1.1) leads to a set of ordinary differential equations for $g_i(\theta)$, the equation for $g_0(\theta)$ corresponding to the Stokes approximation. Formally the terms multiplied by the positive natural powers of $\ln R$ may be added to the series (2.4.1), but the most interesting possibility is to add the term $\nu R g(\theta)$. In this case the function $g(\theta)$ is determined independently from (2.4.1) by the equation

$$g' \sin \theta - 2g \cos \theta + g^2/2 = c_0 \cos^2 \theta - 2b_0 \cos \theta - c_0 + 2a_0 \tag{2.4.2}$$

(the constants in (2.4.2) and in the previous equation (2.1.16) are related to each other by $a = c_0$, $b = -2b_0$, $c = c_0 - 2a_0$); introducing a new variable $\mu = \cos \theta$ the equation is

$$(1 - \mu^2)g' + 2\mu g - g^2/2 = -c_0\mu^2 + 2b_0\mu + c_0 - 2a_0. \tag{2.4.3}$$

Thus the stream function of the form $\psi = \nu R g(\theta)$ is an exact solution of the full Navier-Stokes equation; a variety of problems in the class is determined by different values of the integration constants in (2.4.2). The full formulation of the solution class (2.1.6) and the equation (2.4.2) was made by Yatseyev [73] and by Squire [66, 67], yet the first particular solution of (2.4.2) was obtained by Landau in 1944 [27].

Landau proceeded from the Navier-Stokes equation in the form (1.1.16)

Solutions in spherical coordinates 75

where the tensor components of (1.1.17) in spherical coordinates for axisymmetric case are:

$$\Pi_{RR} = p + |\mathbf{B}|^2/2\mu_0 + \rho v_R^2 - B_R^2/\mu_0 - 2\nu\rho \frac{\partial v_R}{\partial R},$$

$$\Pi_{\theta\theta} = p + |\mathbf{B}|^2/2\mu_0 + \rho v_\theta^2 - B_\theta^2/\mu_0 - 2\nu\rho \left(\frac{1}{R} \frac{\partial v_\theta}{\partial \theta} + \frac{v_R}{R} \right),$$

$$\Pi_{\varphi\varphi} = p + |\mathbf{B}|^2/2\mu_0 + \rho v_\varphi^2 - B_\varphi^2/\mu_0 - 2\nu\rho \left(\frac{v_R}{R} + \cotan \theta \frac{v_\theta}{R} \right),$$

(2.4.4)

$$\Pi_{R\theta} = \rho v_R v_\theta - B_R B_\theta/\mu_0 - \nu\rho \left(\frac{1}{R} \frac{\partial v_R}{\partial \theta} + \frac{\partial v_\theta}{\partial R} - \frac{v_\theta}{R} \right),$$

$$\Pi_{\theta\varphi} = \rho v_\theta v_\varphi - B_\theta B_\varphi/\mu_0 - \nu\rho \left(\frac{1}{R} \frac{\partial v_\varphi}{\partial \theta} - \cotan \theta \frac{v_\varphi}{R} \right),$$

$$\Pi_{R\varphi} = \rho v_R v_\varphi - B_R B_\varphi/\mu_0 - \nu\rho \left(\frac{\partial v_\varphi}{\partial R} - \frac{v_\varphi}{R} \right),$$

and the components of tensor divergence are [16]:

$$(\text{div } \Pi)_R = \frac{1}{R^2} \frac{\partial}{\partial R} (R^2 \Pi_{RR}) + \frac{1}{R \sin \theta} \frac{\partial}{\partial \theta} (\sin \theta \, \Pi_{R\theta}) -$$

$$- \frac{1}{R} \Pi_{\theta\theta} - \frac{1}{R} \Pi_{\varphi\varphi},$$

$$(\text{div } \Pi)_\theta = \frac{1}{R^2} \frac{\partial}{\partial R} (R^2 \Pi_{\theta R}) + \frac{1}{R \sin \theta} \frac{\partial}{\partial \theta} (\sin \theta \, \Pi_{\theta\theta}) +$$

$$+ \frac{1}{R} \Pi_{R\theta} - \frac{1}{R} \cotan \theta \, \Pi_{\varphi\varphi},$$

(2.4.5)

$$(\text{div } \Pi)_\varphi = \frac{1}{R^2} \frac{\partial}{\partial R} (R^2 \Pi_{\varphi R}) + \frac{1}{R \sin \theta} \frac{\partial}{\partial \theta} (\sin \theta \, \Pi_{\varphi\theta}) +$$

$$+ \frac{1}{R} \Pi_{\varphi R} + \frac{1}{R} \Pi_{\varphi\theta} \cotan \theta.$$

Landau set up a problem to find the solutions describing flows whose momentum flux is invariant through any closed surface surrounding the coordinate origin. This is the case (cf. Section 2.3) if the velocity decreases as R^{-1} from the coordinate origin (the corresponding stream function $\psi = \nu R g(\theta)$):

$$v_R = \frac{\nu}{R \sin \theta} g'(\theta), \qquad v_\theta = -\frac{\nu}{R \sin \theta} g(\theta). \tag{2.4.6}$$

Substituting (2.4.6), $v_\varphi = 0$, $\mathbf{B} = 0$ in (2.4.4), (2.4.5), and div $\Pi = 0$ we can obtain (2.4.2). Yet Landau restricted the problem by seeking the solution with all the components of Π, except Π_{RR}, equal to zero, and came to the equation (2.4.2) with $a_0 = b_0 = c_0 = 0$. Indeed, Yatseyev [73] showed that for the velocity field (2.4.6) with the differential relationships (2.4.2)

$$\Pi_{\varphi\varphi} = \frac{2\nu^2 \rho}{R^2} \left(\frac{b_0 \cos \theta - a_0}{\sin^2 \theta} \right),$$

$$\Pi_{\theta\theta} = \frac{2\nu^2 \rho}{R^2} \left(\frac{a_0 - b_0 \cos \theta}{\sin^2 \theta} - c_0 \right), \tag{2.4.7}$$

$$\Pi_{R\theta} = \frac{2\nu^2 \rho}{R^2} \left(\frac{c_0 \cos \theta - b_0}{\sin \theta} \right),$$

i.e. these components are zero for $a_0 = b_0 = c_0 = 0$ ($\Pi_{\theta\varphi} = \Pi_{R\varphi} = 0$ due to $v_\varphi = 0$). Using the Landau solution to (2.4.2)

$$g = \frac{2 \sin^2 \theta}{A - \cos \theta} \tag{2.4.8}$$

the component Π_{RR} can be expressed as

$$\Pi_{RR} = \frac{4\nu^2 \rho}{R^2} \left[\frac{(A^2 - 1)^2}{(A - \cos \theta)^2} - \frac{A}{A - \cos \theta} \right]. \tag{2.4.9}$$

The constant A, determining the flow intensity, is uniquely related to the axial momentum flux across the surface $R = $ const. In the general case it is expressed as

$$F_0 = 2\pi \int_0^\pi (\Pi_{RR} \cos \theta - \Pi_{R\theta} \sin \theta) R^2 \sin \theta \, d\theta$$

and for the problem of Landau

$$F_0 = 2\pi \nu^2 \rho \left[\frac{32}{3} \frac{A}{A^2 - 1} + 4A^2 \ln \frac{A - 1}{A + 1} + 8A \right]. \tag{2.4.10}$$

Landau's solution (2.4.8) conforms with the boundary conditions of zero velocity sources on the axis $g(0) = g(\pi) = 0$ and vanishing viscous stress on the axis due to the symmetry: $\sin \theta \, g''|_{\theta=0} = \sin \theta \, g''|_{\theta=\pi} = 0$, and this may be

Solutions in spherical coordinates

interpreted as a jet flow from the end of a thin tube into the unbounded space filled with the same liquid. The first two conditions mean that the mass flux through a closed surface surrounding the origin is equal to zero, in accordance with the physical meaning of the stream function (cf. Section 1.5). A solution of the problem with a finite mass flux was obtained by Rumer [42], who used the approximation for the velocity field obtained by adding the first term of the series (2.4.1).

For the intense jet ($F_0 \to \infty$) from (2.4.10) it follows that A tends to unity as $A = 1 + a^2/2$ ($a = 32\pi\rho v^2/3F_0$), and for small angles $\theta \sim a$ we obtain the velocity distribution

$$v_\theta = -\frac{4v\theta}{a^2 + \theta^2}, \qquad v_R = \frac{8va^2}{(a^2 + \theta^2)^2}$$

which is similar to that obtained by Schlichting [46] in the boundary layer approximation. The difference between the two solutions is felt merely in the outer part of the jet, as was shown by Kashkarov [25, 70].

If, from among the boundary conditions mentioned, only those at $\theta = 0$ are satisfied, then the right side of (2.4.2) is equal to $-a_0(1 - \cos\theta)^2$ and $b_0 = c_0 = a_0$. For this case an analytic solution can also be constructed [73]:

$$g = -2v\,\frac{\sin^2\theta}{1 + \cos\theta}\left[n\tanh(nx + A) + \frac{1}{2}\right], \qquad a_0 > -\frac{1}{2};$$

$$g = -2v\,\frac{\sin^2\theta}{1 + \cos\theta}\left[\frac{1}{2} - n\tan(nx + A)\right], \qquad a_0 < -\frac{1}{2}; \qquad (2.4.11)$$

$$g = -2v\,\frac{\sin^2\theta}{1 + \cos\theta}\left[\frac{A}{1 + Ax} + \frac{1}{2}\right], \qquad a_0 = -\frac{1}{2},$$

where $x = \ln(1 + \cos\theta)$, $n = \frac{1}{2}\sqrt{|1 + 2a_0|}$. The presence of the as yet undetermined constant a_0 permits us to extend the feasible forms of flow within the class considered. Thus, Yatseyev obtained the solution for a jet emanating from a semiaxis $\theta = \pi$. Squire [67] applied the impermeability condition on the plane $\theta = \pi/2$ ($g(\pi/2) = 0$) and obtained the solution for a jet 'emanating' from a point source on the plane wall (Fig. 2.2b). Kashkarov [25] generalized Squire's solution for a case of impermeable cone $\theta = \theta_0$ (Fig. 2.2c). Note that the no slip condition on this surface cannot be imposed since it leads to the trivial solution $g \equiv 0$. This means physically that the presence of the solid wall infringes the condition of constant momentum flux through the surface $R = $ const: the momentum is wasted in overcoming friction at the surface.

If both the semiaxes $\theta = 0$ and $\theta = \pi$ are assumed to be the sources of the θ-component of the velocity, then, depending on the intensity of the sources, a conical jet of arbitrary cone angle θ_0, issuing from a ring source, can be described (Fig. 2.2d) [68]. The velocity maximum in these jets takes place on the cone surface θ_0. For sources of equal strength the solution to the radial jet

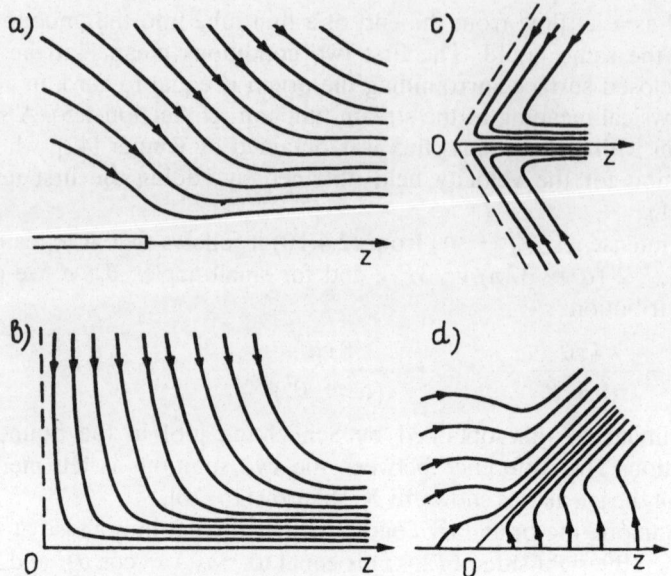

Fig. 2.2. Flow schemes feasible in the class (2.4.6): (a) a jet issuing at the end of a thin tube with a fluid slip at the tube's surface; (b) a jet from a hole in an impermeable plane; (c) a jet from an apex of impermeable cone; (d) a conical jet.

problem was obtained by Loitsyanskiy [32] in the boundary layer approximation. For this case the right-hand part of (2.4.2) is equal to the constant $-2a$.

For the cases listed above ($a_0 = b_0 = c_0 = 0$; $a_0 = b_0 = c_0$; $b_0 = c_0 = 0$; $a_0 \neq 0$) the boundary layer approximation solutions can be constructed [50]. To do this, it is more convenient to use a cylindrical analogue instead of the equation (2.4.2):

$$(1 + \eta^2)\psi' + \eta\psi + \tfrac{1}{2}\psi^2 = -2b_0\eta\sqrt{1 + \eta^2} + 2a_0(1 + \eta^2) - c_0, \qquad (2.4.12)$$

where $\eta = z/r = \cotan\theta$, $\psi(\eta) = -g(\theta)/\sin\theta$. (More about relations between the solutions of the class considered in spherical and cylindrical coordinates is given in Section 4.2.)

Consider, for instance, the derivation of Schlichting's round jet solution from the Yatseyev-Squire problem. In (2.4.12) let $a_0 = b_0 = c_0$ and introduce a variable $t = \eta^{-1} = r/z$ (the jet axis coincides with the z-axis ($t = 0$)), then (2.4.12) is

$$-(1 + t^2)\psi' + \frac{1}{t}\psi + \frac{1}{2}\psi^2 = -a_0\left(\frac{2}{t^2}\sqrt{1 + t^2} - \frac{2}{t^2} - 1\right). \qquad (2.4.13)$$

According to the conceptions of the boundary layer theory, the scales of longitudinal (along jet axis) and transverse dimensions of jet are related as $\sqrt{Re}$

Solutions in spherical coordinates 79

(the *Re* number is constructed by use of the jet momentum F_0), therefore the variable

$$t_0 = \sqrt{Re}\, t \tag{2.4.14}$$

is introduced. Also the stream function for large *Re* is expanded in the series

$$\psi = \sqrt{Re}\, \psi_0 + \psi_1 + \frac{1}{\sqrt{Re}} \psi_2 + \ldots \tag{2.4.15}$$

Substituting (2.4.14) and (2.4.15) in (2.4.13) and passing to the limit $Re \to \infty$, we obtain the equation

$$t_0 \psi_0'' - \psi_0' - \tfrac{1}{2} t_0 \psi_0'^2 = 0,$$

which, after the substitution $\psi_0 = -F/t_0$, finally becomes the Schlichting equation [46]:

$$t_0 F' + 2F + \tfrac{1}{2} F^2 = 0. \tag{2.4.16}$$

Consequently, the round jet solution can be constructed for any high *Re* number. Note also that equation (2.4.16) does not contain the constant a_0, which is present in the right-hand side of (2.4.13). This means that the axial flow (2.4.16) — Schlichting's jet — for high *Re* satisfactorily describes either the flow of Landau, or the flow of Squire, or the flow of Kashkarov. The constant a_0 appears only in the next term of the expansion (2.4.15), reflecting the flow conditions in the outer part of the jet.

Similarly (but without the variable *t*) the solution of Loytsyanskiy [32] for the radial jet may be obtained in the boundary layer approximation. In this case, as we have already mentioned, the right-hand part of (2.4.12) contains only the constant $2a_0$, which is equal to the maximum velocity $rv_r/v = \psi'(0) = 2a_0$ on the jet axis ($\theta = \pi/2$) and determines the asymptotic behaviour of the function for large η:

$$\psi = -2a_0 \frac{\ln \eta}{\eta} + A_1 \frac{1}{\eta} + \ldots$$

Setting $\eta = \sqrt{Re}\, \eta_0$ and using the series (2.4.15), we obtain, in the limit as $Re \to \infty$, the equation

$$\psi_0'' + \tfrac{1}{2} \psi_0'^2 = 2a_0,$$

the solution of which is known [32]: $\psi_0 = \sqrt{a_0}\, 2 \tan \eta \sqrt{a_0}$. The following corrections can also be constructed in analytic form. Thus, the first approximation gives $\psi_1'' + \psi_0' \psi_1' = 0$ and $\psi_1 = D \cosh^2 \sqrt{16 a_0}$ [50]. An exact solution for the radial jet also exists for any high *Re* number.

For arbitrary values of a_0, b_0, c_0 the solution of equation (2.4.2) can be expressed by hypergeometric functions [73], yet it is not convenient for applications. Nevertheless, we may conclude that, in the general case, equation (2.4.2) describes a fluid motion between two coaxial cones $\theta = \theta_1$ and $\theta = \theta_2$ with a common apex, on the surfaces of which the values of tangential and normal components of velocity may be set. Moreover, as Morgan [36] showed, the simultaneous conditions of no slip and impermeability on the surfaces $g(\theta_1, \theta_2)$ = $g'(\theta_1, \theta_2) = 0$ lead to the unique solution $g \equiv 0$. Thus the flow between the two coaxial cones may be determined by assuming either normal velocity, or slip velocity on at least one of the conical boundaries θ_1, θ_2. From this Morgan concluded that the class of solutions with the velocity field (2.4.6) is not the axisymmetric analogue of the plane diverging Jeffrey-Hamel flow [22, 23], which also has velocity components that are inversely proportional to the distance from the source. The major difference, according to Morgan, is the following: for the plane flow the condition of no slip is satisfied on both diverging walls and the mass flux remains constant for any distance from the source, whenever for the axisymmetric flow, the flux across the surface of a spherical sector is proportional to $\psi_1 - \psi_2 = \nu R[g(\theta_1) - g(\theta_2)]$, i.e. to the distance from the apex of the cones (the requirements for the boundary conditions were formulated above).

Of course, a complete analogy should not be expected. Nevertheless, the Hamel flow and the flows determined by equation (2.4.2) have many common features. This is especially clear from their analysis in cylindrical coordinates. As is shown in [50], if the velocity field is set in cylindrical coordinates in the form

$$v_z = \frac{\nu}{r} w(\varphi, \eta), \qquad v_r = \frac{\nu}{r} f(\varphi, \eta), \qquad (2.4.17)$$

where r is the cylindrical radius, $\eta = z/r = \cotan \theta$, then both the flows can be obtained supposing the functions w and f to depend on either one of the two variables φ or η. Thus, Hamel flow has $w(\varphi), f(\varphi)$. Moreover, the v_r velocity component is determined independently of the component v_z and in this sense the flow may be considered as planar (the equation for $w(v_z)$ contains $f(v_r)$). Yet if $w = w(\eta), f = f(\eta)$, we come to the velocity field (2.4.6). Therefore both kinds of the flows belong to the same class of solutions (2.4.17), permitting us to exclude one of the three variables, namely the variable r.

Furthermore, for the plane Hamel flow there is an integral characteristic of mass flux per unit length of line source located at the intersection of plane walls (at the z-axis). As was shown in Section 2.3 (cf. also [50, 55]), the same characteristic may also be applied to the solution of equation (2.4.2). In this case the no slip and impermeability conditions can be satisfied on one conical wall (according to what was said above, this is possible, assuming an injection velocity on the second 'cone' $\theta = 0$, but the injection is also a flux along a unit length of line source). The solution to the problem of a line source in the presence of a rigid conical wall, based on the equation (2.4.2), was first studied in [50, 55] and later in [20]. It will be discussed in greater detail in Section 4.4.

In problems of the jet type, which have the characteristic jet momentum, the no slip conditions, previously mentioned, cannot be satisfied, and, at first glance, equation (2.4.2) is not suitable to describe jets in the presence of a rigid wall. Nevertheless, in a recent work [47], Schneider suggests an acceptable resolution, viz. at the limit $Re \to \infty$ the relatively thin jet acts as a line sink on the axis $\theta = 0$ for the surrounding bulk part of the fluid. Moreover, the fluid

Solutions in spherical coordinates

entrainment turns out to be constant at a unit length of the sink either for a laminar, or a turbulent jet, or also for a laminar heated jet:

$$Q_0 = 2\pi \int_R^{R+1} v_\theta R \sin\theta \, dR = -2\pi v g(\theta_0).$$

Consequently, the exact solution is suitable to describe the entrained flow of the outer jet part, and it satisfies the no slip conditions. Schneider found a uniformly valid solution at $Re \to \infty$ matching the external flow to Schlichting's round jet. The reverse problem of boundary layer flow on a circular cone (a plane is the particular case) with the velocity field (2.4.6) is considered in [12] for Re high enough to use the boundary layer approximation.

Squire [66] offers an interpretation of the jet flow as a result of a point force of the δ-function type located at the coordinate origin (the procedure is common in hydrodynamics of jets or wake flows in the boundary layer approximation). This interpretation arouses interest into the time development of the fluid flow from rest after instantaneously applying a point force F at the origin. The problem of Landau-Squire jet development has been considered by Sozou and Pickering [64]. Since the force F is applied impulsively, the initial data of the problem do not contain a representative time scale, hence the time variable t can be made dimensionless by using only the dimensional quantities — coordinate R and viscosity v: $\lambda^2 = tv/R^2$. The flow development is described by the stream function

$$\psi = vRg(\theta, \lambda), \tag{2.4.18}$$

where $g(\theta, \lambda)$ depends on the two nondimensional variables. The function g is determined by solving an elliptical partial differential equation of the fourth order [64]. Boundary conditions set for the variable λ reflect mutual connection of t and R, so the function (2.4.18) for $\lambda \to 0$ corresponds to the stream function at great distance $R \to \infty$ in a fixed time t and also to the initial phase of motion for $t \to 0$ and a fixed distance R. Thus for $\lambda = 0$ we may assume a state of absence of motion before the force F is applied. The boundary condition for $\lambda \to \infty$ corresponds to the steady flow at $t \to \infty$ given by the Landau's solution (2.4.8). This means that for $R \to 0$ and any $t > 0$ there exists a domain, surrounding the point where the force is applied, from which the steady Landau-Squire jet spreads into the surrounding fluid.

According to [64] at initial time ($\lambda \to 0$) the flow field is that of a dipole oriented in the direction of the applied force:

$$\psi = \frac{F}{4\pi\rho} \frac{t \sin^2\theta}{R}, \tag{2.4.19}$$

i.e. the initial flow is irrotational and is independent of viscosity, which would be anticipated due to the absence of rigid walls. Vorticity spreads from the point where the force is applied with a finite velocity. For small F a linear equation determines the flow, and the solution [60] is symmetric relative to the midplane $\theta = \pi/2$. For higher F the nonlinear numerical solution [64] is essentially asymmetric after the initial flow development period, and the higher the applied force F, the shorter is the time of flow development to the steady state.

Reference [11] presents a study of particle trajectories in the unsteady flow of a developing Landau-Squire jet. Particle trajectories in the meridional plane are determined by the equations

$$\frac{dR}{dt} = v_R, \qquad \frac{d\theta}{dt} = \frac{v_\theta}{R}. \qquad (2.4.20)$$

After substitution of the stream function (2.4.18) and of the similarity variables, the set (2.4.20) can be solved analytically for the cases of the beginning flow of dipole type (2.4.19), of the steady jet (2.4.8), and of slow motion $F \ll 1$ [60]. Depending on the Reynolds number $Re = \sqrt{F/\rho v^2}$ two bifurcation points in the solution of (2.4.20) are found where the particle path structure changes abruptly. The transition is from one unsteady state to another unsteady laminar state. The method for analysing the transition to different states of a system may also be applied to other nonparallel flows, and it is not related to the small disturbance analysis of the common stability theory.

The stability analysis for small disturbances of a Landau-Squire jet for high Re numbers was given in [5]. The round jet is shown to be unstable to the inviscid spiral perturbation. However, the critical Re number in [5] is not determined. In [19] a value of Re is determined for which a new inflection point appears in the jet's profile and, according to the Rayleigh theory, instability may occur.

Until now the flow was considered merely in meridional planes, i.e. $v_\varphi = 0$. Adding the azimuthal velocity (2.1.8), the set of equations (2.1.12), (2.1.13) without electromagnetic forces modifies to give

$$\tfrac{1}{2}(1 - \mu^2)(g^2)''' + 2\ell\ell' = (1 - \mu^2)^2 g^{IV} - 4\mu(1 - \mu^2)g''', \qquad (2.4.21)$$

$$g\ell' = (1 - \mu^2)\ell'', \qquad (2.4.22)$$

where $\mu = \cos\theta$, $g = g(\mu)$, $\ell = \ell(\mu)$.

It is instructive, to gain a better perception of the physical situation with a rotational motion of form (2.1.8), to assume $g = 0$ in equation (2.4.22) and to bound the fluid with a plane $\theta = \pi/2$. Then the solution of (2.4.22) is $\ell = b_1\mu + b_2$. If the plane is assumed to be a free surface, then the absence of stress leads to $\ell'(0) = 0$ and $\ell = b_2$. This is the solution of a constant circulation, determined by a vortex line on the symmetry axis. The azimuthal velocity $v_\varphi = vb_2/R \sin\theta$ corresponds to the potential vortex. Such a rotation does not influence meridional flow since $\ell\ell' = 0$ in (2.4.21). However, for the rigid plane the no slip condition $\ell(0) = 0$ gives $\ell = b_1\mu$ and $\ell\ell' \neq 0$ in the meridional motion equation (2.4.21). Thereby the vortex line, $\ell(1) = b_1$, normal to the rigid plane, generates a flow in meridional planes (for details see Section 2.5).

The action of a vortex line normal to the plane was described by Goldshtik [17] by use of the equations (2.4.21), (2.4.22). He found that the solution existed only for finite values of the tangential Reynolds number $Re = \Gamma/v = b_2 \leq 8$. Later the paradoxical result of the solution's nonexistence for the

Solutions in spherical coordinates

higher Re was checked by many authors, and the value of the critical Re number was redetermined more exactly by direct numerical solution [21, 26, 48] (see also an extended study [74]). The nonexistence of a solution is attributed to the effect of nonlinear interaction of the rotation and intense axial flow at critical Re, which can lead to the vortex breakdown phenomenon and destroy the self-similarity of the flow [29].

The rotating Landau jet was studied by Tsuker [69], who represented the azimuthal velocity in the form of a series $v_\varphi = v\ell_1(\theta)/R \sin \theta + \ell_2(\theta)/R^2 + \ldots$ (and the stream function, too). The boundary conditions for the functions $\ell_i(\theta)$ are $\ell_i(0) = \ell_i(\pi) = 0$. Solution of the equation (2.4.22) for $\ell_1(\mu)$ ($\mu = \cos \theta$) is

$$\ell_1 = C_1 \int_{-1}^{\mu} \left(\exp \int \frac{g \, d\mu}{1 - \mu^2} \right) d\mu + C_2,$$

which for the conditions $\ell_1(\pm 1) = 0$ gives $\ell_1 \equiv 0$. Thence the jet rotation may be taken into account only by the second term; however, this is outside the scope of the class of exact solutions. Tsuker's solution for $F \to \infty$ goes to the solution of Loitsyanskiy [31] obtained for the boundary layer approximation. An interesting property of this solution is that, with the increase of rotation, the radial velocity maximum is displaced from the axis to a conical surface.

Consider now the possibility of including additional physical factors in the flow models based on the exact solution class. Squire [66] studied heat transfer in the jet, supposing that a heated jet, issuing from a small aperture, could be described by adding to the force F a heat source of strength q in the coordinate origin. The heat transfer equation

$$\rho c_p \left(v_R \frac{\partial T}{\partial R} + \frac{v_\theta}{R} \frac{\partial T}{\partial \theta} \right) = \kappa \nabla^2 T, \qquad (2.4.23)$$

where κ is the thermal conductivity coefficient and c_p is the specific heat, satisfies the temperature distribution $T = R^n F(\theta)$ if the velocity components are in the form (2.4.6). The heat flux from the source across a surrounding spherical surface

$$q = 2\pi c_p \rho \int_0^\pi \left(v_R T - \kappa \frac{\partial T}{\partial R} \right) R^2 \sin \theta \, d\theta$$

remains constant for any radius R if $n = -1$, i.e.

$$T = \frac{1}{R} F(\theta). \qquad (2.4.24)$$

However, the presence of the heat source destroys the similarity of equation of motion (2.1.1) since it contains the buoyancy force of magnitude $|-\beta \mathbf{g} T| \sim R^{-1}$ (β is the thermal expansion coefficient and $\mathbf{g}$ = gravity). Consequently, if the velocity field is determined by (2.4.6), a new temperature distribution (2.4.24) can be determined; the reverse influence of the buoyancy force on the flow seems to be impossible in the similarity formulation. For small q this influence can be determined approximately by expanding the stream function [40] $\psi = \nu R g_1(\theta) + R^3 g_3(\theta) + \ldots$

Note that an equation similar to (2.4.23) also determines a passive scalar transport in the jet (it was studied by Rumer [43]).

New possibilities of the class of exact solutions considered here appear with the introduction of the magnetic field. The first description of this class in magnetohydrodynamics was made by Wu [72], who considered a case with a meridional magnetic field B_R, B_θ ($B_\varphi = 0$). The full formulation of the class of exact solutions in magnetohydrodynamics, including all components of velocity and magnetic field, is described in [39, 54] and is analysed in detail in the monograph [50].

The set of equations of the class is presented in Section 2.1, and the feasible external meridional magnetic fields in Section 2.2. Naturally, each of the problems mentioned can be considered with any of these fields. From among the problems that have been solved note the MHD-analogue of the Landau jet considered in [54, 72] for the magnetic field shown in Fig. 2.1e. The studies reveal that the effect of the magnetic field is to deform the jet's stream lines in such a way as to diminish the slope between the flow lines and the magnetic lines of force; then the interaction of flow and magnetic field is reduced. Another feature of the solution is the decrease of the total jet momentum flux with increasing Hartmann number. This is easily explained by the damping effect of the induced electromagnetic force. Consequently, if the condition is set to conserve the momentum flux for the increasing field, then the initial momentum (without the field) has to be increased.

The flows of Landau and Yatseyev-Squire for high electric conductivity of the medium are investigated in [39, 49]. In this case the initial form of the external magnetic field does not matter, since any initial magnetic field assumes the form determined by the fluid flow due to the freezing-in of the magnetic lines of force. The flows in these studies are also considered with the azimuthal magnetic field of a linear current on the symmetry axis. It was found that the magnetic field $B_\varphi = \mu_0 I/R \sin \theta$ acts to focus the jet on the symmetry axis.

A growing interest in the class of solutions in magnetohydrodynamics has developed owing to its potential applications to the flow induced by an electric current externally supplied to the fluid (the schemes of current supply are discussed in Section 2.2). The significance of these problems to the applied sciences has given rise to a stream of publications, which will be discussed in the following sections and chapters.

2.5. Electrically induced vortex flow in a cone

Let us consider the situation depicted in Fig. 2.3 with the aim of demonstrating the qualitative features of flow induced by an electric current radially diverging from a point source. Apart from this, suppose an additional motive source due

Solutions in spherical coordinates 85

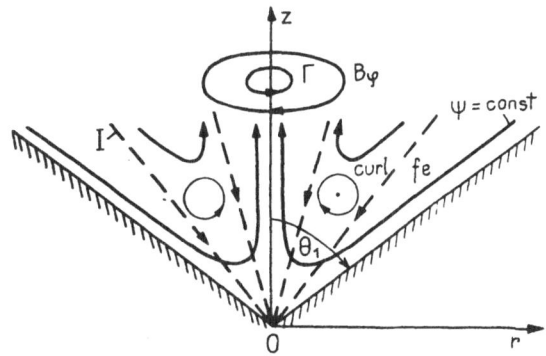

Fig. 2.3. Electrically induced vortex flow in a cone with the vortex line.

to a vortex line* of intensity Γ on the symmetry axis (as discussed in Section 2.3, the class of exact solutions permits us to use this characteristic). To make the problem more tractable the function ℓ is determined in Stokes' approximation, which means that the nonlinear term $g\ell'$ in (2.2.14) must be negligible. Apart from this, let $k = 0$ in (2.2.14), which presupposes the absence of the external meridional magnetic field. Then the velocity of rotation is determined by the equation

$$\left(\frac{\ell'}{\sin \theta}\right)' = 0 \qquad (2.5.1)$$

and thereby the boundary conditions: of no slip on the cone surface $\theta = \theta_1$: $v_\varphi(\theta_1) = 0$, and of the vortex line present on the symmetry axis: $\lim_{\theta \to 0} v_\varphi = \Gamma/2\pi R \sin \theta$. For the function ℓ the conditions are respectively: $\ell(\theta_1) = 0$ and $\ell(0) = \Gamma/2\pi\nu = Re$. Then the solution is

$$\ell = \frac{Re}{\cos \theta_1 - 1} (\cos \theta_1 - \cos \theta). \qquad (2.5.2)$$

Introducing a new variable $\mu = \cos \theta$, the velocity components have the form

$$v_R = -\frac{\nu}{R} g'(\mu), \qquad v_\theta = -\frac{\nu}{R} \frac{g(\mu)}{\sqrt{1-\mu^2}}. \qquad (2.5.3)$$

Then substituting (2.5.2) in the form $\ell = Re(\mu - \mu_1)/(1 - \mu_1)$, $\mu_1 = \cos \theta_1$,

---

* Vortex line — yet another example of velocity field singularity, in addition to the point, dipole, and so forth. This is the line around which the circulation Γ (see definition in Section 2.3) of an infinitesimal closed contour remains a finite quantity. In the field surrounding a straight vortex line the potential flow is induced of the form $v_\varphi = \Gamma/2\pi R \sin \theta$ which is singular on the symmetry axis. The azimuthal magnetic field of an infinitely thin wire carrying an electric current has a similar form (cf. Section 1.7 and Table 1.1).

and taking into account $A = B = (1 - \mu_1)^{-1}$ (2.2.11) for the electric current flowing from the apex into the cone, we obtain the equation for meridional flow

$$(1 - \mu^2)g' + 2\mu g - \tfrac{1}{2}g^2 = a\mu^2 - b\mu + c + A_1(1 + \mu)^2 \ln(1 + \mu) - $$
$$ - A_2(1 - \mu)^2 \ln(1 - \mu), \qquad (2.5.4)$$

where

$$A_1 = \frac{S - Re^2 \tfrac{1}{2}(1 - \mu_1)}{(1 - \mu_1)^2}, \qquad A_2 = \frac{Re^2}{2(1 - \mu_1)^2}.$$

To determine the constants a, b, c and the integration constant of (2.5.4) the following conditions are used: the no slip conditions on the cone's surface $v_R(\mu_1) = v_\theta(\mu_1) = 0$, the absence of a normal velocity source at the symmetry axis $v_\theta(1) = 0$, and the radial velocity profile regularity $(\partial v_R/\partial \theta)|_{\theta=0} = 0$. For the function g these conditions transform respectively to

$$g(\mu_1) = g'(\mu_1) = 0, \qquad (2.5.5)$$
$$g(1) = 0, \qquad (2.5.6)$$
$$\lim_{\mu \to 1} \sqrt{1 - \mu^2}\, g''(\mu) = 0. \qquad (2.5.7)$$

It is natural to assume that velocity v_R on the axis is finite

$$\lim_{\mu \to 1} g'(\mu) = \text{const} < \infty. \qquad (2.5.8)$$

Applying the conditions (2.5.6) and (2.5.8) to the equation (2.5.4), and the conditions (2.5.6)–(2.5.8) to the differentiated equation (2.5.4), we obtain

$$a - b + c + 4A_1 \ln 2 = 0,$$
$$-2a - b + (4 \ln 2 + 2)A_1 = 0, \qquad (2.5.9)$$

which means, in fact, that the right-hand side of (2.5.4) has a double zero at $\mu = 1$. The third relation between the coefficients a, b and c follows from (2.5.5) applied to equation (2.5.4):

$$a\mu_1^2 - b\mu_1 + c + A_1(1 + \mu_1)^2 \ln(1 + \mu_1) + $$
$$ + A_2(1 - \mu_1)^2 \ln(1 - \mu_1) = 0. \qquad (2.5.10)$$

Expressing b and c by a, and A_1, A_2 with (2.5.9), (2.5.4) is rewritten in the form

$$(1 - \mu^2)g' + 2\mu g - \tfrac{1}{2}g^2 = a(1 - \mu)^2 + A_1[2 - (4 \ln 2 + 2)\mu + $$
$$ + (1 + \mu)^2 \ln(1 + \mu)] - $$
$$ - A_2(1 - \mu)^2 \ln(1 - \mu). \qquad (2.5.11)$$

For $S = Re = 0$, i.e. $A_1 = A_2 = 0$, and $a \neq 0$ we have the formulation of Squire's problem [68] for the finite momentum jet emanating in a conical domain. From (2.5.11) it follows that the no slip conditions (2.5.5) cannot be

Solutions in spherical coordinates

satisfied in this case. This is related to the fact that in Squire's problem (or in the Landau problem) only one of the conditions (2.5.5) can be fulfilled, e.g. the impermeability of the surface $\mu = \mu_1$: $g(\mu_1) = 0$. Yet if $S \neq 0$ and $Re \neq 0$, the no slip conditions give

$$a = -\frac{A_1}{(1-\mu_1)^2}[2 - (4\ln 2 + 2)\mu_1 + \\ + (1+\mu_1)^2 \ln(1+\mu_1)] + A_2 \ln(1-\mu_1). \quad (2.5.12)$$

From this it follows that the boundary conditions at a rigid boundary can be satisfied only in the presence of a current ($S \neq 0$) or a vortex line ($Re \neq 0$), and no other integral constraints (e.g. momentum) can be satisfied. The momentum may be prescribed (in other words, it is possible to find, other than (2.5.12), a relation between a, A_1, A_2, and the set value of F) only in the case where the conditions (2.5.5) are not satisfied simultaneously. The reason is, as has been noted in Section 2.4, that in the presence of a rigid boundary (or, contrastingly, friction at the wall) the momentum is wasted to overcome the friction, i.e. it does not remain constant on all the surfaces $R = $ const. Thence, the assumption that the momentum is a flow characteristic would destroy the similarity of flow.

Qualitative features of the flow can be demonstrated by solving the equation (2.5.4) in the Stokes approximation. Neglecting the term $g^2/2$ in (2.5.4), it is easily solved:

$$g = (1 - \mu^2)\left[-\frac{a}{1+\mu} - (2\ln 2 + 1)\frac{A_1}{1-\mu^2} + \right. \\ + \frac{A_2}{2}\ln\left|\frac{1+\mu}{1-\mu}\right| + A_1 \frac{\ln(1+\mu)}{1-\mu} + \\ \left. + A_2 \frac{\ln(1-\mu)}{1+\mu} + A_1 \frac{\mu}{1-\mu^2} + D\right], \quad (2.5.13)$$

where the integration constant

$$D = \frac{a}{1+\mu_1} + (2\ln 2 + 1)\frac{A_1}{1-\mu_1^2} - \\ - \frac{A_2}{2}\ln\frac{1+\mu_1}{1-\mu_1} - A_1 \frac{\ln(1+\mu_1)}{1-\mu_1} - \\ - A_2 \frac{\ln(1-\mu_1)}{1+\mu_1} - A_1 \frac{\mu_1}{1-\mu_1^2}, \quad (2.5.14)$$

and the constant a is determined by (2.5.12).

Note, in the particular case when $S = 0$, $\mu_1 = 0$ the solution (2.5.13) becomes the solution of Goldshtik [17], obtained in 1960 by analysing a tornado-type vortex flow at a plane surface (the tornado vortex is modeled by a vortex line on the symmetry axis). Assuming in the solution (2.5.13) that $Re = 0$ and $\mu_1 = 0$, it becomes the solution of Lundquist's problem [34] for the flow induced by an electric current radially diverging from a point in a semi-infinite liquid medium. The linear solution for $Re = 0$ and an arbitrary μ_1 is presented in [59]. The solution of the complex problem ($Re \neq 0$, $S \neq 0$, and μ_1 arbitrary) is reproduced from [56].

The linearity of the problem permits us to consider separately the meridional flow induced by an electric current and by a vortex line. Stream lines $\psi/\nu = Rg(\mu) = $ const are shown in Fig. 2.4 for the case $\mu_1 = 0$ and (a) $Re = 0$, $S = 2$ with $A = B = 1$, i.e. current is supplied to the origin from below; (b) $Re = \sqrt{2}$, $S = 0$; profiles of velocity Rv_R/ν are shown in Fig. 2.5. From the figures we

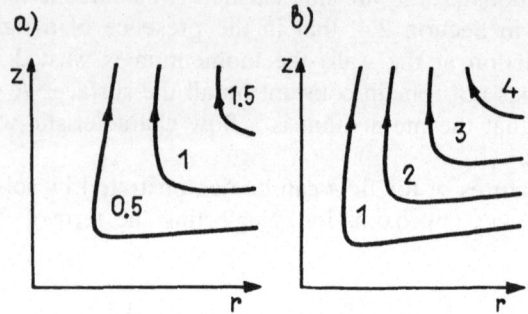

Fig. 2.4. (a) Stream lines in the electrically induced flow and (b) in the flow induced by the vortex line.

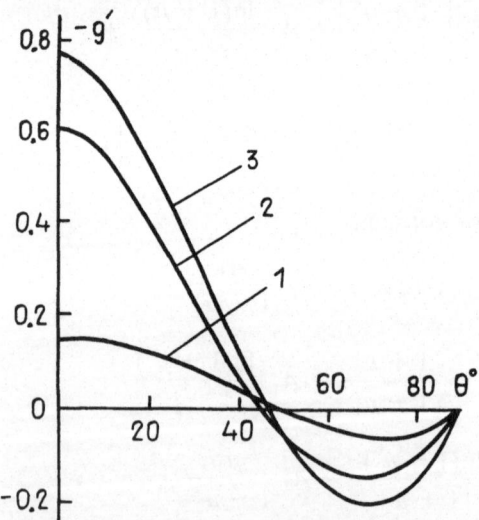

Fig. 2.5. Radial velocity profiles: (1) for electrically induced flow, (2) for vortex line induced flow, (3) for joint action of current and vortex line.

Solutions in spherical coordinates

may conclude that the meridional flows induced either by current or vortex line are similar: the clearly evident jet propagates along the symmetry axis in the direction from the cone's vertex; the fluid is entrained in the jet along the cone's surface.

The reason for the similarity of the meridional flows lies in the qualitatively similar nature of the force distributions responsible for the flows, though the forces themselves are different in nature and even in direction. For the electric current case it is the electromagnetic force. Interaction of the radial electric current and its self-magnetic azimuthal field yields the electromagnetic force which is orthogonal to the electric current lines. The force is nonuniform along the current line. Indeed, from (2.2.7)–(2.2.9) it follows that

$$j_R = \frac{I}{2\pi} \frac{1}{R^2(1 - \cos\theta_1)},$$

$$B_\varphi = \frac{\mu_0 I}{2\pi R \sin\theta} \frac{1 - \cos\theta}{1 - \cos\theta_1}, \qquad (2.5.15)$$

$$\mathbf{f}_e = \mathbf{j} \times \mathbf{B} = -\frac{\mu_0 I^2}{4\pi^2(1 - \cos\theta_1)^2} \frac{1}{R^3} \frac{1 - \cos\theta}{\sin\theta} \mathbf{i}_\theta,$$

i.e. $\mathbf{f}_e$ increases as R^{-3} approaches the point source along the line $\theta = $ const (Fig. 2.6a). Yet most important is the fact that the force $\mathbf{f}_e$ is rotational:

$$\text{curl } \mathbf{f}_e = \frac{\mu_0 I^2}{2\pi^2(1 - \cos\theta_1)^2} \frac{1}{R^4} \frac{1 - \cos\theta}{\sin\theta} \mathbf{i}_\varphi \neq 0.$$

Because of this fact $\mathbf{f}_e$ cannot be balanced by the potential pressure forces, but only by forces of a rotational nature, i.e. related to a viscous fluid motion (see Section 3.1). Since curl $\mathbf{f}_e > 0$ and it is directed azimuthally, the fluid flow, driven by the electromagnetic force, lies in the meridional planes, and it is directed according to the sign of curl $\mathbf{f}_e$ (Fig. 2.6a).

The cause of meridional flow in the presence of a vortex line is also the

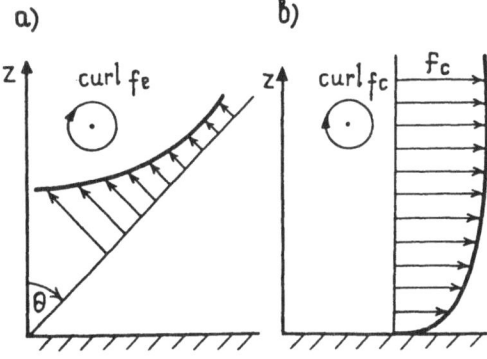

Fig. 2.6. Generation of meridional flow by rotational electromagnetic (a) and centrifugal (b) forces.

rotationality, but of the centrifugal force. According to (2.1.8) and (2.5.2), the expression for the centrifugal force is

$$\mathbf{f}_c = \rho \frac{v_\varphi^2}{R \sin \theta} \mathbf{i}_r = \frac{\rho v^2 Re^2}{(1 - \cos \theta_1)^2} \frac{(\cos \theta - \cos \theta_1)^2}{r^3} \mathbf{i}_r, \qquad (2.5.16)$$

where r is the cylindrical radius, with $\mathbf{f}_c$ in the direction of the cylindrical radius. Consider now the change of $\mathbf{f}_c$ for $r = $ const, varying θ: from (2.5.16) it is easy to see that $\mathbf{f}_c$ decreases monotonously from a maximum at $\theta = 0$, i.e. far from the plane, to zero on its surface (Fig. 2.6b). The latter is due to the friction at the surface and, as in the previous case

$$\operatorname{curl} \mathbf{f}_c = \frac{2\rho v^2 Re^2}{(1 - \cos \theta_1)^2} \frac{\cos \theta - \cos \theta_1}{R^4 \sin \theta} \mathbf{i}_\varphi \neq 0$$

and it is positive in the whole fluid domain. According to the sign of curl $\mathbf{f}_c$, the meridional flow is induced, similar to the previous case. Thus, in spite of the different origin of the forces and, moreover, of their different directions (cf. Fig. 2.6), the induced meridional flows are qualitatively similar, since the curls of the forces have the same sign.

Let us now evaluate the velocity v_R on the symmetry axis $\mu = 1$ for $Re = 0$. Differentiating (2.5.13), we obtain from (2.5.3)

$$\left. \frac{R}{\nu} v_R \right|_{\mu = 1} = -g'(1) = -a - A_1(2 + \ln 2) + 2D$$

$$= -\frac{S}{(1 - \mu_1)^2} \left(2 + \frac{3 + \mu_1}{1 - \mu_1} \ln \frac{1 + \mu_1}{2} \right). \qquad (2.5.17)$$

From this there follows the significant dependence of the electrically induced flow intensity, which can be measured by the velocity magnitude, on the symmetry axis $v_R|_{\mu = 1}$, on the geometric conditions of current flow (in the present case, from the angle of current divergence). According to Fig. 2.7, constructed after (2.5.17), for the constant current density and variable cone angle $(S(1 - \mu_1)^{-2} = $ const$)$ the flow intensity decreases with the decrease of θ_1 down to zero at $\theta_1 = 0$. Yet if the total current is set constant ($S = $ const), then the flow intensity decreases more slowly at $\theta_1 \to 0$ ($\lim_{\mu \to 1} Rv_R/\nu = 1/24$). The dependence of flow intensity on the angle of current divergence is important for various applications (some of them being discussed in Chapter 8).

Finally, let us estimate the limits of application for the Stokes solution (2.5.13). Assuming Reynolds number $Re = Rv_R(1)/\nu$ and supposing the Stokes solution is valid up to $Re \sim 1$, from (2.5.17) we obtain

$$\frac{S}{(1 - \mu_1)^2} \left(2 + \frac{3 + \mu_1}{1 - \mu_1} \ln \frac{1 + \mu_1}{2} \right) \approx 1.$$

In particular, for $\mu_1 = 0$ (plane surface) $0.0794 S = 1$, i.e. the linear solution is valid for $S \lesssim 10$.

Solutions in spherical coordinates

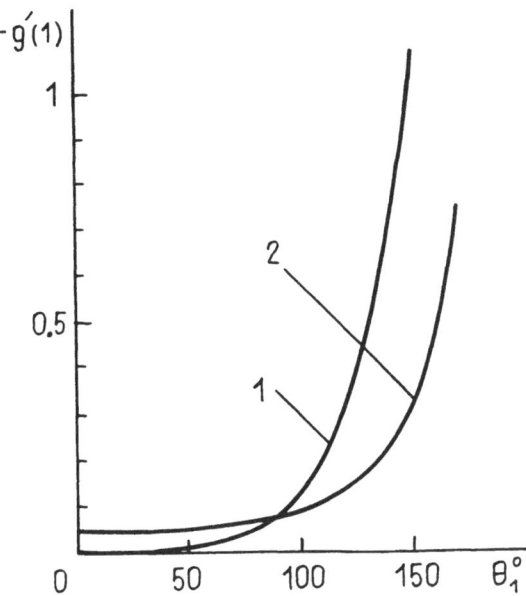

Fig. 2.7. Velocity on the symmetry axis vs. the cone's angle: (1) $S/(1 - \mu_1)^2 = $ const, (2) $S = $ const.

2.6. Gas flow in an electrical arc

The method used in Section 2.5 may be applied to the estimation of a flow arising in electrical arcs. The electrical arc is a type of electrical discharge with a relatively high current strength (of the order of 10^1–10^3 A). It is commonly represented by three regions: a cathode zone, an anode zone, and a cylindrical domain between these. To visualize the regions, consider an example of a typical arc between two carbon electrodes. For the total current magnitude of 40 A the current density in the arc column ~100 A/cm², at the cathode ~500, and at the anode ~40 [44] (Fig. 2.8).

Suppose a cathode spot is located at the apex of a cone $\theta = \theta_2$ ($\cos \theta_2 = \mu_2$), the arc burning zone at the cathode is close in form to a cone, and is limited by the cone $\theta \leqslant \theta_1$ ($\cos \theta_1 = \mu_1$) (region I in Fig. 2.9). Suppose also (this is quite a strong assumption) that the physical properties of the gas (viscosity, density, and so forth) are equal in the discharge zone and in the surroundings; but in the discharge zone, suppose that the gas is electrically conducting, ensuring electric current flow. The current direction coincides with the direction of the spherical radius, the current being uniformly distributed along a spherical surface in the region I. Thus the equations of motion for the regions I and II differ merely by the presence of electromagnetic force in the region I and its absence in the region II.

As in the previous problem, in (2.2.8) we set $A = B = (1 - \mu_1)^{-1}$. Then in the region I gas flow is determined by the same equation (2.5.4) assuming there $Re = 0$, and in the domain II by

$$(1 - \mu^2)g_2' + 2\mu g_2 - \tfrac{1}{2}g_2^2 = a_2\mu^2 - b_2\mu + c_2. \tag{2.6.1}$$

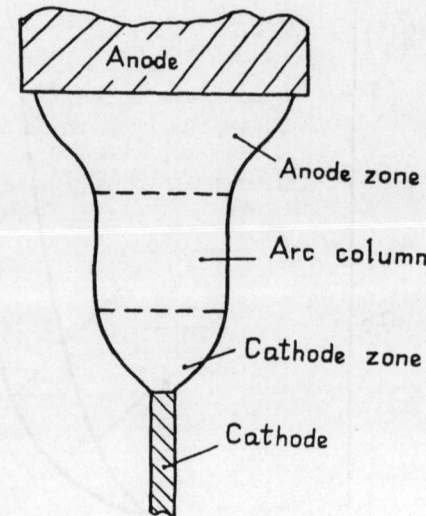

Fig. 2.8. Typical zones of electrical arc.

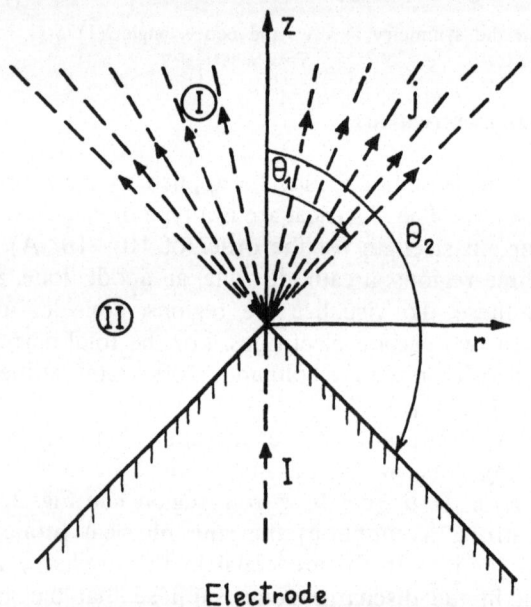

Fig. 2.9. Computational scheme of electrical arc.

Boundary conditions for the function g_1 on the axis $\mu = 1$ are analogous to (2.5.6)–(2.5.8); the function g_2 must satisfy the conditions (2.5.5) on the wall $\mu = \mu_2$. Apart from these, the continuity conditions must be satisfied on the common boundary $\mu = \mu_1$ for the function and its derivatives to the third order:

$$g_1(\mu_1) = g_2(\mu_1), \qquad g'_1(\mu_1) = g'_2(\mu_1), \\ g''_1(\mu_1) = g''_2(\mu_1), \qquad g'''_1(\mu_1) = g'''_2(\mu_1). \tag{2.6.2}$$

Solutions in spherical coordinates

The last of these conditions expresses continuity of pressure, and the other expresses continuity of the velocity field.

The conditions listed are sufficient to determine explicitly the constants a_i, b_i, c_i ($i = 1, 2$), and also two integration constants of the equations (2.5.4) and (2.6.1) [51].

The solution in the Stokes approximation is

$$g_1 = -a_1(1 - \mu) + \frac{S}{(1 - \mu_1)^2} [\mu - 2\ln 2 - 1 +$$
$$+ (1 + \mu)\ln(1 + \mu)] + D_1(1 - \mu^2), \qquad (2.6.3)$$

$$g_2 = -a_1(1 - \mu) + \frac{S}{(1 - \mu_1)^2}\left[(1 + \mu)\ln(1 + \mu_1) + \right.$$

$$+ \left(\frac{3}{2} - \frac{1}{4}(1 - \mu_1)^2\right)\mu - 2\ln 2 - \mu_1 - \frac{1}{2} +$$

$$\left. + \frac{1}{8}(1 - \mu_1)^2(1 - \mu^2)\ln\frac{1 + \mu}{1 - \mu}\right] + D_2(1 - \mu^2), \qquad (2.6.4)$$

where

$$D_2 = \frac{1}{(1 - \mu_2)^2}\left\{a_1(1 - \mu_2) - \frac{S}{(1 - \mu_1)^2}\left[(1 + \mu_2)\ln(1 + \mu_1) + \right.\right.$$

$$+ \left(\frac{3}{4} + \frac{1}{4}(1 - \mu_1)^2\right)\mu_2 - 2\ln 2 - \mu_1 - \frac{1}{2} +$$

$$\left.\left. + \frac{1}{8}(1 - \mu_1)^2(1 - \mu_2^2)\ln\frac{1 + \mu_2}{1 - \mu_2}\right]\right\},$$

$$D_1 = D_2 + \frac{S}{(1 - \mu_1)^2(1 - \mu_1^2)}\left[\frac{7}{8} - \frac{\mu_1}{2} + \frac{\mu_1}{4}(1 - \mu_1)^2 + \right.$$

$$\left. + \frac{1}{8}(1 - \mu_1)^2(1 - \mu_1^2)\ln\frac{1 + \mu_1}{1 - \mu_1}\right],$$

$$a_1 = -\frac{S}{(1 - \mu_1)^2(1 - \mu_2)^2}\left[(1 + \mu_2)^2\ln(1 + \mu_1) - \right.$$

$$\left. - 2\mu_2(2\ln 2 + \mu_1) + \frac{1}{2}(1 - \mu_1)^2 + \frac{3}{2}(1 + \mu_2^2)\right].$$

Fig. 2.10a shows stream lines in the arc, constructed by the solution (2.6.3), (2.6.4) for $\theta_1 = 45°$ and $\theta_2 = 135°$ (the dashed line separates the regions of conducting and nonconducting gases). According to the figure, the electromagnetic forces in the arc cause cold gas entrainment into the arc burning zone. In this zone the gas flow is of jet type, propagating along the symmetry axis from the cathode spot (from the vertex of the cone) (curve 1 in Fig. 2.11).

Numerical solution of the nonlinear problem stated above shows the effect of nonlinear terms in the equation of motion (the effect is pronounced for higher S), which leads us to the velocity concentration at the symmetry axis, though on the cone surface differences are minor between the Stokes and the nonlinear solutions (Fig. 2.11). This is evident by the comparison of stream lines in Fig. 2.10 corresponding to $S = 9$, especially in Fig. 2.10b, in which, by the dashed line, the linear stream line $\psi = 9$ is added for comparison with the nonlinear stream line $\psi = 9$ (full line).

The difference in intensity of the linear and nonlinear flows is demonstrated in Fig. 2.11 showing the distributions of normalized radial velocity g'/S. The nonlinear solution for the same S gives a substantially more intense flow than the Stokes solution. The reasons for such behaviour of the nonlinear solution will be discussed below.

The solution for the electrical arc also may be obtained without the use of the last condition in (2.6.2), i.e. dropping the demand of pressure continuity at the discharge edge. This approach may be motivated by the discontinuity of

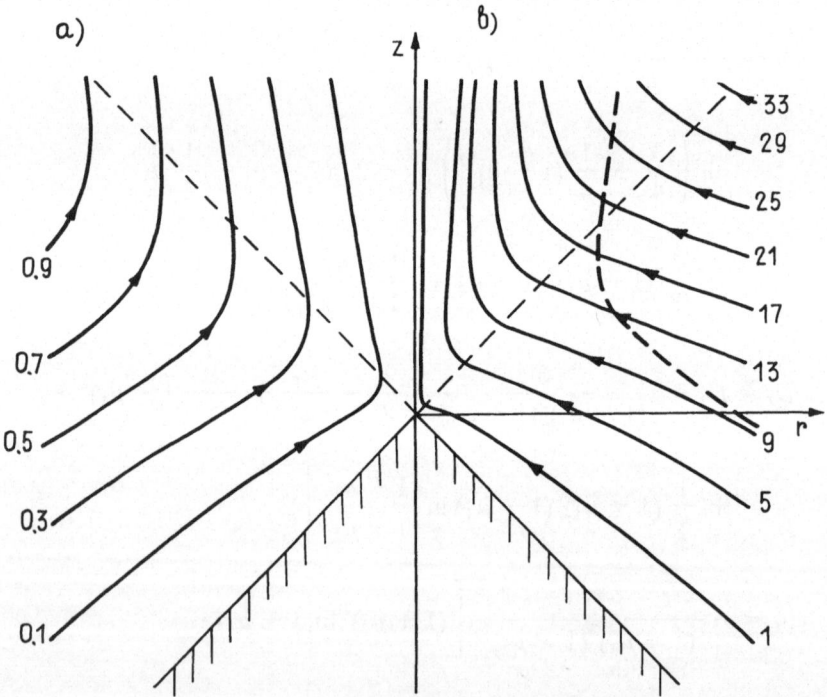

Fig. 2.10. Stream lines in an electrical arc according to the linear (a) and nonlinear (b) solutions.

Solutions in spherical coordinates

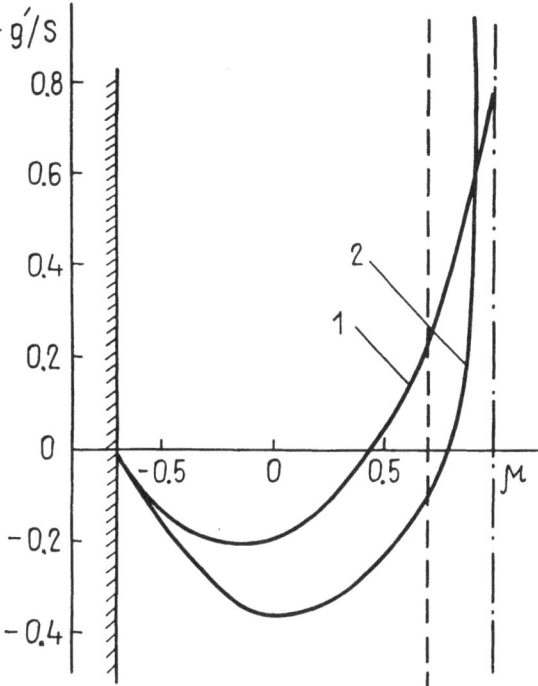

Fig. 2.11. Distribution of the normalized radial velocity in the arc: (1) linear solution; (2) nonlinear solution. (The position of solid electrode is indicated by a cross-hatched line, the discharge edge by a dashed line.)

the electromagnetic force (it is maximum at the inner side of the discharge boundary, and zero at the external side). The solution shows [51] that in this case the pressure is distributed in the following way: in the axial zone the pressure is higher than atmospheric; approaching the boundary of discharge it falls below atmospheric; and crossing the boundary it 'jumps' to nearly atmospheric pressure.

Note that the coefficients a_i, b_i, c_i in the right-hand side of equations (2.5.4), (2.6.1) may also be determined by the above method for larger numbers of alternating electrically conducting and nonconducting zones. Thus, for instance, inside the conical discharge there may be located a nonconducting axial zone. This scheme corresponds to the arc structure with an axial magnetic field [30] and to the structure of an electrically induced vortex in atmosphere (cf. Section 4.5).

2.7. Problems of the nonlinear solution

It is no accident that the electrically induced flow is considered in Section 2.5 jointly with the secondary flow induced by a vortex line. The reason is, as Goldshtik first showed [17], that the solution of (2.5.4)–(2.5.8) exists and is unique (for $S = 0$ and $\mu_1 = 0$) only for $Re \leq 4.8096$. For $Re > 8$ a bounded solution does not exist*, i.e. the meridional flow, induced by a vortex line normal to the plane surface, becomes singular. The singularity first appears on the symmetry axis where the condition (2.5.8) that v_R be finite is infringed.**

Since the meridional flow of the vortex line and the electrically induced flow are qualitatively similar, moreover, the mathematical formulation of the meridional flow problem is the same (the right-hand side of (2.5.4) merely varies with either $S = 0$ or $Re = 0$), then the singularity may also be expected in the electrically induced flow.

Indeed, the numerical solution of nonlinear problem (2.5.4)–(2.5.8) in [57, 58] and also in [50] showed that for the case of the current diverging from a point on the plane ($\theta_1 = \pi/2$), the radial velocity on the symmetry axis initially increased linearly with the parameter S, but further on it increased at a much higher rate than the linear (Fig. 2.12). Moreover, at $S = S_{cr} = 150$ the velocity became unbounded.

To estimate the consequences of this fact, note that the value of parameter $S = 150$ corresponds to the total current in mercury ($\rho = 13.6 \times 10^3$ kg/m^3, $\nu = 10^{-7}$ m^2/s) equal to only 0.9 A. If we take into account that the magnitude

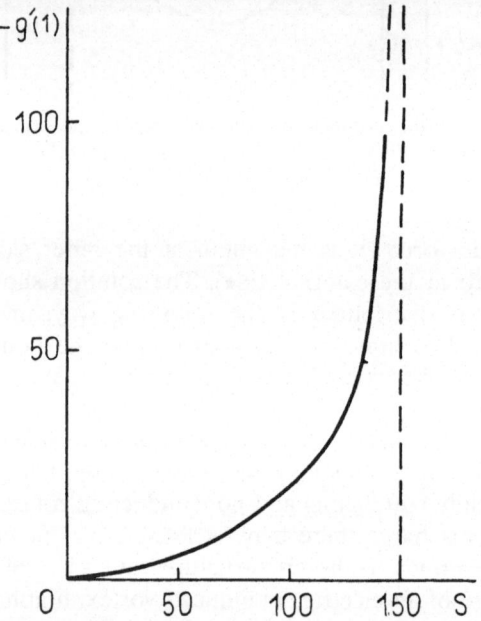

Fig. 2.12. Axial velocity dependence on the parameter S.

---

* The problem was solved numerically in [21]. For the flow in a cone of angle 2α at the apex the authors found the condition for the solution's nonexistence to be $Re > 16 \cot(\theta_1/2)$, which is double the value of Goldshtik's result.

** Particularly, from this it follows that the solution for boundary layers in [13, 71], proposed to model processes in vortex chamber, tornado vortex, etc., are questionable.

Solutions in spherical coordinates 97

of current in a liquid conductor in experiments and in real high-current technology may reach values as high as $10^3 - 10^4$ A (with $S = 10^8 - 10^{10}$), then the quite limited potential of the model to describe the phenomena quantitatively becomes clear.

The authors of [37] determined the value of S_{cr} to be substantially higher for the current discharge in a cone of angle $2\theta_1 < \pi$ at the apex. So, for $\theta_1 = \pi/6$, $S_{cr}(1 - \cos \theta_1)^{-2} = 1.5 \times 10^5$. In the following paper [38] the authors obtained an approximate formula for $S_{cr}(\theta_1)$:

$$S_{cr} = 2.56 \cdot 10^3 (1 - \cos \theta_1)^2 (\theta_1)^{-6.3}; \qquad (2.7.1)$$

they also estimated the Stokes solution to be valid up to approximately $S = 0.1 S_{cr}$; the formula (2.7.1) is valid for the parameter S_{cr} based on the total current I_{cr}, yet if we want to conserve the current density j_{cr} when changing the cone's angle, then

$$S_{0cr} = \frac{S_{cr}}{(1 - \cos \theta_1)^2} = \frac{\mu_0 I_{cr}^2}{4\pi^2 \rho v^2 (1 - \cos \theta_1)^2}$$

$$= \frac{\mu_0 j_{cr}^2 R^4}{\rho v^2} = 2.56 \times 10^3 (\theta_1)^{-6.3}. \qquad (2.7.2)$$

Now estimate a critical current value, e.g. for $\theta_1 = \pi/6$. According to (2.7.1) we have $I_{cr} = 39$ A, i.e. it is substantially higher than for $\theta_1 = \pi/2$, but not high enough to use the model in practice.

The most complete study of the dependence of S_{cr} on the geometric conditions of current discharge and on the fluid flow domain is made in [7]. These results (Fig. 2.13) contain the particular values obtained in [37, 38, 50, 57]. The data of Fig. 2.13b represent the various forms of arc discharge, when the gas is

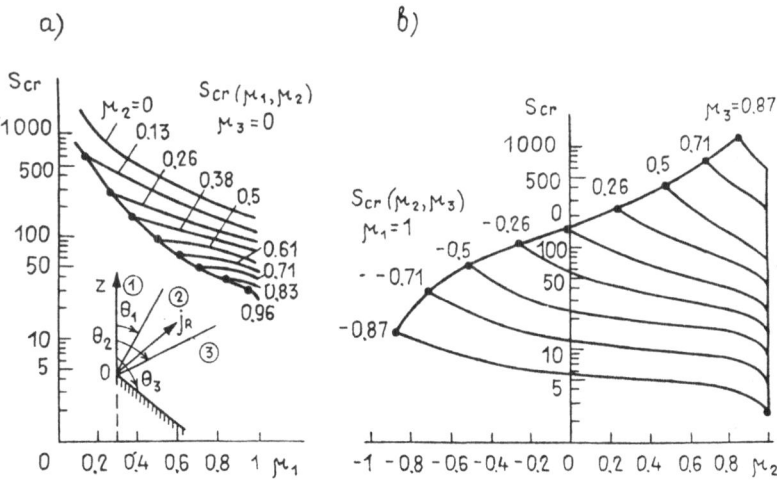

Fig. 2.13. Dependence of the critical parameter S_{cr} on the flow geometry and position of the discharge zone.

divided in two zones — conducting and nonconducting (in a particular case the latter is absent — the curve $\mu_2 = \mu_3$) — and Fig. 2.13a represents the various forms of tornado model (cf. Section 4.5), which has an additional axial zone of nonconducting gas (scheme 3 in Table 2.1).

Analysis of the results represented in Fig. 2.13b reveals that, for the fixed rigid boundary (μ_3 = const), narrowing of the discharge zone decreases the value of S_{cr}. If the total current is assumed to be fixed, this may be attributed to the growing current density in the process of contraction. A more definite conclusion may be reached if the discharge zone is fixed (μ_2 = const). As follows from the same figure, decrease of the rigid cone domain ($\mu_3 \rightarrow -1$) leads to a substantial decrease in S_{cr}. Conversely, the decrease of the fluid domain leads to the increase of S_{cr}. In particular, when the electric current penetrates the whole fluid ($\mu_2 = \mu_3$), contraction of the discharge ($\mu_3 \rightarrow 1$) also increases S_{cr}.

Consequently, viscous friction on the rigid cone's surface plays a substantial role in limiting the velocity field breakdown. If viscous effects are totally ignored, the singularity of velocity on the axis takes place for any $S > 0$, as has been demonstrated by Shercliff [53] who analysed a discharge in an inviscid fluid. It is tempting to conclude that the appearance of S_{cr} or Re_{cr} in the model (2.5.4)–(2.5.8) may be related to an insufficiently effective viscous mechanism of dissipation in the pure meridional flow converging to the symmetry axis flow, or to some other unexplained factors. One of the latter factors in a pure electrically induced flow ($Re = 0$) could be related to the flow induced electric currents which, as is well known, damp the fluid flow. Next, the very statement of the problem contains a singularity at the point electrode, since in this case the electromagnetic force is singular in the coordinate origin [35] (although in the equation of motion (2.5.4) the singularity has cancelled out). Finally, the breakdown may be related to the unbounded flow domain. We shall examine these factors in the listed order below, but before this we shall consider the influence of another integral flow characteristic on the value of S_{cr} (Section 2.8). The first hypothesis, concerning the viscous effects, will be discussed in the next Chapter.

Note also that the nonlinear problem of flow at a point electrode was studied in a nonstationary regime of the flow development in time from the rest state after an instantaneous switch-on of the electric current [62]. Since statement of the problem does not contain a representative time interval and a length scale, the dimensional time and distance can be combined in one nondimensional similarity variable $T = (\nu t)^{1/2}/R$, and the stream function can be determined in the form $\psi = \nu R g(\theta, T)$. Then the nonstationary equations of motion, after the elimination of pressure, are reduced to a set of two nonlinear second order partial differential equations (or one of the fourth order). The computational procedure in [62] is based on the assumption that a steady state is reached exponentially in time, and then a boundary condition for the similarity variable $T \rightarrow \infty$ is the previously known stationary solution.

In consequence of this assumption the developing flow field exists merely up to the value $S = S_{cr}$. In the interval of feasible S the magnitudes of the stream function initially increase with T, reach a maximum, and then fall off to the

Solutions in spherical coordinates 99

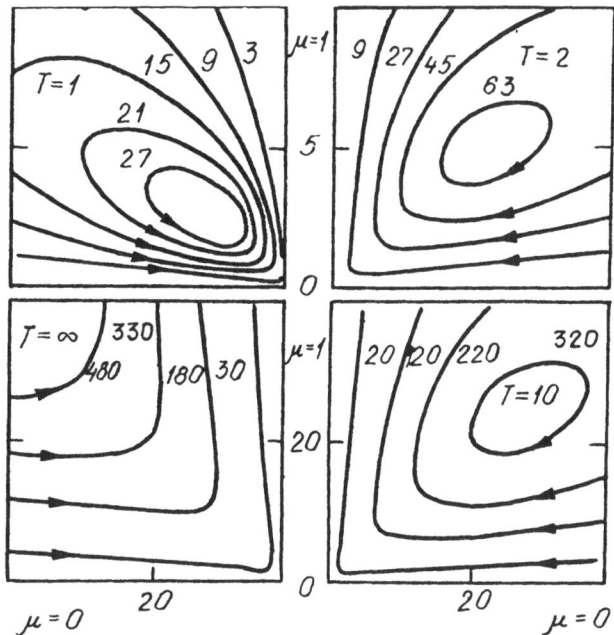

Fig. 2.14. Time development of the electrically induced flow at a point electrode.

steady state. With the increase of S the maximum magnitude of ψ increases, and it is reached for a smaller T. The flow field for a very small T is of a toroidal vortex form with closed stream lines (Fig. 2.14). With the growth of T the centre of vortex moves to infinity, where also the stream lines close. The steady state is reached in a half-sphere of radius R in a time of order R^2/ν. However the higher the value of S, the slower is the rate at which the vortex centre moves to infinity.

2.8. Landau-Squire flows in the presence of a radially diverging electric current

Other flow characteristics, not associated with electric current, may also substantially influence the value of critical S. As mentioned in Section 2.3 the exact solution class permits us to consider the flows which are driven by a hydrodynamic momentum flux originating from the point. This class of flows (in the absence of current) is known as Landau-Squire flows (see Section 2.4). It would be of interest to examine a simultaneous action of the hydrodynamic momentum and of the electric current. The situation may find application, for instance, in a plasmatron where an electric arc burns in a flow of injected gas.

In the case of an immersed jet issuing from the vertex of a cone [27, 28], the solution satisfies the integral constraint of momentum flux conservation on all the surfaces $R = $ const. As we shall demonstrate, the jet momentum constraint is equivalent to a given velocity slip on the cone's surface. To see this, consider the problem of Landau, for which the equation (2.5.4) has a zero value on the right-hand side and $\theta_1 = 0$ (or $S = 0$), $\theta_2 = \pi$ in Fig. 2.9. The solution of the equation is (2.4.8) and, respectively,

$$g'(\mu) = 2\left[\frac{1-A^2}{(A-\mu)^2} + 1\right]. \tag{2.8.1}$$

The radial velocity of the 'cone's' surface $\mu = -1$ is proportional to

$$g'(-1) = \frac{4}{A+1} = \alpha, \tag{2.8.2}$$

i.e. the slip velocity $\sim \alpha$ is related by the constant A to the jet momentum F_0 (2.4.10). Naturally, a similar relation also occurs for an arbitrary cone angle.

Thus, the proposed problem, with the addition of the electric current discharge, is defined by the equations (2.5.4) and (2.6.1) with boundary conditions similar to (2.5.5)–(2.5.8) and (2.6.2), but the second condition (2.5.5) is $g'_2(\mu_2) = \alpha$. The problem with the two parameters S and α was solved numerically [10] for the case of a discharge bounded by $\theta_1 = 45°$ ($\mu_1 = 0.707$) and the surface of the electrode varied from $\theta_2 = 150°$ ($\mu_2 = -0.866$) to $\theta_2 = 180°$ ($\mu_2 = -1$). Note also that $\alpha > 0$ corresponds to the positive momentum (jet from a source), and $\alpha < 0$ corresponds to the jet to a sink.

Figure 2.15a shows stream lines in the jet-source for $S = 0$ and $\theta_2 = 150°$. The flow resembles the electrically induced flow in the electrical arc model (cf. Fig. 2.10). Therefore it may be expected that a joint action of the positive momentum and the electric current would intensify the flow, particularly at the

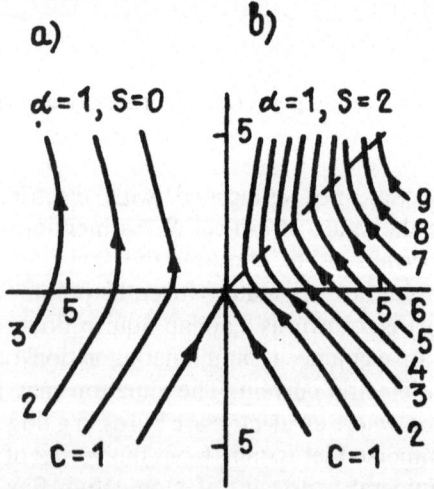

Fig. 2.15. Stream lines in the source-jet: (a) in the absence of electric current, (b) in the case of integral action of hydrodynamic momentum and electric current.

Solutions in spherical coordinates 101

symmetry axis (Fig. 2.15b*). This means also that for $\alpha > 0$ the value of S_{cr}, at which the radial velocity grows to infinity, must decrease, and the higher the value of α, the faster is the decrease. This is illustrated in Fig. 2.16. For the limiting value $\alpha = 2.61$ the critical value $S_{cr} = 0$ for given $\theta_2 = 150°$.

The existence of a limiting value of $\alpha > 0$ for a given θ_2 is easy to see, for example, in Landau's problem. Indeed, for $A \to 1 + 0$ the jet momentum $F_0 \to +\infty$ (2.4.10) and the radial velocity (2.8.1) on the axis $\mu = 1$ also grows to infinity ($g'|_{\mu \to 1} \to \infty$). Corresponding to $A = 1$ is the limiting value $\alpha_\ell = 2$ (2.8.2). Numerical solutions for the particular values of α indicate that α_ℓ increases with decreasing θ_2. So, for $\theta_2 = 175°$ $\alpha_\ell = 2.1$, for $\theta_2 = 150°$ $\alpha_\ell = 2.61$, etc. (Table 2.2).

Table 2.2. Values of jet momentum $\bar{F}$ for different slip coefficients α and angles θ_2.

α	$\theta_2 = 150°$	175°	180° (the Landau problem)
1	2.50	3.28	3.12
2	9.71	1.67×10^3	∞
2.6	0.17×10^4	∞	∞
−1	−1.63	−1.76	—
−2	−3.05	−6.72	—
−30	−55.01	−45.50	—

It is of interest to note that, according to the Landau solution, for $A = 1$ the velocity component $Rv_\theta/\nu = -g/\sqrt{1-\mu^2}$ is infinite on the axis $\mu = 1$, as follows from (2.4.6) and (2.4.8), i.e. the symmetry axis turns out to be a line sink. This situation is used by Schneider [47] to construct the solution for a jet in the presence of rigid boundary (Section 2.4). Thus for $\alpha = \alpha_\ell$ the flow changes its structure and assumes the form depicted in Fig. 2.17a. The similar result for $\theta_2 < \pi$ is represented in Fig. 2.17b.

If the momentum is negative (Landau's problem $F_0 < 0$, $A < -1$, $\alpha < 0$), the fluid flow (Fig. 2.18a) is in opposition to the electrically induced flow. Consequently, the electrically induced flow must be weakened by the action of the momentum sink, and one would expect an increase of the critical S with decreasing $\alpha < 0$. Stream lines for this case are represented in Fig. 2.18b, and the dependence of $S_{cr}(\alpha)$ in Fig. 2.16. For a certain ratio of S to α the flow is divided into two zones: in the axial zone the flow is closer to the electrically induced one (the electromagnetic forces are concentrated there), and the sink flow takes place closer to the conical electrode. Similar results prior to [10] were obtained by Sozou [58] for the particular case $\theta_1 = \theta_2 = \pi/2$.

* The value of $S = 2$ corresponds to a current in the plasma of $I = 1.5$ A. Hence the substantial flow intensification, as in Fig. 2.15b, caused by such a small current, is more likely to be due to a proximity of $S = 2$ to the critical value of S for given α and θ_2. In other words, although the intensification is too high to be expected in an experiment, the qualitative result — jet intensification with the passage of the current — is, evidently, correct.

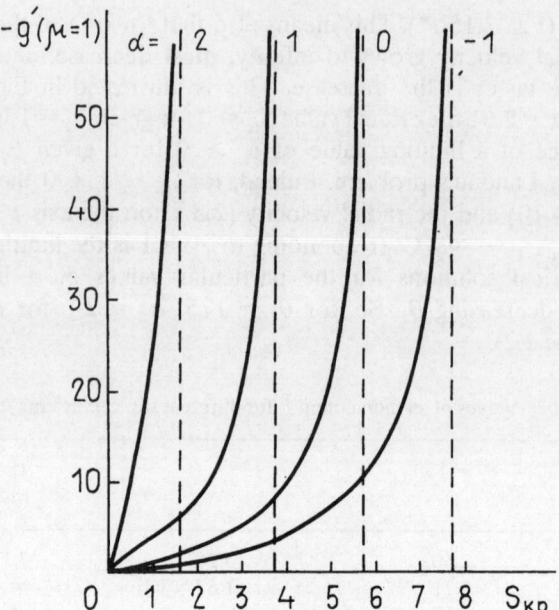

Fig. 2.16. The effect of hydrodynamic momentum on S_{cr}.

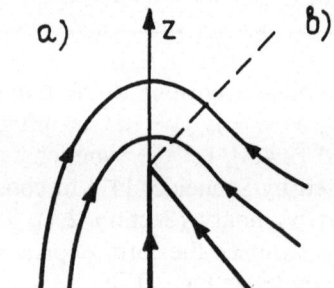

Fig. 2.17. The flows at critical slip velocities.

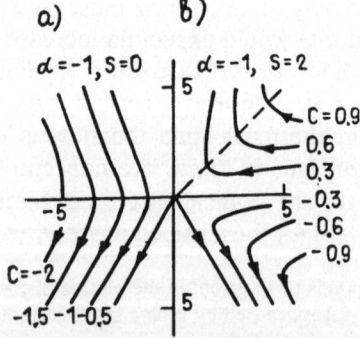

Fig. 2.18. Stream lines in the sink-jet: (a) in the absence of electric current, (b) in the case of integral action of negative momentum and electric current.

Solutions in spherical coordinates 103

The action of electromagnetic force increases the total hydrodynamic momentum flux. Figure 2.19 demonstrates the dependence of the total momentum on S for a particular a. Dependences $F(a, S) = \bar{F}2\pi\rho v^2$ for $S = 0$ and $\bar{F}(S)$ for $a = 0$ are shown in Fig. 2.20.

As noted in Section 2.3, apart from the jet momentum, the vortex line Γ on the symmetry axis can be added to the flow characteristic parameters; the results of Section 2.5 indicate that, with increasing Γ, the values of S_{cr} are expected to decrease.

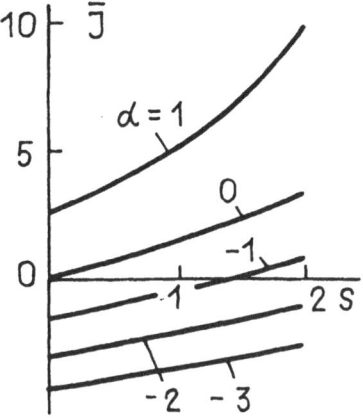

Fig. 2.19. The nondimensional total momentum vs. the parameter S for different slip velocities.

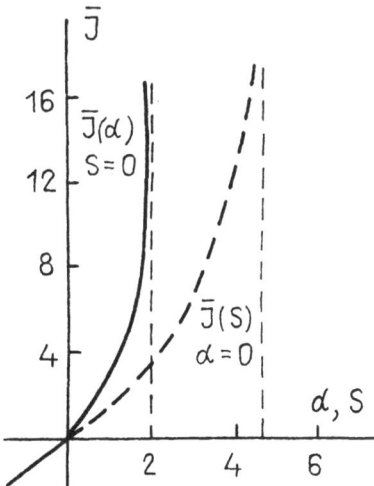

Fig. 2.20. Effect of the parameters S and a on the total momentum.

2.9. Effect of the induced electric current on the flow at a point current source

An additional circumstance, not included in the model (2.5.4)—(2.5.8) which, in principle, may affect the mere fact of the existence of S_{cr} is associated with an effect of induced currents on the fluid motion. It is well known that the currents induced by a fluid flow usually act so as to damp the motion, thereby it is quite logical to suppose that, by taking into account the induced current, an unlimited increase of velocity may be prevented in the problem under consideration.

Prior to examining the specific results concerning the induced current effect, let us state some general considerations about it. In the scheme proposed by Lundquist [34] an applied electric current flows along a spherical radius, therefore its redistribution under the action of fluid flow can take place due to the radial component of the induced electric field $\mathbf{E}_i$ (or the current density $\mathbf{j}_i = \sigma \mathbf{E}_i$). From Ohm's law $\mathbf{E}_i = \mathbf{v} \times \mathbf{B}$, and $\mathbf{E}_i$ is always normal to the velocity vector. Glancing at the flow pattern represented in Figs. 2.4a or 3.9a, it is easy to conclude that near the axis and at the plane the radial component of $\mathbf{E}_i$ must be close to zero and the maximum magnitude of $\mathbf{E}_i$ is reached in an intermediate interval. Taking into account the direction of $\mathbf{E}_i$ (it is always opposed to the direction of the externally applied current density), it becomes clear that the most considerable reduction of the unperturbed current density will take place in this very domain. Under the constraint of a constant total electric current applied externally to the fluid this means that this current deficiency in the intermediate domain has to be compensated by an excess density at the axis and the plane. In other words, for the case of intense interaction of flow and electrodynamic field, the applied current should pass mainly along the symmetry axis and the plane bounding the fluid domain (Fig. 2.21) [53].

The second consideration concerns estimates the validity of the low magnetic Reynolds number ($Re_m \ll 1$) approximation. The first estimate was made by Lundquist [34]; it was based on his Stokes solution of the fluid flow. To estimate

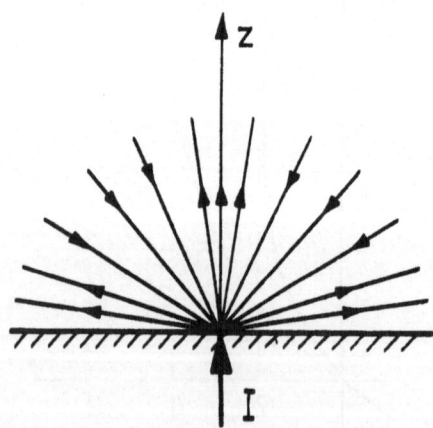

Fig. 2.21. Effect of the fluid motion on the distribution of electric current diverging in a half-space.

Solutions in spherical coordinates

$Re_m = vL/v_m = \mu_0 \sigma vL$ let us use the result (2.5.17) and a representative distance from the current source R:

$$Re_m = \mu_0 \sigma v_R|_{\theta=0} R = 0.079 \mu_0 \sigma \frac{\mu_0 I^2}{4\pi^2 \rho v^2} = 0.079 S\beta.$$

The requirement $Re_m \ll 1$ in this case means the total current must satisfy the condition

$$I \ll I_0 = \frac{2\pi v}{\mu_0} \sqrt{\frac{\rho}{0.079\sigma}}.$$

For mercury ($v = 10^{-7}$ m²/s, $\rho = 10^4$ kg/m³, $\sigma = 10^6$ Ω^{-1}m^{-1}) $I_0 = 900$ A. However, the critical current magnitudes, computed in Section 2.7 for the respective S_{cr} are much smaller than the above I_0 and obviously satisfy the condition $Re_m \ll 1$. Nevertheless the nonlinear solution leads to flow breakdown for the specified currents.

Consider the problem from a different perspective; namely, let us estimate the magnitude of Batchelor number $\beta = v\mu_0\sigma$ for which the effect of induced electric current may be expected. For this purpose we use equation (2.2.1) in which the magnetic field enters in the form of expansion by powers of β (recall that the first term in the expansion represents the unperturbed magnetic field L_0 and the second, L_1, represents the field of flow induced electric current); its right-hand part has terms containing $L_0 L_1$ that are proportional to $S\beta$, yet the terms of unperturbed magnetic field with L_0^2 are proportional to S. From (2.2.5) $L_1 \sim g$, i.e. it is induced by the motion; then if $g \sim S^n$, both terms in (2.2.1) are of the same order if

$$S \sim S\beta S^n. \tag{2.9.1}$$

For Stokes flow $n = 1$ and the relation (2.9.1) is

$$S\beta \sim 1 \tag{2.9.2}$$

i.e. for $S \sim 10^1$ the effect of induced current is felt beginning with $\beta \sim 10^{-1}$ and higher; however for liquid metals $\beta \sim 10^{-6}$—10^{-7} and lower. Of course, for S close to the critical value the velocity increases at a higher rate than that according to the Stokes solution (cf. Fig. 2.12), i.e. $n > 1$ and the values of β are lower, but the effect would be questionable. In the order of magnitude estimates it is more reasonable to consider values of S not very close to S_{cr}.

In any case, (2.9.1) indicates that for higher β the effect of induced currents is enhanced; therefore the final answer to the problem of S_{cr} may be sought by the use of a numerical solution. Sozou and English [61] numerically solved the set (2.1.14) and (2.1.16) for $G = 0$, $\ell = 0$; the boundary conditions for L were the same either for $\beta = 0$ or $\beta > 0$ (cf. Section 2.2).

Fig. 2.22 represents the numerical results for the current density distribution $IL'/2\pi R^2 \sin\theta$ (the left part of the figures) and the stream lines (right part) for quite high $\beta = 1$ and different S. The results indicate that the inclusion of induced currents increases the range of S in which the solution for the velocity

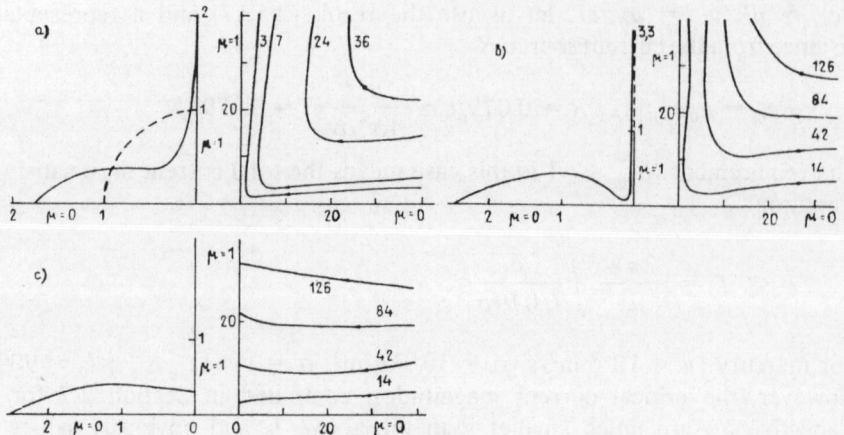

Fig. 2.22. The distribution of current density (left part of figures) and stream lines (right part) for different S: (a) $S = 100$, (b) 350, (c) 406.5.

field exists. However, the effect of induced currents only increases the value of S_{cr}, but does not eliminate the fact of its existence. Thus, for $\beta = 0$ $S_{cr} = 150$ and for $\beta = 1$ $S_{cr} = 406.5$. Similarly to the case $\beta = 0$, when a critical S is reached for $\beta > 0$, it leads to a radical change of the velocity field with the appearance of the velocity singularity at the axis (Fig. 2.22c).

For a fixed β the current density is redistributed with the increase of S. If for $\beta = 0$ the current density is uniform (in Fig. 2.22a the dashed circle of radius 1), then for $\beta > 0$ the supplied current is concentrated at the plane and the axis, and, as Fig. 2.22b indicates, the greater part of it passes near the plane. This should lead to an increase of the rotational part of the electromagnetic force, since the divergence of electric current lines is responsible for it. This, in addition, may decrease S_{cr} due to the intensification of the flow. It is interesting to note, also, that the growth of the current density at the symmetry axis is nonmonotonous, reaching a maximum $3.3I/2\pi R^2$ for $S = 350$, and it then falls off, and for $S = 406.5$ it is only $0.3I/2\pi R^2$.

A similar redistribution of the current is observed for increasing β and fixed S: an increase of β is favourable to S_{cr} growth, yet the higher β is, the lower is the growth rate of S_{cr} [61].

Thus, although the induced currents reduce the growth rate of the axial velocity with increasing S, these do not eliminate the breakdown of the velocity field. Even assuming $\beta = 1$, we merely obtain a threefold increase in S_{cr}. For real media ($\beta = 10^{-6} - 10^{-7}$ and lower) the increase of S_{cr} is slight and addition of the equation (2.1.14) to account for the induced currents does not improve the mathematical model (2.5.4)–(2.5.8).

The paper [61] also studies a joint action of electric discharge, hydrodynamic momentum and induced currents. In summary the qualitative result is the following: by increasing β the electrically induced flow is retarded, and the flow approaches the pure flow due to the momentum source.

Solutions in spherical coordinates

2.10. Electrically induced flows at finite size electrodes

The velocity field singularity for $S \geq S_{cr}$ may be related to the singularity of physical quantities at the coordinate origin (this view was particularly expressed by Cowley [14]). Consider now a situation which eliminates the singular point $R = 0$. Of course, the problem in this case is not self-similar and does not belong to the class of exact solutions considered. Nevertheless, the discussion of flows at finite electrodes is immediately related to the limiting case of the point electrode.

Let the electric current flow in the half-space filled with an electrically conducting fluid, from the hemispherical electrode of radius $R = R_0$ in the coordinate origin (Fig. 2.23). The surface $\theta = \pi/2$ is assumed to be a free surface.

Introducing nondimensional quantities $R = R^*/R_0$, $\psi = \psi^*/\nu R_0$ (where the asterisk denotes the dimensional quantities) and a new variable $\mu = \cos\theta$, the equation (1.10.15) for $\psi_2 = 0$, $v_\varphi = 0$, and B_φ in the form (2.2.11) for $\theta_1 = \pi/2$ is expressed as

$$\mathbf{i}_\varphi \cdot R\sqrt{1-\mu^2}\,\nabla\psi \times \nabla\left(\frac{E^2\psi}{R^2(1-\mu^2)}\right) - E^4\psi = 2S\frac{1-\mu}{R^3}, \quad (2.10.1)$$

where

$$E^2 = \frac{\partial^2}{\partial R^2} + \frac{1-\mu^2}{R^2}\frac{\partial^2}{\partial \mu^2}.$$

The equation (2.10.1) must be solved with the boundary conditions:

impermeability and no slip on the surface of electrode

$$\psi(1,\mu) = \frac{\partial\psi}{\partial R}(1,\mu) = 0, \quad (2.10.2)$$

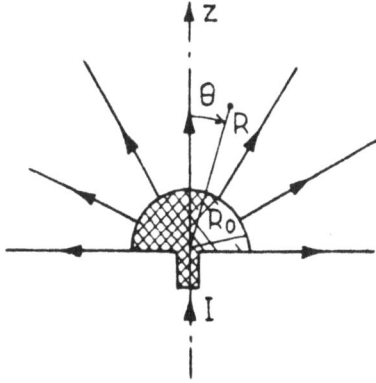

Fig. 2.23. Model of hemispherical electrode.

zero normal velocity and zero tangential stress on the free surface

$$\psi(R, 0) = \frac{\partial^2 \psi}{\partial \mu^2}(R, 0) = 0, \qquad (2.10.3)$$

absence of source at the symmetry axis

$$\psi(R, 1) = 0. \qquad (2.10.4)$$

Assume also that the velocities fall off at infinity no slower than R^{-1} (cf. 2.1.7):

$$\lim_{R \to \infty} R^{-1} \psi(R, \mu) = g(\mu). \qquad (2.10.5)$$

The solution to the problem (2.10.1)–(2.10.5) is constructed in a form of power series expansion by the parameter S [8]:

$$\psi = S\psi_1 + S^2\psi_2 + \ldots . \qquad (2.10.6)$$

To determine the functions ψ_i we then have the equations

$$-E^4 \psi_1 = 2 \frac{1 - \mu}{R^3}, \qquad (2.10.7)$$

$$E^4 \psi_2 = \mathbf{i}_\varphi \cdot R\sqrt{1 - \mu^2}\, \nabla \psi_1 \times \nabla \left(\frac{E^2 \psi_1}{R^2(1 - \mu^2)} \right), \qquad (2.10.8)$$

etc.

According to the condition (2.10.5), the solution in the external domain ($R \geqslant 1$) contains decreasing powers of R:

$$\psi_1 = R[-(1 + \mu)\ln(1 + \mu) + (2\ln 2 - \tfrac{1}{2})\mu + \tfrac{1}{2}\mu^2] +$$

$$+ \sum_{n=1}^{\infty} J_{2n+1}(\mu)[B^1_{2n+1} R^{-2n} + D^1_{2n+1} R^{-2n+2}], \qquad (2.10.9)$$

where $J_k(\mu)$ are Gegenbauer functions (see Section 1.7).

The Gegenbauer functions of odd index k satisfy the conditions (2.10.3). For the purpose of fulfilling (2.10.2) we define the first line of (2.10.9) to be antisymmetric for negative μ and expand it by $J_{2n+1}(\mu)$. Then (2.10.9) can be expressed

$$\psi_1 = \sum_{n=1}^{\infty} J_{2n+1}(\mu)(a_{2n+1} R + B^1_{2n+1} R^{-2n} + D^1_{2n+1} R^{-2n+2}), \qquad (2.10.10)$$

where

$$a_{2n+1} = \frac{(-1)^n}{\sqrt{\pi}} \frac{(4n + 1)\Gamma(n + \tfrac{1}{2})}{2n(2n + 1)(2n - 1)(n + 1)\Gamma(n + 1)}$$

(where $\Gamma(x)$ is a gamma function).

Solutions in spherical coordinates 109

The conditions (2.10.2) determine

$$B^1_{2n+1} = a_{2n+1} \frac{2n-1}{2}, \qquad D^1_{2n+1} = -a_{2n+1} \frac{2n+1}{2}.$$

The coefficients a_{2n+1} decrease,

$$a_3 = -\frac{5}{24}, \qquad a_5 = \frac{3}{160}, \qquad a_7 = -0.0048\ldots,$$

at a rate fast enough to retain only the first terms of the series (2.10.10) in the following computation of ψ_2. Thus, if we are to retain the term $n = 1$, the error in the magnitude of ψ_2 is of the order of $|a_3 a_5| \approx 10^{-3}$ and the equation (2.10.8) transforms to

$$E^4 \psi_2 = -\frac{6}{7} a_3^2 \left[J_3(\mu) \left(R^{-3} - \frac{15}{2} R^{-4} + 9R^{-5} + \frac{7}{2} R^{-6} - 6R^{-7} \right) + \right.$$

$$\left. + J_5(\mu)(20R^{-3} - 66R^{-4} + 54R^{-5} + 7R^{-6} - 15R^{-7}) \right].$$

Its solution is

$$\psi_2 = -\frac{6}{7} a_3^2 \left[J_3(\mu) \left(B_3^2 R^{-2} + D_3^2 + \frac{R}{24} - \frac{1}{4} \left(\ln 2 - \frac{1}{6} \right) - \right. \right.$$

$$\left. - \frac{3}{8} R^{-1} - \frac{1}{20} R^{-2} \ln R - \frac{1}{24} R^{-3} \right) + J_5(\mu) \left(B_5^2 R^{-4} + \right.$$

$$\left. + D_5^2 R^{-2} + \frac{1}{18} R - \frac{33}{140} + \frac{3}{8} R^{-1} + \right.$$

$$\left. \left. + \frac{1}{18} \left(R^{-2} \ln R - \frac{5}{14} R^{-2} \right) + \frac{3}{16} R^{-3} \right) \right], \qquad (2.10.11)$$

and the conditions (2.10.2) give

$$B_3^2 = \frac{29}{240}, \qquad D_3^2 = \frac{51}{240}, \qquad B_5^2 = -\frac{311}{80 \cdot 63 \cdot 2},$$

$$D_5^2 = -\frac{3343}{80 \cdot 63 \cdot 3}.$$

Note that solutions (2.10.10), (2.10.11) contain the highest positive power of $R = R^*/R_0$ equal to one. If the dimensional radius R^* is fixed when passing to the limit $R_0 \to 0$, then the dimensional stream function $\psi^* = \nu R_0 \psi$ will contain only the terms proportional to the first power of R:

$$\lim_{R_0 \to 0} \psi^* = \lim_{R_0 \to 0} \nu R_0 \psi \left(\frac{R^*}{R_0}, \mu, S \right)$$

$$= \lim_{R_0 \to 0} \nu R_0 \left[\frac{R^*}{R_0} g(\mu, S) + o\left(\frac{R^*}{R_0} \right) \right] = \nu R^* g(\mu, S),$$

i.e. the solution transforms to the self-similar exact solution of a point current source on a free surface. The same conclusion may be reached if R_0 is fixed and proceeding with $R^* \to \infty$.

The similarity solution of the problem with a point electrode on a free surface was obtained by Lundquist [34] in the Stokes approximation. Solution of the same nonlinear problem by Sozou and Pickering [63] demonstrated that with the free surface the value of $S_{cr} = 47$, i.e. three times lower than for the problem with the rigid surface ($S_{cr} = 150$). Consequently, the presence of critical S in the problem with a free surface means that the above solution with a finite electrode, eliminating the singular point $R = 0$, has the same singularity, but the velocity field breakdown is moved to a domain of high R. The conclusion, of course, is independent of the limitation of the solution procedure, whether retaining only the two terms of series (2.10.6), or setting $n = 1$ and constructing (2.10.11), because the limiting form of the solution is preserved. The limitations of the procedure affect the precision of the computed ψ,

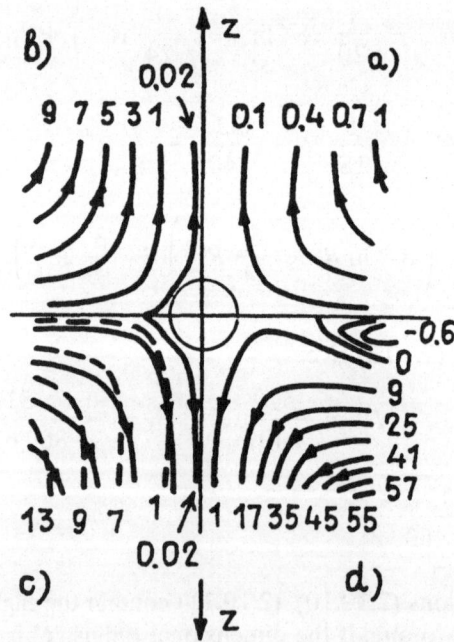

Fig. 2.24. Flow at the finite size electrode for (a) $S = 5$, (b) 50, (c) 75, (d) 250.

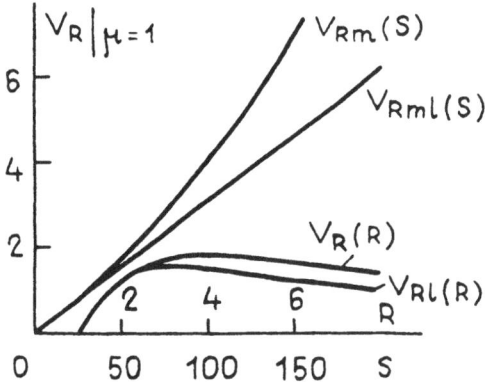

Fig. 2.25. Radial velocity on the symmetry axis vs. the distance from origin, $v_R(R)$, and maximum of the velocity vs. parameter S, $v_{R_m}(S)$. (Index ℓ refers to the linear solution.)

especially for higher S. However, we may suppose the solutions (2.10.10) and (2.10.11) to be valid near the electrode for a moderately high S.

Consider now the numerical results of the solution. As expected, for small S (Fig. 2.24) the stream lines constructed by the linear (2.10.10) and the nonlinear (2.10.10) + (2.10.11) solutions are practically indistinguishable. Appreciable differences begin with $S = 50$. Fig. 2.24c displays, for comparison, the linear stream lines (dashed lines) and nonlinear stream lines (full lines). Evidently, the nonlinearity retards the flow at the free surface and intensifies at the axis as compared to the linear flow. For $S = 250$ even a zone of reversed flow is formed at the free surface, which, however, may be related to insufficient accuracy of the numerical solution.

Fig. 2.25 demonstrates the behaviour of the nondimensional velocity v_R on the axis $\mu = 1$. By the linear solution (2.10.10) v_R has a maximum at $R = 2.9$; by the nonlinear solution the velocity maximum is shifted slowly to higher R with increasing S. Thus for $S = 50$ the maximum v_{R_m} occurs at $R = 3.6$. Concerning the behaviour of $v_{R_m}(S)$, as displayed in the figure, beginning with $S \approx 50$ v_{R_m} increases at a higher rate than the linear one. This indicates a similarity to the exact solution with a point current source.

2.10.1. Flows in hemispherical containers

Sozou and Pickering [63] investigated the flow in a bounded region with a point electric current source (Fig. 2.26). The full nonlinear equation of motion in partial derivatives (2.10.1) with the boundary conditions (2.10.2)–(2.10.4) is solved numerically. Yet, since the problem is set up for the internal domain of hemispherical shape ($R^* \leq R_0$), the constraint at infinity (2.10.5) must be replaced by a condition $R \to 0$. The condition is a numerical matching of the solution in the main flow region with the solution near a point source which is assumed to be the self-similar solution of the point source problem in the semi-infinite domain with free surface, determined by the same authors. (The transient Stokes flow was investigated in [1].)

The singularity at $S_{cr} = 47$ is expected to be present in the total solution due

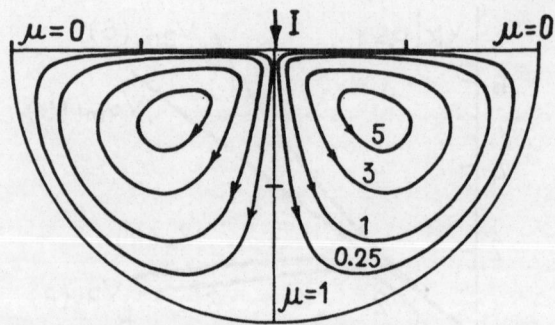

Fig. 2.26. Stream lines in the hemisphere with a point electrode and free surface ($S = 5$).

to the matching. Yet the most surprising result appears to be the impossibility of determining a regular solution even for S much lower than S_{cr}. It was found that in a region close to the spherical surface ($R^*/R_0 > 0.4$), the increase of S from 7.5 to 8.5, i.e. by 13 per cent, increased the velocity on the axis ($R^* v_R^*/\nu S$) normalized by S by 30 per cent. Yet an increase of S from 8.5 to 9.25, i.e. by 9 per cent, increased the velocity more than 70 per cent. Since $v_R|_{R^* = R_0} = 0$, then, increasing S in the axial region near the spherical surface, the velocity gradient became very high, precluding the authors' obtaining a solution by the numerical method used. Moreover, the authors suggested that the velocity field breaks down in this region in the interval $9.25 < S < 10$.

A similar tendency was found by the authors of [9] for an electrically induced flow between two concentric hemispherical electrodes of finite size and with the free surface $\mu = 0$. The finite difference solution of the problem shows that the velocity growth at $R^* = R_2^*$, $\mu = 1$ leads to a sharp increase of the radial component of the velocity gradient for high enough S, and the computational procedure based on a constant sized mesh division ceases to converge. Nevertheless, inclusion of the finite sized electrodes permits us to obtain the solution for higher S. Thus for $R_1 = 0.1$ satisfactory results can be obtained up to $S = 6.3 \times 10^3$, and for $R_1 = 0.5$ — even up to $S = 1.5 \times 10^5$ (for details of the problem see Section 3.7). Consequently, the possibility of obtaining the solution for higher values of the parameter S depends on the relative current density at the small electrode, which, in turn, determines the rotational part of the electromagnetic force.

In the same context it is of interest to analyse the computational results of Atthey [4] for a hemispherical container with the current distribution on the free surface, which is relevant to a welding pool. The following model is assumed by the author (Fig. 2.27). A metallic plate of thickness b contains a hemispherical pool of radius a filled by liquid metal. An electric current is applied to the free surface of metal, where the current density is assumed to be of Gaussian distribution:

$$j_z = \frac{I}{\pi r_0^2} \exp(-r^2/r_0^2),$$

where r_0 is a parameter controlling the current density maximum and its

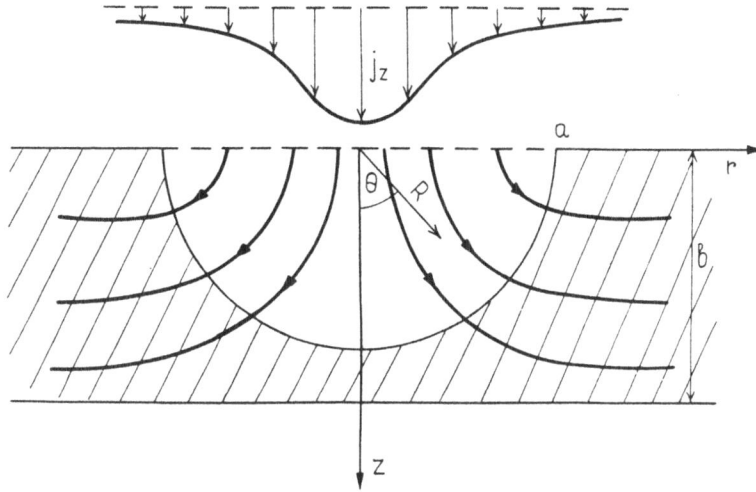

Fig. 2.27. Scheme of the current passage in the model of arc melting.

steepness. This current distribution is close to the real distribution in the arc welding process (see also Section 8.1). The plane $z = b$ is assumed to be nonconducting, i.e. the current is withdrawn to the periphery of the metallic plate. On changing the position of the nonconducting plane, the divergence of the electric current lines is altered in the liquid metal pool, and, thereby, the rotationality of electromagnetic force.

The numerical procedure [4] is stable up to $S \sim 10^5$. From comparison of the results presented in Fig. 2.28 it follows that the magnitudes of the stream function normalized by S, $\Psi = \psi/avS$, for 1.15×10^5 are slightly smaller than for $S = 0.115$; however, behaviour of the nondimensional radial velocity on the

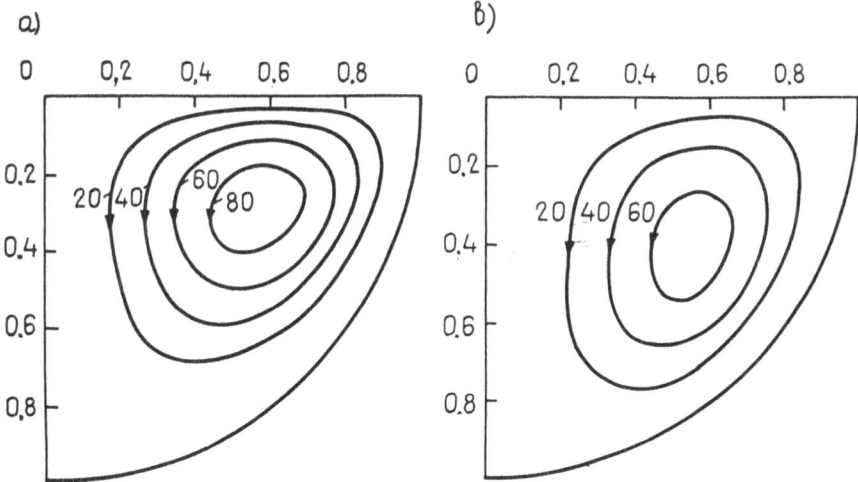

Fig. 2.28. Stream lines in the hemisphere with the distributed current supply for (a) $S = 0.115$, (b) 1.15×10^5.

symmetry axis, normalized by S, $av_R/\nu S|_{\theta=0}$ indicates the tendency to increase the radial gradient near $R = a$ (Table 2.3). It follows also from the table that, on decreasing the parameter r_0, the magnitudes of the velocity are increased. The computations confirm, as may be expected, that the flow intensity falls off with the increase of the plate's thickness b.

A finite-sized current supply to the fluid was also modeled by other methods. Thus in the papers [3, 4] the situation was considered in which a point current source was shifted upstream along the symmetry axis out of the hemisphere, and an additional point current sink was provided on the axis below the hemisphere. For varied positions of the source and sink, different flow configurations were obtained, even giving the reverse of the flow shown in Fig. 2.28.

Table 2.3. Values of the normalized velocity $10^3 av_R/\nu S$ on the axis (for $b = a$).

R	$S = 11.5$	11.5×10^2	11.5×10^4	11.5×10^4	11.5×10^4	11.5×10^4
0.1	4.39	4.20	0.97	0.40	1.93	2.83
0.2	5.52	5.54	1.50	0.64	2.73	3.61
0.3	5.87	6.12	1.92	0.82	3.39	4.48
0.4	5.43	5.90	2.12	0.90	3.67	4.74
0.5	4.39	5.00	2.14	0.91	3.77	4.96
0.6	3.13	3.70	1.98	0.84	3.53	4.83
0.7	1.87	2.28	1.68	0.69	3.27	4.43
0.8	0.87	1.09	1.23	0.47	2.55	3.57
0.9	0.22	0.28	0.61	0.19	1.69	2.53
$r_0 = 0.8$	0.8	0.8	1.2	0.6	0.5

2.10.2. Flows in a spheroidal container

The flow in a spheroidal container with a finite sized electrode was investigated in the Stokes approximation in [65], the nonlinear problem was solved in [2] and independently in [52]. The container was of the form of the lower half of an oblate spheroid: in [52] its semiaxis along the radius was $r/a = 1.64$ and along the symmetry axis it was $z/a = 1.3$; in [2] these were varied. The electric current was supplied to the horizontal plane surface $z = 0$ by a circular planar electrode of radius a, the second electrode being the equipotential, rigid wall of the spheroid.

Four variants of the boundary conditions for velocity on the horizontal surface are analysed in [52]: (a) the surface is free; (b) the central part — circle of radius a — is free, the remaining part being rigid; (c) the central part is rigid, and the remainder is free; (d) the whole surface is rigid. The applications for the variants (a) and (b) could be various situations in the arc melting of metal; the variants (c) and (d) describe situations in slag melting. The problem is solved in

Solutions in spherical coordinates 115

the coordinates of an oblate spheroid ξ, η, φ (see Section 1.4 — No. 4), which are related to the cylindrical coordinates z, r, φ by the expressions

$$z = a \sinh \xi \cos \eta, \qquad r = a \cosh \xi \sin \eta, \qquad \varphi = \varphi;$$

the following new variables are introduced:

$$\lambda = \sinh \xi, \qquad \mu = \cos \eta.$$

The numerical solutions (finite-difference in [52] and semianalytical series expansion in [2]) show, for all the variants and for the whole investigated range of the parameter $S = \mu_0 I^2/2\pi^2 \rho v^2$, that the flow is of the form of a toroidal vortex circulating in meridional planes, and in the axial region the flow is always directed from the circular electrode. The stream lines $\psi = $ const, which are very similar for all the variants, give little insight into the flow details. More interesting information may be obtained from the distribution of vorticity $\mathbf{w} = $ curl $\mathbf{v}$. For a pure meridional flow one has merely the φ-component of vorticity

$$w_\varphi = -\frac{1}{H_3} E^2 \psi$$

$$= \frac{1}{\lambda^2 + \mu^2} \left(\sqrt{1 + \lambda^2} \frac{\partial}{\partial \lambda} v_\mu \sqrt{\lambda^2 + \mu^2} + \right.$$

$$\left. + \sqrt{1 - \mu^2} \frac{\partial}{\partial \mu} v_\lambda \sqrt{\lambda^2 + \mu^2} \right). \tag{2.10.12}$$

From the whole set of equal vorticity lines $w_\varphi = $ const we consider the behaviour of the zero vorticity line $w_\varphi = 0$. This line is of particular interest for the following reason. A hydrodynamic boundary layer may be defined as the region in which the viscosity takes effect. The boundary (conventional) of the region is determined by the displacement thickness (for flow behind a body), by the jet's half-width (for jet type flows), and so on. It is difficult to determine a boundary layer's thickness for the flows in closed volumes. We propose to use for this purpose the zero vorticity line; the line separates, for instance, the high vorticity layer near the rigid boundary from the bulk of fluid. Of course, such an interpretation is not valid everywhere, because the lines $w_\varphi = 0$ also include the symmetry axis (the z-axis) and the free surface.

The lines $w_\varphi = 0$ for the variant (a) are shown in Fig. 2.29a. The overall tendency of the boundary layer's behaviour at the spheroidal wall with the increase of S (with the relevant increase of velocity) agrees with the assumption of boundary layer theory: with the increase of velocity the thickness of the layer decreases. An increase of the thickness is observed only in the vicinity of the line joining the rigid and free horizontal surfaces. The boundary layer behaviour is similar at the spheroid's wall for all the variants.

When the rigid ring is present on the horizontal surface (Fig. 2.29b) the boundary layer at the spheroidal wall joins the ring-shaped layer. On increasing

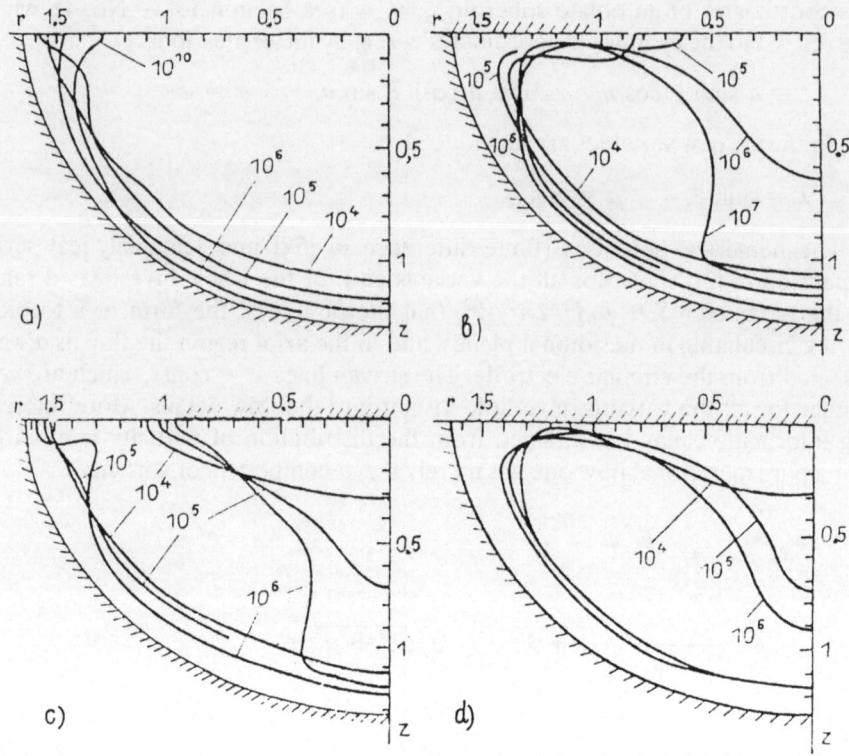

Fig. 2.29. Lines of zero vorticity in the hemispheroidal container for different S and for the flows: (a) with the free horizontal surface, (b) free is the central part, (c) the central part is rigid, (d) the whole horizontal surface is rigid.

S, the zone of viscous effects at the horizontal surface moves to the symmetry axis, and the zone's volume grows quite abruptly near the axis. The 'boundary layer' growth in the zone, free from rigid walls, with the increase of velocity is obviously related to the converging flow near the horizontal surface. The extra vorticity accumulated in the fluid at the rigid walls is convected to the axis.

If the rigid part of the horizontal surface is merely the circular electrode, the joining of boundary layers is not observed (Fig. 2.29c) (although reliable numerical results for this variant were obtained up to $S = 10^6$), and the axial zone of the 'erupted' vorticity originates from the central rigid electrode. The zero vorticity lines for the case in which the whole horizontal surface is rigid (variant (d)) are shown in Fig. 2.29d. Comparing the axial layers in Figs. 2.29c and d, it is evident that for equal S the vorticity zone in the case (c) is more extended than in the case (d). Obviously, this may be explained by the presence of the sharp edge where the rigid electrode contacts the free surface. The axial zones of vorticity for the cases of rigid central electrode (c, d) grow with increasing S in opposition to the boundary layer at the spheroidal wall.

The velocity components could be of extreme magnitudes in the vicinity of zero vorticity line (2.10.12). This assumption is checked by constructing the

Solutions in spherical coordinates 117

profiles of velocity components normal to the coordinate lines for $S = 10^6$; the coordinate lines are selected as those which intersect the zero vorticity lines. The profile of the velocity component normalized by $\sqrt{S}$ on the coordinate line $\lambda = 0.3$ (in meridional section this is the ellipse with the semiaxes of length 0.3 along the z-axis and 1.04 along the r-axis) is presented in Fig. 2.30. It is evident from the figure that an extremum is always located on the symmetry axis $\mu = 1$ and at the free surface $\mu = 0$. If the surface $z = 0$ is rigid, an extremum also appears. The most unexpected result is the second extremum in the axial zone, which is especially clear for the cases (c) and (d); this agrees with the above assumption of a possible velocity extremum on the zero vorticity line (indeed, the position of the zero vorticity line, marked by the dots, when it intersects the coordinate line $\lambda = 0.3$ is close to the location of the extremum; exact coincidence should not be expected, since in the expression (2.10.12) also enters the derivative of the second velocity component). This suggests that for certain situations the greatest magnitude of velocity may be located not on the symmetry axis, as it is for all the previous solutions, but it can be located off the axis. Similarly, the situation is possible in which the greatest velocity is off the free surface.

Thus, a direct relation is found between the development of additional extremal velocities and the position of zero vorticity. The meaning of the second extremum close to the symmetry axis remains to be clarified. Fig. 2.31 reveals the growth dynamics of this extremum with the current for the case (b). For S up to 10^5 the maximum velocity magnitude is located on the symmetry axis. For higher values of S the maximum is moved off the axis and the velocity dip at the axis increases. The second factor is the size of the horizontal rigid surface. The ratio of the radii r_1 (electrode) and r_2 (horizontal surface of the container) is $r_1/r_2 = 0.6$ for Fig. 2.29c. When the ratio is set as $r_1/r_2 = 0.75$ then for $S = 10^{12}$ a slow reversed flow is even observed in the axial zone [52]; the tendency is suggested in Fig. 2.31 by the dip in the growth with S. Such a

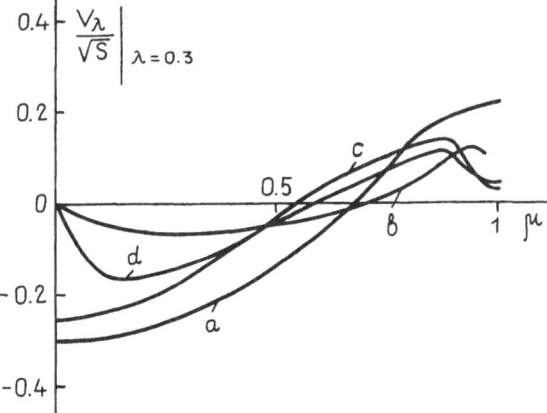

Fig. 2.30. Velocity profiles along the coordinate line $\lambda = 0.3$ in the different flow cases (a, b, c, d). The dots on the curves mark the intersections with the zero vorticity line.

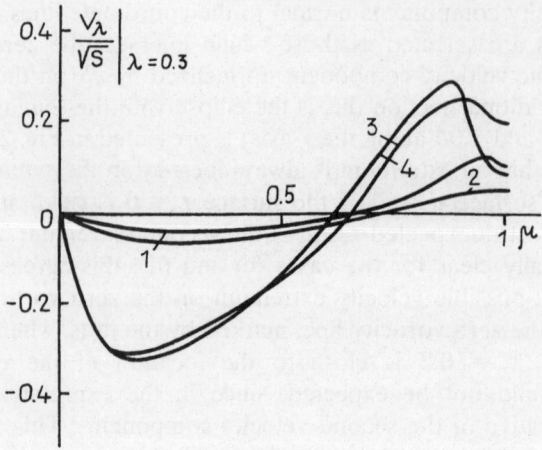

Fig. 2.31. Velocity profiles along the line $\lambda = 0.3$ in the case (b).

reversed flow was not found in [2] because of the generally lower S considered, yet the authors found another kind of reversed flow for r_1/r_2 close to unity in the vicinity of the electrode rim for small S [65], which disappeared for higher S.

It is of interest to determine the limits of application of the Stokes (linear) approximation. With this aim the maximum velocity v_λ ($\mu = 1$) varying with the parameter S is evaluated. In the Stokes regime (for equal orders of magnitude of viscous and electromagnetic terms in the equation of motion) the velocity at any point is proportional to S ($v_{\lambda_{max}} = k_1 S$). Hence in the coordinates of Fig. 2.32 the Stokes regime corresponds to the horizontal section of curves $v_{\lambda_{max}}(S)$. It is evident that, for all the cases, the Stokes approximation is valid up to $S = 10^4$. The onset of the nonlinear regime is for $S = 10^5-10^7$ depending on the specific situation, and it is characterized by the relation $v_{\lambda_{max}} = k_2 \sqrt{S}$. For quantitative estimates the magnitudes of k_1 and k_2 are evaluated for the respective cases: (a) $k_1 = 2 \times 10^{-3}$, $k_2 = 0.4$; (b) $k_1 = 1.75 \times 10^{-3}$, $k_2 = 6.3 \times 10^{-2}$; (c) $k_1 = 1.1 \times 10^{-3}$, $k_2 = 3.6 \times 10^{-2}$; (d) $k_1 = 10^{-3}$, $k_2 = 2.3 \times 10^{-2}$.

The numerical solution for relatively high S in [2, 52] is, of course, related to the relatively great size of the upper electrode. Nevertheless, for each computed situation there exists an upper magnitude of S at which the computational procedure becomes unstable, especially for $r_1/r_2 \rightarrow 0$. Similarly to the results [4], the axial velocity derivative increases in magnitude very rapidly with S in the vicinity of the container's bottom. The general tendency of the critical S is the following: the greater the surface area of the rigid horizontal surface, the lower is the value of S at which the numerical procedure becomes unstable; the ring-shaped lid on the horizontal surface is more favourable to increasing S than the central rigid circle.

The material presented above revealed that, taking account of the flow induced electric currents and the finite sized electrodes, the limited flow region did not in principle solve the problem of the critical parameter S. It seems likely

Solutions in spherical coordinates

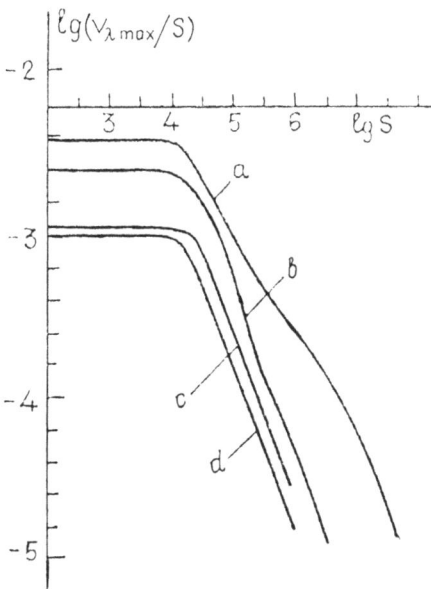

Fig. 2.32. To determine the linear and nonlinear flow regimes.

that the velocity field breakdown, found in the different models, is related to the specific properties of the flow investigated, the most typical of which is the convergence of stream lines to the symmetry axis and the corresponding intense flow in this region. This problem will be considered in the following chapter.

3

Electrically induced vortex flow at a point electrode and azimuthal rotation

The previous chapter dealt with the flows arising when an electric current radially diverges through the fluid from a small electrode. In spite of the simplicity of the physical model with a point current source, which was successfully combined with the similarity solution of equations of motion and magnetic field, it was found to be impossible to describe the flows for magnitudes of electric current higher than a certain critical magnitude. The source of failure may be sought in the similarity of the equations, which greatly restrict the form of the motions. However it should be recalled that the similarity is not an artificial assumption to describe the flows at a point electrode, it is derived by the dimensional analysis of the given physical quantities entering the problem [26]. Moreover, as was demonstrated in Section 2.10, the description of analogous flows without the assumption of similarity, e.g. in a closed hemispherical container, also led to serious difficulties. Consequently, the cause of difficulties should be sought in the physical statement of the problem, which probably does not take into account an essential mechanism limiting the growth of velocities in the flow when the critical magnitude of electric current is reached. The limiting mechanism cannot be related to the flow induced electric current since, even for excessive magnitudes of electrical condictivity (by 6—7 orders of magnitude compared to real materials) the critical magnitude of the electric current is not appreciably increased. This mechanism cannot be the mechanism of turbulent dissipation for the following reasons. If the flow Reynolds number is estimated by the formula (2.5.17) (i.e. $Re = 0.079S$), which is valid up to $S = 15$ (Stokes regime), then the upper value $S = 15$ corresponds to the current in mercury $I = 0.1$ A, and the respective $Re = 1.2$. It seems unlikely to expect turbulence after a mere threefold increase of the electric current (to 0.9 A for $S = 150$).

With the aim of finding a mechanism which would be able to prevent the unlimited growth of velocities, we now consider some integral properties of the electrically induced flows and then, using this knowledge, we attempt to construct a model of flow at a point source of electric current, which would be free from the disadvantages mentioned above.

3.1. Integral properties of the flows driven by rotational electromagnetic forces

The flows considered in this book differ from ordinary hydrodynamic flows, because they are driven by rotational body forces instead of being the result of imposed inhomogeneous boundary conditions of velocity or pressure. A detailed analysis of the action in a fluid of rotational electromagnetic forces can be found in reference [28]; here we shall procede in a different way, analysing the integral relations.

A natural characteristic of vortical motion is the vorticity $\mathbf{w} = \text{curl } \mathbf{v}$. Let us seek a relation between steady flow of the total vorticity $\int w^2 \, dV$ in a fluid volume V with the vortical body forces $\mathbf{f}$ which sustain the motion. Given the equation of stationary motion

$$\mathbf{v} \times \mathbf{w} = \frac{1}{\rho} \nabla P + \nu \nabla \times \mathbf{w} - \frac{1}{\rho} \mathbf{f}, \qquad \nabla \cdot \mathbf{v} = 0;$$

$$P = p + \rho v^2/2, \tag{3.1.1}$$

we express the quantity w^2 by the use of vector identities and (3.1.1):

$$w^2 = \mathbf{w} \cdot \text{curl } \mathbf{v} = \nabla \cdot (\mathbf{v} \times \mathbf{w}) + \mathbf{v} \cdot (\nabla \times \mathbf{w})$$

$$= \nabla \cdot (\mathbf{v} \times \mathbf{w}) + \frac{1}{\nu} \mathbf{v} \cdot \left(\mathbf{v} \times \mathbf{w} - \frac{1}{\rho} \nabla P + \frac{1}{\rho} \mathbf{f} \right)$$

$$= \nabla \cdot (\mathbf{v} \times \mathbf{w}) + \frac{1}{\nu \rho} [-\nabla \cdot (P\mathbf{v}) + \mathbf{v} \cdot \mathbf{f}].$$

The total vorticity in the volume V, bounded by a surface S_0, then, is equal to

$$\int_V w^2 \, dV = \frac{1}{\nu \rho} \int_V \mathbf{v} \cdot \mathbf{f} \, dV + \oint_{S_0} \left(\mathbf{v} \times \mathbf{w} - \frac{1}{\nu \rho} \nu P \right) dS_0, \tag{3.1.2}$$

after the use of Gauss's theorem.

In ordinary hydrodynamics, usually, $\mathbf{f} = 0$, and the total vorticity is generated at the surface S_0. If $\mathbf{f} \neq 0$, it is easy to imagine a situation where the surface integral is equal to zero, e.g. for the homogeneous boundary conditions $\mathbf{v} = 0$ on the rigid walls S_0 of a closed volume. In this case the total vorticity is quantitatively equal to the rate of work done by the forces $\mathbf{f}$ in the fluid, i.e. the power of these forces. On the other hand, the scalar product of the vector equation (3.1.1) by $\mathbf{v}$, and integration over the volume V with the surface S_0, which is impermeable to the fluid ($v_n = 0$), after analogous manipulations gives

$$\int_V \mathbf{v} \cdot \mathbf{f} \, dV = \nu \rho \int_V \mathbf{v} \cdot (\nabla \times \mathbf{w}) \, dV, \tag{3.1.3}$$

i.e. the integral power of the forces $\mathbf{f}$ is totally consumed in overcoming viscous friction [20]. Combining (3.1.2) and (3.1.3), we can conclude that the total vorticity, created by the rotational body forces in a closed volume, is quantitatively equal to the rate of work done by viscous stresses.

Consider now the flow arising when the electric current diverges axisymmetrically from a point electrode. Since the force **f** in this case has only the meridional component f_θ (3.3.2), fluid particles travel in meridional planes. In the unbounded domain the stream lines close at infinity where the velocity falls off as R^{-1} (2.1.7), and the pressure falls off as R^{-2} (2.1.17). We integrate the equation of motion (3.1.1) along the closed contour C, consisting of a stream line ψ = const and the line which connects the stream line's ends at $R \to \infty$ [20]. Part of the integral along the line at infinity is zero, because the velocity and the pressure decrease sufficiently fast with the distance from the origin. The integral along the line C gives

$$\oint_C \mathbf{f} \cdot d\mathbf{l}_c = \nu\rho \oint_C (\nabla \times \mathbf{w}) \cdot d\mathbf{l}_c, \qquad (3.1.4)$$

since $d\mathbf{l}_c \times \mathbf{v} = 0$ according to definition of stream line, and $\oint \nabla P \cdot d\mathbf{l}_c = 0$. This means that the circulation of electromagnetic forces along the stream line is balanced only by viscous stresses.

One may derive some useful conclusions from the relations (3.1.2)–(3.1.4). Note first that the potential forces (curl **f** = 0) can be included in the expression for pressure P, and therefore they do not contribute to the integral vorticity (3.1.2) in a closed volume. Further, the relations (3.1.3)–(3.1.4) indicate that attempts to construct asymptotic models of vortical flows at high velocities, i.e. for the parameter $S \to \infty$, may prove to be a complicated matter. Thus, the assumption that the flow is described in the first approximation by equations that neglect viscosity is in contradiction to (3.1.3)–(3.1.4). Indeed, the solution for an inviscid fluid [27] leads to singularities in the velocity field. If we attempt to make the inviscid model consistent with the relations (3.1.3) and (3.1.4) by introducing viscous boundary layers, then all the stream lines, beginning in the inviscid domain, should cross the viscous layer to balance the circulation of electromagnetic forces by viscous forces. However, the condensation of the stream lines in the boundary layer, indicating the high intensity of flow, may then infringe the usual assumption in the boundary layer theory of a thin layer at high Reynolds number.

In the problem with the radial electric current diverging from a point source the flow with an increase of the parameter S takes the form of a narrow jet centred on the symmetry axis (Section 2.7), and high viscous stresses act in the jet's shear layer. Nevertheless, the numerical solution for this flow showes that the balance of vorticity (3.1.2) in this case is not attained when the value of the parameter $S = S_{cr}$ is exceeded, and singularities appear in the flow field. Obviously, viscous forces in the free shear layer of the axial jet are insufficient to compensate the vorticity created by the action of electromagnetic forces. The effect of viscous forces may be increased by placing a rigid wall near the axis; for instance, a narrow rigid cone surrounding the axis. The solution of such a methodological problem in Section 3.2 does not show the symptoms of unlimited velocity growth at a finite parameter S, yet the solution for high S becomes extremely sensitive to small disturbances of the boundary conditions.

The effect of viscous forces would be even higher, if the fluid flow is reversed, because then the flow is directed to the plane with the electrode, and a radial jet along the rigid plane is formed instead of the axial jet. The reversed flow can be created by two methods: with an immersed electrode (Section 3.3), when the curl of electromagnetic force changes its sign, or by applying an external magnetic field $\mathbf{B}_0$, by which the fluid acquires an additional differential rotation around the symmetry axis. As will be shown in Section 3.5, the differential rotation drives a meridional flow diverging from the symmetry axis. Further investigation of the interaction of azimuthal rotation and meridional flow (Section 3.6, 3.7) will show a way to resolve the problem of the breakdown of the solution at $S = S_{cr}$ for the flow at a point electrode. In this flow, converging to a narrow axial jet, even a small azimuthal disturbance is amplified to a finite state of azimuthal rotation by mechanical energy transfer from the meridional flow to the rotation. The amplification mechanism, based on the nonlinear interaction of meridional and rotational motion, involves additional viscous forces (see Section 3.8), which thereby limit the vorticity growth in agreement with the integral relations (3.1.2) and (3.1.3).

3.2. A model demonstrating the effect of viscosity

In the previous paragraph we found out that viscosity had to play a key role in limiting the growth of vorticity in a fluid, which was generated by electromagnetic forces. Therefore, let us attempt to alter the problem with a point electrode (Sections 2.5–2.7) in such a way as to increase the action of viscous forces in the axial jet. This may be achieved if a narrow, rigid cone is placed on the symmetry axis (Fig. 3.1) [4]. Then, to the viscous forces in the jet's free shear layer will be added the friction forces at the rigid surface due to the jet flow surrounding the conical core. In this situation the electric current does not penetrate the inner cone, and the constants defining the magnetic field (2.1.5)

$$B_\varphi = \frac{\mu_0 I}{2\pi R} \frac{B - \mu A}{\sqrt{1 - \mu^2}} \tag{3.2.1}$$

are specified as $A = 1/\mu_1$ and $B = 1$, where $\mu_1 = \cos \theta_1$ is on the cone's surface.

The equation of motion (2.2.15), after the change of variable to $\mu = \cos \theta$, is

$$(1 - \mu^2)g' - g^2/2 + 2\mu g = SK(\mu, S),$$
$$K = a_0\mu^2 - b_0\mu + c_0 + \tfrac{1}{2}A[(A + B)(1 + \mu)^2 \ln(1 + \mu) +$$
$$+ (A - B)(1 - \mu)^2 \ln(1 - \mu)]; \tag{3.2.2}$$
$$a_0 = a/S, \quad b_0 = b/S, \quad c_0 = c/S,$$

and it is solved for the boundary conditions of zero velocities ($g = g' = 0$) at

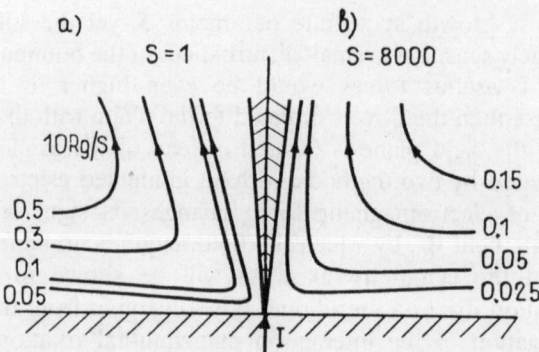

Fig. 3.1. Stream lines $Rg/S = C$ for the flow with an axial cone in (a) linear, and (b) nonlinear regimes.

the rigid walls $\mu = 0$ and $\mu = \mu_1$. These conditions give two relations to determine the constants a_0, b_0, c_0: $K(0, S) = 0$, whence $c_0 = 0$, and from $K(\mu_1, S) = 0$ we can express $b_0(a_0)$, but the constant a_0 which, in the general case, depends on the parameter S, is determined numerically by solving the boundary value problem (3.2.2) with the specified conditions on the rigid walls.

The problem is solved by the Runge-Kutta shooting method, and the stream lines $\psi/S = $ const for the small value of the parameter $S = 1$ and for the conical angle $\theta_1 = 5°$ are presented in Fig. 3.1. With increasing S the flow concentrates more at the cone's wall forming a jet along its surface. With the increase of S the maximum of the jet velocity, normalized by S ($v_R/S = -vg'_m/RS$), reaches a maximum magnitude and then decreases (Fig. 3.2). This behaviour of the velocity differs essentially from the case without the rigid cone, when the velocity ($v_R \sim -g'(\theta = 0)$ in Fig. 3.2) increases without limit at the critical value $S = S_{cr}$.

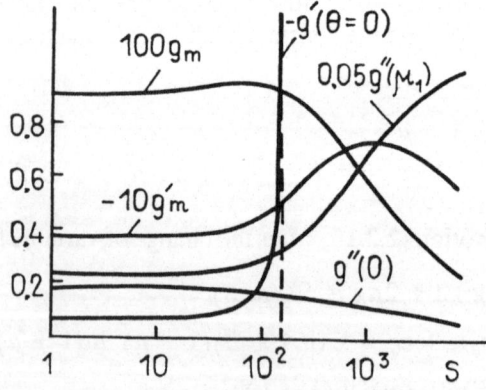

Fig. 3.2. The velocities and friction vs. S.

Consider now the variation of viscous friction on the rigid walls expressed by the component $\sigma_{R\theta}$ of viscous stress acting on the surface $\theta = \text{const}$:

$$\frac{1}{v^2 \rho S} \sigma_{R\theta} = \frac{1}{vS}\left(\frac{1}{R}\frac{\partial v_R}{\partial \theta} + \frac{\partial v_\theta}{\partial R} - \frac{v_\theta}{R}\right)$$

$$= \frac{\sqrt{1-\mu^2}}{SR^2} g''(\mu).$$

To express the friction for the flow without the central cone, i.e. $\mu_1 = 1$, the boundary conditions (2.5.6) and (2.5.7) are set on the axis instead of the zero velocity conditions, whence $a_0 = -2$, $b_0 = 4\ln 2 - 2$, $c_0 = 0$, and the friction normalized by S $g''(0)/S = 1 - b_0 = 3 - 4\ln 2$ is constant on the plane $\mu = 0$. If the cone of nonzero angle θ_1 is present, the friction on the plane $\mu = 0$, i.e. $g''(0)$, according to Fig. 3.2, decreases with S. However, on the cone itself the viscous stress $g''(\mu_1)$ grows with an increase in parameter S. The growing contribution of viscous forces at the cone's surface ensures that the integral relations (3.1.3) and (3.1.4) are satisfied, and the velocity remains finite in the flow.

The growth of friction with the increase of S leads to another interesting phenomenon, viz. a small change in the boundary conditions upstream, i.e. at the surface $\mu = 0$, should lead to much greater flow changes downstream, i.e. at the cone $\mu = \mu_1$. This is suggested by the convergence of the numerical solution procedure: thus, for small S the constant a_0, relating the conditions upstream and downstream, can be determined with an accuracy of several significant digits, but on increasing S the number of digits must be increased to ensure the convergence. For $S \geqslant 8000$ the limiting 17 digit precision of the ES-1022 computer is insufficient to achieve convergence in the iterations. Now, we may conclude that this extreme sensitivity to the perturbation of boundary conditions is also a sort of flow breakdown.

3.3. Flow at an immersed electrode

We may expect that difficulties of numerical solution for converging flows would not appear for a reversed flow. Then the fluid flows along the symmetry axis to the plane and diverges along it. In this case a small perturbations in the axial region would lead to even smaller perturbations downstream at the plane.

To this end, consider a situation in which an isolated conical electrode is immersed in an electrically conducting liquid, the tip of the conical electrode is touching the plane $\mu = 0$, which is either a rigid wall or a free surface of the liquid, and from the small non-isolated area at the cone's tip an electric current flows into the liquid. An idealized variant of the problem can be considered in the class of exact solutions (2.1.7) [4]. The problem's definition differs from the previous Section only in that the electric current is supplied to the point source on the plane along the axis $\theta = 0$ (see Table 2.1, No. 4). A narrow coaxial cone $\theta = \theta_1$ models the immersed electrode supplying the current $I = I_2$ (Fig. 3.3a). Now the magnetic field (3.2.1) is specified by the constants

$$A = \frac{1}{\mu_1}, \qquad B = 0. \tag{3.3.1}$$

Fluid flow is determined by the stream function from (2.1.7) and the equation (3.2.2) with the boundary conditions of zero velocities on the rigid walls $\mu = \mu_1$ and $\mu = 0$ ($g = g' = 0$), or the conditions $g = g'' = 0$ on the plane $\mu = 0$ if it is supposed to be a free undeformed surface. Since now the electromagnetic force in the fluid

$$f_{e\theta} = \frac{\mu_0 I^2}{4\pi^2 R^3} \frac{A(A\mu - B)}{\sqrt{1 - \mu^2}} \tag{3.3.2}$$

is positive, and, $(\text{curl } \mathbf{f}_e)_\varphi < 0$, the motion is directed from the axial zone to the plane. In this case the flow takes the form of an axisymmetric wall jet or a jet along the free surface. Note that the jet along the free surface is analogous to the radial jet of Loitsyanskiy (Section 2.4), because the conditions $g = g'' = 0$ in the centre of the jet coincide and the difference is only in the method by which the jet is generated.

Fig. 3.3 shows the stream lines $\psi/S = $ const in the fluid with the free surface $\mu = 0$ for the parameter $S = 1$. In agreement with the previous suppositions, the numerical solutions for such reversed flow can be obtained for practically any magnitude of S, including the magnitudes which correspond to high electric currents, e.g. 10^3–10^4 ampères in mercury. Thus in Fig. 3.4a the stream lines are presented for $S = 10^9$ and $\theta_1 = 5°$, in Fig. 3.5a they are shown for the rigid wall $\mu = 0$ and $S = 10^7$. We see in the figures that for high S a radial jet is formed, in which the fluid is entrained. Let the reader note that in these problems, the extreme sensitivity of the numerical solution to the precision in determination of the constants a_0, b_0, c_0 in equation (3.2.2) is not observed. Minor difficulties arise for S higher than 10^7, yet these are related to the presence of large parameter in the equation and are eliminated by taking more

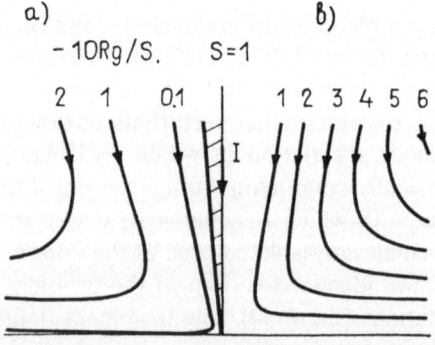

Fig. 3.3. Stream lines at the immersed electrode ($S = 1$): (a) $\theta_1 = 5°$, (b) $\theta_1 = 0$.

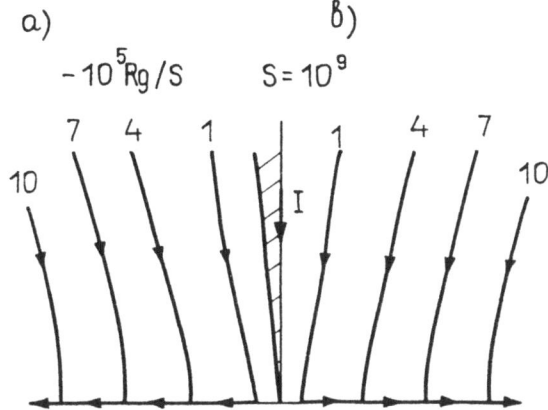

Fig. 3.4. Stream lines of the flow with the free surface ($S = 10^9$) (a) for the finite size electrode, (b) for the infinitely thin wire.

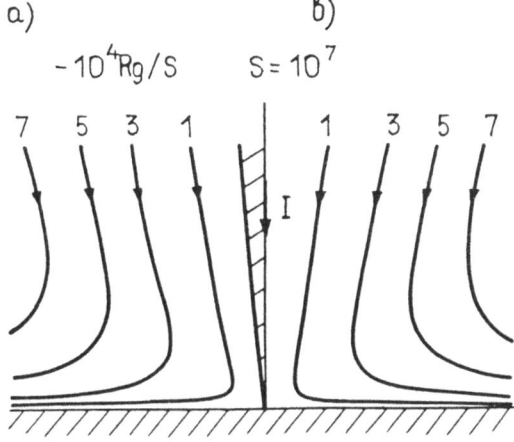

Fig. 3.5. Stream lines of the flow with the rigid plane ($S = 10^7$) (a) for the finite size electrode, (b) for the infinitely thin wire.

mesh points in the finite difference solution near the boundaries $\mu = 0$ and $\mu = \mu_1$, where boundary layers are developing, as will be shown in Section 3.4.

The solution depends also on the cone's angle θ_1. The limiting case of decreasing $\theta_1 \to 0$ corresponds to an infinitely thin wire supplying the current along the axis. It carries a finite current I_2, and the electromagnetic force (3.3.2) there is singular, $\sim 1/\sin \theta$, which must be compensated by a respective singularity of pressure and viscous stress. Nevertheless, the problem with $\theta_1 = 0$ is attractive, because in this case the constants a_0, b_0, c_0 can be determined immediately from the boundary conditions without numerical iterations. It follows from equation (3.2.2) that at the axis $\mu = 1$ the function $g \sim (1 - \mu^2) \ln(1 - \mu)$, consequently, the velocity components are finite and the

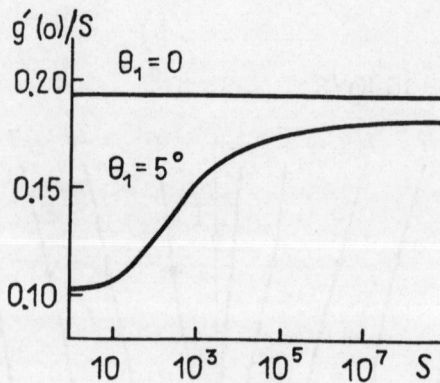

Fig. 3.6. Dependence of the velocity at free surface vs. the parameter S for the finite and infinitely thin electrodes.

boundary conditions (2.5.6), (2.5.7) at the axis are satisfied. For the free surface case we obtain

$$a_0 = -\ln 2 - \tfrac{1}{2}, \qquad b_0 = 0, \qquad c_0 = \tfrac{1}{2} - \ln 2. \tag{3.3.3}$$

Comparison of the stream lines in Fig. 3.3 demonstrates that the rigid cone slows down the motion for small S, but, with increasing S, the rigid wall near the axis is felt to an increasingly smaller extent, excluding the narrow axial region. Thus, in Figs. 3.4 and 3.5 the stream lines for high S are only slightly shifted relative to the flow with the central cone. To estimate the effect of the cone's rigid surface on the flow, let us evaluate the velocity on the free surface $\mu = 0$

$$v_R = -\frac{\nu}{R} g'(0) = -\frac{\nu}{R} Sc_0. \tag{3.3.4}$$

For the case of an infinitely thin central wire the velocity, according to (3.3.3) and (3.3.4), grows at a linear rate with S (note, the velocities in the flow's inner part grow at a slower rate, this is clearly demonstrated by the asymptotic solution in Section 3.4). For $\theta_1 \neq 0$ the surface velocity is determined by numerical solution, because the dependence of $c_0(S)$ is not known *a priori*. The numerically found dependence of $g'(0)$ vs. S for $\theta_1 = 5°$ (Fig. 3.6) shows that the velocity for high S tends asymptotically to the linear value for $\theta_1 = 0$, whence we may conclude that for high currents the approximation of a thin wire electrode yields at least the correct surface flow.

3.4. Asymptotic solution for high S

A high magnitude of S (10^6–10^9) for real flows in the laboratory and in practice is a natural prerequisite for constructing asymptotic solutions for $S \to \infty$. Since the numerical results for the reversed electrically induced flow (Section 3.3) indicate the existence of the solution for high S at least for some variants of the flow, we attempt to construct the asymptotic solution [5].

For high values of S the stream function $g(\mu, S)$ will also be of high magnitude, thus equation (3.2.2) suggests the expansion in a form of regular perturbation by S:

$$g(\mu, S) = S^{1/2}g_1(\mu) + g_2(\mu) + S^{-1/2}g_3(\mu) + \ldots \qquad (3.4.1)$$

The first approximation is determined from the equation

$$-\tfrac{1}{2}(g_1)^2 = K(\mu),$$

where $K(\mu)$ is assumed to be independent of S. The function g_1 is the solution for an inviscid fluid, which has been studied in [27]. This solution infringes the viscous conditions on the fluid boundaries; moreover, it gives a singular axial velocity. Obviously the next terms of the series (3.4.1) cannot correct the situation. Therefore we should add to the expansion (3.4.1) additional expansions in those flow regions where (3.4.1) ceases to be valid, namely, in narrow layers at the plane $\mu = 0$ and at the symmetry axis.

The correctness of the division into regions is supported by experience of the numerical solution of equation (3.2.2), which revealed that the leading term for high S was nonlinear in the bulk of the flow, except for the narrow regions at boundaries where the linear (viscous) terms reach the nonlinear term in magnitude. Increasing the number S, the difference in magnitudes of the terms also increases in the inner region and the viscous layers decrease in thickness. Thus, we have the problem with boundary layers, separated by the region in which the fluid behaves as if it were inviscid.

For a mathematical description let the flow be divided into three regions: 1 — a narrow region at the axis, 2 — bulk of the fluid between the viscous layers, 3 — a thin layer at the plane $\mu = 0$. It is necessary to introduce stretched variables $x_1(\mu, S)$ and $x_3(\mu, S)$ in the regions 1 and 3 to describe satisfactorily the rapid change of the function at the boundaries for high S. The direct expansion (3.4.1) stays in the region 2, and the deficient boundary conditions are replaced by matching the expansions in the boundary layers [34].

The expansion of $g(\mu, S)$ for $S \to \infty$ in each region is denoted in the following way:

$$g^i = \sum_{j=1}^{J} D_j^i(S) g_j^i(x_i), \qquad (3.4.2)$$

where i is the region number, j is the order of expansion, and x_i is an independent variable in the ith region, which is fixed in the limiting process $S \to \infty$.

The expansions for particular problems depend considerably on the specific boundary conditions, therefore we shall restrict our study to the flows which contain the symmetry axis, i.e. either with an infinitely thin wire on the axis $\theta = 0$, or a current supply from the external side of the plane boundary along $\theta = \pi$. For these cases the coefficients in the right-hand part of equation (3.2.2) can be expressed explicitly, e.g. in the form (3.3.3), and the right-hand part of (3.2.2) can easily be expanded in each domain. When the solution in each of the three regions are to be obtained, they will need to be matched in such a way as to have a smooth solution for the whole flow. We shall demonstrate this procedure for the two situations of current supply mentioned above.

Region 2 (direct expansion)

In this region the independent variable is not deformed: $x_2 = \mu$. Substituting (3.4.2) with two terms ($J = 2$) in (3.2.2), we obtain:

$$-\tfrac{1}{2}(g_1^2)^2 = K(\mu), \qquad D_1^2 = S^{1/2};$$
$$(1 - \mu^2)(g_1^2)' - g_1^2 g_2^2 + 2\mu g_1^2 = 0, \qquad D_2^2 = 1, \qquad (3.4.3)$$

whence, using (3.4.3), we find

$$g_2^2 = 2\mu + (1 - \mu^2)\frac{K'}{2K}.$$

Region 3

Introduce the stretched variable $x_3 = M = \mu/\Delta_3(S)$. After the substitution of (3.4.2) with two terms ($J = 2$), the equation (3.2.2) transforms to

$$(1 - (\Delta_3 M)^2)\left(\frac{D_1^3}{\Delta_3}(g_1^3)' + \frac{D_2^3}{\Delta_3}(g_2^3)'\right) - \frac{1}{2}(D_1^3 g_1^3)^2 -$$
$$- D_1^3 D_2^3 g_1^3 g_2^3 + 2\Delta_3 M D_1^3 g_1^3 = SK(\Delta_3 M), \qquad (3.4.4)$$

where the differentiation is with respect to M. To evaluate $K(M)$ for $S \to \infty$ we apply the Taylor series expansion of $K(\mu)$ at $\mu = 0$:

$$K(M) = c_0 + \Delta_3 M(AB - b_0) + \tfrac{1}{2}(\Delta_3 M)^2(2a_0 + 3A^2) + \ldots$$

For a rigid wall $M = 0$ the conditions $g = g' = 0$ must be satisfied, whence $c_0 = 0$, and for a free surface $M = 0$ the conditions $g = g'' = 0$ lead to $b_0 = AB$.

Consider the boundary layer at the free surface. Then $c_0 \neq 0$ and, comparing the terms in the equation (3.4.4) for $S \to \infty$, we see that

$$\Delta_3 = S^{-1/2}, \qquad D_1^3 = S^{1/2}, \qquad D_2^3 = S^{-1/2};$$
$$(g_1^3)' - \tfrac{1}{2}(g_1^3)^2 = c_0.$$

The velocity on the free surface is determined by the first approximation only, and it coincides with the exact expression (3.3.4). The direction of the velocity is determined by sign of the constant c_0. If $c_0 > 0$, the solution for g_1^3 is of the form

$$g_1^3 = -\sqrt{2c_0}\,\tan(\sqrt{c_0/2}\,M).$$

Such a highly oscillating function in the interval $M \in [0, \infty)$ has no physical meaning, which corresponds to the nonexistence of solutions for high S in the flows directed along the plane to the axis (see Section 2.7). Thus, the assumption that the boundary layer forms in the converging flow is not justified.

If $c_0 < 0$, the solution

$$g_1^3 = \sqrt{-2c_0}\,\frac{1 - \exp(\sqrt{-2c_0}\,M)}{1 + \exp(\sqrt{-2c_0}\,M)} \tag{3.4.5}$$

is the first approximation of the boundary layer solution in diverging flow.

For the rigid wall $c_0 = 0$, and the expansion of $K(\mu)$ begins with the second term. Comparison of the terms in (3.4.4) leads, respectively, to

$$\Delta_3 = S^{-1/3}, \quad D_1^3 = S^{1/3};$$
$$(g_1^3)' - \tfrac{1}{2}(g_1^3)^2 = M(AB - b_0).$$

Thus, thickness of the boundary layer is of the order of $S^{-1/2}$ at the free surface, and $S^{-1/3}$ at the rigid wall.

Region 1

In the axial region the stretched variable $x_1 = T = (1 - \mu)/\Delta_1$ is introduced. Substituting again (3.4.2) in (3.2.2), we obtain

$$-(2 - \Delta_1 T)T[D_1^1(g_1^1)' + D_2^1(g_2^1)'] - \tfrac{1}{2}(D_1^1 g_1^1)^2 - D_1^1 D_2^1 g_1^1 g_2^1 +$$
$$+ 2(1 - \Delta_1 T)(D_1^1 g_1^1 + D_2^1 g_2^1) = SK(\Delta_1 T). \tag{3.4.6}$$

Taking into account the boundary conditions $g = \sqrt{T}g'' = 0$ on the axis $T = 0$, we have in the expansion

$$K(\Delta_1 T) = [a_0 - b_0 + c_0 + A(A + B)2 \ln 2] +$$
$$+ \Delta_1 T[-2a_0 + b_0 - A(A + B)(2 \ln 2 + 1)] +$$
$$+ \tfrac{1}{2}(\Delta_1 T)^2[\ln(\Delta_1 T)A(A - B) + 2a_0 +$$
$$+ A(A + B)\ln 2 + 3A^2] + \ldots$$

the expressions in the first and second brackets equal zero. These conditions, together with the boundary conditions in the region 3 determine all the three constants a_0, b_0, c_0.

Comparing the terms in (3.4.6) for $S \to \infty$ we assume that D_1^1 is not a growing function of S, otherwise the flow in the first approximation would be inviscid with a singularity on the axis $T = 0$, therefore $D_1^1 = \text{const}$. Let $D_1^1 = 1$. Then from (3.4.6) we find for $A - B \neq 0$ that Δ_1 is determined by the expression

$$-S(\Delta_1)^2 \ln \Delta_1 = 1 \quad \text{or} \quad \Delta_1 = (\tfrac{1}{2}S \ln S)^{-1/2} + \ldots,$$

and the stream function in the first approximation by the equation

$$-2T(g_1^1)' - \tfrac{1}{2}(g_1^1)^2 + 2g_1^1 = \tfrac{1}{2}A(A - B)T^2.$$

If the electric current is supplied along the axis $\mu = 1$, then $A = 1$, $B = 0$ and the solution is

$$g_1^1 = -T\,\frac{C + e^{T/2}}{C - e^{T/2}}, \qquad (3.4.7)$$

where C is a constant. This solution does not contain the logarithmic singularity $g''(T) \sim \ln T$, which must compensate for the electromagnetic force at the infinitely thin wire (Section 3.3). The singularity appears only in the second approximation with $D_2^1 = S(\Delta_1)^2 \sim (\ln S)^{-1/2}$ and hence exerts little influence on the flow at $S \to \infty$.

The case with $(A - B) = 0$ corresponds to the current supply from below along $\mu = -1$ and leads to similar, highly oscillating functions as in the region 3. Apart from this, evaluating $K(\mu)$ with the already known constants a_0, b_0, c_0, we find that $K(\mu) > 0$ in the region 2 and therefore equation (3.4.3) has no real solution.

For the case of the flow diverging along the plane we have found the asymptotic representations for the stream function in the three regions. These expansions should be related to each other, which leads to a need to match them in the intermediate subregions. Let us apply a suitable matching principle, proposed by Shivamoggi [29]: "[The n-term formal Laurent series expansion of the outer expansion about the inner boundary written in terms of the inner variable] = [The n-term formal outer limit of the inner expansion]".

We obtain for the first approximation:

at the axial region boundary

$$g^2(t \to 0) = T = g^1(T \to \infty) = T;$$

at the region 3 boundary for $c_0 \neq 0$

$$g^2(\mu \to 0) = -S^{1/2}\sqrt{-2c_0} = g^3(M \to \infty) = -S^{1/2}\sqrt{-2c_0},$$

for $c_0 = 0$

$$g^2(\mu \to 0) = -S^{1/3}\sqrt{2M(b_0 - AB)} = g^3(M \to \infty)$$
$$= -S^{1/3}\sqrt{2M(b_0 - AB)}.$$

Of course, the matching is formal in this case, since the constants a_0, b_0, c_0 are initially set the same in all three regions.

For the problem with a free surface the first approximation in all regions is expressed explicitly, hence a composite expansion can be constructed that is uniformly valid in all regions. This is done by adding the three expansions, but then we would have double common parts, which must be subtracted, i.e. we subtract the external limits of the expansions in the regions 1 and 3:

$$g_1 = D_1^3 g_1^3 + D_1^2 g_1^2 + D_1^1 g_1^1 - [D_1^3 g_1^3]_1^2 - [D_1^1 g_1^1]_1^2$$

(indices on the brackets denote a one-term external limit, i.e. in the region 2).

The asymptotic composite solution for high S numbers of the problem with immersed electrode and free surface can be expressed explicitly by the use of (3.4.3), (3.4.5), and (3.4.7):

$$S^{-1/2} g_1 = \sqrt{-2c_0}\,\frac{1 - \exp(\mu\sqrt{-2Sc_0})}{1 + \exp(\mu\sqrt{-2Sc_0})} -$$

$$- \sqrt{2}\left[-a_0\mu^2 + b_0\mu - c_0 - \frac{1}{2}(1+\mu)^2 \ln(1+\mu) - \right.$$

$$\left. - \frac{1}{2}(1-\mu)^2 \ln(1-\mu)\right]^{1/2} + \sqrt{-2c_0}, \qquad (3.4.8)$$

where the arbitrary constant in (3.4.7) is set $C = 0$. The next approximations can also be constructed, however their expressions are either quite lengthy or they cannot be expressed explicitly. The problem with a rigid cone surrounding the central axis demands expansions of $a_0(S)$, $b_0(S)$, and $c_0(S)$ and hence is more difficult to analyse.

The final expression (3.4.8) describes the diverging flow for high magnitudes of the parameter S and shows that velocities everywhere in the flow are proportional to $S^{1/2}$, i.e. grow at a linear rate with the current I, except in the region of the radial jet along the free surface, where $v_R \sim S$. The solution for the diverging flow exists for high S owing to the fact that vorticity is transported to the plane $\mu = 0$ where the viscous boundary layer grows. By contrast, in the converging flow vorticity is transported to the symmetry axis where, as we have seen, a viscous layer cannot be formed, and we failed to obtain a solution (cf. Section 2.10 where the numerical solution in the spheroidal container with a finite electrode shows boundary layer' eruption).

3.5. Electrically induced flow with differential rotation

One of the possibilities to limit vorticity growth in the axial region of converging flow, which is mentioned in Section 3.1, is to apply an external magnetic field, owing to which the fluid is made to rotate relative to the symmetry axis in addition to the meridional electrically induced flow. The external field permits us to control the electrically induced flow and the method is used for the magnetic control of electric arcs [13, 18], electroslag [12], and other processes [22] of electrometallurgy. Often for this purpose use is made of the field at a solenoid's end, which is placed coaxially to the current supplying electrode externally to the fluid.

For a mathematical description of the flow at a point current source an external magnetic field $\mathbf{B}_0$ should be of a form suitable to retain the flow's similarity [7]. Due to the equations (2.2.4) and (2.1.9) the external magnetic field inversely proportional to the distance from the origin

$$\mathbf{B}_0 = \frac{k\mu_0 I}{R} \left(C, \frac{C\mu + D}{(1 - \mu^2)^{1/2}}, 0 \right) \tag{3.5.1}$$

can be added to the field B_φ (3.2.1) of the radial current. If the constants are set $C = -1$, $D = 1$, the magnetic force lines are parabolic in a meridional plane (see Fig. 2.1e), and the constant k determines the relative magnitude of the external field to the self-magnetic field. The external field (3.5.1) interacting with the radial electric current (2.5.15) yields the azimuthal force:

$$f_{e\varphi} = j_R B_{0\theta} = \frac{k\mu_0 I^2}{2\pi R^3} A \frac{C\mu + D}{\sqrt{1 - \mu^2}}, \tag{3.5.2}$$

which drives the fluid in rotation. The velocity field

$$\mathbf{v} = \left(-\frac{1}{R^2} \frac{\partial \psi}{\partial \mu}, -\frac{1}{R\sqrt{1 - \mu^2}} \frac{\partial \psi}{\partial R}, v_\varphi \right), \tag{3.5.3}$$

according to the statements in Section 2.1, is sought in the similarity form:

$$\psi = RI \left(\frac{\mu_0}{\rho} \right)^{1/2} g(\mu), \quad v_\varphi = \frac{I}{R} \left(\frac{\mu_0}{\rho} \right)^{1/2} \frac{\Omega(\mu)}{\sqrt{1 - \mu^2}}, \tag{3.5.4}$$

which permits us to reduce the equation of motion to a set of ordinary differential equations:

$$\frac{1}{2}(1 - \mu^2)(g^2)''' + 2\Omega\Omega' = \frac{1}{R_S}(1 - \mu^2)(2g'' + ((1 - \mu^2)g'')') +$$

$$+ A(A\mu - B)\frac{1}{2\pi^2}, \tag{3.5.5}$$

$$g\Omega' = \frac{1}{R_S}(1 - \mu^2)\Omega'' + kA(C\mu + D)\frac{1}{2\pi}, \tag{3.5.6}$$

where the Reynolds number R_S is introduced, and is related to S by

$$R_S = \frac{I}{\nu}\sqrt{\frac{\mu_0}{\rho}} = 2\pi S^{1/2}.$$

Consider the effect of rotation, arising after the application of the external magnetic field, on the original meridional electrically induced flow. We choose a situation for analysis in which the point electrode is located on the free surface ($\mu = 0$) of fluid filling the half-space $0 \leq \mu \leq 1$ (Fig. 3.7). In the first approximation we neglect the effect of meridional flow on rotation, i.e. in (3.5.6)

Electrically induced vortex flow at a point electrode and azimuthal rotation 135

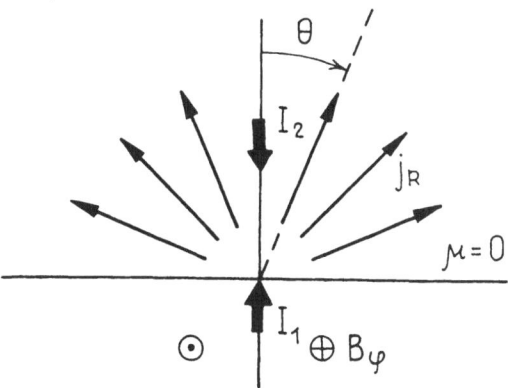

Fig. 3.7. Scheme of the problem with different current supplies to a point electrode on the plane $\mu = 0$.

the left term $g\Omega'$ vanishes. For the boundary conditions of zero tangential stress on the free surface ($\Omega'(0) = 0$) and zero velocity on the axis ($\Omega(1) = 0$) it then follows

$$\Omega = -\frac{kR_S}{2\pi} A[(1 + \mu)\ln(1 + \mu) + 1 - 2\ln 2 - \mu]. \quad (3.5.7)$$

The solution reveals that the rotation is more intense at the plane $\mu = 0$ and it decreases with distance from it. Such a differential rotation, where different fluid layers above the plane rotate with varying velocities, is driven by the electromagnetic force (3.5.2) decreasing in magnitude with the distance from the surface. Respectively decreases also the centrifugal force $f_c = v_\varphi^2/R \sin \theta$, thereby a meridional flow is generated, the fluid being withdrawn from the axis near the free surface and entrained, due to continuity, along the axis from the distant layers.

The differential rotation, defined by (3.5.7), generates the flow diverging along the plane $\mu = 0$, which is similar to the reversed electrically induced flow at an immersed electrode (Section 3.3). The curl of centrifugal force in the equation (3.5.5), $2\Omega\Omega' \leq 0$, corresponds by this analogy to the curl of electromagnetic force for the immersed electrode ($A = 1$, $B = 0$), i.e. it maintains the diverging flow. If the current is supplied along the axis $\mu = -1$ from the external side ($A = 1$, $B = 1$), the centrifugal forces will reduce the electrically induced flow converging to the axis (Section 2.6) and may even reverse the motion if the applied field (3.5.1) is sufficiently great in magnitude. Fig. 3.8 illustrates this with a sequence of stream lines increasing the external field in magnitude in the electric arc model [7].

Electrically induced flows in closed containers in the presence of external axial magnetic field were studied in [11, 19 and others]. The obtained solutions reveal the action of differential rotation in the flows bounded by walls similar to the considered above simple cases.

If was suggested in [30] that the differential rotation, by retarding the

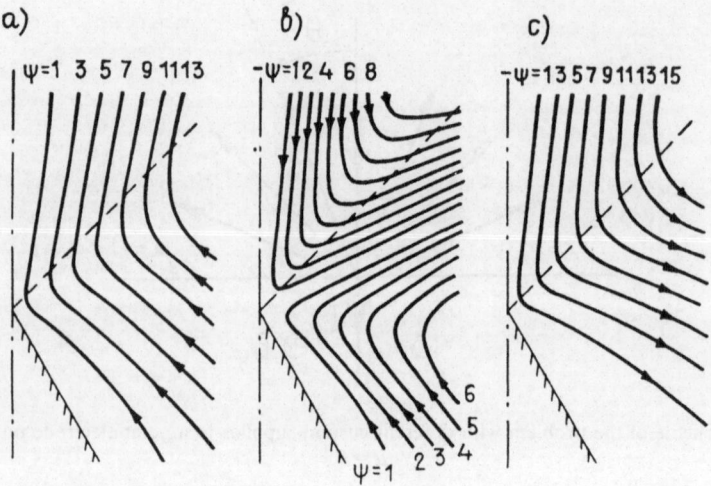

Fig. 3.8. Stream lines $Rg = C$ for $S = 2$ and (a) $k = 0$, (b) $k = 16$, (c) $k = 24$.

converging electrically induced flow at a point electrode, could increase the critical parameter S at which the flow breakdown occurs. A high enough magnitude of the external field (3.5.1) can preclude the flow breakdown at all.

3.6. Growth of azimuthal disturbance in the electrically induced flow at a point electrode

The flows at a point electrode with oppositely directed motions are essentially different. As we have seen, the flow diverging from the axis of symmetry develops smoothly with the increase of current I, and the narrow viscous layers are formed, separated by the region of inviscid flow, when $I \to \infty$ (Sections 3.3, 3.4). The flow breakdown occurs in the flow converging to the axis when the total supplied current I exceeds a certain critical magnitude (Section 2.7).

Previously we have discussed possible causes of the breakdown of the solution at which the velocity in the converging flow grows to infinity. The results of the analysis in Section 3.5 suggest that the difficulties in the model of converging flow at a point electrode may be overcome by setting the fluid in differential rotation with a sufficiently high azimuthal velocity. In [8] it is demonstrated that the state of intense rotation may be reached without any increase of external magnetic field, because the converging flow itself has a propensity to spin-up, similar to the flow above a washbasin. Indeed, the stream lines represented in Fig. 3.9 a resemble the flow to a sink remote from the plane surface, yet the converging sink flow is well known to reach a state of intense swirl resulting from even a small rotational disturbance.

Prior to a theoretical analysis of the flow's spin-up, consider the following experiment substantiating what is stated above. In the experimental model (Fig. 3.10) the electric current is supplied to a small water-cooled electrode (0.8 cm

Electrically induced vortex flow at a point electrode and azimuthal rotation 137

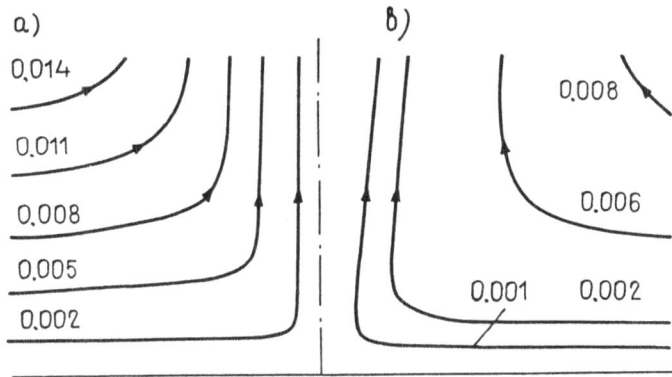

Fig. 3.9. Stream lines of the flow (a) without the rotation ($k = 0$, $R_s = 42$), (b) with the rotation ($k = 0.01$, $R_s = 44$).

in diameter) which is situated in the centre of the free surface of mercury filling a hemispherical copper container (36 cm in diameter) which also serves as the second electrode.

When the electric current $I = I_1$ is supplied from above to the free surface, the flow converging to the symmetry axis is set up; it visually resembles the motion on surface when the fluid enters a sink at the bottom of the container. The flow, observed for the magnitudes of electric current $I \geqslant 15$ A is accompanied by rotation which increases on approaching the central axis. The intensity of swirl also increases with the electric current. For the total current 1200 A the photograph in Fig. 3.10a clearly exhibits azimuthal tracks of graphite particles, added to aid visualization of the motion. In this axisymmetric situation there are no substantial external azimuthal forces which can sustain the observed rotation; however, a small magnitude force is present, arising from the interaction of the radial electric current and the vertical component of the Earth's magnetic field. The perturbing external field is approximately 0.5 gauss in magnitude, while the self-magnetic field in the fluid is of the order of 500 gauss for a current of $I = 1000$ A. Consequently, the rotational disturbance from the weak external field is amplified in a certain way by the intense converging meridional flow.

The same observations reveal that the rotation does not arise in the reversed flow when the electric current $I = I_2$ is supplied from below to the free surface (Fig. 3.10b) up to the maximum current in the experiment $I = 1500$ A, and, even after an artificial mechanical spin-up, it soon dies out and only the radial diverging flow remains visible on the surface. The photograph in Fig. 3.10b shows a surface bulge at the small electrode, where the stream lifts up; the dark ring in the periphery consists of the graphite particles transported out by the radial flow. Thus, for equal external conditions the divergence along the surface flow does not rotate, while the converging flow is set in the rotation resembling the vortex above a fluid sink.

To analyse theoretically the flow's sensitivity to rotation, return now to the similarity equations (3.5.5), (3.5.6) for the problem with a point electrode. The

Fig. 3.10. Motion on the surface of mercury in the hemisphere: (a) the electric current $I_1 = 1200$ A is supplied from above, the converging flow is set in rotation; (b) current $I_2 = 1200$ A is supplied from below, the diverging flow does not rotate.

disturbing external field in this case may be assumed to be the field (3.5.1) with the constant $k \ll 1$; then the azimuthal force (3.5.2) is also proportional to k. Let the plane $\theta = \pi/2$ (see Fig. 3.7) be undeformed free surface. The boundary conditions for the equations of viscous flow are

$$g(0) = g''(0) = \Omega'(0) = 0 \quad \text{on the surface} \quad \mu = 0,$$
$$g(1) = \sqrt{1 - \mu^2}g''(1) = \Omega(1) = 0 \quad \text{on the axis} \quad \mu = 1. \tag{3.6.1}$$

Solution of the boundary value problem (3.5.5), (3.5.6), (3.6.1) is characterized by rapid changes of the functions $g(\mu, R_s)$ and $\Omega(\mu, R_s)$ in proximity to the critical value of R_s. This set up special requirements to the approximation of numerical solution. Therefore in [8] a variant of Galerkin method was used, which is preferable in accuracy by comparison with finite-difference methods [23]. For realization of the method it is supposed that the functions $g(\mu)$ and $\Omega(\mu)$ can be expanded in converging series of the form $\Sigma a_n I_n(\mu)$, where the functions I_n are the eigenfunctions for the linearized equations of creeping motion, and can be expressed by Legendre polynomials: $I_n = (P_{n-2} - P_n)/(2n - 1)$. In the computations the series of I_n were truncated at $n = 41$ (for details of the numerical procedure see [8]).

The expansions assumed immediately give a solution of the linear Stokes equation for the slow flow if we set the left-hand parts of the equations (3.5.5), (3.5.6) equal to zero. In this case the functions g and Ω are determined independently and $g \sim R_s$, $\Omega \sim kR_s$. With the aim of showing the nonlinear growth of the velocity field, when the nonlinear terms are retained, we introduce normalized functions

$$g_n = \frac{g}{R_s}, \quad \Omega_n = \frac{\Omega}{kR_s}. \tag{3.6.2}$$

Consider the nonlinear problem (3.5.5), (3.5.6), (3.6.1) in the case where the electric current $I = I_1$ is supplied from the external side to the fluid's surface (Fig. 3.7). Note that if the azimuthal force is neglected, i.e. it is set at $k = 0$ in (3.5.6), then, as was shown in Section 2.4, we obtain a trivial solution $\Omega = 0$, and the axial velocity will go to infinity at $R_s \to R_{cr} = 43$ ($S_{cr} = 47$) (curve 1 in Fig. 3.11). The flow stream lines in proximity to R_{cr} are shown in Fig. 3.9a.

Now introduce an azimuthal disturbance in the form of (3.5.2). For a fixed small parameter k, and increasing R_s, the disturbance is proportional to kI^2. In this case it may arise due to an asymmetry of the supplying wire. When the parameters k and R_s are small, the normalized axial velocity $g'_n(1)$ and the

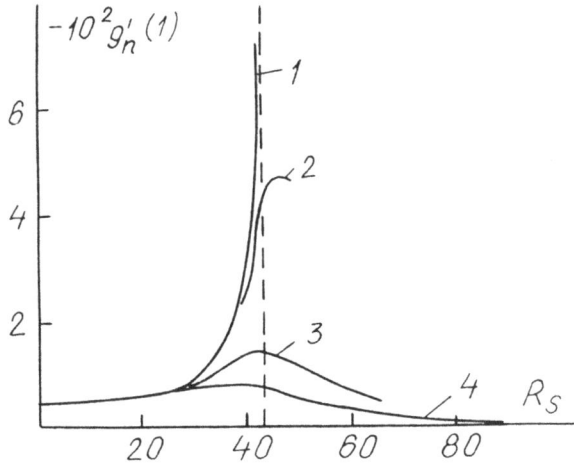

Fig. 3.11. Axial velocity vs. R_s for (1) $k = 0$, (2) 0.001, (3) 0.005, (4) 0.01.

azimuthal velocity on the free surface $\Omega_n(0)$ are constant (Figs. 3.11, 3.12). With increasing R_s, the meridional velocities grow, as does the azimuthal velocity. Increased centrifugal forces at the free surface, where $\Omega(\mu)$ has a maximum, counteract the converging flow along the plane $\mu = 0$, which is sustained by the electromagnetic force due to the self-magnetic field. This results in a decrease of meridional convection; the velocity at R_{cr} remains finite (the stream lines with azimuthal disturbance are represented in Fig. 3.9b). Moreover, the axial velocity $g'_n(1)$ reaches a maximum and then decreases with increasing R_s (curves 2—4 in Fig. 3.11). The variation of azimuthal velocity $\Omega_n(0)$ is similar (curves 2, 3 in Fig. 3.12).

If $k \sim 1$, the relative magnitude of the azimuthal velocitie's maximum is small and does not attract special attention. This situation corresponds to the results in [11, 19, 30]. More interesting are small values of k. Then the solution shows that the maximum magnitude of Ω_n rapidly rises with increasing R_s in the vicinity of R_{cr} (Fig. 3.12), and $g'_n(1)$ still remains finite (Fig. 3.11). This result means that the small disturbance ($\sim k$) of azimuthal motion is amplified by the action of intense meridional flow.

A question arises as to whether the function Ω_n would grow infinitely if $k \to 0$. If this were to be true, then an infinitesimal perturbation can be amplified to a level sufficient to interact with the meridional flow and limit increase of the axial velocity at $R_s \to R_{cr}$. The variation of the inverse value to $\Omega_n(0)$ is indicative in this case, and it is depicted in Fig. 3.13 for $k \to 0$ and different R_s. When $R_s < R_{cr}$, the magnitude of Ω_n^{-1} remains finite for $k = 0$, what means $\Omega = kR_s\Omega_n = 0$, i.e. the azimuthal velocity is zero. Yet for $R_s \geqslant R_{cr}$ the numerical results indicate $\Omega_n^{-1} \to 0$ for $k \to 0$ and it may be expected that Ω tends to a finite value, i.e. the azimuthal velocity does not vanish.

Consider the behaviour at $k \to 0$ in greater detail. For finite k the quantity $g'_n(1)$ remains finite for R_s exceeding R_{cr} (Fig. 3.11). Hence, the finite disturbance eliminates the singularity of axial velocity. If we let k go to zero, solving the problem for consecutively smaller k, the asymptotic variation of the numerical solution at R_{cr} indicates that $\Omega_n \sim \text{const}/k$ (Fig. 3.13), i.e. $\Omega \sim \text{const}$.

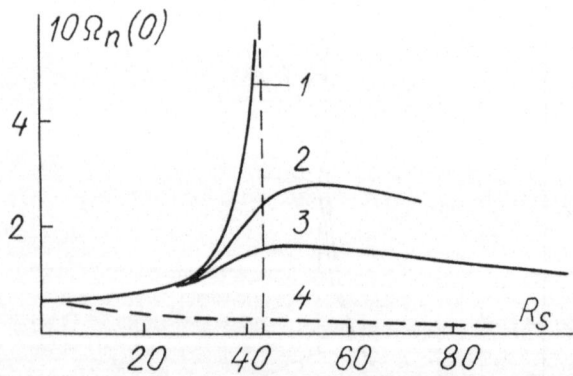

Fig. 3.12. Azimuthal velocity on the free surface vs. R_s for (1) $k = 0.001$, (2) 0.005, (3) 0.01 (all three for $I = I_1$); (4) 0.01 (for $I = I_2$).

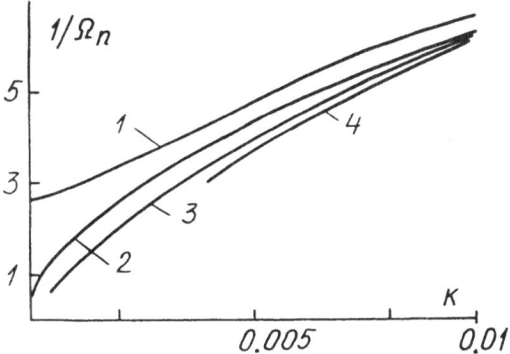

Fig. 3.13. Behaviour of the quantity inverse to azimuthal velocity ($\Omega_n^{-1}(0)$) with the decrease of perturbation parameter $k \to 0$: (1) $R_s = 40$, (2) 42, (3) 45, (4) 50.

Figure 3.14, demonstrating the variation of the inverse of axial velocity for R_{cr} when the solution without the rotation cannot be obtained, also strongly suggests that $g'_n(1)$ remains finite for $k \to 0$. Nevertheless, the limit $k = 0$ cannot be reached, because then in the axisymmetric flow a source of angular momentum vanishes ($f_{e\varphi} = 0$). For $k = 0$ (3.5.6) can be integrated

$$\Omega' = \text{const} \exp \int_0^\mu \frac{R_s g(t)}{1-t^2} \, dt,$$

and for the boundary conditions (3.6.1) $\Omega \equiv 0$. Thus, at least an infinitesimal, but nonzero, perturbation is necessary to eliminate the singularity of axial velocity for $R_s \geqslant R_{cr}$.

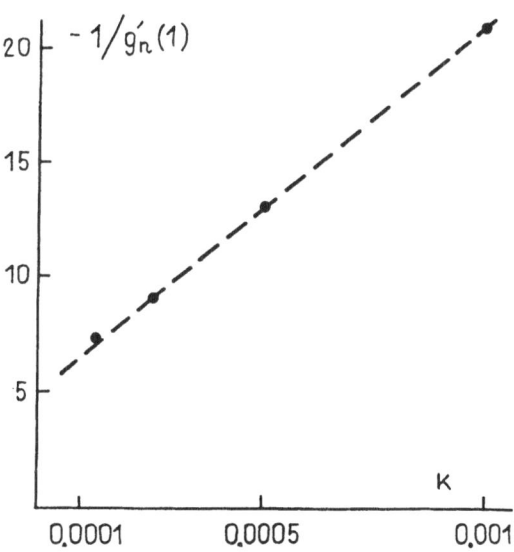

Fig. 3.14. Behaviour of the quantity inverse to axial velocity for $R_s = 43 \geqslant R_{cr}$.

Consider some more properties of the flow with rotation. Components of vorticity in this flow are of the form

$$\text{curl } \mathbf{v} = I \left(\frac{\mu_0}{\rho} \right)^{1/2} \left[-\frac{1}{R^2} \Omega'(\mu), 0, \frac{\sqrt{1-\mu^2}}{R^2} g''(\mu) \right].$$

In the absence of rotation the magnitude of the azimuthal vorticity grows infinitely at the symmetry axis for $R_s \to R_{cr}$, and viscous diffusion of vorticity in this case cannot preclude the flow breakdown. The sensitivity to azimuthal disturbance leads to a rapid increase of radial vorticity at the axis of rotation (Fig. 3.15), yet the increased viscous diffusion of this component makes the increase of the total vorticity bounded, as the analysis of the rotation amplification mechanism in Section 3.8 will reveal.

The amplification of rotation is clearly evident in Fig. 3.16, according to which the angular momentum related to the applied azimuthal force in unit

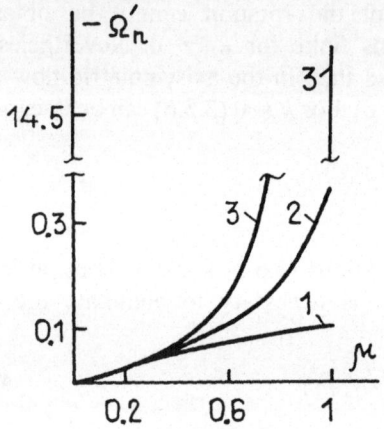

Fig. 3.15. Distribution of radial vorticity for $k = 0.001$ (1) $R_s = 1$, (2) 30, (3) 43.

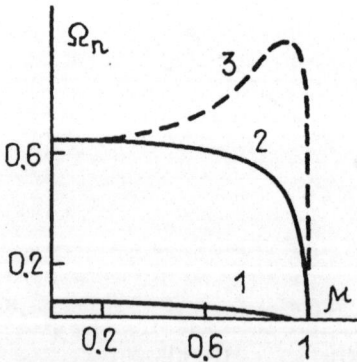

Fig. 3.16. Angular momentum $\Omega_n(\mu)$ (full line) and azimuthal velocity $\Omega_n(\mu)/\sqrt{1-\mu^2}$ (dashed line) for $k = 0.001$: (1) $R_s = 1$ (the curves for both quantities coincide), (2) 43, (3) 43.

volume, Ω_n, increases significantly in intense converging flow ($R_s = 43$), and the increase of azimuthal velocity, which is proportional to $\Omega_n/\sqrt{1-\mu^2}$ demonstrates formation of a rotating vortex.

The amplification of rotation is specific to the converging flow. The diverging flow, which is generated by supplying the electric current $I = I_2$ by an immersed electrode along $\theta = 0$, does not increase the rotation intensity (curve 4 in Fig. 3.12), and a small disturbance in the form of (3.5.2) does not significantly alter the meridional flow. In this case we even observe a retardation of rotation due to the transport of the momentum to infinity.

The fact that the converging flow at a point electrode can be maintained only when coupled with rotation has an important meaning in relation to the well-known phenomenon of a rotating vortex formation in converging flows, observed in nature and technology. This will be discussed in greater detail in Section 3.8.

3.7. Intensification of rotation in a closed volume

Some uncertainty still remains as to what extent the self-similar flow at a point electrode can be related to a real flow in a closed volume with electrodes of finite size. A model suitable for the investigation is flow in a hemispherical container [6]. In the model, electric current radially diverges from a small hemispherical electrode $R = R_1$ to conducting walls of the hemispherical container $R = R_2$ (Fig. 3.17), and the distribution of current density remains the same as in the point-source problem: $j_R = IA/2\pi R^2$. Hence the magnetic field is determined by (3.2.1) in the noninduction approximation and the

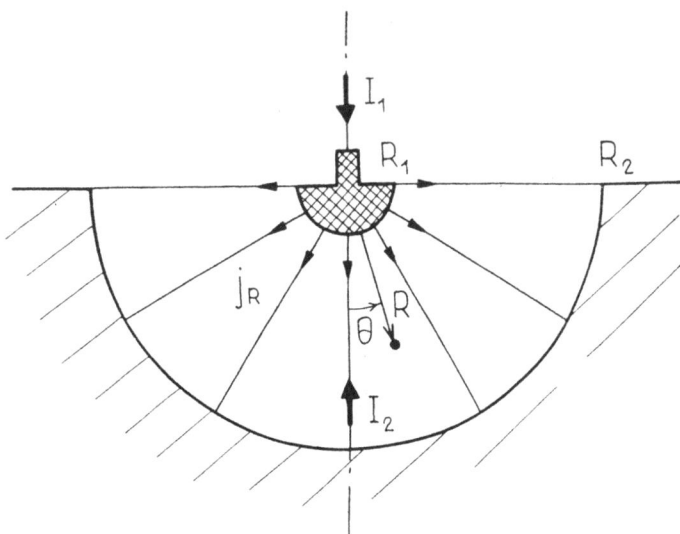

Fig. 3.17. Scheme of the electric current passage in a hemispherical container.

electromagnetic force is given by (3.3.2). An azimuthal disturbance of motion will be generated by the external field (3.5.1), to retain analogy with the flow at point electrode.

The equations of axisymmetric motion (2.1.1), (2.1.2) can be rewritten in a nondimensional form by the use of new variables: the coordinate $R^* = R_2 R$, the stream function $\psi^* = R_2 I \sqrt{\mu_0/\rho}\, \psi(R, \theta)$, the rotation function $\Omega^* = I\sqrt{\mu_0/\rho}\, \Omega(R, \theta)$, and the force $f_e^* = \mu_0 I^2/R_2^3 \cdot f_e$, where stars denote the old dimensional variables; $\psi_1 = I(B - A\cos\theta)/2\pi$, $\psi_2 = \mu_0 kIR(1 - \cos\theta)/2\pi$:

$$(\psi_\theta \xi_R - \psi_R \xi_\theta)/\sin\theta + 2\xi(\psi_R R \cotan\theta - \psi_\theta)/R\sin\theta -$$
$$- 2\Omega(\Omega_R R \cotan\theta - \Omega_\theta)/R\sin\theta$$
$$= R_s^{-1} R^2 E^2 \xi + R^3 \sin\theta\, \mathbf{i}_\varphi \cdot \nabla \times \mathbf{f}_e;$$
$$\xi = -E^2\psi; \qquad (3.7.1)$$
$$(\psi_\theta \Omega_R - \psi_R \Omega_\theta)/\sin\theta = R_s^{-1} R^2 E^2 \Omega + R^3 \sin\theta\, \mathbf{f}_e \cdot \mathbf{i}_\varphi,$$

where $E^2 = \partial^2/\partial R^2 + (\partial^2/\partial\theta^2 - \cotan\theta\, \partial/\partial\theta)/R^2$; $R_s = I\sqrt{\mu_0/\rho}/\nu$; the indices R, θ denote the respective partial derivatives. Now the flow depends on three nondimensional parameters: the Reynolds number R_s, the relative disturbance magnitude k, and the ratio of electrode's radii R_1/R_2, and also from the kind of current supply, i.e. I_1 or I_2 (see Fig. 3.17). The equations were solved by the finite-difference alternating direction implicit method for the boundary conditions of impermeability and no slip on the rigid walls, impermeability and zero stress on the free surface $\theta = \pi/2$ (for details see [6]).

If $f_{e\varphi} = 0$, the meridional volume force (3.3.2) induces motion in meridional planes without azimuthal rotation, hence $\Omega = 0$ is fixed. According to the numerical solution the flow is of the form of a toroidal vortex with the stream lines $\psi/R_s = $ const displayed in Fig. 3.18a for the case of the current $I = I_1$ supply. The values of axial velocity in Table 3.1 demonstrate that the flow intensity increases with the decrease of the small electrode's radius R_1. This is related to the increase of the electromagnetic force applied to the fluid in the hemisphere

$$\mathbf{F}_e = \int \mathbf{f}_e^* \, dV = -\frac{\mu_0 I^2}{4\pi} \ln \frac{R_2}{R_1} \mathbf{i}_\theta\left(\frac{\pi}{2}\right),$$

and to a reduction of friction losses.

For a fixed ratio R_1/R_2 the nondimensional axial velocity $v_R(0) = \psi_\theta/R^2 \sin\theta$ grows linearly with R_s approximately up to $R_s = 10$. Increasing R_s further, the maximum of v_R/R_s decreases[*], yet approaching the external hemisphere there is a region where the magnitude of axial velocity increases comparing to the magnitude at smaller R_s. This reminds us of the velocity growth in the hemispherical flow with a point electrode (Section 2.10). Since at the wall $R^* = $

---

[*] This ratio can be treated as the ratio of dimensional velocity v_R^* to a quantity directly proportional to the applied force $|\mathbf{f}_e^*| \sim I^2$.

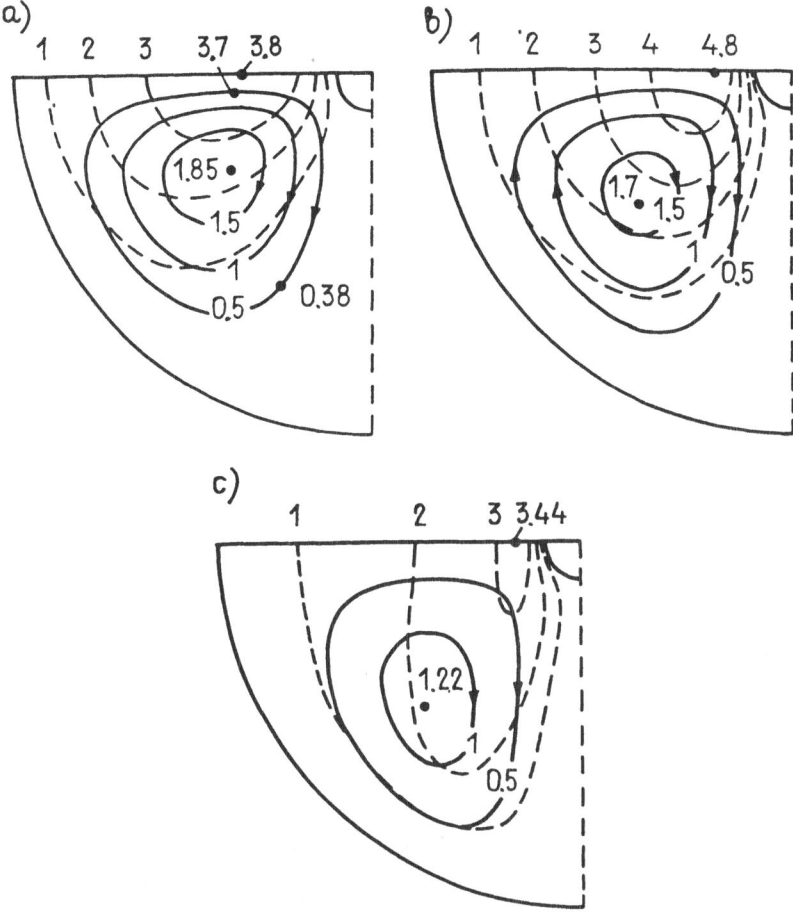

Fig. 3.18. Stream lines $10^4 \psi/R_s$ (full lines) and lines of constant angular momentum $10^2 \Omega_n$ (dashed lines) for $R_1/R_2 = 0.1$, $k = 0.01$: (a) $R_s = 10$, (b) 75, (c) 150.

R_2 the velocity must be zero, the gradient of velocity in this region increases very fast (see Table 3.1). The high velocity gradients deteriorate convergence of the numerical procedure, what precludes to obtain the solution for higher R_s. The local increase of radial velocity in the vicinity of external hemisphere is supported also by the independent results [1], where a similar flow in hemisphere with the current flowing from an electric arc, contacting the free surface of fluid, is computed by the leapfrog method of DuFort-Frankel (see Section 2.10).

The results presented in Tables 2.3 and 3.1 give evidence that the region, where v_R/R_s grows, increases in size decreasing the ratio of representative size of current supply area (R_1) to container size (R_2), i.e. for $R_1/R_2 \to 0$. In the limiting case $R_1/R_2 = 0$ the flow at a point electrode is determined in the class of exact solutions (3.5.4), which leads to the flow breakdown for the critical value of $R_s = 43$. For $R_1/R_2 = \text{const} \neq 0$ the solution remains bounded, but its

Table 3.1. Values of the nondimensional axial velocity $10^2 v_R/R_s$.

R_s	10	150	10	100	10	100
k	0	0	0	0	0.01	0.01
$R = 0.05$	—	—	0	0	0	0
0.1	0	0	1.26	0.85	1.26	0.81
0.15	0.35	0.17	1.41	1.07	1.41	0.93
0.2	0.56	0.30	1.29	1.09	1.29	0.91
0.25	0.63	0.38	1.11	1.06	1.11	0.86
0.35	0.55	0.44	0.77	0.96	0.77	0.73
0.5	0.33	0.42	0.42	0.82	0.42	0.57
0.7	0.13	0.32	0.14	0.59	0.14	0.34
0.8	0.07	0.23	0.07	0.40	0.07	0.20
0.9	0.01	0.08	0.01	0.15	0.01	0.06
1	0	0	0	0	0	0
R_1/R_2	0.1	0.1	0.05	0.05	0.05	0.05

calculation for high R_s is impossible due to the formation of the region with very high velocity gradients.

Consider now modification of the flow in a hemisphere by a disturbance in the form of a small azimuthal force $f_{e\varphi}$. Introduce the azimuthal force (3.5.2), proportional to a small quantity k, into equation (3.7.1). The force (3.5.2) drives azimuthal rotation, and (3.3.2) drives meridional flow, both of which are independent in linear regime for small R_s. Thus, comparison of the axial velocities in Table 3.1 for $R_s = 10$ without rotation ($k = 0$) and with rotation ($k = 0.01$) demonstrates full identity in the interval of computation accuracy. In the linear regime the dimensional velocity v_φ^* grows with R_s in proportion to the applied force $f_{e\varphi} \sim kI^2$, hence we define, similarly to (3.6.2), the function $\Omega_n = \Omega/kR_s$, which represents a change of fluid particle's angular momentum compared to Stokes regime.

For high R_s the nonlinear terms in (3.7.1) significantly affect the flow configuration. Meridional stream lines get closer to the symmetry axis, and the angular momentum Ω_n attains its maximum also closer to the axis (Fig. 3.18). The velocity of rotation $v_\varphi \sim \Omega_n/R \sin \theta$ increases significantly due to the convection of angular momentum to the axis ($R \sin \theta \to 0$), and a rotating vortex is formed in the axial region with a tenfold increase in angular velocity $\omega = v_\varphi/R \sin \theta$ (Fig. 3.19 represents growth of the axial quantity $R^2 \omega_\eta = \lim_{\theta \to 0} \Omega_n/\sin^2 \theta = \Omega''(0)/2$). According to the widely accepted explanation of the rotating vortex formation, angular momentum is transported from surroundings, and it is proportional to the applied azimuthal force. Yet the computational results presented in Fig. 3.20 show that the normalized by the applied force angular momentum $\Omega_n(R_s)$ increases, i.e. the momentum Ω^* grows at a higher rate than the applied force $f_{e\varphi}^*$. The increase of Ω_n is higher, the smaller the ratio R_1/R_2 and, respectively, the more intense is the meridional flow. The

Electrically induced vortex flow at a point electrode and azimuthal rotation 147

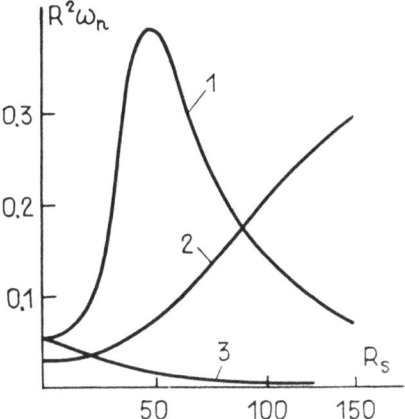

Fig. 3.19. Maximum of $R^2\omega_n$, the quantity proportional to angular velocity on the axis $\theta = 0$, for $k = 0.01$: (1) $R_1/R_2 = 0$ $(I = I_1)$, (2) 0.1 $(I = I_1)$, (3) 0 $(I = I_2)$.

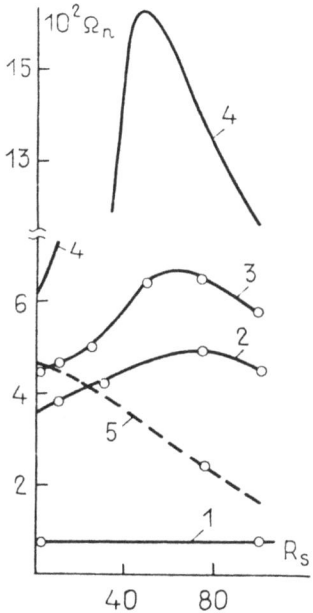

Fig. 3.20. The maximum momentum $\Omega_n(R_s)$ on the free surface $\theta = \pi/2$ for $k = 0.01$: (1) $R_1/R_2 = 0.5$, (2) 0.1, (3) 0.05, (4) 0 [all for the converging flow with $I = I_1$], (5) 0.05 [for the diverging flow with $I = I_2$].

flow with a point electrode $R_1/R_2 = 0$ is evidently the limiting case of maximum increase of Ω_n (Fig. 3.20). (We shall discuss the mechanism of rotating vortex formation in Section 3.8.)

The increased velocity of rotation modifies also the meridional flow. Since the maximum centrifugal forces are located at the free surface, they counteract the converging flow, which is sustained by the electromagnetic force $f_{e\theta}$, and intensity of the meridional flow is weakened; this is demonstrated by reduction of the axial velocity in Table 3.1 ($R_s = 100$, $k = 0$ and $k = 0.01$). Moreover, the rotation inhibits growth of the higher velocity gradients at the outer sphere.

The intensification of rotation is closely related to the flow converging along the free surface. The reverse flow can be generated supplying electric current to the small hemisphere along an immersed isolated wire (I_2 in Fig. 3.17). For the infinitely thin wire along the axis $\theta = 0$ in the expression for magnetic field (3.2.1) the constants $A = 1$, $B = 0$ (3.3.1), and the meridional force (3.3.2) is directed oppositely to the former case. The problem with a point electrode ($R_1/R_2 = 0$) is solved in Section 3.3; the flow with finite electrodes ($R_1/R_2 \neq 0$) also does not contain the region of high velocity gradient which hinders computation of the converging flow. A weak rotation in the diverging flow is not amplified, it is rather weakened, as indicated in the curves 5 in Fig. 3.20, and 3 in Fig. 3.19. The weakening of rotation may be explained by convection of the momentum Ω to the container's side walls, where it dissipates due to friction.

3.8. Mechanism of rotation intensification in an axisymmetric vortex

A remarkable feature of the converging meridional flows described in Sections 3.6 and 3.7 is that of acquiring a relatively high swirl velocity as a result of a small azimuthal disturbance. The spin-up feature is similar to such well-known flows as washbasin-flow, convective flows generating tornado-like vortices in atmosphere, and air entering a jet-engine on a grounded airplane. Also, in electrometallurgy, for instance in electroslag melting, a spontaneous rotation of a liquid slag is observed when electromagnetic forces maintain the meridional circulation.

Development of a rotating vortex is closely related to certain properties of the flow radially converging to a symmetry axis. Thus in a sink vortex fluid enters the sink through a narrow vertical region surrounding the vortex funnel, and it is entrained along horizontal layers. Morton [21] offers considerations, based on integral properties of the equations of motion, by which the vortex can exist only if a certain value of flow rate to the sink is exceeded, i.e. the converging flow is sufficiently intense; when exceeding another, higher value of flow rate the vortex disappears, because any disturbance is carried out into the sink. The first critical value of flow rate may be interpreted as a stability limit of the converging flow to rotational perturbation.

This point of view is supported by experimental results for sink flow [15, 16], and for the convective flow above a localized heat source [33]; according to these the rotation arises when a critical intensity of the converging flow is exceeded.

Let us attempt to discover how the rotation is amplified in an axisymmetric swirling vortex. Usually these vortices are explained by the convection of disturbance vorticity to the axis and stretching of the vorticity lines in the axial region or, in other words, by conservation and convection of the angular momentum Ω to the axis, where the azimuthal velocity $v_\varphi \sim \Omega/r$, and the angular velocity $\omega \sim \Omega/r^2$ increase rapidly with decrease of the distance r from the axis. Yet several facts indicate the substantial effect of an additional mechanism in maintaining the vortex. Thus, observing the rotating vortex, one may note its quasi-stable state even for an occasional change of the seeding Ω far from the axis, which would lead, according to the above explanation, to a $1/r$ times increased change in the rotation at small distance r from the axis; however the occasional short-time perturbation is dissipated in the intense vortex. Moreover, as it was shown in Sections 3.6 and 3.7, the momentum Ω increases with the intensity of converging flow, instead of the plain conservation (see Fig. 3.20). This may be explained by an increased effect of viscous friction in the highly rotating axial region and the associated momentum transfer by viscous diffusion to the vortex periphery, where in a steady state the diffused out momentum would add to the seeding momentum.

The effect of viscosity can be demonstrated in an instructive experiment described by Lavrentyev, Shabat [17]: a cylindrical vessel containing fluid is mounted on rolls and at an initial moment is set to rotate around the symmetry axis. After a sink is opened in the centre of the vessel's bottom, the swirling vortex is generated; at this moment the velocity of vessel's rotation begins to increase, which means the increase of angular momentum per unit volume (Ω) with the fluid running out. The authors explain the acceleration of the vessel by the transfer of viscous friction from the axial vortex region to the container's walls. The explanation would be correct if the momentum transported by viscous diffusion were to be added to the initial Ω at the side wall, yet it does not take any account of the intense convective momentum transport to the sink, which is usually assumed to be much larger than the diffusive flux in fluids of small viscosity, such as water for instance.

The interaction process of convection and viscous effects in a rotating vortex are studied in Burger's model [2, 9], which describes the vortex above a plane maintained by the converging meridional potential flow. Other extensive studies of the vortex formation are numerical [14, 24, 36]. However, the flows considered in Sections 3.6 and 3.7 are of special interest for the analysis of rotating vortex development. These models differ from the previous because the meridional flow is controlled by the volume force (3.3.2), and the source of rotation is the independent force (3.5.2). In this situation the boundary conditions for the closed flow in a hemispherical container are uniquely determined (homogeneous), and there are no difficulties to pose conditions at inflow and outflow boundaries, which usually affect the flow significantly.

To study the mechanism, by which angular momentum in the vortex core is growing at a higher rate than the applied azimuthal force (Section 3.7), consider now the angular momentum conservation in the vortex core. An equation describing transport of the momentum $m_z = \rho r v_\varphi = \rho \Omega^*$ in a fluid is obtained by multiplying the azimuthal component of the Navier-Stokes equation by the distance r from the rotation axis. Using the divergent form of the equations [35], the change of Ω^* in the fluid is determined by the equation

$$\rho\Omega^*_t + \rho\,\text{div}(\mathbf{v}^*\Omega^*) = \nu\rho\left[\text{div grad}\,\Omega^* - 2\,\text{div}\left(\frac{\Omega^*}{r}\mathbf{i}_r\right)\right] + rf^*_{e\varphi} \quad (3.8.1)$$

in dimensional form. It can be rewritten in the nondimensional form using the representative quantities defined in Section 3.7, and the nondimensional time variable $t = \nu t^*/R_2^2$. We integrate (3.8.1) further over a control volume V with a surface S_0, and use Gauss's theorem; the resulting integral form of the momentum conservation equation is

$$\frac{1}{2\pi k}\left\{\frac{1}{R_s}\int_V \Omega_t\,dV + \oint_{S_0}\Omega\mathbf{v}\cdot d\mathbf{S}_0\right.$$

$$= \frac{1}{R_s}\oint_{S_0}\left[\left(\Omega_R - \frac{2\Omega}{R}\right)\mathbf{i}_R + \right.$$

$$\left.+ \frac{1}{R}(\Omega_\theta - 2\Omega\cot\theta)\mathbf{i}_\theta\right]\cdot d\mathbf{S}_0 + \left.\int_V R\sin\theta f_{e\varphi}\,dV\right\}, \quad (3.8.2)$$

or more briefly

$$\Delta + K = D + M_e,$$

where Δ, K, D, M_e denote the consecutive integrals in (3.8.2) with the respective multipliers.

Choosing the control volume shown in Fig. 3.21 by the dashed lines, we represent the surface integrals in (3.8.2) in the form of momentum fluxes

$$\Delta + K_R + K_\theta = D_R + D_\theta + D_{R1} + M_e.$$

Here Δ is the momentum rate of change in time; K_R, K_θ are the momentum convective transport out of the volume V in the direction of externally oriented normal, i.e. across the surfaces $R = \text{const}$ and $\theta = \text{const}$ respectively (Fig. 3.21); D_R, D_θ, D_R are the momentum transport by viscous diffusion into the volume V across the respective surfaces; M_e is the electromagnetic force moment.

Consider the electrically induced flow in a hemispherical container (Section 3.7). For the steady flow ($\Delta = 0$) the electromagnetic moment M_e, acting on the whole fluid volume in the hemisphere, is compensated merely by the integral viscous friction on the rigid walls of the container ($\mathbf{v}$ is zero on the walls), i.e. (3.8.2) is transformed to

$$M_e = -D_R - D_{R1},$$

because $\mathbf{v} = 0$ on the walls, $\mathbf{v}\cdot d\mathbf{S}_0 = 0$ on the free surface $\theta = \pi/2$. Since the nondimensional M_e is not changed with increase of the current I, then, in spite

Electrically induced vortex flow at a point electrode and azimuthal rotation 151

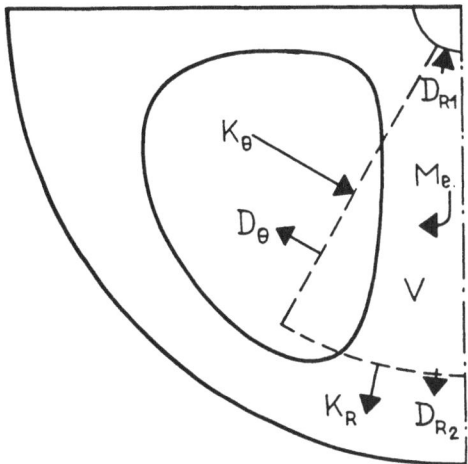

Fig. 3.21. Scheme of the angular momentum Ω balance in the axial conical region ($R_1/R_2 = 0.1$, $R = 0.8$, $\theta = 27°$) for $k = 0.01$, $R_s = 150$ of the steady flow.

of the meridional flow intensification increasing R_s, the integral nondimensional moment of friction is not changed with the increase of R_s*.

However, with the increase of R_s the momentum Ω is substantially redistributed in the fluid volume, and the maximum magnitude of Ω can grow significantly (Fig. 3.20). Since Ω grows in the axial region (see Fig. 3.18), the rotation velocity $v_\varphi = \Omega/r$ there increases sharply, and the rotating vortex is formed. Consider the momentum transport balance (3.8.2) for a vortex region, which, to be specific, is confined in the cone $\theta = 27°$ with the ends at the rigid walls R_1 and $R = R_2 = 1$ (see Fig. 3.21). For this case $K_R = 0$ and the convective momentum exchange by the meridional circulation is given by K_θ.

If convection is weak ($R_s = 1$), K_θ is practically zero, and, after the instan-

---

* The mechanical energy equation for a similar situation was analysed Bershadskiy [3]. On multiplying Navier-Stokes equation by ($\mathbf{e}_\varphi v_\varphi$) and integrating over the hemispherical domain, we obtain by use of continuity and the previously set boundary conditions:

$$\frac{1}{2} \frac{\partial}{\partial t} \int_{R_1}^{R_2} R^2 \, dR \int_0^{\pi/2} v_\varphi^2 \sin\theta \, d\theta$$

$$= \nu \int_{R_1}^{R_2} R \, dR \int_0^{\pi/2} v_\varphi E^2(R \sin\theta \, v_\varphi) \, d\theta -$$

$$- \int_{R_1}^{R_2} R \, dR \int_0^{\pi/2} v_\varphi^2 (v_R \sin\theta + v_\theta \cos\theta) \, d\theta + \int_{R_1}^{R_2} R^2 \, dR \int_0^{\pi/2} f_\varphi v_\varphi \sin\theta \, d\theta.$$

The left side of the equation represents the rate of change of azimuthal motion's kinetic energy. The first term on right is a part of energy dissipation in a unite time, and the second term contains the expression in brackets equal to the cylindrical radial velocity: $v_r = v_R \sin\theta + v_\theta \sin\theta$. Thus, a converging flow ($v_r < 0$) may increase the energy of rotation; the diverging flow decreases the energy.

taneous application of M_e, the momentum increase $\Delta = M_e$ for the initial time moment (Fig. 3.22a). With the increase of time a growing portion of M_e is consumed by the viscous friction $-D_{R_2}$ and $-D_{R_1}$ at the rigid walls, the small momentum part is added by the viscous diffusion D_θ from the periphery into the axial region.

In the case of intense convection ($R_s = 150$) the momentum transport

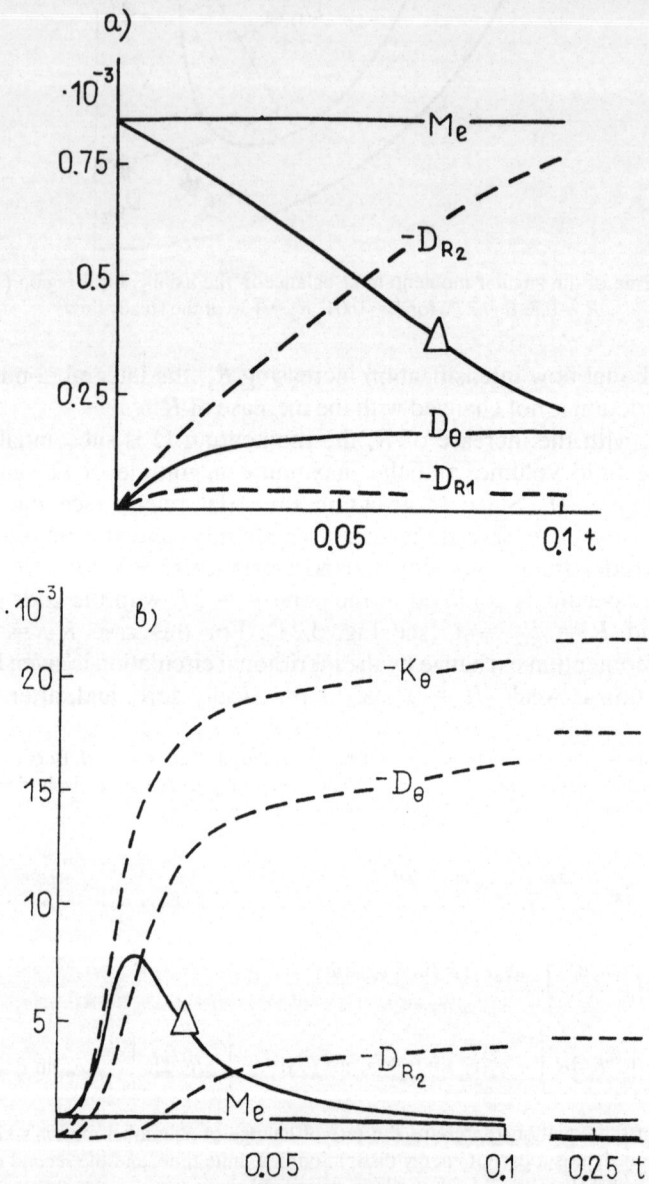

Fig. 3.22. Time development of the momentum transport and the incremental rate in the axial region ($R_1/R_2 = 0.1$, $R = 1$, $\theta = 27°$) after the current $I = I_1$ switch on at the time moment $t = 0$ for $k = 0.01$: (a) $R_s = 1$, (b) $R_s = 150$.

balance is essentially different. Note, first of all, the scale of the Fig. 3.22b is contracted 20 times relative to the previous case (Fig. 3.22a) to represent the momentum change in the same size figure. (Let us recall that Fig. 3.22 represents, according to (3.8.2), the nondimensional momentum change which is normalized by the quantity proportional to the meridional force $(f_{e\theta} \sim I^2)$ sustaining the motion). After the electric current switch-in, both the meridional flow and the rotation begin to grow from the rest. At the initial, very small time interval, when the meridional circulation is still weak, the momentum incremental rate Δ is again determined by the electromagnetic moment M_e in the volume V. With the time increase the convective momentum flux into the axial region $(-K_\theta)$ increases very rapidly, which causes further growth of the momentum incremental rate Δ and, respectively, the velocity of rotation Ω/r. At this the viscous friction of the highly rotating volume V increases in proportion to the incoming convective momentum, yet this means the viscous diffusion of momentum $D_\theta < 0$ and it is directed from the axial vortex region outside.

As shown in Fig. 3.22b, the momentum transport by viscous diffusion upstream the converging flow is the main factor compensating the huge convective momentum flux into the axial region. Thereby the momentum is not transported out of the axial region by the meridional recirculation ($K_\theta < 0$) and it can be accumulated in the region. The momentum increase rate Δ starts to fall only after the viscous friction $-D_{R_2}$ and $-D_{R_1}$ on the rigid walls step in action. This mechanism provides an explanation for the higher momentum growth rate in the vortex core region compared to the applied force $f_{e\varphi}$ generating the rotation: the momentum is not brought out by the meridional convection, it is set back by the upstream viscous diffusion, thereby accumulating in the vortex. The same mechanism of momentum accumulation may explain the acceleration of the rotating container with the fluid being discharged into the sink at its bottom: the fluid flow into the sink practically does not carry out the angular momentum, it is accumulated in the vortex, and is transferred by viscous friction to the container's walls.

An even more interesting possibility is suggested by the similarity solution for the limiting case $R_1/R_2 \to 0$ of electrically induced flow. From Fig. 3.12 it follows that the lower is the applied azimuthal force $f_{e\varphi}$, the higher is the momentum increase in the situation of the intense converging flow. This implies that even an arbitrary small, but nonzero, stationary perturbation will be accumulated at the symmetry axis and will lead to the rotating vortex formation. This may explain the vortex development at the critical Reynolds number in the experiments previously referred to [8, 15, 16, 33].

The analysis of the rotation intensification revealed that the flow at point electrode is in a certain sense optimal to the rotation development in the hemispherical container, since the rotation is amplified to the higher level, the smaller the electrode's surface area and the larger the fluid flow domain (Section 3.7). Therefore the exact solution for the meridional flow model not taking into account the instability to rotation led to the velocity field breakdown at the critical Reynolds number. Only by taking the spin-up process into account were we able to describe the converging flow due to the axisymmetric electric current diverging from a small electrode.

4

Flows with cylindrical symmetry

A new approach to the study of electrically induced flows is opened by the use of cylindrical coordinates. First, substantially more variants of axisymmetric electric current distributions (compared to the spherical case) are available, which, at least partly, have immediate practical implications. Second, in cylindrical coordinates there are several methods for the separation of variables in the Navier-Stokes and Maxwell equations that allow us to construct solutions in a similarity form. In choosing the coordinate system the role played by the geometrical form of the liquid conductor domain is by no means unimportant; a cylindrical form often occurs in practical high-current technology.

4.1. External electric current and magnetic field in cylindrical coordinates

In order to discover some of the feasible electric current distribution schemes consider the equation (1.10.19) which, in cylindrical coordinates, takes the form

$$\frac{\partial^2 \psi_1}{\partial z^2} + r \frac{\partial}{\partial r} \frac{1}{r} \frac{\partial \psi_1}{\partial r} = 0. \qquad (4.1.1)$$

Separating the variables in (4.1.1), i.e. assuming

$$\psi_1 = N(z)M(r), \qquad (4.1.2)$$

we obtain the equations for determining $N(z)$ and $M(r)$:

$$N'' - n^2 N = 0; \qquad (4.1.3)$$

$$r \frac{\partial}{\partial r} \frac{1}{r} M' + n^2 M = 0, \qquad (4.1.4)$$

where n is the separation constant. In the trivial case, $n = 0$, the solution is

$$\psi_1 = (Az + B)(Cr^2 + D). \qquad (4.1.5)$$

Similarly, separating the variables in the potential equation

$$\frac{1}{r} \frac{\partial}{\partial r} r \frac{\partial \Phi}{\partial r} + \frac{\partial^2 \Phi}{\partial z^2} = 0, \qquad (4.1.1a)$$

Flows with cylindrical symmetry

and assuming that the separation constant is equal to zero, we obtain

$$\Phi = (A_1 z + B_1)(C_1 \ln r + D_1). \quad (4.1.5a)$$

The pairs of particular solutions (4.1.5), (4.1.5a): $\psi_1 = $ const and $\Phi = $ const, $\psi_1 = z$ and $\Phi = \ln r$, $\psi_1 = r^2$ and $\Phi = z$ are related to each other and correspond to the current lines and to the equipotential lines represented in Table 1.1 (schemes No. 1, 3, and 5, respectively). The particular solution $\psi_1 = -zr^2$ with the potential $\Phi = z^2 - r^2/2$, which does not follow from (4.1.5a), correspond to the scheme No. 7. Some combinations of these solutions, e.g. No. 11 (with the limiting case No. 7) or No. 16 can be found in the Table. For a nonzero separation constant the solutions to (4.1.1) and (4.1.1a) will be discussed in Chapter 5. In particular, some of these permit us to consider a class of flows driven by a current that periodically varies in space.

Apart from the solutions (4.1.2) similarity solutions to the equation (4.1.1) may be sought. Let us seek the solutions in the form:

$$\psi_1 = A r^\alpha f_1(\eta), \quad (4.1.6)$$

where $\eta = B z^\beta r^\gamma$ and $A, B, \alpha, \beta, \gamma$ are constants. Substituting (4.1.6) in (4.1.1) we obtain

$$r^2 \beta [\beta \eta (\eta f_1')' - \eta f_1'] + z^2[(\alpha - 2)(\alpha f_1 + \gamma \eta f_1') + \\ + \gamma \eta (\alpha f_1 + \gamma \eta f_1')'] = 0. \quad (4.1.7)$$

It follows that similarity of the equation (4.1.1) occurs in the cases where either $\beta = 0$, or $\alpha = \gamma = 0$, or $\alpha = 2$, $\gamma = 0$. All these cases lead to the particular solutions from (4.1.5). The last possible variant is $\beta = 1$, $\gamma = -1$, i.e. $\eta = z/r$. Then (4.1.7) and (4.1.6) transform to

$$-\eta f_1' + \eta (\eta f_1')' + \eta^2[(\alpha - 2)(\alpha f_1 - \eta f_1') - \eta (\alpha f_1 - \eta f_1')'] = 0; \quad (4.1.8)$$

$$\psi_1 = A_1 r^\alpha f_1\left(\frac{z}{r}\right). \quad (4.1.9)$$

Recalling the relation between the cylindrical and spherical coordinates ($r = R \sin \theta$, $z/r = \cotan \theta$), we conclude that the solution (4.1.9) corresponds to the solution (1.7.10) of the equation (1.7.2) in spherical coordinates. Thus for $\alpha = 0$ we have from (4.1.8) $(1 + \eta^2) f_1'' + 3\eta f_1' = 0$ with the solution $f_1 = (A\eta/\sqrt{1 + \eta^2}) + B$, which corresponds to the situations shown in Table 2.1. For $\alpha = 1$ we have $(1 + \eta^2) f_1'' + \eta f_1' - f_1 = 0$ with the solution $f_1 = A\eta + B\sqrt{1 + \eta^2}$. The electric current lines, determined by this solution, coincide with the magnetic lines of force shown in Fig. 2.1.

4.2. Similarity solutions

According to the results of the previous section the variants of the electric current function, for which the equation (4.1.1) is ordinary, reduce to the following cases:

(1) $\psi_1 = A_1 f_1(Br)$;
(2) $\psi_1 = A_1 z f_1(Br)$;
(3) $\psi_1 = A_1 f_1(Bz)$; (4.2.1)
(4) $\psi_1 = A_1 r^2 f_1(Bz)$;
(5) $\psi_1 = A_1 r^\alpha f_1\left(\dfrac{z}{r}\right)$,

in the first four cases expressions for the functions f_1 follow from (4.1.5). It is evident from (4.2.1) there are three ways to define the similarity variable: $\eta = r/a$, $\eta = z/a$, $\eta = z/r$, where a is a representative length. Obviously, the similarity solutions of the equations (1.10.11)–(1.10.14), written in cylindrical coordinates:

$$\dfrac{\partial \psi}{\partial z} r \dfrac{\partial}{\partial r}\left(\dfrac{E^2\psi}{r^2}\right) - \dfrac{1}{r}\dfrac{\partial \psi}{\partial r}\dfrac{\partial E^2\psi}{\partial z} - \dfrac{\partial v_\varphi^2}{\partial z} + \nu E^4 \psi$$

$$= -\dfrac{\mu_0}{\rho}\dfrac{1}{r^2}\dfrac{\partial \psi_1^2}{\partial z} + \dfrac{1}{\rho\mu_0}\left[\dfrac{\partial \psi_2}{\partial z} r \dfrac{\partial}{\partial r}\left(\dfrac{E^2\psi_2}{r^2}\right) - \dfrac{1}{r}\dfrac{\partial \psi_2}{\partial r}\dfrac{\partial E^2\psi_2}{\partial z}\right];$$

$$\dfrac{1}{r}\dfrac{\partial \psi}{\partial z}\dfrac{\partial rv_\varphi}{\partial r} - \dfrac{1}{r}\dfrac{\partial \psi}{\partial r}\dfrac{\partial rv_\varphi}{\partial z} + \nu E^2 rv_\varphi$$

$$= \dfrac{1}{\rho}\dfrac{1}{r}\left(\dfrac{\partial \psi_2}{\partial z}\dfrac{\partial \psi_1}{\partial r} - \dfrac{\partial \psi_2}{\partial r}\dfrac{\partial \psi_1}{\partial z}\right); \qquad (4.2.2)$$

$$E^2\psi_1 = \sigma r\left[\dfrac{\partial \psi_2}{\partial z}\dfrac{\partial}{\partial r}\left(\dfrac{v_\varphi}{r}\right) - \dfrac{\partial \psi_2}{\partial r}\dfrac{\partial}{\partial z}\left(\dfrac{v_\varphi}{r}\right) + \right.$$

$$\left. + \mu_0 \dfrac{\partial \psi}{\partial r}\dfrac{\partial}{\partial z}\left(\dfrac{\psi_1}{r^2}\right) - \mu_0 \dfrac{\partial \psi}{\partial z}\dfrac{\partial}{\partial r}\left(\dfrac{\psi_1}{r^2}\right)\right];$$

$$E^2\psi_2 = \dfrac{\mu_0 \sigma}{r}\left(\dfrac{\partial \psi_2}{\partial z}\dfrac{\partial \psi}{\partial r} - \dfrac{\partial \psi_2}{\partial r}\dfrac{\partial \psi}{\partial z}\right),$$

Flows with cylindrical symmetry 157

where

$$E^2 = \frac{\partial^2}{\partial z^2} + r\frac{\partial}{\partial r}\frac{1}{r}\frac{\partial}{\partial r},$$

should be sought in the classes of functions depending on the same similarity variables. Consider consecutively the three variant ways to define the variables.

Variant I

For the case $\psi = Af(r/a)$ and $\psi_1 = A_1 f_1(r/a)$ (in the noninduction approximation $\psi_1 = Cr^2 + D$, as follows from (4.1.5)) the first equation (4.2.2) reduces to

$$E^4\psi = r\frac{\partial}{\partial r}\frac{1}{r}\frac{\partial}{\partial r}r\frac{\partial}{\partial r}\frac{f'}{r} = 0,$$

the solution of which $f' = rv_z/A = C_1 r^2 + C_2 \ln r + C_3$, describes the laminar flow in an annular tube. The uniform longitudinal current $j_z = rC$, $j_r = 0$ (No. 5 in Table 1.1) in this case is associated with the irrotational electromagnetic force $\partial \psi_1/\partial z = 0$, which does not affect the velocity field.

Setting in the first equation of (4.2.2) $\psi = vzf(\eta)$, $\psi_2 = 0$, $\psi_1 = A_1 zf_1(\eta)$, $\eta = r/a$, we obtain the equation describing electrically induced flow in the electrodynamic approximation:

$$f^{IV} - 2r^{-1}f''' + 3r^{-2}f'' - 3r^{-3}f' + r^{-1}ff''' - r^{-1}f'f'' - 3r^{-2}ff'' +$$
$$+ r^{-2}f'^2 - 3r^{-3}ff' = -\frac{2\mu_0 A_1^2}{\rho v^2 a^2}r^{-2}f_1^2, \qquad (4.2.3)$$

where $f_1 = Cr^2 + D$.

If the azimuthal velocity and the meridional magnetic field (specified by the function $\psi_2 = A_2 zf_2(\eta)$) are also added, this class of the similarity solutions determines the velocity and magnetic fields by the expressions:

$$\mathbf{v} = \{v_z, v_r, v_\varphi\} = \left\{\frac{v}{a}\frac{z}{r}f'(\eta), -\frac{v}{r}f(\eta), \frac{v}{a^2}z\Omega(\eta)\right\};$$

$$\mathbf{B} = \{B_z, B_r, B_\varphi\} = \left\{\frac{A_2}{a}\frac{z}{r}f'_2(\eta), -\frac{A_2}{r}f_2(\eta), \mu_0 A_1 zf_1(\eta)\right\}.$$
(4.2.4)

The set of equations (4.2.2) then takes the form:

$$f^{IV} - 2r^{-1}f''' + 3r^{-2}f'' - 3r^{-3}f' + r^{-1}ff''' - r^{-1}f'f'' -$$
$$- 3r^{-2}ff'' + r^{-2}f'^2 + 3r^{-3}ff' - 2\Omega^2$$
$$= -\frac{2\mu_0 A_1^2 a^2}{\rho v^2}r^{-2}f_1^2 + \frac{A_2^2}{\rho\mu_0 v^2}(r^{-1}f_2 f_2''' -$$
$$- r^{-1}f'_2 f''_2 - 3r^{-2}f_2 f''_2 + r^{-2}f'^2_2 + 3r^{-3}f_2 f'_2); \qquad (4.2.5)$$

$$r\left[\frac{1}{r}(r\Omega')\right]' + r^{-1}[f(r\Omega)' - f'r\Omega] = \frac{A_1A_2a}{\rho v^2} r^{-1}(f_2f_1' - f_2'f_1); \quad (4.2.6)$$

$$f_1'' - r^{-1}f_1' = \beta(2r^{-3}ff_1 + r^{-2}f_1f' - r^{-2}f_1'f) +$$

$$+ \frac{\sigma A_2 v}{A_1 a}\left[f_2\left(\frac{\Omega}{r}\right)' - f_2'\frac{\Omega}{r}\right]; \quad (4.2.7)$$

$$f_2'' - r^{-1}f_2' = \beta r^{-1}(f_2f' - f_2'f). \quad (4.2.8)$$

The set of equations (4.2.5), (4.2.6) for $f_1 = 0$ and $f_2 = 0$, was obtained by Smirnov [52], and the equations for $\Omega = 0$ are presented in [61]. It should be noted that the class of similarity solutions considered has rarely been used in hydrodynamics. In [61] it was used to analyse nonrotating flow in a porous tube and in [52] for rotating flows in a tube with permeable and impermeable walls and with a closed end. Additionally, in [52], the solution obtained was interpreted as a flow reversal in the dead end of the tube. Flow in the tube with an accelerating surface velocity was considered in [8], and the flow development in a porous tube in [9], the latter also discusses the similarity assumption.

It is interesting to note that, specifying the functions ψ and ψ_2 in the form $\psi = vzf(\eta)$, $\psi_2 = A_2zf_2(\eta)$, the similarity of the initial equations also remains for the azimuthal velocity and the electric current function in the form $v_\varphi = C\Omega(\eta)$, $\psi_1 = A_1f_1(\eta)$. However, in this case the curl of the centrifugal force $(\partial/\partial z)(v_\varphi^2/r)$ and the curl of the electromagnetic force for the self-magnetic field (see 1.6.15) are identically zero and do not affect the meridional flow. Examples of the flows with azimuthal velocity $v_\varphi = C\Omega(\eta)$ may be found in [15, 26] and in the review [16].

Electrically induced flows are not considered in this class of solutions. Thus, we merely restrict ourselves to the statement of the class and express a hope that interesting problems of electrically induced flows will be solved in near future.

Variant II

For the similarity variable $\eta = z/a$ the only possible structure of the functions ψ and ψ_1 is $\psi = Ar^2f(z/a)$, $\psi_1 = A_1r^2f_1(z/a)$. Substituting these expressions in the first equation of (4.2.2) we obtain the equation in the electrodynamic approximation

$$-f^{IV} + \frac{2Aa}{v} ff''' = \frac{2\mu_0 A_1^2 a^3}{\rho v A} f_1f_1', \quad (4.2.9)$$

where $f_1 = Az + B$.

In general case with the azimuthal velocity $v_\varphi = Cr\Omega(\eta)$ and the external

Flows with cylindrical symmetry

meridional magnetic field, determined by $\psi_2 = A_2 r^2 f_2(\eta)$, the velocity field and the magnetic field is defined by

$$\mathbf{v} = \{v_z, v_r, v_\varphi\} = \left\{2Af(\eta), -\frac{A}{a} rf'(\eta), Cr\Omega(\eta)\right\};$$

$$\mathbf{B} = \{B_z, B_r, B_\varphi\} = \left\{2A_2 f_2(\eta), -\frac{A_2}{a} rf'_2(\eta), \mu_0 A_1 rf_1(\eta)\right\}, \quad (4.2.10)$$

and the set of equations (4.2.2) takes the form:

$$-f^{IV} + \frac{2Aa}{\nu} ff''' + 2\frac{C^2 a^3}{\nu A} \Omega\Omega'$$
$$= \frac{2\mu_0 A_1^2 a^3}{\rho\nu A} f_1 f'_1 + \frac{2A_2^2 a}{\mu_0 \rho\nu A} f_2 f'''_2; \quad (4.2.11)$$

$$\Omega'' + \frac{2Aa}{\nu}(f'\Omega - f\Omega') = \frac{2A_1 A_2 a}{\rho\nu C}(f'_2 f_1 - f_2 f'_1); \quad (4.2.12)$$

$$f''_1 = -\frac{2A_2 \sigma Ca}{A_1} f_2 \Omega' + 2A\mu_0 a\sigma f'_1 f; \quad (4.2.13)$$

$$f''_2 = 2\mu_0 \sigma Aa(ff'_2 - f'f_2). \quad (4.2.14)$$

The equation (4.2.11) may be integrated once:

$$-f''' + \frac{2Aa}{\nu}\left(ff'' - \frac{1}{2} f'^2\right) + \frac{C^2 a^3}{\nu A}\Omega^2 - \frac{\mu_0 A_1^2 a^3}{\rho\nu A} f_1^2 -$$
$$- \frac{2A_2^2 a}{\mu_0 \rho\nu A}\left(f_2 f''_2 - \frac{1}{2} f'^2_2\right) = \lambda \quad (4.2.11a)$$

The integration constant λ determines the pressure distribution

$$(p - p_0) = \frac{2\rho\nu A}{a}\left[\frac{\lambda}{4}\frac{r^2}{a^2} + F\left(\frac{z}{a}\right)\right], \quad (4.2.11b)$$

which can be easily obtained from the *r*-component of the Navier-Stokes equation. The equation to determine the function $F(\eta)$ is obtained from the z-component of the equations of motion:

$$f'' - \frac{2Aa}{\nu} ff' - F' = 0.$$

The set of equations (4.2.11)–(4.2.14) was derived by Sichev in 1960 [51].

For $f_1 = 0$ and $f_2 = 0$ (i.e. in the absence of a magnetic field) the similarity equations (4.2.11), (4.2.12) are widely used in hydrodynamics to study rotating fluids and also axisymmetric flows at a critical point on plane wall; historically the first study was made by Karman [25] of the flow at an infinite rotating disc. The Karman class was generalized to the case of heat convection in [37]. Batchelor applied the solution class to the flow between two parallel coaxially rotating planes [4]. Recently Karman's class has drawn attention by the numerical studies of the bifurcating and nonunique solutions [27].

In magnetohydrodynamics the Karman class, the generalization of which is the class of exact solutions of the 'layer type' introduced by Lin [29], is mainly applied to the analysis of the flows at a front critical point with external magnetic field. As follows from the solution of the equation (4.2.2) with zero right-hand side, which coincides with the solution (4.1.5) for ψ_1 (for this class $D = 0$ should be fixed), the feasible configurations of the external field also include the practically important case of uniform axial magnetic field ($A = 0$, $\psi_2 = Cr^2$, $B_z = 2C$, $B_r = 0$). The flow at a front critical point with this field was considered by Kakutani [24] and by Axford [2] (the review see [2]), the rotating flow of conducting liquid above a stationary infinite disc in approximation of boundary layer has been described in [14, 28] and elsewhere. From among the other studies note the study [55], which considers the flow between rotating and stationary discs in a uniform field. Electrically induced flows of this class will be considered in Section 4.3.

Variant III

As follows from Section 4.1 the similarity solutions for the electric current function $\psi_1 = A_1 r^\alpha f_1(\eta)$, $\eta = z/r$ may be constructed for any α. For the full set of Navier-Stokes and Maxwell equations it can be shown* that the similarity solutions require $\alpha = 1$ for the stream function (also for the magnetic function), and $\alpha = 0$ for the electric current function. Consequently, the flow class is defined by the functions (we specify $A = -\nu$):

$$\psi = -\nu r f(\eta), \qquad \psi_1 = A_1 f_1(\eta), \qquad \psi_2 = -A_2 r f_2(\eta),$$

---

* For instance, by equating the r powers in the inertial and viscous terms in the equation of motion.

Flows with cylindrical symmetry

the corresponding velocity and magnetic fields are

$$\mathbf{v} = \left\{ -\frac{v}{r}(f - \eta f'), \frac{v}{r} f', \frac{v}{r} \Omega \right\};$$

$$\mathbf{B} = \left\{ -\frac{A_2}{r}(f_2 - \eta f'_2), \frac{A_2}{r} f'_2, \frac{\mu_0 A_1}{r} f_1 \right\}. \tag{4.2.15}$$

The set of equations (4.2.2) in this case is rewritten in the form:

$$(1 + \eta^2)^2 f^{IV} + 10\eta(1 + \eta^2)f''' + (21\eta^2 + 6)f'' + 3\eta f' - 3f +$$
$$+ (1 + \eta^2)ff''' + 3(1 + \eta^2)f'f'' + 3\eta ff'' + 3\eta f'^2 - 3ff' + 2\Omega\Omega'$$
$$= \frac{\mu_0 A_1^2}{\rho v^2} f_1 f'_1 + \frac{A_2^2}{\rho \mu_0 v^2} [(1 + \eta^2)f_2 f'''_2 + 3(1 + \eta^2)f'_2 f''_2 +$$
$$+ 3\eta f_2 f''_2 + 3\eta f'^2_2 - 3f_2 f'_2]; \tag{4.2.16}$$

$$(1 + \eta^2)\Omega'' + 3\eta\Omega' + f\Omega' = \frac{\mu_0 A_1 A_2}{\rho v^2} f_2 f'_1; \tag{4.2.17}$$

$$(1 + \eta^2)f''_1 + 3\eta f'_1 = -\beta(2f'f_1 + ff'_1) + \frac{\sigma v A_2}{A_1}(2f'_2\Omega + f_2\Omega'); \tag{4.2.18}$$

$$(1 + \eta^2)f''_2 + \eta f'_2 - f_2 = -\beta(ff'_2 - f'f_2). \tag{4.2.19}$$

The equation (4.2.16) can be integrated three times. The final result is

$$(1 + \eta^2)f' + \eta f + \frac{1}{2}f^2 +$$
$$+ \eta\sqrt{1 + \eta^2} \int \frac{2\eta^2 + 1}{\sqrt{1 + \eta^2}} \Omega^2 \, d\eta - (2\eta^2 + 1) \int \eta\Omega^2 \, d\eta$$
$$= 2b\eta\sqrt{1 + \eta^2} - 2a(1 + \eta^2) + c +$$
$$+ \frac{\mu_0 A_1^2}{2\rho v^2} \left[\eta\sqrt{1 + \eta^2} \int \frac{2\eta^2 + 1}{\sqrt{1 + \eta^2}} f_1^2 \, d\eta -$$
$$- (2\eta^2 + 1) \int \eta f_1^2 \, d\eta \right] + \frac{A_2^2}{2\rho\mu_0 v^2} f_2^2. \tag{4.2.20}$$

The considered class of solutions is closely related to the class of exact solutions in spherical coordinates (Section 2.1). Indeed, comparing the fields (2.1.7)–(2.1.9) to (4.2.15) with $\eta = \cotan \theta$ it is easy with the substitutions

$$f(\eta) = -\frac{g(\theta)}{\sin \theta}, \quad \Omega(\eta) = \ell(\theta), \quad f_1(\eta) = L(\theta), \quad f_2(\eta) = -\frac{G(\theta)}{\sin \theta}$$

to reduce the equations (4.2.17)–(4.2.20) to the equations (2.1.13)–(2.1.16). This means that all the problems in spherical coordinates can also be considered in cylindrical coordinates. Yet this is not the central argument for the advantage of equations (4.2.17)–(4.2.20). We use them for the following two reasons. The first is that some problems are more naturally set in cylindrical coordinates, and this could lead to the statement of new problems in ordinary and magnetic hydrodynamics. Second, a lot of axisymmetric problems in boundary-layer theory have been solved in cylindrical coordinates. This permits us to apply the known results to the analysis of the asymptotic behaviour of the solutions to the above equations, and in special cases to study the flows in exact formulation, i.e. for arbitrary Reynolds numbers (see Section 2.4). We shall not discuss this class in detail, since it is identical to the Landau-Yatseyev-Squire class discussed in Chapter 2. In addition to the papers cited in the review in Section 2.4, the set of equations (4.2.16)–(4.2.19) has been applied to the flows with a line mass source or a sink in a conical domain [48]. The analogous problems in spherical coordinates were considered in [17] (line source (sink) normal to a plane), [13] (vortex line and line source), [39] (effect of suction at the plane wall). Variants of the magnetohydrodynamic problem with the line source were studied in [48, 49] (the source with radial magnetic field), and [48] (the source in the azimuthal magnetic field of a line wire, including the problem of magnetic field generation). The electrically induced flow with the line source (sink) is considered in Section 4.4.

4.3. Electrically induced flow between two parallel walls

4.3.1. Flow in unbounded layer

We begin the study of electrically induced flows in the Karman solution class (i.e. for the similarity variable $\eta = z/a$) with the simplest situation in which the flow is driven by the electric current specified by the function $\psi_1 = A_1 r^2 f_1(\eta) = A_1 r^2 (z/a)$ (cf. (4.1.5) with $B = D = 0$). The current distribution (case No. 7 in Table 1.1) is the analogue of inviscid flow at a front critical point.

Suppose the flow domain is bounded by two parallel planes separated from each other at a distance a (Fig. 4.1); there is neither an external magnetic field, nor a rotation ($f_2 = 0$, $\Omega = 0$). We also neglect the induced by motion electric current. The latter assumption permits us to determine the function $f_1 = \eta = z/a$. As a representative current quantity it is assumed that the total current I crossing the plane $z = a$ inside the circle $r = a$:

$$I = -2\pi \int_0^a j_z \, r \, dr = -2\pi \int_0^a \left(\frac{1}{r} \frac{\partial \psi_1}{\partial r} \right)_{z=a} r \, dr = -2\pi a^2 A_1,$$

Flows with cylindrical symmetry 163

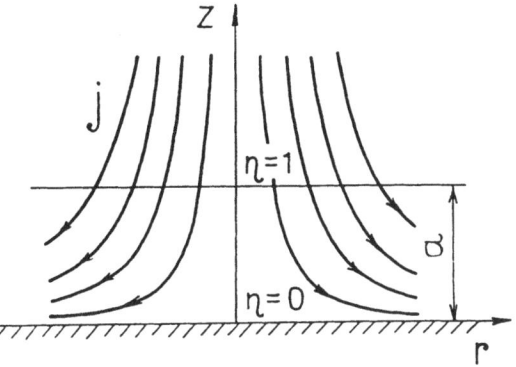

Fig. 4.1. Scheme of the current passage at the critical point $r = 0$, $z = 0$.

from this $A_1 = -I/2\pi a^2$. The same current also penetrates the cylindrical surface $r = a$, $0 \leq z \leq a$:

$$I = 2\pi a \int_0^a j_r \, dz = 2\pi a \int_0^a \left(-\frac{1}{r} \frac{\partial \psi_1}{\partial z}\right)_{r=a} dz = -2\pi a^2 A_1.$$

The similarity constant A in the expression for stream function $\psi = Ar^2 f(z/a)$ is specified by use of the distance a and the kinematic viscosity v: $A = v/2a$. Substituting A, A_1 and a in (4.2.11) and (4.2.11a), we arrive at

$$ff''' - f^{IV} = S\eta; \tag{4.3.1}$$

$$2f''' - 2ff'' + f'^2 + S\eta^2 - 2\lambda = 0, \tag{4.3.1a}$$

where $S = -\mu_0 I^2/\pi^2 \rho v^2$. The boundary conditions for viscous flow are (see (4.2.10))

$$f(0) = f'(0) = 0; \tag{4.3.2}$$

$$f(1) = f'(1) = 0, \tag{4.3.3}$$

at rigid walls $\eta = 0$, $\eta = 1$. If, for instance, the surface $\eta = 1$ is free, the conditions (4.3.3) should be replaced by

$$f(1) = f''(1) = 0, \tag{4.3.4}$$

expressing the zero normal velocity and zero tangential stress. The flow model may be applied to electroslag and arc melting in the regime when the current flows to the lateral wall of a crystallizer mould.

The first attempts to solve the boundary value problem (4.3.1a)–(4.3.4) were made using the power series expansion [6]:

$$f = \sum_{n=0}^{\infty} \frac{f^{(n)}(0)\eta^n}{n!} \tag{4.3.5}$$

containing two undetermined parameters $f''(0)$ and $f'''(0)$. These parameters were determined in [6] by the iterative procedure, the initial values being found from the Stokes approximation:

$$f''(0) = -\tfrac{4}{120}S, \qquad f'''(0) = \lambda = \tfrac{6}{40}S \qquad (4.3.6)$$

for the boundary conditions (4.3.3), and

$$f''(0) = -\tfrac{7}{120}S, \qquad f'''(0) = \lambda = \tfrac{9}{40}S \qquad (4.3.7)$$

for (4.3.4).

The flow stream lines $r = C/\sqrt{f}$ and velocities are presented in Fig. 4.2. The flow is in the form of a toroidal vortex with the motion at the centre directed downward (i.e. in the direction of current density decrease). The flow in Karman's class is characterized by the vertical velocity v_z, being independent of the radius, and the radial velocity v_r growing linearly with r (cf. (4.2.10)).

Comparing the expressions (4.3.6) and (4.3.7) related to the friction and pressure at the plane $z = 0$ (constant λ determines the pressure (4.2.11b)), it follows that the flow intensity is higher for the free surface case. The radial velocity on the free surface can be evaluated as

$$v_r(r, z = a, S) = \frac{v}{a^2} r \left[-\frac{1}{2} f'(1) \right] = -\frac{v}{a^2} r \frac{S}{160}.$$

The power series (4.3.5) provides the solution up to approximately $S = 700$. However, in this range of S the solution hardly deviates from the linear Stokes solution. Indeed, the Reynolds number for $S = 700$ is of the order of unity:

$$Re = \frac{vL}{v} = \frac{v_{z_{max}} a}{v} = \psi_{max} \approx S \times 1.3 \times 10^{-3} \approx 1.$$

Yet for higher S the iterative procedure fails to converge to the solution [6].

In an attempt to extend the solution for higher S and to study the nonlinear effects a semianalytic method was applied [5], which was proposed in a series of papers by Van Dyke (see the review [57]). The method is based on high order

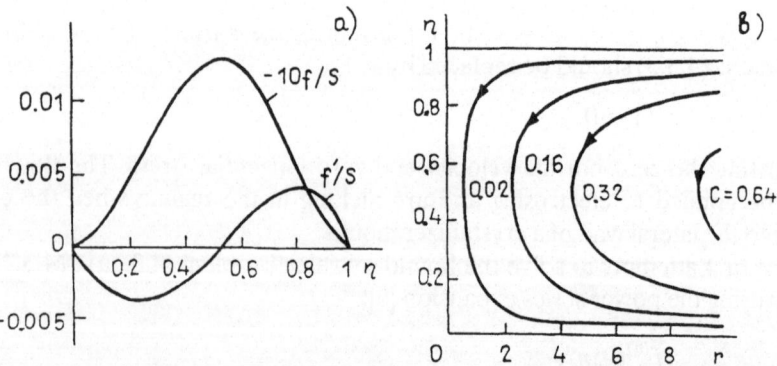

Fig. 4.2. (a) Axial ($f = av_z/v$) and raidal ($f' = -2a^2 v_r/vr$) velocities, (b) stream lines for $S = 100$.

Flows with cylindrical symmetry 165

perturbation series calculated with the computer. Reaching the convergence limiting *Re* number, the structure of the solution is analysed in the artificially extended complex plane of the *Re* parameter; on this basis the perturbation series are rebuilt taking into account the newly found complex singularities which precluded the derivation of a solution using the earlier direct perturbation series. Reference [57] discusses successful applications of the method in hydrodynamics.

The present problem contains the parameter S, an analogue to the Re number, therefore a solution is sought in the form

$$f = \sum_{n=1}^{\infty} \varphi_n(\eta) S^n, \tag{4.3.8}$$

where φ_n are functions of η and do not depend on S. This is an advantage of the method compared to the former one, since $\varphi_n(\eta)$ are computed once and for all S. In addition, $\varphi_n(\eta)$ can be expressed in a closed form, i.e. the coefficients of the series (4.3.8) do not depend on the accuracy of the iterative process.

Substituting (4.3.8) in (4.3.1) and equating the coefficients at like powers of S, we find first the linear Stokes solution ($n = 1$) $\varphi_1 = \Sigma_{m=0}^{5} B_m^1 \eta^m$, $B_4^1 = 0$, $B_5^1 = -1/120$, and for the boundary conditions (4.3.2), (4.3.3) $B_0^1 = 0$, $B_1^1 = 0$, $B_2^1 = -1/60$, $B_3^1 = 1/40$, which coincides with (4.3.5) truncated at $n = 5$. The subsequent φ_n in (4.3.8) are determined from the equation

$$\varphi_n^{IV} - \sum_{K=1}^{n-1} \varphi_K \varphi_{n-K}''' = 0, \quad n \geq 2$$

the solution to which is also the power series

$$\varphi_n = \sum_{m=2}^{6n-1} B_m^n \eta^m, \tag{4.3.9}$$

the coefficients B_m^n for $m \geq 5$, $n \geq 2$ are determined recursively, knowing B_m^n for smaller n [7]. B_2^n and B_3^n are determined by the boundary conditions.

The actual numerical procedure was restricted to the perturbation order $n = 20$. The sequences of the coefficients B_2^n, B_3^n, B_8^n are used to detect independently the convergence radius $S^* = 3500$, however the nearest singularity in the complex S plane is not located on the real axis, i.e. there is in principle no physical limitation to the convergence radius.

The numerical results demonstrate for $S \leq S^*$ that the solution is still close to linear ($n = 1$). Nevertheless, nonlinear tendencies are experienced. Figs. 4.3a, b show the velocity components normalized by S and varying with S. With increasing S there is an initial growth of the velocities compared to the Stokes solution yet, after the maximum is reached, the normalized velocities decrease. The friction at the wall $z = 0$ (Fig. 4.3c) grows monotonously with S, which indicates a substantial growth of velocity gradient at the wall. Coupled to the

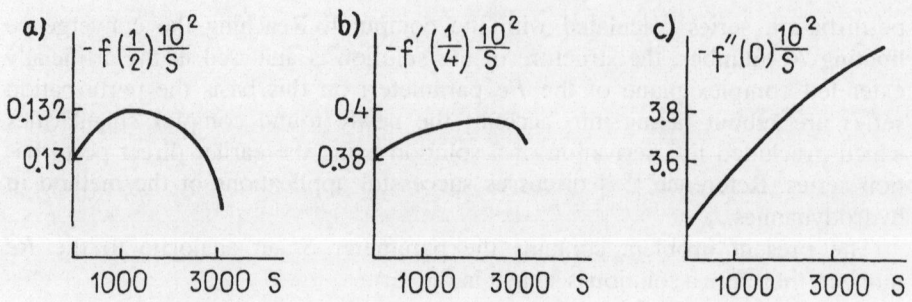

Fig. 4.3. Velocity components and friction vs. the parameter S. (Normalized by S.)

fact that the velocity grows more slowly than the gradient at the wall, this suggests that the boundary layer at $z = 0$ is beginning to develop.

The physical mechanism of the boundary layer formation at the wall $z = 0$ is obvious. The vorticity curl $\mathbf{v} = -\nu/2a^3 r\varphi''\mathbf{i}_\varphi$, generated by both the electromagnetic force and the viscous interaction, is transported by diffusion from the wall inside the fluid, but convection, directed to the wall $z = 0$, retains the vorticity at the wall. At the other wall $z = a$ both convection and diffusion are directed away from the wall, hence the boundary layer cannot grow. Consequently, the flow at high S cannot be described by perturbations of an inviscid flow and boundary layers.

The eruption of vorticity by convection (in this example from the wall $z = a$) indicates the deficiency of viscous effects in limiting the vorticity growth in the flow at the wall $z = a$. As was demonstrated by analysing the electrically induced flow at a point electrode (Sections 3.1, 3.2), this may lead in the extreme case to unlimited velocity growth at a finite S. The present flow compensates the friction deficit at the wall $z = a$ by the increase at the wall $z = 0$, thus the crisis in the flow is eliminated.

4.3.2. Flow in a radially constrained layer

We have at our disposal a possibility to estimate the relationship of the similarity solution (4.3.8) to a more real situation where the flow domain is bounded by the cylindrical wall $r = b$ [5]. For this purpose the first equation (4.2.2) was solved numerically. The expression for the electromagnetic force, determined by the function ψ_1, was the previous $\psi_1 = Ar^2z/a$, the boundary conditions being $\mathbf{v} = 0$, $\psi = 0$ on the walls and zero vorticity at the symmetry axis: $\omega = r^{-1}E^2\psi = 0$. The solution procedure involved iterations with the use of the relaxation parameter $0 < \alpha \leq 1$ for ω at the wall Γ: $\omega^{n+1}|_\Gamma = (1 - \alpha)\omega^n|_\Gamma + \alpha\omega_T$, where n is the iteration number; ω_T was derived from the second order difference approximation of $E^2\psi = r\omega$.

The computed stream lines $\psi = C$ in a meridional plane are shown in Fig. 4.4. The flow domains are transformed by $r = r_k/k$ to allow comparison of the flows at different ratios $k = a/b$ of the height to the radius of the container. The new parameter $S_k = S/k^4$ is also introduced, and it has the following meaning:

Flows with cylindrical symmetry

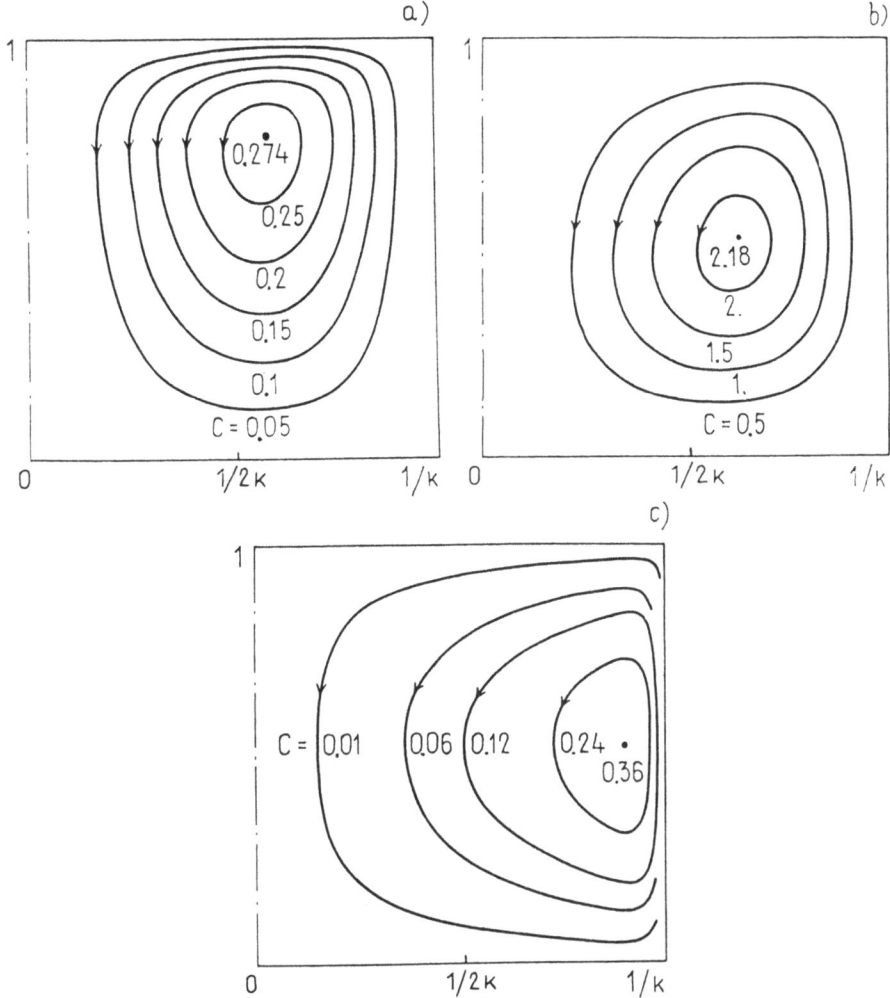

Fig. 4.4. Stream lines $\psi = C$ in the cylindrical container (a) $S_k = 2.5 \times 10^4$, $k = 5$, (b) $S_k = 2.5 \times 10^4$, $k = 1$, (c) $S_k = 700$, $k = 0.1$ ($\psi = 10^{-2} C$).

the flows with the equal S_k, but different ratios k, are driven by equal total currents supplied across the top $z = 1$, yet the equality of S means the same current densities at the top. Comparing Figs. 4.4a and b it is evident that the flow intensity, indicated by the maximum magnitude of ψ, for equal total currents is higher for $k = 1$ than for $k = 5$ (i.e. in the radially stretched domain). The numerical results also show that the vortex centre is stationary, being located with the increase of S, yet its position depends significantly on the stretching of the domain (i.e. on k): on stretching the domain in r (decreasing k), the vortex centre is shifted to the lateral wall $r = k^{-1}$, the stream lines for the stretched domain tend to be more similar to the flow in the radially unbounded layer (Fig. 4.4c).

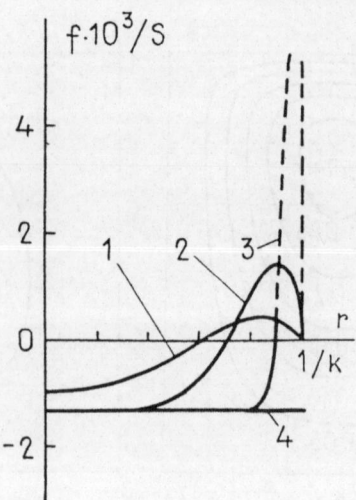

Fig. 4.5. Axial velocity in the plane $\eta = 0.5$ vs. the radial distance: (1) $k = 1$, $S = 5 \times 10^3$, (2) $k = 0.5$, $S = 1.56 \times 10^3$, (3) $k = 0.1$, $S = 0.07$, (4) $k = 0$ (similarity solution).

Fig. 4.5 demonstrates the similarity of velocity field in the growing radial portion of the container with the stretching. The axial velocity obtained by the similarity solution is independent of r (straight line 4 in Fig. 4.5). For $k = 0.1$ approximately 4/5 of the flow region coincides with the similarity flow; only in the vicinity of lateral wall does the flow in the finite container quite abruptly change its direction to close the stream lines. The flows behave in the same manner for the whole interval of S considered in this study.

4.3.3. Other solutions in Karman's class

The problem of fluid flow in a front critical point with the far field of the form

$$v_z = -2v_0 \frac{z}{a}, \qquad v_r = v_0 \frac{r}{a}$$

was investigated in [2] for the two cases of azimuthal magnetic field:

(1) the field associated with the electric current supplied to the fluid through the electrically conducting plane;
(2) and to the current flowing onto the dielectric plane.

(The two cases differ merely by the coefficients A and B specifying the expression for unperturbed field $f_1 = Az + B$.) Since in the present situation we have the representative velocity v_0 instead of the length scale, then, in contrast to the previous problem, the constants in (4.2.9) are specified in the form $A = -v_0$, $a = \nu/v_0$. Apart from this, the problem is set up in the full formulation including the induced currents, i.e. with the addition of equation (4.2.13) with $\Omega = 0$. For this choice of the constants the coefficient in right-hand side of (4.2.13) has the meaning of the Batchelor number:

$$2A\mu_0 a\sigma = -2\mu_0 \nu\sigma = -2\beta.$$

Flows with cylindrical symmetry

However, the solution in [2] was obtained for $\beta = 1$ only, i.e. for conductivities which are unreal in available experimental conditions. Hence we do not discuss the results.

For the same reason the results of [40], in which the rotating plane with the azimuthal magnetic field is investigated, will not be discussed here. However, note that the set of equations for this case can be derived from (4.2.11)–(4.2.13) assuming there

$$f_2 = 0, \quad A = -(\nu\omega)^{1/2}, \quad a = (\nu/\omega)^{1/2}, \quad C = \omega,$$

where ω is a given angular velocity of the rotating plane.

4.4. Flow with a line source in a circular cone

This and the following section discuss the flows described by the class of solutions (4.2.15). As we have noted previously, the class is identical to the solutions of (2.1.7)–(2.1.9) and therefore, to avoid repetition, we shall consider the new flows which are naturally expressed in cylindrical coordinates.

The flow discussed below is related to the axisymmetric flow in a diffuser [48]. The scheme is shown in Fig. 4.6: the line source of radial velocity coincides with the z-axis, and the source ends at the apex of a circular rigid cone, which in the particular case ($\gamma_0 = 0$) is the plane surface.

In addition to the pure hydrodynamic flow investigated in [17, 48], a point electric current source is placed in the cone's apex. The electric current field is determined by (4.2.18) assuming $\beta = 0, \Omega = 0$:

$$f_1 = \left(1 - \frac{\alpha}{\sqrt{1+\alpha^2}}\right)^{-1}\left(1 - \frac{\eta}{\sqrt{1+\eta^2}}\right), \quad (4.4.1)$$

where $\alpha = \tan \gamma_0$.

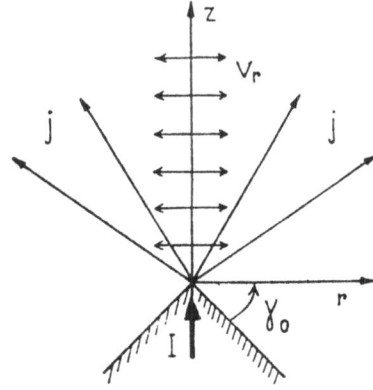

Fig. 4.6. The flow model with a line source of velocity and a point source of electric current.

On substituting (4.4.1) in (4.2.20) and evaluating the integrals on the right-hand side, the following problem is formulated:

$$(1 + \eta^2)f' + \eta f + \tfrac{1}{2}f^2 = 2b\eta\sqrt{1 + \eta^2} - 2a(1 + \eta^2) + c +$$
$$+ S[(1 + 2\eta^2 + 2\eta\sqrt{1 + \eta^2})(\tfrac{1}{2}\ln(1 + \eta^2) -$$
$$- \ln(\eta + \sqrt{1 + \eta^2})) + 2\eta^2 + \eta\sqrt{1 + \eta^2}], \quad (4.4.2)$$

where

$$S = \frac{\mu_0 I^2}{2\rho v^2 \left(1 - \dfrac{\alpha}{\sqrt{1 + \alpha^2}}\right)^2}.$$

The boundary conditions are the impermeability and no slip on the cone's surface

$$f(\alpha) = f'(\alpha) = 0, \quad (4.4.3)$$

the prescribed source strength (the flow rate Q per unit length)

$$\lim_{r \to 0} \int_0^{2\pi} rv_r \, d\theta = Q = 2\pi v f'(\infty) = 2\pi v Re,$$

or

$$f'(\infty) = Re, \quad (4.4.4)$$

and the finiteness of the axial velocity on the symmetry axis

$$\lim_{r \to 0} v_z = \lim_{r \to 0} \frac{v}{r}(\eta f' - f) = \lim_{\eta \to \infty} \frac{v}{z}\eta(\eta f' - f) = \frac{v}{z}D. \quad (4.4.5)$$

The constants a, b, c in (4.4.2) are determined by these boundary conditions. The first of these (4.4.3) gives

$$2ba\sqrt{1 + \alpha^2} - 2a(1 + \alpha^2) + c$$
$$= -S[(1 + 2\alpha^2 + 2\alpha\sqrt{1 + \alpha^2})(\tfrac{1}{2}\ln(1 + \alpha^2) -$$
$$- \ln(\alpha + \sqrt{1 + \alpha^2})) + 2\alpha^2 + \alpha\sqrt{1 + \alpha^2}], \quad (4.4.6)$$

In order to apply the conditions (4.4.4), (4.4.5), consider the asymptotic form of f for $\eta \to \infty$:

$$f = A_0\eta + A_1\frac{\ln \eta}{\eta} + A_2 + A_3\frac{1}{\eta} + \dots. \quad (4.4.7)$$

Then (4.4.4) determines $A_0 = Re$, and (4.4.5) gives $A_1 = 0$, $A_2 = 0$, $A_3 =$

Flows with cylindrical symmetry 171

$-D/2$. On substituting (4.4.7) in (4.4.2) and passing to the limit $\eta \to \infty$, one can obtain the following relations between the constants:

$$2b - 2a = 2Re + \tfrac{1}{2}Re^2 + S(4 \ln 2 - 3),$$

$$b - 2a + c = Re\left(1 - \frac{D}{2}\right) + 2S(4 \ln 2 - 3).$$

(4.4.8)

From these it follows that

$$2b = \left(a^2 + \frac{1}{2} - a\sqrt{1 + a^2}\right)^{-1}\left[Re\left(1 + 2a^2 - \frac{D}{2}\right) + \right.$$

$$+ \frac{Re^2}{2}a^2 + S\left\{(4 \ln 2 - 3)(2 + a^2) + \right.$$

$$+ (1 + 2a^2 + 2a\sqrt{1 + a^2})\left(\frac{1}{2}\ln(1 + a^2) - \right.$$

$$\left.\left.- \ln(a + \sqrt{1 + a^2})\right) + 2a^2 + a\sqrt{1 + a^2}\right\}\Bigg],$$

(4.4.9)

$$2a = 2b - 2Re - \frac{1}{2}Re^2 - S(4 \ln 2 - 3),$$

$$c = b - \frac{1}{2}Re^2 - Re\left(1 + \frac{D}{2}\right) + S(4 \ln 2 - 3).$$

It is evident from the expressions (4.4.9) that the constants a, b, c are determined by the constant D, related to the nondimensional axial velocity (4.4.5) and generally unknown prior to the solution (a, S, and Re values are set). This situation requires us to set an estimated value for D in (4.4.9), then integrate (4.4.2) and find the function (4.4.5) for $\eta \to \infty$. Then the value of D is corrected until it coincides with the prescribed condition (4.4.5); if this is satisfied, the solution is found.

However, in some cases, such a long-winded procedure is not necessary. As follows from (4.4.8) or (4.4.9), the constant D enters only by the combination ReD. Hence for $Re = 0$, i.e. in the absence of a line source the constants are determined, and D is obtained as the result of numerical integration of (4.4.2) for the prescribed S. This case was studied in Section 2.5 for EVF at a point electric current source. Another example of flow in which the axial velocity is set by physical considerations, and thereby the value of D, is the flow at the line source in the absence of electric current ($S = 0$). In this case it is sensible to set

$D = 0$. Then the constants a, b, c are completely determined by tne geometric configuration (i.e. α) and by the source strength (Re).

The pure hydrodynamic flow ($S = 0$) was investigated in [7] and first set up in [48]. The main results are the following. If $D = 0$ is set in (4.4.5), then three kinds of solution are possible in the plane (α, Re): (1) there is a region in the plane where the solution exists, satisfying all the set conditions, including also the asymptotic requirement (4.4.5); (2) for the negative Re, which means the line sink, only those solutions are realized with the asymptotic behaviour $f'(\infty) = -Re - 4$; (3) for negative α there is a region where the finite solution does not exist.

The specific properties of the first kind of flows are the following. For $\alpha \geq 0$ the profile of radial velocity has an extremum near the cone's surface. If $\alpha < 0$, a detached reversed flow may develop, the detachment begins when

$$\alpha = -\frac{2}{Re} \frac{1}{\sqrt{1 + 4/Re}}.$$

From this relation it follows that for high Re the flow at the plane is close to a critical state: an infinitesimal deviation of the surface inclination towards negative α leads to flow detachment. For high Re the boundary layer solution is possible [48]. This flow at the plane surface has practically zero friction.

The most interesting results are obtained for $\alpha = 0$. Fig. 4.7 shows the regions in the plane (D, Re) where the solution of the problem (4.4.2)–(4.4.5) exists. For $-2 < Re < 0$ the solution of the first kind exists, yet for each Re the value of $D < 0$ is uniquely determined; for $Re < -2$ the solution is of the

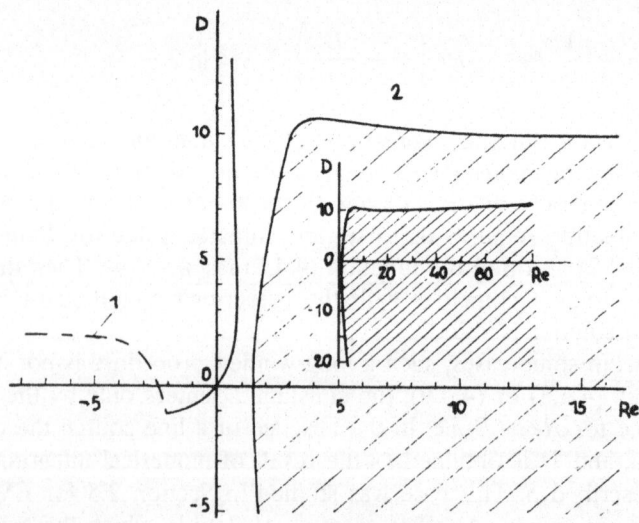

Fig. 4.7. The regions where the solution of the problem (4.4.2)–(4.4.5) exists in the case $\alpha = 0$.

Flows with cylindrical symmetry 173

second kind. In the interval $0 < Re < 0.847$ the solution for each Re is realized for two different D. If $Re \to 0+$, $D \to \infty$, this agrees with the nonexistence of a solution for Squire's problem (see Section 2.4) with the rigid plane surface. For $Re > 1.4$ a continuous interval exists in which the value of feasible axial velocity D lies (the shaded region in Fig. 4.7), but in the interval $0.847 < Re < 1.4$ the solution does not exist. The situation where the solution does not exist for small Re numbers, although it does exist for high Re, is unusual in hydrodynamics.

Now let us proceed with the analysis of mutual interaction of the electrically induced flow, driven by the electric current from the point electrode, and the flow due to the line source. The problem is determined by four independent parameters: α, Re, D, and S; this complicates the analysis. Therefore we restrict the investigation to the particular problem with $\alpha = 0$ (plane surface) and $D = 0$ (zero axial velocity).

The stream lines for the joint action of the line source and the electrically induced flow are shown in Fig. 4.8. It is evident from the figure that a conical jet is formed in the fluid, the slope of which is determined by the zero stream line $\psi = C = 0$. The slope depends on the values of Re and S: by increasing Re for fixed S the line is shifted to the plane $\eta = 0$, yet if Re is fixed, then increasing S the zero stream line moves to the symmetry axis.

Let us now find out how the strength of the line source (Re number) affects the critical value of parameter S, which in the present case for $Re = 0$ is $S_{cr} = 150$ (Section 2.7). As was shown in Chapter 3, the flow breakdown at S_{cr} is the result of a high concentration of vorticity, which is convected by the converging flow to the axis, and the respective deficiency of viscous effects in the narrow axial jet. The presence of a line source, which convects the vorticity from the axis to a larger fluid volume, should increase the value of S_{cr}. Indeed, the data collected for $Re > 1.4$ may be summarized by the approximate formula for $S_{cr}(Re)$: $S_{cr} = 150 + 45.2\,Re^{1.12}$.

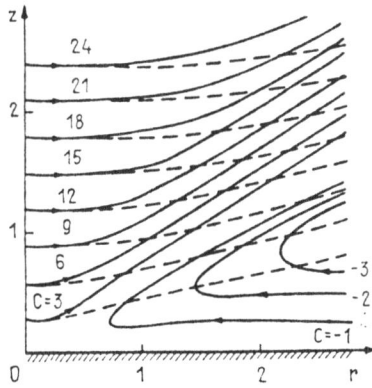

Fig. 4.8. The flow with the line source and point electrode for $Re = 10$, $S = 400$ (the dotted lines represent $S = 0$).

It is straightforward to evaluate the friction on the plane wall. On differentiating (4.4.2) by the use of the expressions for constants (4.4.9) we have

$$(1 + \alpha^2)f''(0) = \frac{Re}{\sqrt{1 + \alpha^2}}[(4 + Re)(\alpha\sqrt{1 + \alpha^2} + \alpha^2) + 2 - D] +$$

$$+ S\left\{\frac{4}{\sqrt{1 + \alpha^2}}(1 + 2\alpha^2 + 2\alpha\sqrt{1 + \alpha^2})\left[\frac{1}{2}\ln(1 + \alpha^2) - \right.\right.$$

$$\left. - \ln(\alpha + \sqrt{1 + \alpha^2})\right] + 2(4\ln 2 - 3)\left(\frac{2 + \alpha^2}{\sqrt{1 + \alpha^2}} + \alpha\right) +$$

$$+ \frac{\alpha}{1 + \alpha^2}(5 + 6\alpha^2 + 6\alpha\sqrt{1 + \alpha^2})\right\}.$$

For $\alpha = 0$ $f''(0) = Re(2 - D) + 4S(4\ln 2 - 3)$, if $D = 0$ also, then the friction is determined by the two parameters Re and S:

$$f''(0) = 2Re + 4S(4\ln 2 - 3). \tag{4.4.10}$$

(4.4.10) suggests that the friction on the wall may be effectively affected by the electric current. Particularly, for

$$S = \frac{Re}{2(3 - 4\ln 2)}$$

the friction is zero, and, on exceeding the S value, a reversed flow develops at the wall.

4.5. Magnetohydrodynamic model of tornado

Let us now give a short review of this natural phenomenon and discuss the mathematical models set forth to describe it. A full description of tornadoes may be found in the remarkable monography written by Nalivkin [38], for a recent review and the numerical theories see [23]. Here we shall present the facts concerning the tornadoes which are necessary to specify them as a special form of atmospheric vortices and to meet the requirements posed by a mathematical model.

4.5.1. Structure and action of a tornado

A tornado may be described as a highly rotating narrow funnel descending from a thunderstorm cloud. Its lateral size may be from several meters up to some hundred meters. Usually the diameter of tornado increases on approaching the cloud, and it resembles a bent slender cone with a tip at the ground. The funnel is made visible by water vapour and dust entrained by the rotating air (Fig. 4.9).

Flows with cylindrical symmetry

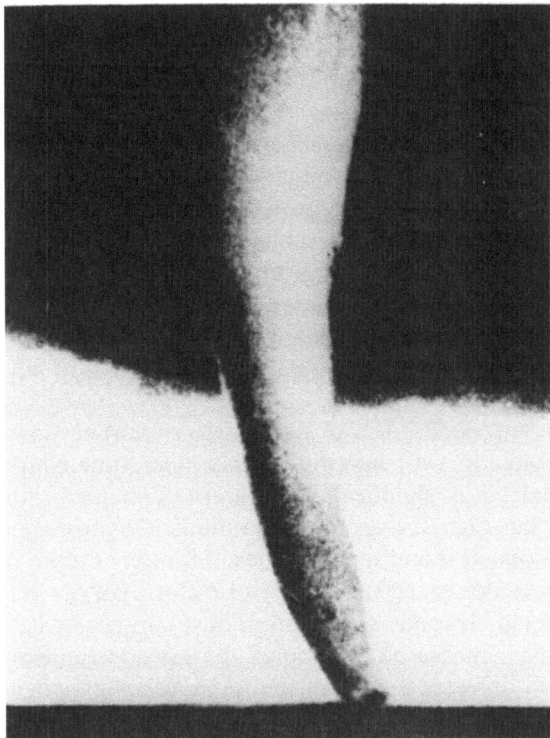

Fig. 4.9. Photograph of a tornado [20].

The height of tornado is supposed to reach the cloud's bottom and it can reach 2—3 km. There are rare events when the vortices form between the clouds. In this case their length is determined by the distance between the clouds.

The tornado is transported relative to the ground with the velocity of the cumulus cloud feeding it. The mean velocity is approximately 50 km/h, yet in some instances it reaches 200 km/h. The lifetime of a tornado varies from several minutes up to several (6—7) hours.

There are different stages in the tornado's lifetime. At the initial stage the cumulonimbus cloud develops a protrusion (or some protrusions). The protrusion extends, sometimes becoming invisible, and reaches the ground. At this moment, usually coinciding with the beginning of the tornado's destructive action, the cone is filled with dust, water droplets, and various debris, and it becomes clearly visible. At the end phase the tornado brightens, narrows, and it is drawn back into the cloud. Often the tip of a mature tornado leaps at the ground. It permits a rare chance to glance into the very heart of tornado.

From the existing photographs and evidence of eyewitnesses it can be concluded that the tornado's inner cavity is similar to the storm's 'eye' — a region where the motion of air is much slower, and sometimes it is quiescent.

Characteristic of the tornado cavity is the downward-directed air flow. In some instances the air flow is intense enough to cause a squeezing effect, which is manifested by things being squeezed into the ground. Another feature of the tornado's cavity is a high electrical activity inside it. The witnesses describe a continuous light of discharges between the walls of the cavity and note the smell of ozone and nitrogen oxides. The cavity of the funnel is surrounded by walls of highly rotating air in spiral paths. This is the most dangerous part of the tornado: the air velocity here reaches hundreds of meters per second, and sometimes even the speed of sound. In the walls the air flow is directed upwards, causing objects on the ground to be sucked into the tornado and sometimes to be lifted up to considerable heights.

The rotational velocity and the vertical velocity in the tornado are highest in the lower part of the funnel and decrease with height. This possibly explains the small height to which heavy objects are lifted. The rotational velocity also varies sharply in the horizontal direction: the velocity rapidly falls off when withdrawing from the wall of the funnel.

A tornado is a specific form of the variety of vortices in the atmosphere. It has much in common with hurricanes, dust and snow whirls and fire whirls; at the same time it is significantly different. Tornado is distinguished from hurricanes by the high energy concentration: if a hurricane is spread over thousands of kilometers, and the velocities in it rarely exceed 50 km/h, the scale of a tornado does not exceed hundreds of meters, yet the velocities in it reach the speed of sound. Tornado differs from dust vortices by its lifetime: if for the dust vortices this is measured in minutes, the tornado can exist for hours. Their energies are also different — that of the tornado being many times greater than that of dust vortices. In addition, the tornado is inseparable from the thunderstorm cloud, which is not a necessary feature for hurricanes, and even more so for the dust vortices which can arise out of a clear sky. It seems to us that the relation to thundercloud is the most characteristic feature of tornadoes, making them a special vortical form in the atmosphere.

Consequently, any model pretending to describe the tornado must satisfy the following basic requirements. First, an energy source must be specified which is capable of feeding the tornado for a long time interval. This, in particular, must take into account the relationship with the cloud. Second, a mechanism must be specified which transforms the energy of the source to the kinetic energy of the tornado vortex. And, last, the model should contain a possibility to explain the individual differences between the tornadoes (the suction effect or the squeezing effect, and so forth).

4.5.2. Historic reference

Descriptions of tornado vortices were known in antiquity (Athens — 200 years B.C., Lucretius [31] — sixties B.C.), the first physical models appeared in the 17th century [3], and mathematical models were developed only in the last two decades.

The physical models of tornadoes may be classified into two groups: the thermodynamic and the electrical.

The thermodynamic theories are based on the assumption that the vortex is initiated by a temperature difference between the ground level and the cloud level, particularly because the hurricane models based on such a theory yield acceptable results. However, Vonnegut [59] estimated that for such highly concentrated energy as is found in the tornado system, reaching the speed of sound, the required temperature difference was of the order of several hundred degrees. Though in the natural environment (volcanic eruptions, great fires) such temperature drop is sometimes achieved, the theory seems unrealistic for most of tornadoes.

The explanation for high local velocities in the tornado funnel was set forth by Brandt [38] and Abdullah [1], and it was based on the conservation of angular momentum. High off the ground the velocities are relatively lower and agree with a moderate temperature drop. The disadvantage of the theory was the absence of any connection between the tornado and the thundercloud. This was taken into account by Mihailov [35], Wegner [60], and Brooks [11] assuming that the tornadoes were initiated in the thunderclouds by horizontal spiralling vortices. It is now generally accepted that the lower part of the tornado cloud is a huge ring vortex, a specific tornado cyclone, a protrusion of which, at least in the initial stage, is the tornado.

Linking the two previous assumptions would yield a well-balanced theory, if there were no other controversy. As will be shown later (see also Section 3.5), the differential rotation of air in the conical gap with a maximum angular velocity at the cone's apex is always associated with a downward air flow (this flow direction agrees with the fact that, at the moment when the protrusion contacts the ground, objects on the ground are scattered from the contact area). If in the mature stage the tornado were still to be feeding from the cloud vortex, we will always have objects being pushed from the tornado zone.

An interesting theory, which did not demand a high temperature drop, was constructed by Gutman and Malbahov [20, 21, 33] based on an unstable stratification of the atmosphere. In the unstably stratified atmosphere the motion of air could lead to an additional energy release. Gutman constructed a model of thermal, i.e. vertical spontaneous air flow, based on this phenomenon. Later the model was supplemented by a rotational motion, and it was applied to explain the tornado. This theory led to an infinitely growing vertical velocity with height. To overcome this Gutman and Malbahov assumed that the unstable stratification took place in a layer of finite height above which the atmosphere was stable. In this case the velocity reaches a maximum and then decreases with height. In the latest work by Malbahov [32] the theory was improved by taking account of the ground layer.

For a suitable choice of parameters specifying the state of the atmosphere, the above theory can lead to results agreeing with the tornado measurements. However, the theory cannot explain the association of tornado with the cumulus cloud. Particularly, as we have already mentioned, the tornado initially develops from the cloud, whenever the thermal is associated with the unstable lower atmosphere, i.e. it develops upward. Moreover, the theory does not explain the tornado vortices between the clouds, which can be nearly horizontal. Nevertheless the physical assumptions laid down in the theory may be accepted, and not merely to explain the thermals. The mechanism of the unstably stratified atmosphere can maintain the steady tornado after it has developed from the thundercloud.

The second group of theories explained the tornado's development by reference to electrical phenomena. It is of interest to note that hypotheses of the electrical tornadogenesis appeared historically earlier than the thermodynamic theories. Thus, Francis Bacon [3] expressed an idea that, speaking in modern terms, the tornado could arise due to the air heating by electrical discharges and the subsequent intense convection. In 1840 Peltier [41] wrote: "Everything proves that the tornado is nothing other than a conductor formed of the clouds, which serves as a passage for a continual discharge of electricity . . .". An even more specific statement was made at the same time (1837) by Hare [22]: "After maturely considering all the facts, I am led to suggest that a tornado is the effect of an electrified current of air superseding the more usual means of discharge between the earth and clouds, in those vivid sparks which we call lightning". The hypothesis was later forgotten; only in the fifties of the present century, after the systematic observation of tornadoes had begun, it was considered anew. There is certainly enough electrical energy in the thundercloud to cause and maintain the tornado; we refer to the estimates made by Vonnegut [59]: assuming equal orders of magnitude of the vertical and circular velocities, to accelerate an air column of 100 m in diameter up to 250 m/s it is necessary to have a power of 10^{18} erg/s, or 10^8 kW. Yet the energy of a single lightning discharge, according to Braham [10], is 10^{22} erg (the same order of magnitude estimates of the discharge energy was obtained also by Uman [56]). This energy is sufficient to sustain the tornado for several hours.

Two mechanisms which may be responsible for the transformation of electrical energy into the kinetic energy of air moving in the tornado were discussed: according to the first, air was heated electrically up to the temperature capable of causing intense convection, and according to the second, proposed earlier by Hare, the charged air is accelerated up to a high velocity in a strong electrical field.

The best known type of electrical discharge in the atmosphere is the lightning discharge. Vonnegut [59] suggests a situation in which lightning discharges occurred consecutively and the ions, created by the previous discharge, had no time to recombine until the following discharge. In this case the consecutive discharges could traverse the same path and produce intense heating of a relatively small air volume. By the estimates, ten discharges in a second would be sufficient to heat an air column of diameter 50 m up to 300°C and thereby bring the air velocity up to 200 m/s. Air might also be heated by a steady current in a glow discharge or in an electrical arc [59]. Of course, the electrical theories in such a form were not essentially different from the heat theories: the differences were merely in the heat sources.

A different mechanism for the transformation of the electrical energy into kinetic energy is the acceleration of charged particles in an electrical field. Estimates show that an electrical field of magnitude 10^5 V/m gives rise to a motion of charged water droplets with a velocity which, for comparison, can be attained by convection for a temperature difference of 20°C.

4.5.3. Tornado model

An acceptable hypothesis of tornado genesis and development may be suggested in the framework of modified electrical theory. We shall start by a schematic specification of the basic processes which take place, in our view, in consecutive stages of the tornado's existence, then we shall present computational results concerning mainly the hydrodynamic structure of tornado in the mature stage.

After the appearance of the conical protrusion at the bottom of thundercloud the electrical charges travel from the cloud to the cone's tip under the action of an electric field between the cloud and the ground (the conical form of protrusion, resulting from the action of the electrical field on a concentrated charge in the moving tip, seems to be a natural choice). The resulting electrical field between the tip and the ground increases, and the tip is pulled to the ground. The part of the protrusion at the cloud's bottom is rotating, since the protrusion is part of a tornado cyclone. Supposing that in different parts of the cone's cross section angular momentum is conserved, the velocity on the surface of rotating cone of gas is varying as $v_\varphi \sim r^{-1}$, where r is the distance from the axis of rotation. The angular velocity respectively grows on approaching the cone's apex as r^{-2}, or in other words, there is a differential rotation in the gas cone. This kind of rotation, as will be shown later (see also Section 3.5), induces a secondary flow directed along the cone's axis to the apex. Since, in the lower part of the cloud, there are located predominantly charges of equal sign [56], the differential rotation also causes the convective charge transport from the cloud to the cone's tip.

The downward air flow takes place during the initial stage until the tornado contacts the ground. This form of the air flow can explain the blast effect at the moment of contact, as a result of which objects in the contact area are scattered.

From this moment, according to Peltier, the tornado acts as "a passage for a continual discharge of electricity" from the cloud, and a new factor comes into effect, — viz. the interaction of the discharge current with the self-magnetic field, generating an electrically induced flow. The electrically induced flow for the current flowing in the cone is always directed upwards along the axis (see Chapter 2). In conjunction with the differential rotation, which is responsible for the downdraft in the axial zone, the two kinds of flow determine the variety of hydrodynamic structures in a tornado vortex. Thus, if the centrifugal forces are leading, the downdraft will be observed, together with the 'squeezing' effect; if the leading mechanism is electrically induced, the main flow will be upwards and the lifting of objects from the ground will be observed. Intermediate variants are also possible.

In Chapter 3 we have shown that the converging electrically induced flow can amplify a small azimuthal disturbance until a finite state of rotation is achieved. The nonlinear interaction of the electrically induced flow and rotation results in a kinetic energy transfer from the electrically induced meridional flow to the rotational motion, thus feeding it. Another possibility for maintaining the intense rotation is the direct generation by means of an electromagnetic force.

Indeed, if the current lines are spiralling along the tornado walls, an axial magnetic field arises (similar to the field of a solenoid) which, interacting with the radial current, gives the azimuthal force (Fig. 4.10) additionally increasing the rotation. The terrestrial magnetic field could also take part.

Hence, the electrically induced flow, the nonlinear interaction of meridional and azimuthal motion, and the direct action of azimuthal force constitute the basis for the mechanism transforming the electrical energy of the thundercloud to the kinetic energy of the tornado.

Consider now a hydrodynamic structure of a tornado in the mature stage, using solutions of the class (4.2.15). Since the main kinetic energy is concentrated in the walls of a funnel and, thereby, the air electrification due to friction of the rotating layers takes place there, the tornado model may be represented schematically as consisting of an inner conical nonconducting cavity 1 (Fig. 4.10), a discharge zone 2 confined by the coaxial cones $\eta_1 = \tan \alpha_1$ and $\eta_2 = \tan \alpha_2$, and a nonconducting ground zone 3. The physical properties of the gas, for simplicity, are supposed to be equal everywhere, apart from the discharge zone 2 where the gas is electrically conducting, which ensures the passage of the current I with uniform density on a spherical surface.

Assuming for air, roughly, $\sigma = 10^2$ (Ohm m)$^{-1}$, $\nu = 10^{-5}$ m^2/s, then $\beta = \mu_0 \sigma \nu = 10^{-9}$ and the effect of motion on magnetic field can be neglected. The equations (4.2.18), (4.2.19) for the azimuthal and meridional magnetic fields in this case are

$$(1 + \eta^2)f_1'' + 3\eta f_1' = 0, \tag{4.5.1}$$

$$(1 + \eta^2)f_2'' + \eta f_2' - f_2 = 0. \tag{4.5.2}$$

The electric current flow, represented in Fig. 4.10 by I, is specified by the solution of (4.5.1)

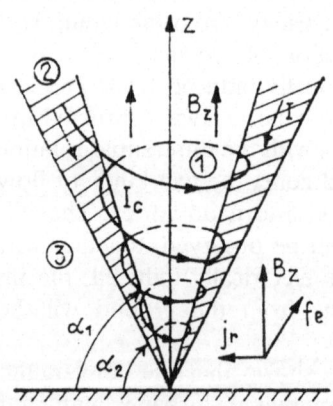

Fig. 4.10. The computational model of tornado and the scheme explaining generation of rotation by the spiralling current passage.

$$f_1 = \left(\frac{\eta_2}{\sqrt{1+\eta_2^2}} - \frac{\eta_1}{\sqrt{1+\eta_1^2}}\right)^{-1}\left(\frac{\eta}{\sqrt{1+\eta^2}} - \frac{\eta_1}{\sqrt{1+\eta_1^2}}\right). \quad (4.5.3)$$

From the solutions of (4.5.2) we choose the solution

$$f_2 = \eta - \sqrt{1+\eta^2} \quad (4.5.4)$$

responsible for the meridional magnetic field whose lines of force lie on rotational paraboloids (see Fig. 2.1e); they are similar to the field in a solenoid of variable radius with spiralling current lines.

The equations of motion are (4.2.16) and (4.2.17) where $S = \mu_0 A_1^2/\rho v^2 = \mu_0 I^2/\rho v^2$, $N = \mu_0 A_1 A_2/\rho v^2$, and the right-hand part is determined by (4.5.3) and (4.5.4) (then the expression in square brackets in the right-hand side of (4.2.16) is identically zero).

The boundary conditions on the surface $z = 0$ are

$$f(0) = 0, \quad f'(0) = 0, \quad \Omega(0) = 0. \quad (4.5.5)$$

At the vortex axis $r = 0$ ($\eta = \infty$) a solution should provide zero radial velocity

$$\lim_{r \to 0} v_r = \lim_{\eta \to \infty} \frac{v}{z} \eta f' = 0, \quad (4.5.6)$$

limited axial velocity (except $z = 0$)

$$\lim_{r \to 0} v_z = \lim_{\eta \to \infty} \frac{v}{z} \eta(\eta f' - f) = \text{const} \frac{v}{z}, \quad (4.5.7)$$

and zero azimuthal velocity

$$\lim_{r \to 0} v_\varphi = \lim_{\eta \to \infty} \frac{v}{z} \eta \Omega = 0. \quad (4.5.8)$$

The numerical solution of the problem set up was obtained in [53], and the solution of the linearized problem earlier in [50]. If $f_2 f_1' = 0$ is inserted in (4.2.17), it is easy to obtain

$$\Omega' = C \exp\left(-\int \frac{3\eta + f}{1+\eta^2} d\eta\right),$$

i.e. the derivative of the azimuthal velocity function Ω does not change sign in the whole interval considered, and for the conditions (4.5.5) and (4.5.8) only a trivial solution may be obtained $\Omega \equiv 0$. Thus, a nonmonotonous behaviour of Ω with zeroes on the axis and the plane, may be obtained upon introducing a disturbing force. In the present problem this is the electromagnetic force arising from the current interaction with the meridional magnetic field.

Neglecting the description of the numerical method, which may be found in [53], we shall discuss the main results. Note, first, that the parameter S is directly responsible for the intensity of the meridional flow (for f), and the parameter N for intensity of the rotation. Additionally these parameters influence the flow by the nonlinear interaction (the term $f\Omega'$ in (4.2.17)).

Fig. 4.11 shows the distributions of axial ($zv_z/v = \eta(\eta f' - f)$) and azimuthal ($zv_\varphi/v = \eta\Omega$) velocities in horizontal sections $z = $ const. For $S = 10$, $N = 100$ (full line 1) the contribution of centrifugal force is substantially higher than that of the electromagnetic force due to the azimuthal self-magnetic field, and in the axial zone of vortex the downflow is induced. For $S = N = 50$ (curve 4) both the factors interchange their effect and the upflow becomes dominant. This is clearly evident also in Fig. 4.12 where the stream lines $r = C/f$ are represented. From this it follows that, varying magnitudes of the parameters S and N, a variety of flow configurations may be obtained. Thus for $S = 50$, $N = 100$ (Fig. 4.12c) the downflow exists in the axial zone (cone's cavity), and the intense updraft takes place in the discharge zone denoted in the figures by the dotted lines. This kind of tornado may show both the effects — objects would be lifted at the outer part of the funnel and 'squeezing' would be observed in the cavity zone. For $S = 50$, $N = 92$ (Fig. 4.12d) we have almost quiescent air in the cavity of the funnel. Only upflows would be observed for this kind of tornado. Velocity profiles in the different cross sections of tornado are shown in Figs. 4.11 and 4.13. Inspecting these we can conclude that for a fixed N growth of S leads to retardation of the axial downflow, and increasing S further (or decreasing N) the flow is transformed to upflow. The flow transformation in the axial zone is associated with the corresponding transformation of the groundflow (Fig. 4.13 depicts the radial velocity $f' = rv_r/v$ in sections $r = $ const). Thus for $N \gg S$ the flow scatters objects on the ground (full line 1); for the remaining variants they would be drawn inside the tornado zone.

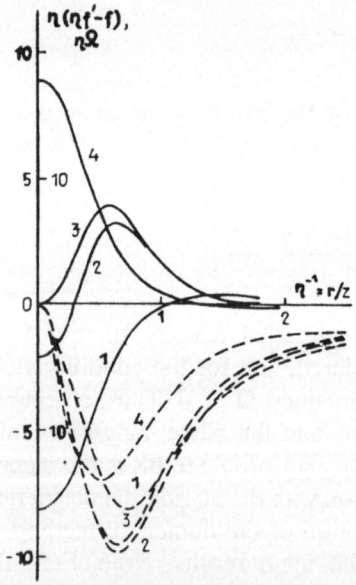

Fig. 4.11. Radial dependence of axial and azimuthal velocities: (1) $S = 10$, $N = 100$; (2) 50, 100; (3) 50, 92; (4) 50, 50 (vertical scale for the curve 4 is increased two times).

Flows with cylindrical symmetry

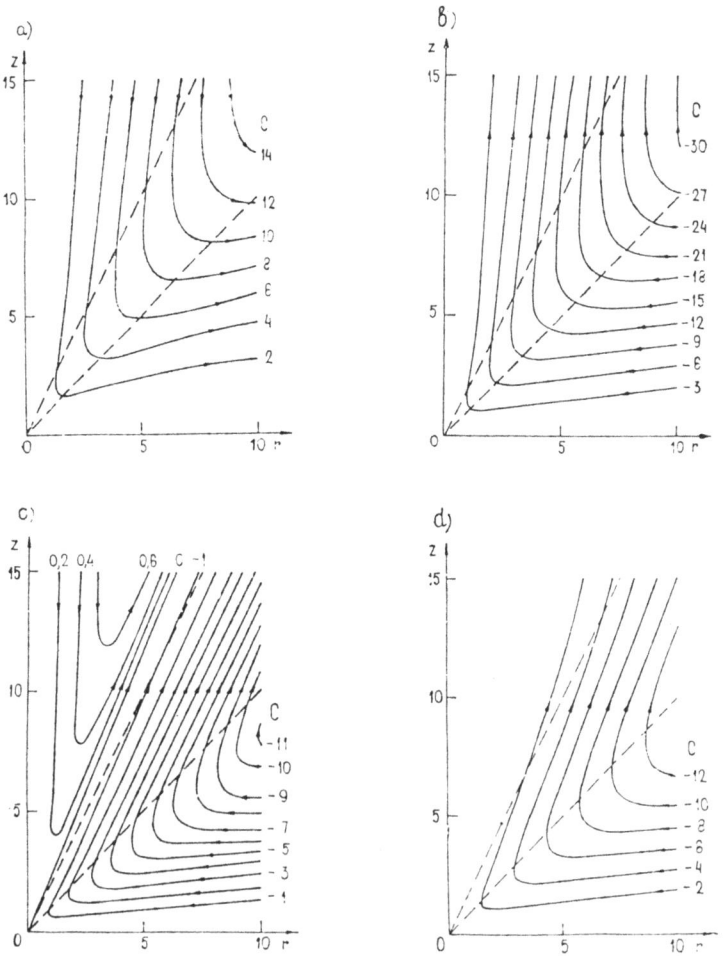

Fig. 4.12. Stream lines in a tornado: (a) $S = 10$, $N = 100$; (b) 50, 50; (c) 50, 100; (d) 50, 92.

Dotted lines in the figures represent the profiles of azimuthal velocity (in Fig. 4.13 the function $\Omega = rv_\varphi/\nu$ has the meaning of azimuthal velocity in sections $r = $ const). Obviously, decreasing the parameter N, responsible for the fluid rotation, and retaining a value of S, the rotational velocity is diminished. The most significant here is the effect of the parameter S for $N = $ const. As expressed by (4.2.17), the parameter S can affect the rotation velocity only by means of the function f of meridional flow. Hence the growth of rotation velocity with the increase of S means the energy transfer from the meridional flow to the rotational motion, and thereby the meridional flow also sustains rotation in the tornado.

A different model for an electrically induced tornado vortex, where the electromagnetic forces generate the meridional up-draught, were considered by Sozou [54]. Rotation in this model is associated with the line vortex which coincides with the axis of electrical discharge. In agreement with the results of

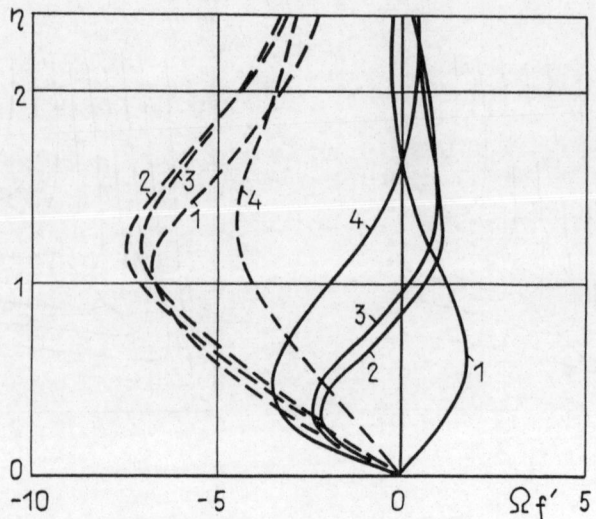

Fig. 4.13. Distributions of radial and azimuthal velocities along the axis: (1) $S = 10$, $N = 100$; (2) 50, 100; (3) 50, 92; (4) 50, 50.

Section 2.5 the line vortex also generates the up-draught, thus the model [54] is able to predict only the flow configuration converging along the ground, instead of the flow variety predicted by [50, 53].

The numerical results presented display merely qualitative features of a tornado, and cannot be applied for a quantitative description mostly due to the small magnitudes of the parameters. The real parameters are significantly higher. For example, for $I = 10^3$ A, $\rho = 1$ kg/m^3, $\nu = 10^{-5}$ m^2/s the value of S is 10^{10}.

In conclusion let us estimate the conductivity of air in the tornado, which is necessary to pass a prescribed current. According to the data in [56], even for clear weather the electric field in the lower atmosphere is 10^2 volts/m. Assuming a typical cross section of the tornado's conducting zone to be 10 m^2, a current of $I = 10^3$ A could pass if the conductivity $\sigma = 1$ (Ohm m)$^{-1}$.

4.6. EVF in a cylindrical container

Similarity solutions describe flows in unbounded or semi-infinite volumes. Real high-current technology (electroslag welding, melting etc.) requires us to consider finite sized current-carrying domains. It is obvious that similarity solutions may only be partly useful in predicting EVF in a finite domain, and it is necessary to solve the equations (4.2.2) numerically. The process of numerical solution requires us to restrict the flow parameters. We shall consider here two factors only: the effect of motion-induced currents on EVF in a cylindrical container, and a possibility of operating the flow by means of an external, axial, uniform magnetic field. Since one may encounter well-known computational

Flows with cylindrical symmetry 185

difficulties for either too high or low values of the parameters, the results for moderate values of the parameters discussed below should be considered as suggesting certain tendencies in the flow behaviour, affected by the previously mentioned factors.

In (4.2.2) let $\psi_2 = 0$, $v_\varphi = 0$ and introduce nondimensional coordinates related to the radius of container a, velocity to the quantity v/a, magnetic field to $\mu_0 I/2\pi a$ (function ψ_1 to $I/2\pi$), where I is the total supplied current. Then the effect of induced currents can be evaluated solving the set of equations:

$$\frac{1}{r}\frac{\partial \psi}{\partial r}\frac{\partial w}{\partial z} - \frac{\partial w}{\partial z}\frac{\partial}{\partial r}\left(\frac{w}{r}\right)$$

$$= \frac{\partial}{\partial r}\left[\frac{1}{r}\frac{\partial}{\partial r}(rw)\right] + \frac{\partial^2 w}{\partial z^2} - \frac{S}{r}\frac{\partial}{\partial z}(B_\varphi^2);$$

$$\frac{\partial}{\partial r}\left[\frac{1}{r}\frac{\partial}{\partial r}(rB_\varphi)\right] + \frac{\partial^2 B_\varphi}{\partial z^2}$$

$$= \beta\left[\frac{1}{r}\frac{\partial \psi}{\partial r}\frac{\partial B_\varphi}{\partial z} - \frac{\partial \psi}{\partial z}\frac{\partial}{\partial r}\left(\frac{B_\varphi}{r}\right)\right],$$

where the vorticity $w = \partial v_r/\partial z - \partial v_z/\partial r = -r^{-1}E^2\psi$ and the magnetic field $B_\varphi = \psi_1/r$ are introduced. The parameters $S = \mu_0 I^2/4\pi^2\rho v^2$ and $\beta = \mu_0 \sigma v$ have the same meaning as previously.

The numerical results [36] are obtained for $r = z = 1$, the radius of the bottom electrode at $z = 0$ (Fig. 4.25) is $r_0 = 0.2$, the radius of the top electrode $(z = 1) - r = 1$. Boundary conditions for the viscous flow are:

$$\psi = \frac{\partial \psi}{\partial z} = 0 \quad \text{for} \quad z = 0, z = 1;$$

$$\psi = \frac{\partial \psi}{\partial r} = 0 \quad \text{for} \quad r = 1;$$

$$\psi = w = 0 \quad \text{for} \quad r = 0.$$

The computational procedure includes the underrelaxation for the function w at the rigid walls (see Section 4.3). For the magnetic field the conditions are: $B_\varphi = 0$, on $r = 0$, $B_\varphi = 1$ on $r = 1$, and $B_\varphi = r$ on $z = 1$ (for a uniform current density on the surface of larger electrode). Concerning the field at the smaller electrode, it would not be generally correct to impose a uniform current density on it, since it is known that the current is condensed at the edges. We assume the current density at the electrode is that for a current passing from a disc of radius r_0 into an infinite fluid. For this purpose consider the equation $E^2\psi_1 = 0$

in the coordinates of an oblate spheroid (1.7.13), the solution of which $\psi_1 = I(1 - \mu)/2\pi$ gives the distribution of current density we seek

$$j_z\Big|_{\substack{z=0\\r\leq r_0}} = j_\lambda|_{\lambda=0} = -\frac{I}{2\pi r_0^2} \frac{1}{\sqrt{\lambda^2 + \mu^2}\sqrt{1+\lambda^2}} \frac{\partial \psi_1}{\partial \mu}\Big|_{\lambda=0}$$

$$= \frac{I}{2\pi^2 r_0^2 \mu} = \frac{I}{2\pi^2 r_0^2 \sqrt{1 - \left(\frac{r}{r_0}\right)^2}},$$

i.e. the current density actually condenses at the electrode's edges. The magnetic field is expressed with (1.6.10):

$$B_\varphi = \frac{\mu_0 \psi_1}{r} = \frac{\mu_0 I}{2\pi} \frac{1-\mu}{r},$$

and at the electrode's surface $z = 0$, $r \leq r_0$ in the non-dimensional form

$$B_\varphi = \frac{1}{r}\left(1 - \sqrt{1 - \left(\frac{r}{r_0}\right)^2}\right).$$

Outside the electrode at $z = 0$ the magnetic field distribution is $B_\varphi = 1/r$ for $z = 0, r \geq r_0$.

The numerical solution [36] is obtained for relatively small S up to 500. Hence to detect an effect of the motion induced currents it is assumed that $\beta = 0.2$.

The flow stream lines are shown in Fig. 4.14; they take the form of a toroidal vortex with the fluid moving in the central zone upwards from the smaller sized electrode. The dependence of the axial velocity, normalized by S (Fig. 4.15),

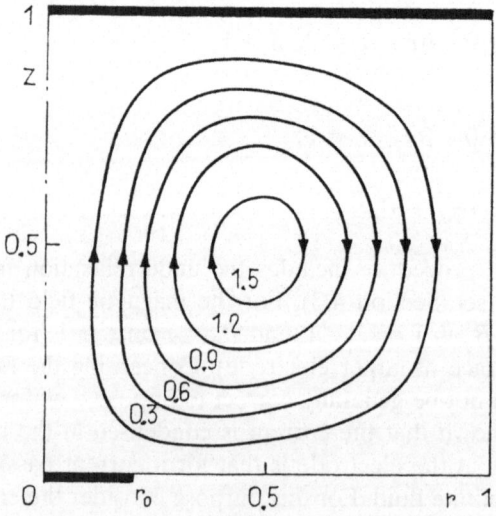

Fig. 4.14. Stream lines for $S = 500$, $\beta = 0.2$.

Flows with cylindrical symmetry

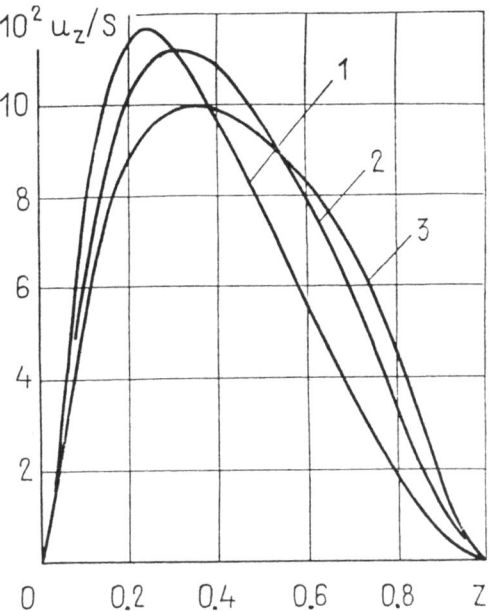

Fig. 4.15. Axial velocity along the symmetry axis for $\beta = 0.2$ and (1) $S = 25$, (2) 250, (3) 500.

demonstrates that with increasing S the velocity maximum is shifted towards the container's midplane, and its magnitude is diminished. This fact indicates that the induced currents make the growth of velocity with S slower than linear.

The effect of an external magnetic field on the electrically induced flow will be demonstrated for the case of uniform axial magnetic field of the magnitude B_0 ($\psi_2 = B_0 r^2/2$). Now the mathematical model is simplified assuming in (4.2.2) that $\sigma = 0$ (or $\beta = 0$):

$$\frac{1}{r}\frac{\partial \psi}{\partial r}\frac{\partial w}{\partial z} - \frac{\partial \psi}{\partial z}\frac{\partial}{\partial r}\frac{w}{r} - \frac{1}{r}\frac{\partial}{\partial z}(v_\varphi^2)$$

$$= \frac{\partial}{\partial r}\left[\frac{1}{r}\frac{\partial}{\partial r}(rw)\right] + \frac{\partial^2 w}{\partial z^2} - \frac{S}{r}\frac{\partial}{\partial r}(B_\varphi^2),$$

$$\frac{1}{r}\frac{\partial \psi}{\partial r}\frac{\partial v_\varphi}{\partial z} - \frac{1}{r^2}\frac{\partial \psi}{\partial z}\frac{\partial}{\partial r}(rv_\varphi)$$

$$= \frac{\partial}{\partial r}\left[\frac{1}{r}\frac{\partial}{\partial r}(rv_\varphi)\right] + \frac{\partial^2 v_\varphi}{\partial z^2} + N\frac{\partial B_\varphi}{\partial z},$$

$$\frac{\partial}{\partial r}\left[\frac{1}{r}\frac{\partial}{\partial r}(rB_\varphi)\right] + \frac{\partial^2 B_\varphi}{\partial z^2} = 0,$$

where $N = aIB_0/2\pi\rho v^2$ is the parameter determining the intensity of fluid rotation.

Let the axial size of the fluid region be increased to $z = 2$. Then all the above boundary conditions stay but, instead of the top wall $z = 1$, the respective conditions are imposed on $z = 2$. There are added boundary conditions for the azimuthal velocity: $v_\varphi = 0$ on the rigid walls and at $r = 0$.

The change of the flow character for the varied parameter N is demonstrated by the stream lines in Fig. 4.16. For $N = 0$ (Fig. 4.16a) the flow is similar to that considered previously. The axial magnetic field causes the fluid to rotate differentially, which in turn generates a secondary toroidal motion opposed to the electrically induced one. In the process of increasing the axial field, initially, two toroidal vortices (Fig. 4.16b) are formed, then the rotation induced secondary flow suppresses the electrically induced flow (Fig. 4.16c). This is supported also by the axial velocity distribution (Fig. 4.17).

It is of interest to follow the variation of rotation with increasing values of the parameter S, which is responsible for the intensity of the electrically induced flow, and for fixed N, responsible for the rotation intensity (since both the parameters contain the current I, then a fixed N with an increase of S means a corresponding decrease of the external magnetic field, so that $IB_0 = $ const). Fig.

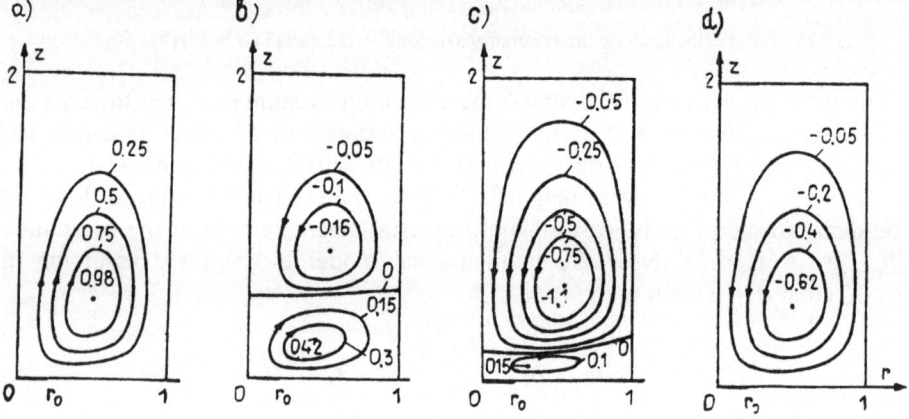

Fig. 4.16. Meridional flow stream lines for (a) $S = 250$, $N = 0$; (b) 250, 250; (c) 250, 500; (d) 0, 250.

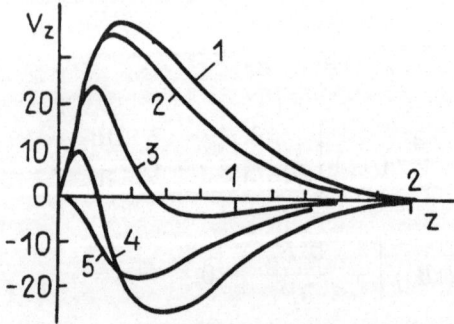

Fig. 4.17. Axial velocity ($r = 0$) for (1) $S = 250$, $N = 0$; (2) 250, 100; (3) 250, 250; (4) 250, 500; (5) 0, 250.

Flows with cylindrical symmetry

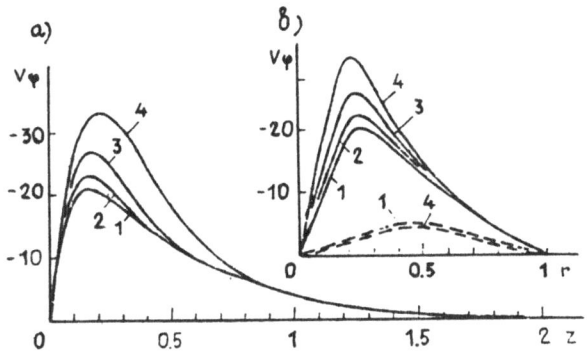

Fig. 4.18. Azimuthal velocity for $N = 250$: (a) along the height for $r = 0.3$; (b) along the radius for $z = 0.1$ (full lines) and $z = 1$ (dashed). (1) $S = 0$, (2) 100, (3) 250, (4) 500.

4.18 shows that, on increasing S, the intensity of rotation is changed: at the smaller electrode v_φ grows significantly, and at the larger electrode it is somewhat retarded (Fig. 4.18b). It is evident that the meridional flow significantly affects the azimuthal rotation.

To evaluate the effect, the following integral quantities are computed: momentum in a meridional plane $K = \int_0^1 \int_0^2 v_\varphi \, dz r \, dr$, total angular momentum $M = \iint v_\varphi \, dz r^2 \, dr$, and kinetic energy of rotation $E = \iint v_\varphi^2 \, dz r \, dr$ for $N = 250$ and varied S. The computed results are presented in Table 4.1, where also the energies in the separate cylindrical layers of height $\Delta z = 0.4$ are represented (their numbers in the table begin from the bottom $z = 0$). The table reveals that all the computed quantities grow significantly with increasing S. The growth of kinetic energy is the most pronounced. This supports the statement, made in the previous section, that the kinetic energy of the meridional electrically induced motion is partly transferred to the azimuthal rotation, feeding it.

For smaller S the quantities in Table 4.1 are subtly decreased (apart from the total energy E). The table indicates that the decrease occurs in the region remote from the smaller electrode, where for $S \sim N$ the electrically induced flow takes place only in the vicinity of the plane $z = 0$ (see Fig. 4.16b). This is the only region in which the energy of the electrically induced flow is supplied to rotation.

In conclusion it may be suggested that the hydrodynamics of the melt can be affected by operating the axial magnetic field; particularly, heat generated at the smaller electrode could be optimally transferred, and droplets from the melting electrode could also be transported to desired areas.

Table 4.1. Integral quantities for different S.

S	E_1	E_2	E_3	E_4	E_5	E	K	M
0	23.92	14.62	2.63	0.28	0.017	41.47	−4.10	−2.31
100	25.72	13.99	2.42	0.25	0.015	42.40	−4.08	−2.29
250	30.23	13.87	2.15	0.23	0.014	46.48	−4.13	−2.28
500	41.60	18.66	2.17	0.21	0.013	62.64	−4.53	−2.40
750	49.48	33.71	4.90	0.37	0.019	88.48	−5.41	−2.72

4.7. Effect of electric current configuration on flow in a cylindrical container

The electrically induced flows, such as those in electroslag technology, are of a quite diverse nature, depending on the specific process. Thus, welding may involve a relatively thin wire electrode, and the depth to which the electrode is immersed in the bath is varied; for electroslag melting the electrode's size is comparable to the diameter of the slag bath, and so forth, yet the most significant feature of the flows in technology is the high magnitude of the electric current.

The flows for high S, relevant to technology, and for a varied depth of the electrode's immersion were numerically simulated by Sandler [45] (Figs. 4.19–4.21). The electrodes considered were of a cylindrical form with isolated lateral surface, i.e. current entered the fluid through its end face. Results are reported for $4.5 \times 10^2 \leq S \leq 4.5 \times 10^6$, the respective current is $6.5 \text{ A} \leq I \leq 650 \text{ A}$. As the author notes, the numerical solution for relatively high S is obtained owing to the use of a nonuniform computational grid with condensation of mesh points in the wall layers.

Let us discuss some quantitative results of the numerical simulation [45]. The bath circulation grows in intensity with the increase of welding current (Figs. 4.19, 4.20); the single loop circulation at small currents (Figs. 4.19a, 4.20a) is replaced by the double loop (Figs. 4.19b, c; 4.20b, c) with relatively weak flow at the side walls. The stream lines in Figs. 4.19 and 4.20 reveal only the flow patterns. Additional information may be obtained from the velocity profiles in the cross-section of the bath. It is evident from Fig. 4.21 that the maximum of the velocity v_z close to the electrode's end lies off the symmetry axis and moves to the axis with distance from the end face (the distance increases with the current). Evidently this flow property indicates a kind of 'wake' in the flow downstream of the electrode.

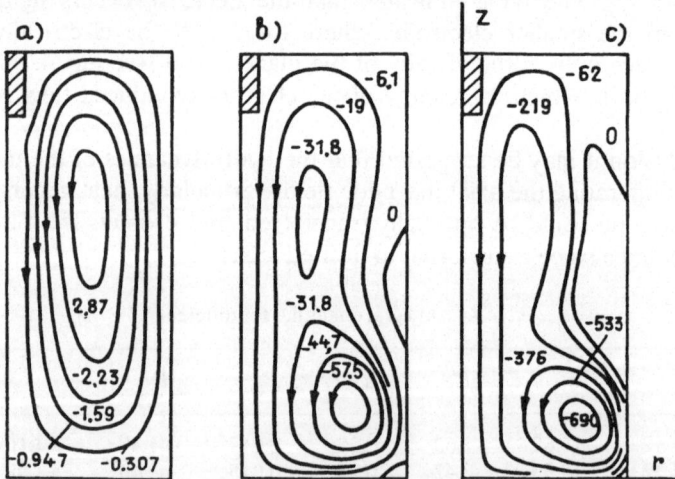

Fig. 4.19. Stream lines $\psi/\nu =$ const in the container of depth 2.48 (all lengths are related to the radius of container R) with the immersed electrode of radius 0.097 and depth of immersion 0.35: (a) $S = 4.5 \times 10^2$; (b) 4.5×10^4; (c) 4.5×10^6.

Flows with cylindrical symmetry 191

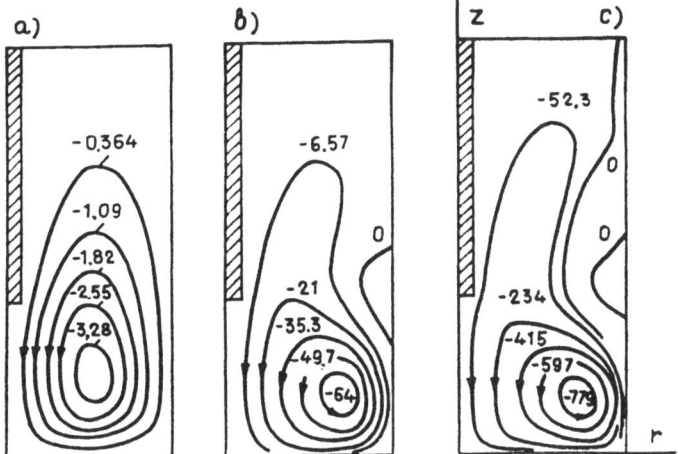

Fig. 4.20. The same as in Fig. 4.19 for the electrode's immersion depth 1.55.

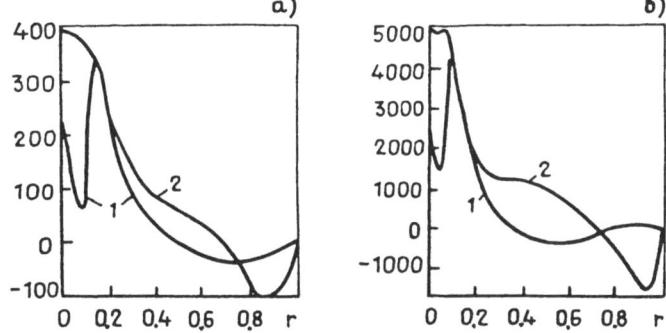

Fig. 4.21. Vertical velocity profiles $(-v_z R/\nu)$ in the horizontal sections $z/R = 2.09$ (curve 1) and 1.6 (curve (2)) for the geometric conditions of Fig. 4.19: (a) $S = 4.5 \times 10^3$; (b) 4.5×10^5.

Specific welding conditions may involve electrodes of different size r_0 relative to the radius R of the slag bath. The effect of the electrode's relative size was simulated by Vlasyuk [58] for electrodes ending at the bath top and for a depth of the bath equal to its radius. The computed stream lines are similar to Fig. 4.14 (for higher S the vortex centre is shifted to the point with the coordinates $z = 0.4$, $r = 0.625$). The velocity field in Figs. 4.22, 4.23 is represented by the equal velocity level lines, i.e. the lines where either the radial velocity component or axial (related to the maximum values) stays constant. It is evident from the figures that the velocities for different r_0/R are not similar because of the different electromagnetic force distributions. For small S (Figs. 4.22b; 4.23b) the most intense motion, independently of r_0/R, is located close to the melted electrode where the points with $v_{z_{max}}$ and $v_{r_{max}}$ are situated.

For high S (Figs. 4.22a; 4.23a) $v_{r_{max}}$ is close to the bottom of bath. The most significant difference is for lines v_z. The flow with a thin electrode ($r_0/R = 0.2$) varies very little over a substantial part of the axial zone, as

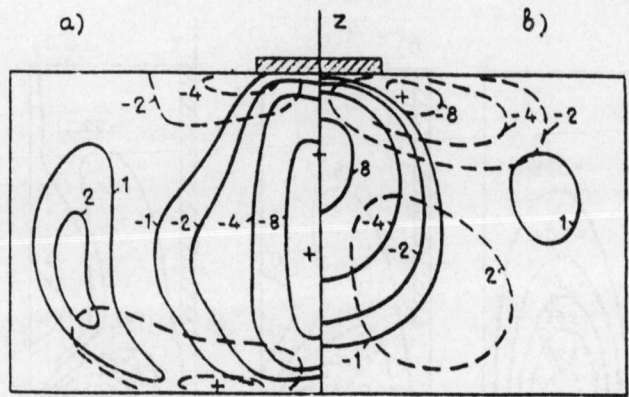

Fig. 4.22. The lines of equal axial (full lines) and radial (dashed lines) velocities in the case of electrode's radius $r_0/R = 0.2$: (a) $S = 10^6$, (b) 10^2.

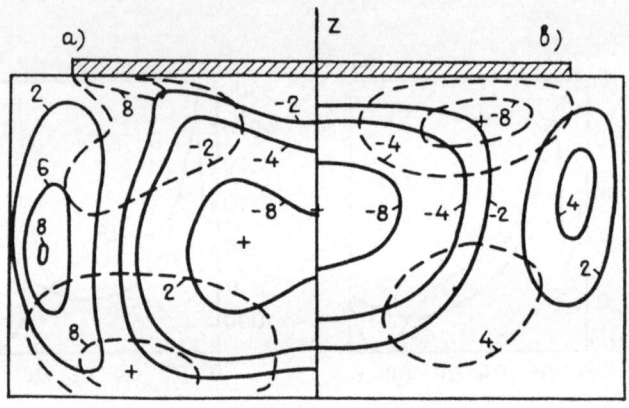

Fig. 4.23. The same as in Fig. 4.22 for $r_0/R = 0.8$.

indicated, e.g. the lines $v_z/v_{z_{max}} = 0.8, 0.4$, and the point where $v_{z_{max}}$ is situated, is practically on the axis. As shown in Fig. 4.24, in this case the velocity profiles vary little with the increase of S for S higher than 10^4. This means that the velocity field related to $|v|_{max}$ stays the same while S is increased, i.e. it becomes self-similar by S.

For the wide electrode ($r_0/R = 0.8$) the downflow enters deeper in the bath with increasing S. The upflow velocity at the side wall also grows. In the process of rearrangement the maximum of velocity v_z moves off the symmetry axis (Fig. 4.25). The displacement reaches a maximum for $S = 10^6$ and for higher S the flow field is again self-similar.

Since, in the situation considered, the electrode does not protrude into the bath, motion of the point with $v_{z_{max}}$ off the axis is not explained by the wake downstream of the electrode. A similar situation was examined in Section 2.10, where the zero vorticity line $w_\varphi = 0$ was used to follow the vorticity transport.

Flows with cylindrical symmetry

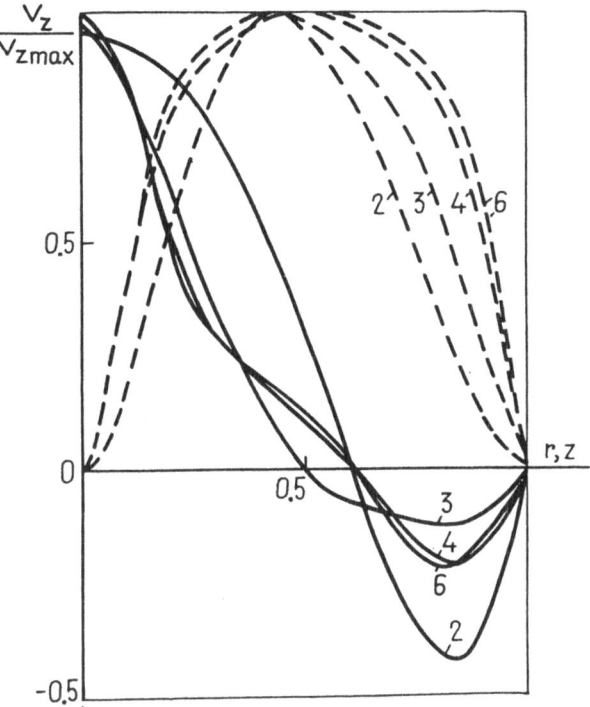

Fig. 4.24. Axial velocity profiles $v_z(r)$ at $z = 0.5$ (full lines) and $v_z(z)$ at $r = 0$ (dashed lines) in the case $r_0/R = 0.2$. The numbers on the lines are equal to log S.

The line is shown in Fig. 4.26 for the two values of r_0/R and agrees with the results for a hemispheroidal container. Note that the bath flow in the axial zone is mainly along the z-axis, then the condition $w_\varphi = 0$ gives approximately $\partial v_z/\partial r = 0$ on this line with the relevant velocity maxima in Fig. 4.25.

As we mentioned earlier, for S higher than a certain value the relative velocity $v/|v|_{max}$ field is nearly constant, independent of S. In this context consider the dependence of $|v|_{max}$ vs. S and r_0/R. It is evident from Fig. 4.27, the plot on a logarithmic scale consists of the two straight lines for $S < 10^3$ and $S > 10^5$, and the transitional curve for $10^3 < S < 10^5$.

For the first region the dependence $v_{max}(S, r_0/R)$ is approximated by the expression

$$v_{max} = \frac{S}{\sqrt{10^{1+5r_0/R}}} \quad \text{for} \quad S < 10^3.$$

The dependence is for Stokes flow where the electromagnetic force is balanced by viscous forces. In the transitional region $10^3 < S < 10^5$ the nonlinear terms are of growing importance; the velocity field is rearranging to the fully nonlinear regime. The last region is of practical importance, and there the dependence is approximated by the expression

$$v_{max} = \sqrt{S} \sqrt[3]{10^{3-5r_0/R}} \quad \text{for} \quad S > 10^5.$$

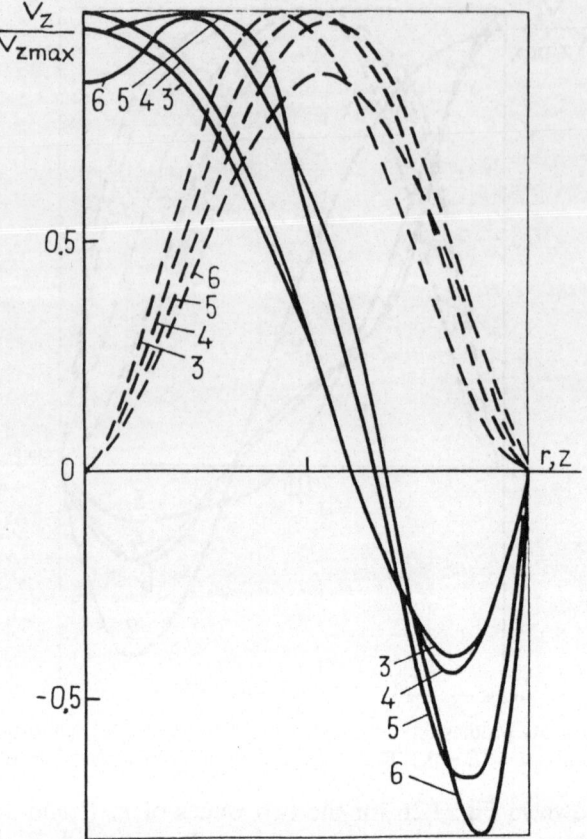

Fig. 4.25. The same as in Fig. 4.24 for $r_0/R = 0.8$.

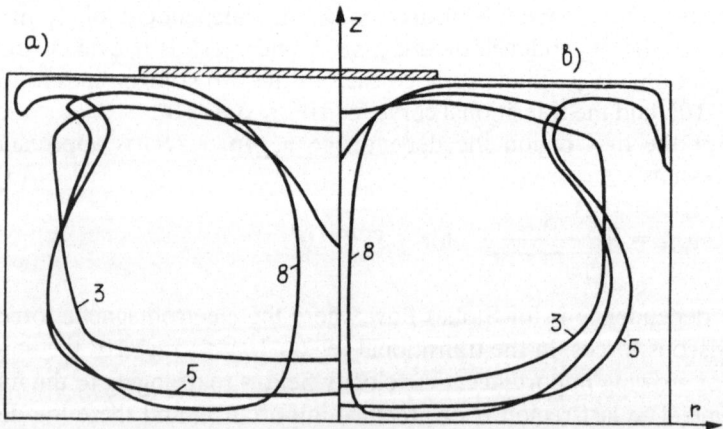

Fig. 4.26. Zero vorticity lines: (a) $r_0/R = 0.6$, (b) $r_0/R = 0.3$. The numbers on the lines are equal to log S.

Flows with cylindrical symmetry

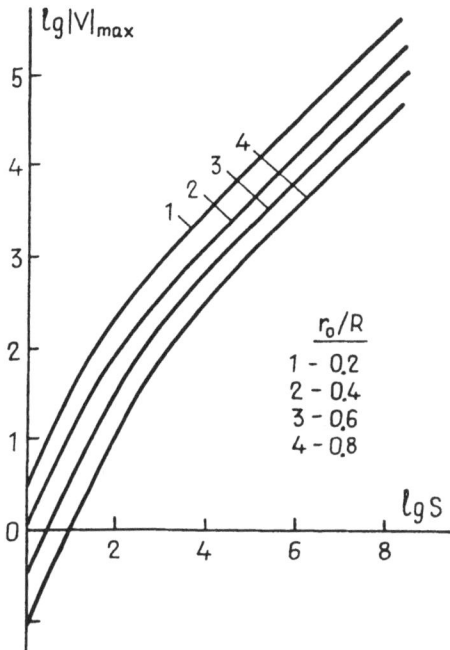

Fig. 4.27. The dependences of maximum velocities on S for different r_0/R.

In this regime the electromagnetic force is balanced in the main part of the flow by inertial forces quadratic in velocity, and the velocity growth is proportional to the current magnitude. The effect of viscosity is felt in the boundary layers. The most important feature of the velocity field after the rearrangement is that it stays undeformed with any further increase of S.

5

Periodic electrically induced flows

Periodic distributions of electric current density can be examined most conveniently and naturally in cylindrical coordinates. The corresponding periodic electrically induced flows constitute a new class of flows, which is interesting from the viewpoint either of novel physical effects or of applications.

5.1. Periodic distributions of current and magnetic field in cylindrical coordinates

Possible configurations of electric current flow, derived from the solutions of the equations (4.1.1) and (4.1.1a) for zero constant of separation n, were considered in the previous chapter. For a nonzero constant of separation $n \neq 0$ the solution of the equations and the corresponding distributions of electric current field and potential can be represented in a general form

$$\psi_1 = rB_\varphi/\mu_0 = r[aI_1(nr) + bK_1(nr)]e^{inz};$$
$$\Phi = i\sigma^{-1}[aI_0(nr) - bK_0(nr)]e^{inz}, \qquad (5.1.1)$$

where a, b, n are arbitrary complex constants.

Linear superpositions of (4.1.5) and the solutions (5.1.1) with different complex coefficients yield a great variety of current flow patterns typical for the cylindrical coordinate system and realizable in cylindrical volumes filled by a liquid conductor. Fig. 5.1 presents some typical distributions of the current and potential following from (5.1.1). The particular solutions (4.1.5) and (4.1.5a) can be added to each of them.

If we consider the possible current distributions in circular tubes and assume the domain to extend from $z = -\infty$ to $z = +\infty$ the condition that **B** and **j** are bounded requires the solutions (4.1.5) with $A = 0$, corresponding to the external and internal field of a circular conductor carrying a uniform current, and the solutions (5.1.1) with real n, representing a periodic dependence along z.

If the current-carrying region contains the symmetry axis $r = 0$, solutions

Periodic electrically induced flows

that are singular on the axis must be omitted, i.e. those with $D \neq 0$ in (4.1.5) and with the function $K_1(nr)$ in (5.1.1). They must be retained for the domain in a form of circular gap between coaxial tubes.

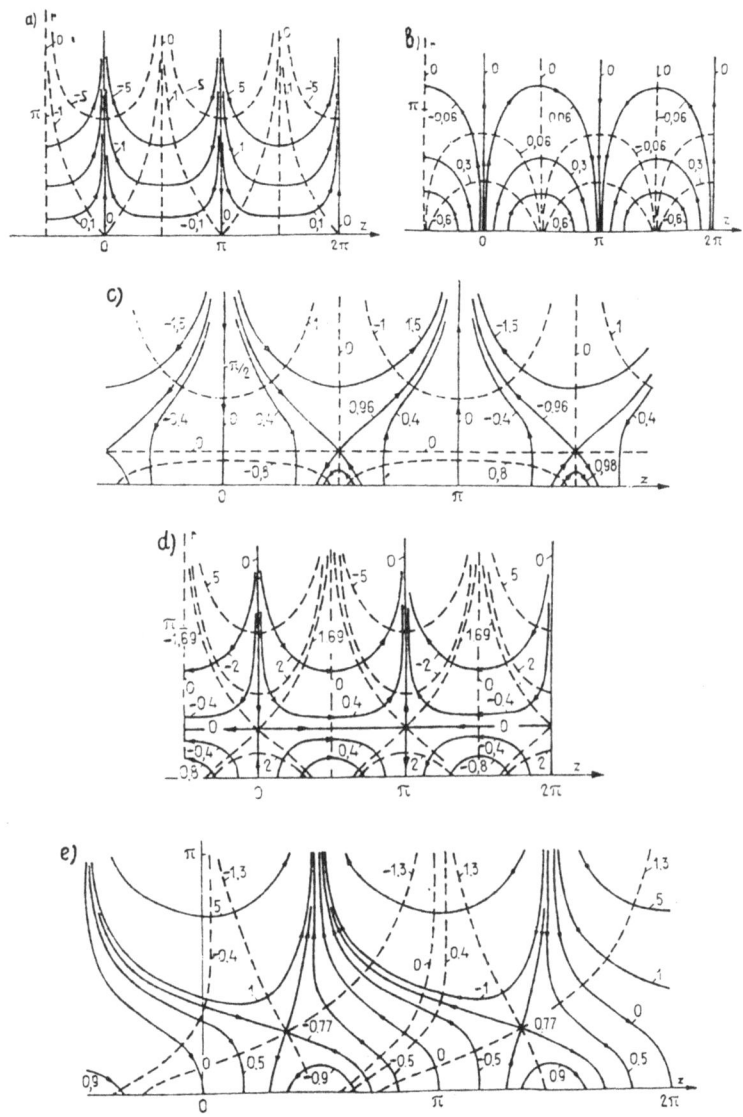

Fig. 5.1. The lines of electric current and of equal potential for

(a) $\psi_1 = -rI_1(r) \sin z$, $\Phi = -I_0(r) \cos z$;
(b) $\psi_1 = rK_1(r) \sin z$, $\Phi = K_0(r) \cos z$;
(c) $\psi_1 = r[I_1(r) - K_1(r)] \sin z$, $\Phi = [I_0(r) - K_0(r)] \cos z$;
(d) $\psi_1 = r[I_1(r) + K_1(r)] \sin z$, $\Phi = [I_0(r) + K_0(r)] \cos z$;
(e) $\psi_1 = rI_1(r) \cos z - rK_1(r) \sin z$, $\Phi = -I_0(r) \sin z - K_0(r) \cos z$.

In the case of real n the solutions (5.1.1) are monotonous along r. If the boundary conditions (e.g. a potential distribution) are given on a nonperiodic surface of rotation, they may be satisfied by using a Fourier integral with the functions e^{inz}. Yet if the bounding surface is periodic along z, the conditions stated on the surface are satisfied using Fourier series of the same functions with n being natural.

In EVF problems an external factor, affecting the flow structure, may be the external magnetic field, induced by currents outside the liquid conductor, which must satisfy the equations div $\mathbf{B} = 0$ and curl $\mathbf{B} = 0$ in the volume of the conductor. From curl $\mathbf{B} = 0$ it follows that the uniquely possible distribution of an irrotational azimuthal field is $B_\varphi = \text{const}/r$ — the field outside an axisymmetric current carrying element. The meridional magnetic field $\mathbf{B}_M = (B_z(z, r), B_r(z, r), 0)$ can be represented by a potential: $\mathbf{B}_M = \text{grad } \Phi_M$. The potential $\Phi_M(z, r)$ is determined by solving the equation $\nabla^2 \Phi_M = 0$. The feasible solutions of the equation in the cylindrical coordinate system are discussed above. Hence the distributions of the irrotational meridional magnetic field correspond to the distributions of the meridional electric current responsible for the rotational azimuthal magnetic field, e.g. the current distributions shown in Fig. 5.1. The external magnetic field is singular at $r = 0$ or $r = \infty$, or $|z| \to \infty$, i.e. where the field sources are located.

5.2. Integral action of electromagnetic force

It is of interest in applied problems of EVF to consider not only the local flow quantities, but also the flow characteristics for the fluid as a whole: the total flow rate through the tube's cross section, mean fluid rotation (angular momentum) relative to the tube's wall, mean pressure drop along the tube, drag coefficient, and so forth. Certain integral characteristics can be evaluated or estimated by applying the conservation laws of momentum and angular momentum to the whole fluid volume.

Different kinds of volume and contact forces act in the fluid volume. We shall express first the total electromagnetic force and the associated total angular momentum in the volume by use of the conservation laws, and then we shall proceed to the evaluation of these quantities by taking into account all the forces acting in the fluid volume. At the same time, let the reader note that, if the conductor is liquid, it is impossible to measure the electromagnetic force and its moment directly. These can be determined by the reaction forces on the current supplying elements and the electromagnetic system generating the external magnetic field.

The relevant integral quantity can be evaluated by integrating over the volume of the conductor the force $\mathbf{f}_e = \mathbf{j} \times \mathbf{B}$ and the moment $\mathbf{m}_e = \mathbf{R} \times \mathbf{f}_e$. The computation can also be performed in a different way with certain advantages. The method is often used in hydrodynamics, and it includes the surface integral of flux density of the relevant quantity.

Periodic electrically induced flows 199

The momentum flux associated with the magnetic field in the vicinity of a point is determined by the Maxwell stress tensor σ^e [8] with the components

$$\sigma_{ik}^e = \frac{1}{\mu_0}\left(B_i B_k - \frac{|\mathbf{B}|^2}{2}\delta_{ik}\right) \quad (5.2.1)$$

where δ_{ik} is the Cronecker δ-symbol (unit second order tensor). Equation (5.2.1) neglects the volume forces due to the electric field.

Electromagnetic force density is related to the stress tensor σ^e by

$$\mathbf{f}_e = \operatorname{div} \sigma^e, \quad (5.2.2)$$

cf. (1.1.16), (1.1.17). Applying Gauss's theorem, the total electromagnetic force is a volume V is expressed as

$$\mathbf{F}_e = \int_V \mathbf{f}_e \, dV = \mathbf{i}_j \oint_S \sigma_{nj}^e \, dS \quad (5.2.3)$$

where S is the closed surface bounding the volume V, n is the unit normal externally oriented to the surface, and summation is implied over the repeated indices.

Similarly, to simplify the computation of the integral moment $\mathbf{M}_e$, a tensor μ^e should be constructed for angular momentum, so that $\mathbf{m}_e = \operatorname{div} \mu^e$. It can be shown that $\mathbf{R} \times \operatorname{div} \sigma^e = \operatorname{div}(\mathbf{R} \times \sigma^e)$ for a symmetric tensor σ^e. From this the tensor $\mu^e = \mathbf{R} \times \sigma^e$ and its components are

$$\mu_{ik}^e = R_j \sigma_{\ell k}^e \varepsilon_{j\ell i} \quad (5.2.4)$$

where $\varepsilon_{j\ell i}$ is the absolute antisymmetric third rank tensor, R_j are components of the radius-vector, and summation is also implied over the repeated indices.

The total momentum of the electromagnetic force acting in the volume V is determined by the integral:

$$\mathbf{M}_e = \int_V \mathbf{m}_e \, dV = \mathbf{i}_j \oint_S \mu_{nj}^e \, dS. \quad (5.2.5)$$

For the axisymmetric distributions of the magnetic field considered here, including both the self-field associated with the currents inside the volume and the external field, the integral action of electromagnetic field onto an axisymmetric volume is expressed by two dynamic parameters: the force F_{ez} acting parallel to the symmetry axis and the component of moment M_{ez} along the same axis. Other components of the vectors $\mathbf{F}_e$ and $\mathbf{M}_e$ are zero due to the axial symmetry.

Note that we are discussing only the integral action of the force on a conductor as a whole, not the effect of the electromagnetic force on the integral motion in a liquid conductor. In this context the results presented are equally valid either for a liquid conductor or a solid one. As we shall see later, the effect of an electromagnetic force on the hydrodynamics of a liquid conductor, entirely or partly enclosed by rigid walls, can lead to an unexpected resultant motion.

Outside the current carrying volume, where the magnetic field is irrotational (curl **B** = 0), the electromagnetic force **f**$_e$ is zero, and we have

$$\text{div } \sigma^e = 0, \qquad \text{div } \mu^e = 0, \tag{5.2.6}$$

thence the surface S surrounding the given volume may be deformed arbitrarily, retaining the result of integration in (5.2.3) and (5.2.5). It is advantageous in cylindrical coordinates to choose the control surface in the form of a cylinder whose lateral surface lies in the space where the electric current is absent (Fig. 5.2). For periodic problems the length of cylinder along z is assumed to be equal to the period h. Note also that the results for periodic problems that are presented below are also valid in situations in which the electric current lines cross the lateral surface, because the inflowing current is equal to the outflowing one and the total current is zero.

We define the electromagnetic momentum flux over the closed surface of the cylinder, i.e. the force acting on a current-carrying body wholly contained within the surface:

$$F_{ez} = \oint_S \sigma^e_{zn} \, dS = \int_{S_1} \sigma^e_{zz} \, dS + \int_{S_2} \sigma^e_{zr} \, dS - \int_{S_3} \sigma^e_{zz} \, dS \tag{5.2.7}$$

The explicit expressions for the Maxwell stress tensor's components follow from (5.2.1):

$$\sigma^e_{zz} = \frac{1}{2\mu_0}(B_z^2 - B_r^2 - B_\varphi^2), \qquad \sigma^e_{zr} = \frac{1}{\mu_0} B_z B_r. \tag{5.2.8}$$

The meridional field components B_z, B_r are associated with the external currents, which do not penetrate the surface S. In (5.2.8) these components enter by the separate terms from the B_φ terms. Therefore, the integral result

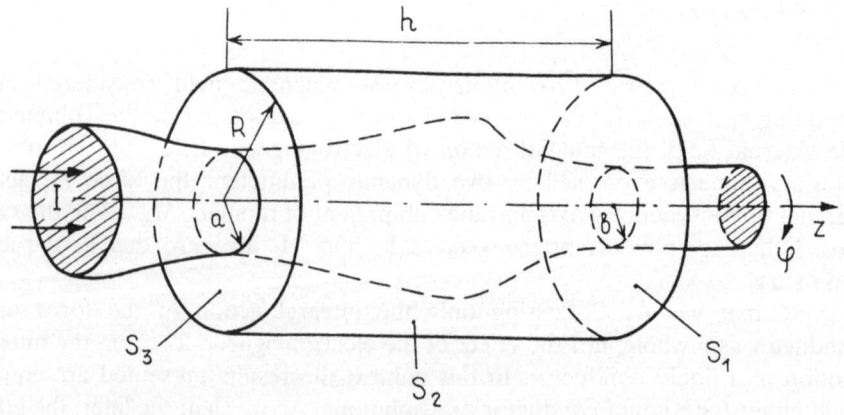

Fig. 5.2. To determine the momentum flux of electromagnetic field.

Periodic electrically induced flows

(5.2.7), in light of (5.2.6), does not depend on the meridional field. Hence, instead of (5.2.7), we have

$$F_{ez} = \frac{\pi}{\mu_0}\left[-\int_0^R B_\varphi^2(z_0, r)r\,dr + \int_0^R B_\varphi^2(z_0 + h, r)r\,dr\right]. \quad (5.2.9)$$

The magnitudes of both the terms in (5.2.9) will be equal if the sections S_1 and S_3 are chosen to lie at a distance equal to the period along z, then the current distribution in the sections is identical. Thereby there are no axisymmetric periodic current distributions including the unsymmetric relative to the z-axis direction change, which can produce the total electromagnetic force acting on the conductor. The total force F_{ez} will arise, for instance, in an expanding conductor. Thus, assuming uniform current densities in the circular sections of radii a and b (see Fig. 5.2), we obtain from (5.2.9)

$$F_{ez} = \frac{\mu_0 I^2}{4\pi} \ln a/b, \quad (5.2.10)$$

where I is the total current.

Similarly the moment M_{ez} can be expressed as the flux of the electromagnetic field's angular momentum across the same surface of the cylinder $S = S_1 + S_2 + S_3$. For this purpose we need components of the tensor (5.2.4)

$$\mu^e_{zz} = \frac{rB_\varphi B_z}{\mu_0}, \quad \mu^e_{zr} = \frac{rB_\varphi B_r}{\mu_0}, \quad (5.2.11)$$

which give a contribution to the z-component of the moment for the specified surface S. Then

$$M_{ez} = \int_{S_1} \mu^e_{zz}\,dS + \int_{S_2} \mu^e_{zr}\,dS - \int_{S_3} \mu^e_{zz}\,dS. \quad (5.2.12)$$

From (5.2.11) it follows that a current-carrying body could be made to rotate only in the presence of an external meridional magnetic field. Further, since the electric current does not cross the lateral surface S_2, the value of rB_φ stays constant on it. In the case of a periodic (with the period h) meridional magnetic field the second integral in (5.2.12) turns out to be zero due to the fact that div **B** = 0. If the current is also periodic in the same volume, then the first and third integrals balance each other so that the total moment is zero.

The moment will arise if the external field is periodic, yet the current is nonperiodic for the length h. For example, the current in the situation shown in Fig. 5.2 with the uniform axial magnetic field $B_z = B_0$ yield the moment

$$M_{ez} = B_0 I(b^2 - a^2)/4.$$

The moment will arise also for a nonperiodic external field and a periodic current, and for both the quantities being nonperiodic.

As we have noted previously, the results are in fact valid both for a rigid and liquid current-carrying body. In the case of a liquid conductor, enclosed in a rigid tube with rigid current supplying electrodes, the electrically induced flow gives rise to viscous stresses and a specific pressure distribution at the walls, which may produce the integral force and torque acting on the walls. In this case the conserved quantities are the total momentum and angular momentum of the fluid—rigid wall integral system, where both conducting and nonconducting wall areas must be taken into account.

To determine the integral force acting on a fluid volume, the tensor Π (1.1.17) can be used; its components in cylindrical coordinates in the axisymmetric case are the following:

$$\Pi_{zz} = p + \frac{B^2}{2\mu_0} + \rho v_z^2 - 2\rho\nu \frac{\partial v_z}{\partial z} - \frac{B_z^2}{\mu_0};$$

$$\Pi_{rr} = p + \frac{B^2}{2\mu_0} + \rho v_r^2 - 2\rho\nu \frac{\partial v_r}{\partial r} - \frac{B_r^2}{\mu_0};$$

$$\Pi_{\varphi\varphi} = p + \frac{B^2}{2\mu_0} + \rho v_\varphi^2 - 2\rho\nu \frac{v_r}{r} - \frac{B_\varphi^2}{\mu_0};$$

$$\Pi_{r\varphi} = \Pi_{\varphi r} = \rho v_r v_\varphi - \rho\nu \left(\frac{\partial v_\varphi}{\partial r} - \frac{v_\varphi}{r} \right) - \frac{B_r B_\varphi}{\mu_0}; \qquad (5.2.13)$$

$$\Pi_{\varphi z} = \Pi_{z\varphi} = \rho v_\varphi v_z - \rho\nu \frac{\partial v_\varphi}{\partial z} - \frac{B_\varphi B_z}{\mu_0};$$

$$\Pi_{zr} = \Pi_{rz} = \rho v_z v_r - \rho\nu \left(\frac{\partial v_z}{\partial r} + \frac{\partial v_r}{\partial z} \right) - \frac{B_z B_r}{\mu_0}.$$

If we seek the integral force, acting on the rigid wall which wholly or partly (in periodic problems) bound the fluid volume, then, in face of the no slip conditions on the wall $\mathbf{v} = 0$, from the tensor Π there remains merely the part Π^0 related to the stress tensor (1.1.19). Moreover, if the wall is curvilinear, it is advantageous to introduce on its surface the natural coordinate system $q_1 = \ell$, $q_2 = n$, $q_3 = \varphi$ with $H_1 = H_2 = 1$, $H_3 = r$ so that the unit vector $\mathbf{i}_2$ is normal to the wall, and it points outside of the fluid (i.e. $\mathbf{i}_2 = \mathbf{n}$). Then the coordinate surface $q_2 = $ const coincides with the rigid wall, and the unit vector $\mathbf{i}_1 = \mathbf{l}$ is tangent to the surface.

The integral momentum flux from the fluid through the rigid surface, i.e. the force acting on the rigid surface, is expressed by the integral $F_z = \int \Pi_{nz}^0 \, dS = 2\pi \int r \Pi_{nz}^0 \, d\ell$. Substituting (1.1.19), taking account of (5.2.13) and the relation between the coordinates ℓ, n and z, r:

$$\mathbf{l} \cdot \mathbf{i}_z = \frac{\partial r}{\partial n}; \quad \mathbf{n} \cdot \mathbf{i}_z = -\frac{\partial z}{\partial \ell},$$

we obtain

$$F_z = \pi \int r^2 \frac{\partial p}{\partial \ell} d\ell - 2\pi\rho\nu \int \frac{\partial r}{\partial n} E^2\psi \, d\ell - 4\pi\rho\nu \int \frac{\partial}{\partial \ell}(rv_r) d\ell +$$

$$+ 2\pi \int \sigma^e_{nz} r \, d\ell - \pi \int \frac{\partial}{\partial \ell}\left[r^2\left(p + \frac{B_\varphi^2}{2\mu_0}\right)\right] d\ell, \quad (5.2.14)$$

where the integration path lies along the rigid wall in a meridional plane. The third integral in (5.2.14) is zero due to the no slip condition; the fourth coincides with (5.2.3) and determines the electromagnetic force. The last integral reduces to $\pi R_0^2 \Delta p$ where Δp is the pressure drop between the surface points separated by the period length, R_0 is the radius of tube in the initial (and final) integration point. The terms with B_φ^2 compensate due to equal values of B_φ in the two end points.

In the Stokes approximation the relation between pressure and stream function is [7]:

$$\frac{\partial p}{\partial \ell} = \frac{\rho\nu}{r} \frac{\partial}{\partial n} E^2\psi; \quad \frac{\partial p}{\partial n} = -\frac{\rho\nu}{r} \frac{\partial}{\partial \ell} E^2\psi.$$

Thence the first two integrals in (5.2.14) can be combined into one:

$$\pi\rho\nu \int r^3 \frac{\partial}{\partial n} \frac{E^2\psi}{r^2} d\ell.$$

Note, finally, that expression (5.2.14) is valid either for electric current lines that pass through the wall or not.

The angular momentum flux is expressed by the tensor μ^0_{ik} with the components $\mu^0_{ik} = R_j \Pi^0_{\ell k} \varepsilon_{ij\ell}$ where $\Pi^0_{\ell k}$ is defined in (1.1.19). Then the torque by which the fluid acts on the wall is expressed as

$$M_z = \oint_S \mu^0_{nz} \, dS = 2\pi \int r^2 \Pi^0_{\varphi n} d\ell$$

$$= 2\pi\rho\nu \int r \frac{\partial}{\partial n}(v_\varphi r) d\ell + \frac{2\pi}{\mu_0} \int r^2 B_n B_\varphi d\ell.$$

5.3. A method used to construct linear solution of periodic EVF in tubes

For the theoretical study of electrically induced flows in tubes we specify the following simplifying assumptions. As usual, axisymmetric flows will be considered. We assume the flows to be periodic along the symmetry axis. This simplifies the problem, since in an infinitely long tube we consider a consecutive series of elements instead of single or several localized elements of nonuniformity. This periodicity assumption permits us to decrease the flow volume that we have to study by imposing periodic boundary conditions on it. Sometimes the interpretation of periodicity may be that of a hydraulic system (tube) closed on itself.

We shall use the electromagnetic approximation in constructing the electromagnetic force. A substantial simplification is also the neglect of inertial terms, i.e. the flow is studied in the Stokes approximation. This may be motivated by the possibility of constructing a closed-form solution for the velocity field.

The Stokes equation for meridional flow in the cylindrical coordinates follows from (4.2.2) if the terms quadratic in ψ are neglected:

$$E^4\psi = -\frac{\mu_0}{\rho v r^2}\frac{\partial}{\partial z}\psi_1^2. \tag{5.3.1}$$

Solution of the equation (5.3.1) may be represented in the form $\psi = \psi_0 + \psi_p$, where ψ_0 is a general solution of the biharmonic equation $E^4\psi_0 = 0$, and ψ_p is a particular solution of the nonhomogeneous equation (5.3.1). The general solution $\psi_0(z, r)$ is

$$\psi_0 = A_0 r^4 + B_0 r^2 + C_0 r^2 \ln r + D_0 +$$

$$+ \sum_n [A_n r^2 I_0(nr) + B_n r I_1(nr) + C_n r^2 K_0(nr) + D_n r K_1(nr)] \cdot e^{inz}. \tag{5.3.2}$$

The constants A_n, B_n, C_n, D_n ($n = 0, \pm 1, \pm 2 \ldots$) are to be determined by boundary conditions. In particular, for the problem containing a symmetry axis the symmetry conditions are set on the axis:

$$v_r = -\frac{1}{r}\frac{\partial\psi}{\partial z} = 0, \quad \frac{\partial v_z}{\partial r} = \frac{\partial}{\partial r}\frac{1}{r}\frac{\partial\psi}{\partial r} = 0 \quad \text{for} \quad r = 0. \tag{5.3.3}$$

Then $C_n = D_n = 0$ for any n.

To determine the particular solution ψ_p a specific distribution of $\psi_1 = rB_\varphi/\mu_0$ must be known. As it was stated above (Section 5.1), the azimuthal magnetic field associated with a meridional electric current periodic along z is represented in a general case by the series:

$$\psi_1 = a_0 r^2 + b_0 + r\sum_n [a_n I_1(nr) + b_n K_1(nr)]e^{inz}. \tag{5.3.4}$$

The curl of electromagnetic force (right-hand part of equation (5.3.1)) then is written in the form

$$\frac{1}{r^2}\frac{\partial}{\partial z}\psi_1^2 = 2i\left(a_0 r + \frac{b_0}{r}\right)\sum_n [a_n I_1(nr) + b_n K_1(nr)]n e^{inz} +$$

$$+ i \sum_{n,\ell \geq n} C_{n\ell}[a_n I_1(nr) + b_n K_1(nr)][a_\ell I_1(\ell r) +$$

$$+ b_\ell K_1(\ell r)](n + \ell)e^{i(n+\ell)z}, \tag{5.3.5}$$

Periodic electrically induced flows

where $C_{nn} = 1$, and $C_{n\ell} = 2$ if $n \neq \ell$. It is evident from (5.3.5) that the right-hand part of (5.3.1) may include only the following terms (with constant multipliers):

$$rI_1(nr)e^{inz}, \quad rK_1(nr)e^{inz}, \quad r^{-1}I_1(nr)e^{inz}, \quad r^{-1}K_1(nr)e^{inz},$$
$$I_1(nr)I_1(\ell r)e^{i(n+\ell)z}, \quad K_1(nr)K_1(\ell r)e^{i(n+\ell)z},$$
$$I_1(nr)K_1(\ell r)e^{i(n+\ell)z}. \tag{5.3.6}$$

The constants n and ℓ in the expressions can be any integers.

In the particular solution $\psi_p(z, r) = \gamma(r)Z(z)$ of the equation (5.3.1), corresponding to each component of (5.3.6), the periodic z-dependence in the form of factor e^{imz} is retained. Part of the solution (5.3.1) which depends on r is found simply by guesswork, since we have no general method to determine the particular solution. The results of this process are classified in Table 5.1, where $f_n(r)$ represents each of the functions (5.3.6) depending on r, $\gamma_n(r)$ — the respective particular solution, i.e. if $\psi_p = \gamma_n(r)e^{inz}$ then $E^4\gamma_n(r)e^{inz} = f_n(r)e^{inz}$.

Thereby the general solution of the equation (5.3.1) can be constructed for any periodic electric current distribution. Full solution of a particular problem

Table 5.1. Particular solutions of the equation (5.3.1).

No.	$f_m(r)$	$\gamma_m(r)$	e^{imz}
1.	$rI_1(nr)$	$r^3(n^2/8)I_1(nr)$	e^{inz}
2.	$rK_1(nr)$	$r^3(n^2/8)K_1(nr)$	e^{inz}
3.	$r^{-1}I_1(nr)$	$\dfrac{r^2}{4n^2}\left[\dfrac{I_0(nr)}{nr} - \ln(nr)I_1(nr)\right]$	e^{inz}
4.	$r^{-1}K_1(nr)$	$\dfrac{r^2}{4n^2}\left[\dfrac{K_0(nr)}{nr} + \ln(nr)K_1(nr)\right]$	e^{inz}
5.	$I_1(nr)I_1(\ell r)$	$(n-\ell)r(n+\ell)^{-3}/4[I_0(nr)I_1(\ell r)/n - I_0(\ell r)I_1(nr)/\ell] -$ $- r^2(n+\ell)^{-2}/2[I_0(nr)I_0(\ell r) + I_1(n\ell)I_1(\ell r)/\ell]$	$e^{i(n+\ell)z}$
6.	$K_1(nr)K_1(\ell r)$	$\dfrac{(n-\ell)r}{4(n+\ell)^3}\left[\dfrac{K_0(nr)K_1(\ell r)}{n} - \dfrac{K_1(nr)K_0(\ell r)}{\ell}\right] -$ $- \dfrac{r^2}{(n+\ell)^2 2}[K_0(nr)K_0(\ell r) + K_1(nr)K_1(\ell r)]$	$e^{i(n+\ell)z}$
7.	$I_1(nr)K_1(\ell r)$	$\dfrac{(n-\ell)r}{4(n+\ell)^3}\left[\dfrac{I_1(nr)K_0(\ell r)}{\ell} + \dfrac{I_0(nr)K_1(\ell r)}{n}\right] +$ $+ \dfrac{r^2}{2(n+\ell)^2}[I_0(nr)K_0(\ell r) - I_1(nr)K_1(\ell r)]$	$e^{i(n+\ell)z}$

depends, first, on the specific form of electric current (or magnetic field (5.3.4)) and on external flow constraints (an applied drop of pressure or flow rate), and second, the specific boundary conditions.

It is instructive to see the method applied in some simple examples [14]. Consider a flow in a circular tube. Then the fields **v** and **B** must be regular at the axis $r = 0$, and the stream function, satisfying (5.3.3), takes the form

$$\psi = A_0 r^2 + B_0 r^4 + \psi_{\text{per}}, \tag{5.3.7}$$

where

$$\psi_{\text{per}} = \sum_n [A_n r^2 I_0(nr) + B_n r I_1(nr) + \gamma_n(r)] e^{inz}$$

is the periodic part and the functions $\gamma_n(r)$ represent the particular solutions from Table 5.1.

The boundary conditions of no slip and impermeability are set on the tube's wall, which is specified by $r = R(z)$. If the coordinates n, ℓ are introduced at the surface, as described in the previous paragraph, then we have

$$v_n = -\frac{1}{r}\frac{\partial \psi}{\partial \ell} = 0, \quad v_\ell = \frac{1}{r}\frac{\partial \psi}{\partial n} = 0 \quad \text{for} \quad r = R(z).$$

From the condition of vanishing normal velocity $\partial \psi / \partial \ell = 0$ it follows that $\psi = $ const along the rigid wall. Recalling the physical meaning of the stream function (see Section 1.5) and setting $\psi = 0$ on the axis we can determine $\psi = Q/2\pi$ by Q — the flow rate over the cross section of the tube.

The second boundary condition is conveniently replaced in our case by the condition $\partial \psi / \partial r = 0$. From the relation between the coordinates $z, r; n, \ell$ on the bounding surface it follows that

$$\frac{\partial \psi}{\partial r} = \frac{\partial \psi}{\partial n}\frac{\partial n}{\partial r} + \frac{\partial \psi}{\partial \ell}\frac{\partial \ell}{\partial r}$$

hence the boundary conditions

$$2\pi\psi = Q \qquad \partial\psi/\partial r = 0 \quad \text{for} \quad r = R(z) \tag{5.3.8}$$

are equivalent to the original ones if $\partial n/\partial r \neq 0$ on the surface $r = R(z)$, i.e. there are no instantaneous expansions or contractions of the tube.

The boundary conditions (5.3.8) determine uniquely the stream function (5.3.7), i.e. the constants A_n, B_n. Substituting (5.3.7) in (5.3.8) and specifying $r = R_0$, we obtain for each harmonic $n \neq 0$ two equations to determine the coefficients $A_n = A_n(R_0)$, $B_n = B_n(R_0)$ (we denote henceforth by $\alpha_n(R_0) = R_0^2 I_0(nR_0)$, $\beta_n(R_0) = R_0 I_1(nR_0)$):

$$\begin{aligned} A_n \alpha_n(R_0) + B_n \beta_n(R_0) + \gamma_n(R_0) &= 0, \\ A_n \alpha'_n(R_0) + B_n \beta'_n(R_0) + \gamma'_n(R_0) &= 0; \end{aligned} \tag{5.3.9}$$

Periodic electrically induced flows

The nontriviality of the solution to the set (5.3.9) is guaranteed for the linearly independent functions α_n, β_n and γ_n:

$$A_n = W[\beta_n, \gamma_n]/W[\alpha_n, \beta_n];$$
$$B_n = -W[\alpha_n, \gamma_n]/W[\alpha_n, \beta_n]; \tag{5.3.10}$$

where

$$W[f_1, f_2] = f_1 f_2' - f_1' f_2.$$

The constants $A_n, B_n \neq 0$ for $n \neq 0$, if $\gamma_n \neq 0$.

The solution determines ψ_{per} — a periodic meridional flow in the form of toroidal vortices of the same structure and intensity, and of an alternating sense of rotation. Such an electrically induced flow can generate neither a transitional flow in the tube, nor a mean pressure gradient along the tube, since both directions along the tube's axis are equivalent. If the external hydraulic circuit does not contain motive sources, and the tube is of a constant cross section, the constants A_0, B_0 should be set equal to zero.

Until now we have discussed a cylindrical tube. A closed form solution $\psi(z, r)$ is also possible for a tube with a non-straight wall if the deformation is small:

$$r = R_0(1 + \varepsilon \xi(z)), \qquad \varepsilon \ll 1, \tag{5.3.11}$$

where R_0 is the mean radius of the tube, and the function $\xi(z)$ determines the deformation of cylindrical surface — a periodic corrugation with the repeated period h: $\xi(z) = \xi(z + h)$; ε is the amplitude of deformation. The solution may be sought in the form of series expansion in powers of the small parameter ε. For this purpose all the constants A_n, B_n are expanded in the series

$$A_n = \sum_m \varepsilon^m A_{nm}, \qquad B_n = \sum_m \varepsilon^m B_{nm}. \tag{5.3.12}$$

To satisfy the boundary conditions (5.3.8) at the boundary (5.3.11) with a required accuracy of ε the functions of r in (5.3.7) are also expanded in Taylor series at the tube's wall, for example

$$a(r) = a(R_0) + a'(R_0) R_0 \varepsilon \xi(z) + \tfrac{1}{2} a''(R_0) R_0^2 \varepsilon^2 \xi^2(z) + \cdots$$

Substituting these expressions in the boundary conditions, and collecting the terms of equal order in ε and the summands of equal Fourier harmonics in z, we obtain the sets of algebraic equations to determine the constants A_{nm} and B_{nm}. The constants A_{n0}, B_{n0} are those determined previously for the straight tube.

A transit flow in the tube is determined by the two constants A_0 and B_0 in (5.3.7), corresponding to the parabolic profile of mean velocity. These can be equivalently replaced by the mean pressure gradient along z: $G = \overline{\partial p/\partial z}^z = 16\rho\nu B_0$ and the total flow rate Q, which is also independent of z. The boundary conditions (5.3.8) for a known ψ_{per} reduce to the set of two equations to determine three constants A_0, B_0, Q. On eliminating A_0 and replacing B_0 by G, we obtain:

$$\frac{\pi R^4}{8\rho v} G + Q = -\frac{R(z)^3}{2}\left(\frac{\psi_{\text{per}}(R(z), z)}{R^2(z)}\right)^{(1)}. \tag{5.3.13}$$

The average of the equation (5.3.13) along z leads to the expression equivalent to the pressure-flow rate $(G(Q))$ characteristic of the tube:

$$\frac{\pi R_0^4 \lambda(\varepsilon)}{8\rho v} G + Q = T(\varepsilon), \tag{5.3.14}$$

where $\lambda(\varepsilon)$ is the variation of hydraulic drag for the corrugated tube relative to the smooth tube of radius R_0, and the nonzero term on the right:

$$T(\varepsilon) = -\frac{1}{2h}\int_0^\pi \left[r^3 \frac{\partial}{\partial r}\left(\frac{\psi_{\text{per}}(r, z)}{r^2}\right)\right]_{r=R(z)} dz \tag{5.3.15}$$

determines the pump effect of the EVF. For the present the linear approximation $G(Q)$-curve determined by (5.3.14) is straight, and its inclination (i.e. the hydraulic drag) is determined by the quantity $\lambda(\varepsilon)$. For the tube in which the pump effect is absent $T(\varepsilon) = 0$ and the line $G(Q)$ goes through the coordinate origin: $G = 0$, $Q = 0$. In the case with the pump effect, $T(\varepsilon) \neq 0$, a nonzero flow rate $Q = T(\varepsilon)$ for $G = 0$ (unloaded regime) or a pressure gradient for zero flow rate (choked regime) are possible, and the curve $G(Q)$ misses the coordinate origin at a distance determined by the quantity $T(\varepsilon)$.

In the corrugated tube vortices in the fluid ($\psi_{\text{per}} \neq 0$) are generated either due to the electric current $I \neq 0$ (EVF) or due to the flow rate resulting from the externally applied pressure drop (secondary flows). In the linear approximation the secondary flow merely increases the hydraulic drag, i.e. it determines $\lambda(\varepsilon)$, yet EVF is capable of generating the pump effect, i.e. $T(\varepsilon)$. We are interested only in the pump effect of EVF in a corrugated tube. To establish the existence of pump effect it is sufficient to determine the first nonzero term T_m in the expansion $T(\varepsilon) = \Sigma_{n=m}^\infty \varepsilon^n T_n$ (if $T_n = 0$ for $n < m$). According to (5.3.13) T_m is determined by the periodic solution $\psi_{\text{per}} = \Sigma_{n=1}^{m-1} \varepsilon^n \psi_n$ (to an accuracy of ε^{m-1}), since, in light of (5.3.15), ψ_n terms with $n \geq m$ does not contribute to T_m. We are not interested in the variation of hydraulic drag due to the corrugations, whence G and Q in (5.3.14) are assumed to be of the order ε^m and, respectively, $\lambda = 1$. This approach is set up due to unwieldy expressions which grow with an accuracy ε and with the number of harmonics.

The function ψ_{per} in (5.3.7) and the term $T(\varepsilon)$ in $G(Q)$ relation (5.3.14) are conveniently normalized by $\mu_0 I^2 R_0/\rho v$. Then instead of (5.3.14) we have

$$\rho v Q + \pi R_0^4 G/8 = \mu_0 I^2 R_0 \varepsilon^m [\delta + O(\varepsilon)], \tag{5.3.16}$$

and the pump effect for a constant mean radius R_0 of tube and a constant amplitude of corrugation $d = \varepsilon R_0$ is given by the quantity $\delta = T_m \rho v / \mu_0 I^2 R_0$, which is determined by the profile of corrugation, i.e. by the function $\xi(z)$ in

Periodic electrically induced flows 209

(5.3.11) and by the ratio of the period to the radius h/R_0. Our aim is to find the quantity δ, then the pump effect of EVF would be demonstrated.

The method can also be applied in the case of an annular tube in which either the outer wall or the inner one is corrugated. For this case the number of algebraic equation sets (5.3.9) is doubled.

Note that a situation may be considered in which the tube's wall coincides neither with the electric current line, nor with the equipotential line. The single requirement is that the periods of the wall's corrugation and of the electric current be comparable, and then the Fourier series expansion method can be applied.

5.4. EVF in a tube with radial current supply

The simplest nonuniform periodic distribution of azimuthal magnetic field is given by the first term of series (5.3.4), i.e. by the single harmonic

$$\psi_1 = ckrI_1(kr) \sin kz, \qquad \Phi = dI_0(kr) \cos kz. \tag{5.4.1}$$

The wave number $k = 2\pi/h$ is introduced, h is the periodic length of the current distribution. This distribution (see Fig. 5.1a) leads to a simple expression for the right-hand term of the Stokes equation, which simplifies the solution and its analysis. To begin with, let us consider a hypothetical situation in which the current and potential distributions (5.4.1) are induced.

According to (5.4.1), the electric current lines ψ_1 = const and the equipotential lines Φ = const extend to infinity in a radial direction and in periodic patterns. Therefore, a surface of the current-carrying fluid volume can be bounded neither by the electric current lines nor by the equipotential lines. The surface can be constructed by an interleaving set of both lines. The sections of the surface are formed by the electric current lines, given implicitly by $rI_1(kr)|\sin kz|$ = const, and the intermediate sections by the equipotential lines, $I_0(kr)|\cos(kz)|$ = const.

The generatrix of the rotational body's surface consists of segments of the current lines and the equipotential lines situated between the neighbouring points of intersection. The current lines represent the boundary section in contact with an insulator, and the equipotential lines represent the contact surface with an ideal conductor — a ring electrode supplying or withdrawing the current (Fig. 5.3). The expression (5.4.1) is then valid inside the central conducting body. The current distribution inside the electrodes does not matter, yet; the only important restriction is that their surface must be equipotential. In practice this may be realized by choosing the material for the electrodes of much higher conductivity than the central current-carrying body. The typical current in the situation considered, passing the narrow section, is $I = 2\pi ckR_0 I_1(kR_0)$ i.e. half of the current is supplied by the single electrode. This specifies the expression for the constant c.

Suppose there is a rigid current-carrying body of rotation with a cavity filled with liquid of the same conductivity as the rigid conductor between the equi-

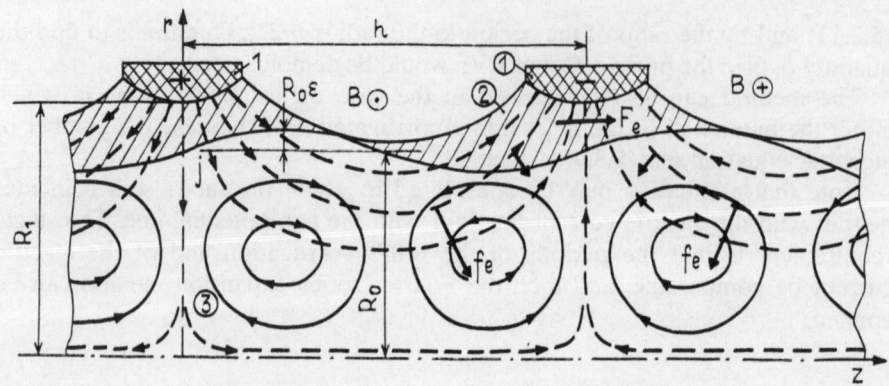

Fig. 5.3. Scheme of current supply to the corrugated tube: 1 — ring electrodes, 2 — electrically conducting wall, 3 — liquid metal. Dashed lines show the electric current lines, full lines — the primary vortices in the fluid.

potential surfaces, and there is no contact resistance between the liquid and rigid conductors. Then the previous distribution of the electromagnetic field is realized.

Consider a situation in which the wall of the tube is slightly deformed in the sinusoidal form:

$$r = R_0(1 + \varepsilon \sin 2k(z - z_0)), \qquad \varepsilon \ll 1, \qquad (5.4.2)$$

with the period of corrugation equal to the period of the electromagnetic force. Generally the corrugation is shifted in phase relative to the electrodes; the shift is specified by the distance z_0. If $|z_0| < h/8$ the configuration of current lines and the electromagnetic force are distributed in the fluid asymmetrically relative to the sinusoid of corrugation. The situation with $z_0 = h/8$ is symmetrical. The greatest deviation from the symmetry is for $z_0 = 0$ (Fig. 5.3).

In the asymmetrical situation a total electromagnetic force acts on the period of the liquid volume, determined by the total momentum flux of electromagnetic field over the liquid conductor surface (5.4.2):

$$F_e = - \frac{\mu_0 I^2}{8 I_1(kR_0)} \varepsilon \cos 2kz_0. \qquad (5.4.3)$$

For $z_0 = 0$ the force is directed oppositely to the z-axis, i.e. to the left in Fig. 5.3. The same force, but in the direction of the z-axis, is applied to the rigid conductor, thus, according to the results of Section 5.2, the total electromagnetic force in the entire current-carrying volume is zero.

The total force applied to the fluid may lead to transit flow in the same direction. However, the final conclusion can be reached by solving the full problem, which includes the determination of viscous friction and pressure distribution [5].

The solution for stream function (5.3.7), with the particular solution $\gamma(r)$

determined by No. 5 from Table 5.1. for $n = \ell$, by the perturbation method with the small parameter ε takes the form

$$\psi = vR_0 kr\{[A_{10} 2krI_0(2kr) + B_{10} I_1(2kr) + \gamma(2kr)] \sin 2kr +$$
$$+ \varepsilon[A_{21} 2krI_0(4kr) + B_{21} I_1(4kr)] \cos 2k(2z + z_0) +$$
$$+ \varepsilon[A_{01} 2kr + B_{01}(2kr)^3] \cos 2kz_0 + O(\varepsilon^2)\}. \qquad (5.4.4)$$

The solution is written to first-order accuracy in ε, and the coefficients A_{ij}, B_{ij} are functions of the nondimensional radius $a = kR_0 = 2\pi R_0/h$ and are determined similarly to (5.3.10). The term in the first line of solution (5.4.4) may be interpreted as a stream function describing periodic toroidal vortices in a cylindrical circular tube ($\varepsilon = 0$), and the term in the second line as the deformation of the vortices due to the wall deformation; the term in the third line describes the flow rate—pressure relation of the flow due to the phase shift of the wavy wall relative to the electrodes.

For the terms of higher ε-order the boundary conditions (5.3.8) give an equation similar to (5.3.16). This relates the two constants A_{01}, B_{01}:

$$A_{01} + 2B_{01} a^2 + \delta(a)/4 = 0. \qquad (5.4.5)$$

Knowing the stream function (5.4.4), the sought-for characteristics of EVF can be evaluated. The central axial velocity is determined by the expression

$$v_z(z, 0) = \frac{v}{R_0} S[(2A_{10} + B_{10}) \sin 2kz +$$
$$+ 2\varepsilon(2A_{21} + B_{21}) \cos 2k(2z + z_0) + 2A_{01}\varepsilon \cos 2kz_0]. \qquad (5.4.6)$$

Similarly to the expression of stream function (5.4.4), the first term in (5.4.6) depends on the period length between the electrodes and determines the main vortices independent of the wall corrugation; the second is related to the vortices of order ε and with the half period; the third relates to the presence of through-flow (this term is independent of z).

The flow rate through any cross-section of the channel

$$Q = -vR_0 2\pi Sa\varepsilon(A_{01} + B_{01} a^2) \cos 2kz_0, \qquad (5.4.7)$$

to the first order of accuracy in ε is independent of z. The pressure distribution in the channel is given by the expression

$$p(z, r) = \frac{\rho v^2}{R_0^2} 2Sa^2 \left\{ \left[-\frac{1}{4} I_0^2(kr) + \left(A_{10} + \frac{1}{4} \right) I_0(2kr) \right] \cos 2kz + \right.$$
$$+ 4\varepsilon[A_{21} I_0(4kr) \sin 2k(2z + z_0) + 2B_{01} 2kz \cos 2kz_0] -$$
$$\left. - \frac{1}{4} I_0^2(kr) \right\}, \qquad (5.4.8)$$

from which the pressure drop Δp on the interval h for any fixed r is evaluated:

$$\Delta p = \frac{\rho v^2}{R_0^2} 64 a^2 B_{01} \varepsilon \cos 2k z_0. \qquad (5.4.9)$$

As is evident from (5.4.7), (5.4.9), the constants A_{01}, B_{01} are related to the pressure drop and the flow rate in the tube. To determine these the constraint of the external hydraulic load must be added to the boundary condition (5.4.5). If the resistance of the hydraulic load is zero, or the section of tube is closed on itself, then $\Delta p = 0$. From (5.4.9) and (5.4.5) it then follows that

$$A_{01} = -\delta(a)/4, \qquad B_{01} = 0. \qquad (5.4.10)$$

The stream lines for this case are shown in Fig. 5.4. The characteristic quantity for $\Delta p = 0$ is the axial velocity:

$$v_z(z, 0) = \frac{v}{R_0} S[-F_1(a) \sin 2kz - \varepsilon F_2(a) \sin 2k(2z + z_0) +$$
$$+ \varepsilon F_3(a) \cos 2k z_0], \qquad (5.4.11)$$

where the nondimensional quantities $F_i(a)$, depending merely on the geometric ratio $h/R_0 = 2\pi/a$, are determined by the coefficients in the expression (5.4.6), and their plots are represented in Fig. 5.5.

The flow rate velocity in the case (5.4.10)

$$v_Q \equiv \frac{Q}{\pi R_0^2} = \frac{v}{R_0} S F_3(a) \varepsilon \cos 2k z_0. \qquad (5.4.12)$$

Thence, for fixed R_0 and h the maximum flow rate corresponds to $z_0 = 0$ — the most asymmetrical state of the corrugation relative to the electrodes. This situation is depicted in Fig. 5.4, and the respective through-flow is in the positive z-axis direction, i.e. in the direction opposite to the electromagnetic force. As is evident from (5.4.12) and Fig. 5.5, the maximum flow rate for fixed other parameters is reached at $h/R_0 \cong 2.1$. This involves the comparison of the

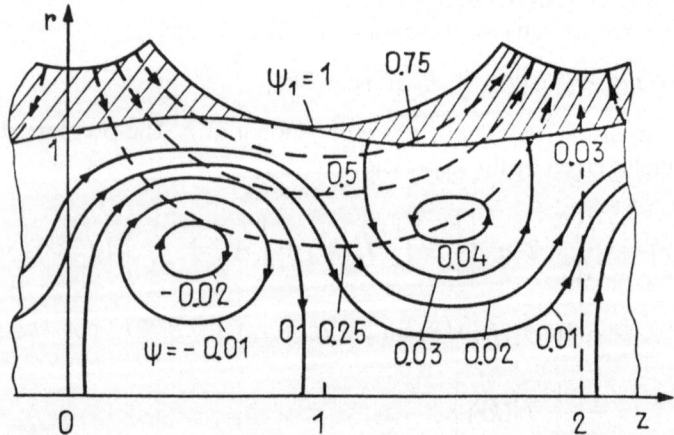

Fig. 5.4. Stream lines 100 ψ for the zero pressure drop regime ($\varepsilon = 0.05$).

Periodic electrically induced flows

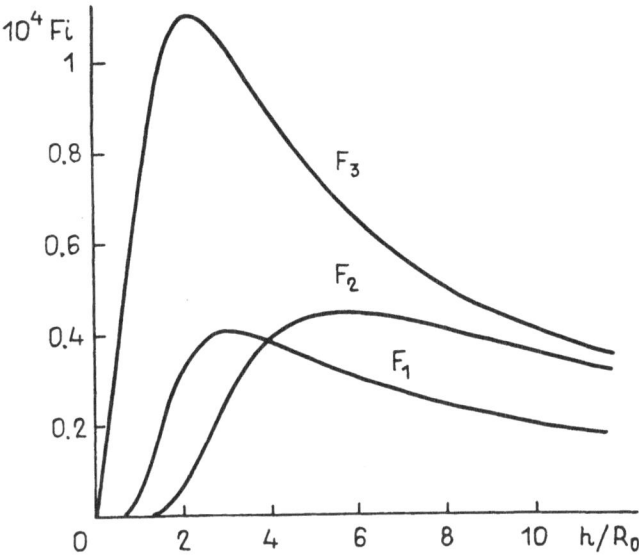

Fig. 5.5. Behaviour of the functions F_i specifying the axial velocity in the zero pressure drop regime.

geometrically non-similar flows (with different ratios h/R_0) at the constant typical current I and the constant height of corrugation ε.

In a choked regime the condition of zero flow rate $Q = 0$ yields

$$A_{01} = \delta/4, \qquad B_{01} = -\delta/4a^2. \qquad (5.4.13)$$

In this situation the stream lines in Fig. 5.6 are closed within the single period of corrugation, and the characteristic quantity — the pressure distribution on the wall — is expressed by (5.4.8) and (5.4.13) after substitution of (5.4.2) and neglecting the terms of accuracy ε^2 and higher:

$$p = \frac{\rho v^2}{R_0^2} S \left[p_1(a) \cos 2kz + p_2(a)\varepsilon \sin 2k(2z + z_0) + \right.$$

$$\left. + p_3(a) \frac{z}{\pi} \varepsilon \cos z_0 \right]. \qquad (5.4.14)$$

The plots of functions $p_i(a)$ are shown in Fig. 5.7. As is evident from (5.4.14), the pressure on average grows linearly along the channel (the last term in (5.4.14)).

The most surprising result of the problem considered is the direction of transit flow (and the respective pressure drop), which is oriented oppositely to the electromagnetic force in the fluid. Indeed, from Fig. 5.3, part of the electromagnetic force $\mathbf{f}_e$ in a half-period h is 'hidden' in the electrically conducting wall 2, and the z-component of the resulting force in the wall is oriented along the z-axis (F_{ez}^+). Since the total force in the entire conducting region

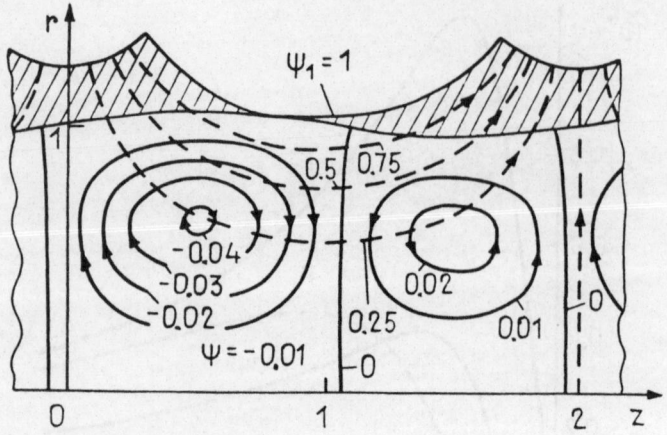

Fig. 5.6. Stream lines 100 ψ in the choked regime ($\varepsilon = 0.05$).

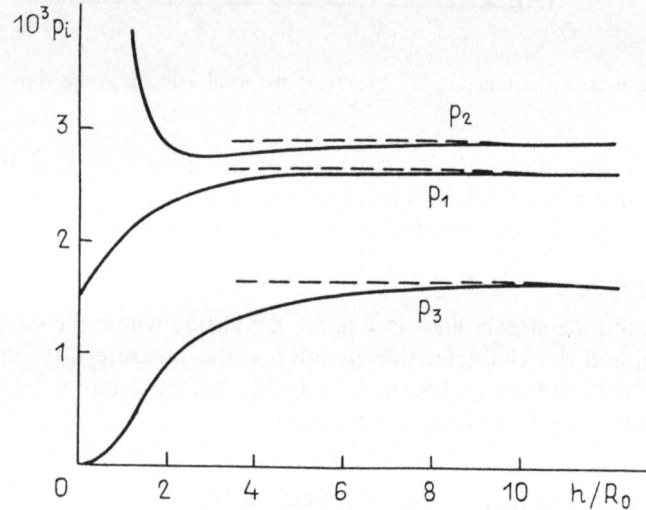

Fig. 5.7. Behaviour of the functions p_i specifying the pressure on the tube's wall in the choked regime. Dotted lines show the asymptotic values.

(fluid—wall) of half-period length is zero, then the force of equal magnitude should act in the fluid and in the opposite direction (F_{ez}^-). Whence the hasty conclusion may be reached that the fluid will move in the direction of the force F_{ez}^- (i.e. opposite to the z-axis). However, the solution shows the fluid as a whole is moving in the z-axis direction, in spite of the force in it. Let us find out what causes this paradox.

Consider the case of choked regime $Q = 0$ (see Fig. 5.6). The balance of the forces acting on the wall includes the electromagnetic force, the friction force on the surface due to the toroidal vortices, and the force due to the pressure difference. The electromagnetic force is definitely positive. The same result is obtained by evaluating the fourth integral in the expression (5.2.14). As for the friction, a more involved analysis is necessary in order to reach a definite conclusion. In a straight tube with ring electrodes the pair of toroidal vortices between the neighbouring electrodes would be fully equivalent: the vortices rotate in opposite senses and their friction on the wall is mutually compensated. The corrugated wall (see Fig. 5.6) contracts the right vortex, and, at the same time, the left vortical torus is able to rotate in a larger volume as compared to the straight tube. Both the vortices have different areas where the respective friction is applied, and they have different intensities, too. These considerations suggest unequal frictional forces on the wall, not compensated, and the resulting force may arise on the wall. Yet the direction of the force is not clear from Fig. 5.6. An even less definite conclusion is reached for the pressure distribution from these qualitative considerations. Though, obviously, the mean pressure over the tube's section increases along the axis, as follows from (5.4.14).

To reach a conclusive answer let us evaluate the integrals in the formula (5.2.14) by use of (5.4.4) and (5.4.8). The results are classified in Table 5.2 where the values of the consecutive integrals in (5.2.14) determine: F_p — the pressure force (first and fifth integrals), F_f — the frictional force (second integral), F_e — the electromagnetic force (fourth integral), and the resulting force F_z in dependence on the ratio h/R. The magnitudes of the forces are normalized by the electromagnetic force $F_e = \rho v^2 \pi^2 S \varepsilon a^2 I_1^2(a/2)$, which allows us to estimate the relative contribution of the forces in the total pressure drop. The magnitude of F_p is higher than F_f, F_e, and F_z, the resulting force F_z is the smallest. The sign of F_p suggests the following scheme for the formation of pressure force on the wall. The toroidal electrically induced flow (see Fig. 5.6) develops a high pressure zone in the sections of the tube where $dR/dz > 0$, and for the opposite wall inclination ($dR/dz < 0$) we see a zone of low pressure. The resulting action of the vortices 'pulls' the wall to the left. The frictional force of the more intense 'left' vortex, moreover acting on a greater area, suppresses the effect of the right, hence the integral frictional force 'pulls' the wall to the right. Since, according to Table 5.2, $|F_p| > |F_f| + |F_e|$, the integral force on the wall is directed to the left (see Fig. 5.6). Respectively, the force in the fluid acts parallel to the z-axis, and in the choked regime the positive pressure drop is developed between the sections $z = 2$ and $z = 0$. If the tube-pump is connected to a hydraulic load of small hydraulic resistance, the pressure drop (the potential energy, in other words) is transformed to the kinetic energy of the transit flow. Thus, the main role in the development of

Table 5.2. Forces acting on the tube's wall.

h/R_0	F_p	F_f	F_e	F_z
0.5	−1.130	0.116	1	−0.014
1.0	−1.276	0.216	1	−0.060
2.0	−1.489	0.310	1	−0.178
5.0	−1.632	0.334	1	−0.297
10.0	−1.657	0.339	1	−0.323

pressure drop (or transit flow) belongs to the pressure exerted by the toroidal vortices. The role of electromagnetic force is merely to generate the vortices of EVF.

5.5. EVF in a tube with longitudinal electric current

The physical mechanism of the transit flow development revealed in the previous section suggests a general principle for producing a flow rate or pressure drop in a tube by the periodic axisymmetric electrically induced flow. A periodic nonuniform distribution of meridional electric current generates in a tube a periodic set of toroidal vortices. In the smooth tube the two neighbouring vortices with opposite vorticities are of equal intensity and do not drive the transit flow. To generate the transit flow it is sufficient to constrict selectively the vortices of equal rotation sense by the corrugated wall. Then the set of vortices is asymmetric relative to the direction of the z-axis. The nonzero integral flux of the z-component of momentum through the boundary fluid—rigid wall of the tube results from the asymmetric pressure distribution and viscous stresses.

The distribution of current flowing along the axisymmetrical conductor would be disturbed if, for instance, the cross section of the conductor is varying lengthwise [16]. The distribution of current is disturbed even more by electrically conducting wall sections embedded in the nonconducting tube, and the disturbance is larger the higher the conductivity of the built-in section relative to the conductivity of the current-carrying fluid. Vortices are developed in the fluid where the current enters and exits the wall. By widening the channel in the location of the first vortex and constricting it in the location of the second, we may expect the transit flow to develop in a real channel (Fig. 5.8).

We replace the exact model represented in Fig. 5.8 by a particular case of the general axisymmetric distribution (5.3.4):

$$\psi_1 = cr(r/2 + \tau I_1(r) \cos z), \qquad (5.5.1)$$

which is the superposition of uniform current and a single harmonic of periodically nonuniform current (c and τ are constants), and this model includes

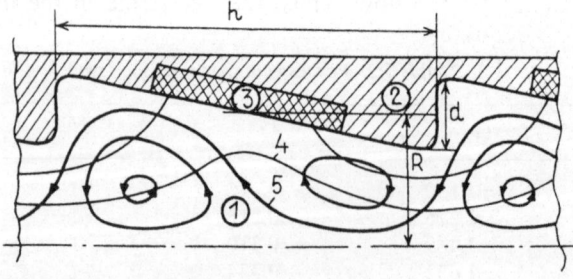

Fig. 5.8. Scheme of the tube: 1 — electrically conducting fluid, 2 — nonconducting wall, 3 — highly conducting sections, 4 — electric current lines, 5 — stream lines.

the basic qualitative features of the exact distribution. The expression (5.5.1) contains the nondimensional coordinates $r = kr^*$, $z = kz^*$, where the starred quantities are dimensional. The respective distributions of electric current lines $\psi_1 = \text{const}$ and equipotential lines $\Phi \sim (z + \tau I_0(r) \sin z) = \text{const}$ are shown in Fig. 5.9a.

The nonuniformity of current grows with the increase of the constant τ and radius r. If the cylinder $r < a$ is considered, then the choice is limited by the highest possible magnitude of $\tau = I_0^{-1}(a)$, when the saddle points formed by the current function and the potential for $z = (2n + 1)\pi$, $r = a$ (see Fig. 5.9a) are located outside the cylinder. Then in the cylinder $r < a$ the current flows everywhere along the axis, $j_z = ck^2(1 + \tau I_0(r) \cos z) > 0$, and it can be produced by applying a potential difference at the ends of the tube. This region contains the current lines passing from $z = -\infty$ to $z = +\infty$, and each of them may serve as the generatrix of the current carrying volume's surface, inside which the current distribution differs from uniformity due to the cross-sectional variation along the axial direction, and the distribution of current and magnetic field is exactly that described by expression (5.5.1). These lines retain the mean distance from the axis and for small r are sinusoidal [16]. The amplitude of the lines increases with r and their form deviates from the sinusoidal. The highest amplitude is reached on the critical electric current line crossing the saddle points at $r = a$ with the inflections of 90°. This line comes to the minimum distance from the axis, being substantially smaller than a. If the electric current line, passing higher than the critical line and extending to $r \to \infty$, is chosen as the generatrix of the current carrying region, then the integral generatrix of the current-carrying tube must include the sections of current lines intercepted by the sections of equipotential lines crossing the saddle points. The current lines form the surface of an insulator, and the equipotential lines form the surface of ideally conducting sections. The integral current across the contact surface of each ideally conducting section is zero (in contrast to the case considered in Section 5.4). Their thickness may be assumed to be small, thence any section of the surface formed by the equipotential line may be considered as a layer of surface currents (Fig. 5.9b). This configuration of the current-carrying tube also satisfies the distribution (5.5.1) inside it.

Consider now the hydrodynamic part of the problem. Let the current-carrying volume consist of the fluid region including the symmetry axis and the rigid region. It is advantageous to assume that the maximum radius of the domain filled by the fluid does not exceed the saddle point ($r \leq a$), and the external surface of the rigid conductor is formed by the electric current lines and equipotential lines tangent to the cylinder $r = a$ (see Fig. 5.9b). Such a mathematical model qualitatively corresponds to the distribution arising when the current passes through the tube filled with mercury, in the nonconducting wall of which are embedded copper sections strongly disturbing the current distribution in mercury ($\sigma_{Cu} \gg \sigma_{Hg}$) (see Fig. 5.8). Then the nondimensional quantity a may be specified as the ratio of two length scales — the mean radius of tube R_0 and the period length h: $a = kR_0 = 2\pi R_0/h$.

The representative current is reasonably specified as the effective magnitude

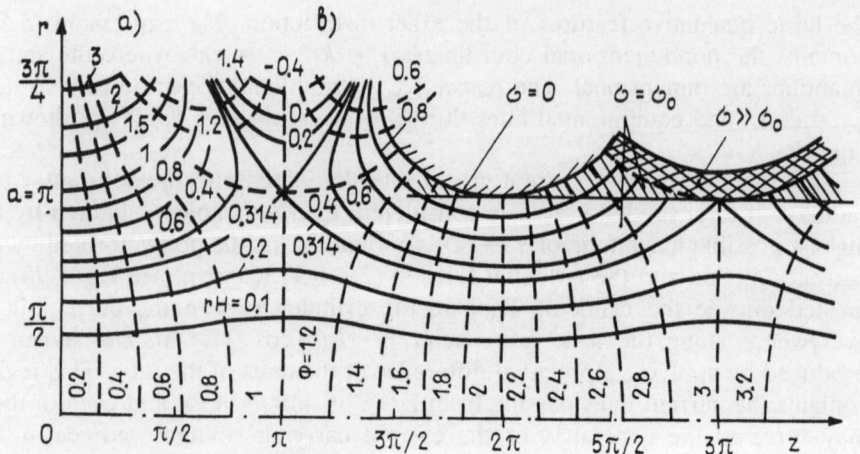

Fig. 5.9. Electric current lines (full) and equipotential lines (dashed) for $a = \pi$ (the right part shows the choice of tube's wall form).

I_{eff} determined by the mean-square density. This means that geometrically similar models are compared for equal Joule losses in them, independent of the specific distribution. For the distribution (5.5.1) this condition amounts to the constant c being expressed as

$$c = I_{\text{eff}}/(\pi a^{3/2}(a + \tau I_1(a))^2).$$

For the stream function in this case we have the equation

$$E^4 \psi = S \frac{rI_1(r) \sin z + I_1^2(r) \sin 2z / I_0(a)}{a^3 (aI_0(a) + I_1(a))} \tag{5.5.2}$$

where

$$S = \mu_0 I_{\text{eff}}^2 / \rho v^2.$$

On taking into account the two spatial harmonics present in (5.5.2), we choose the generatrix of the tube's corrugated wall in the form

$$r = a(1 + \varepsilon_1 \sin(z - z_1) + \varepsilon_2 \sin 2(z - z_2)). \tag{5.5.3}$$

Assuming the amplitude of corrugation to be small $\varepsilon_1, \varepsilon_2 \ll 1$, we will seek the solution of the equation (5.5.2) by applying the perturbation method described previously. To an accuracy of the first order in ε the solution is

$$\psi(z, r) = Sr \left\{ \sum_{\ell=1}^{2} [C_\ell \alpha(\ell r) + D_\ell \beta(\ell r) + \gamma_\ell(\ell r)] \sin \ell z \right.$$

$$+ \sum_{\ell=1}^{4} \sum_{m=1}^{2} [C_{\ell m} \alpha(\ell r) + D_{\ell m} \beta(\ell r)] \cos(\ell z - z_m) +$$

$$\left. + Ar + Br^3 \right\}, \tag{5.5.4}$$

Periodic electrically induced flows

where the following notation is used:

$$\gamma_1(x) = \frac{x^3 I_2(x)}{8a^3(aI_0(a) + I_1(a))} \; ; \qquad \gamma_2(x) = \frac{-x[I_1^2(x/2) + I_0^2(x/2)]}{16a^3(aI_0(a) + I_1(a))} \; ;$$

$$C_\ell = W[\beta, \gamma_\ell]/W[\alpha, \beta]; \qquad D_\ell = W[\gamma, \alpha]/W[\alpha, \beta];$$

$$C_{\ell m} = c_{\ell m}\delta_{3-m}\varepsilon_m\beta/aW[\alpha, \beta];$$

$$D_{\ell m} = -c_{\ell m}\delta_{3-m}\varepsilon_m\alpha/aW[\alpha, \beta]; \qquad (5.5.5)$$

$$\delta_\ell = (C_\ell\alpha'' + D_\ell\beta'' + \gamma_\ell'')/a^2;$$

$$\alpha = \alpha(x) = xI_0(x); \qquad \beta = I_1(x); \qquad x = \ell a;$$

$$c_{11} = c_{31} = 2; \qquad c_{12} = c_{32} = c_{42} = 1/2; \qquad c_{21} = 1/4;$$

$$c_{22} = c_{41} = 0.$$

The first sum in (5.5.4) is the solution for a circular tube and describes the set of primary toroidal vortices. The second sum represents the vortices of different sizes arising when the vortices are deformed as a result of interaction with the wall's irregularities. The last two terms with the coefficients A and B characterize the transit properties of flow — pressure drop along the tube's element and the flow rate Q over the tube's cross section. The interaction of primary vortices with the corrugation is responsible for the origin of these transit properties.

The boundary conditions give the equation connecting the constants A and B:

$$Aa + 2Ba^3 + \tfrac{1}{4}\delta_1\varepsilon_1 \cos z_1 + \delta_2\varepsilon_2 \cos z_2 = 0. \qquad (5.5.6)$$

This equation, according to Section 5.3, serves as a $G(Q)$ characteristic for the tube-pump, and it determines the linear relation between the flow rate Q and the mean pressure gradient along z:

$$\frac{\pi R_0^4 G}{8\rho v} + Q + \pi R_0 vS\left(\frac{1}{2}\delta_1\varepsilon_1 \cos z_1 + 2\delta_2\varepsilon_2 \cdot \cos z_2\right) = 0. \qquad (5.5.7)$$

Efficiency of the pump is determined by the expression in round brackets in equation (5.5.7). The maximum magnitude is reached for $z_1 = z_2 = 0$. The choice of geometrical ratios for the corrugation: h/R, $\varepsilon_1/\varepsilon_2$, depends on the general properties of the functions $\delta_1(a)$ and $\delta_2(a)$. The exact expressions are derived from (5.5.5):

$$\delta_1(a) = \frac{I_0(a) + I_1^3(a)/W[\alpha(a), \beta(a)]}{a[aI_0(a) + I_1(a)]} \; ; \qquad (5.5.8)$$

$$\delta_2(a) = 2I_1(2a)[aI_0(2a)(I_0^2(a) + I_1^2(a)) - I_0(a)I_1(2a)(I_0(a) +$$

$$+ 2aI_1(a)] - I_0(a)I_1(a)\, W[\alpha(2a), \beta(2a)] \div$$

$$\div aI_0(a)[aI_0(a) + I_1(a)]\, W[\alpha(2a), \beta(2a)].$$

On using the representations of modified Bessel functions $I_n(x)$ in the series form and the asymptotic expansion [6], we obtain

$$\delta_1 = \frac{a}{72}\left(1 - \frac{3}{16}a^2 + O(a^4)\right) \quad \text{and}$$

$$\delta_2 = \frac{a}{72}\left(1 - \frac{1}{2}a^2 + O(a^4)\right) \quad \text{for} \quad a \to 0,$$

$$\delta_1 = \frac{1}{4a^2}\left(1 - \frac{3}{2a} + O(a^{-2})\right) \quad \text{and}$$

$$\delta_2 = \frac{1}{8a^3}\left(1 - \frac{5}{4a} + O(a^{-2})\right) \quad \text{for} \quad a \to \infty.$$

From these it follows that the transit flow decays for very short, as well as for very long corrugation. Both the functions in the interval $0 < a < \infty$ have a single maximum (Fig. 5.10). Hence, the maximum of the transit characteristics is ensured by the choice of $\varepsilon_1/\varepsilon_2 > 0$. Let $\varepsilon_1 = \varepsilon$, $\varepsilon_2 = \varepsilon/2$. In this case, in view of the fact that $z_1 = z_2 = 0$, the wall generatrix (5.5.3) contains the first two terms of the Fourier expansion for the function $r = a[1 - \varepsilon(z - \pi)/2]$. Since other harmonics of the expansion in linear approximation do not yield the transit flow for the distribution (5.5.1), then the wall corrugation may be assumed in the form of a sawtooth (see Fig. 5.8). The amplitude of the 'sawtooth' $d = \varepsilon \pi R_0$. The ideally conducting sections of the tube are situated midway along the sloping sections. The function $F = \delta_1/2 + \delta_2$, expressing for this channel the dependence of v_Q and G, is also shown in Fig. 5.10, and it contains a maximum at $h = 4.8R_0$.

The transit flow in the tube of the form shown in Fig. 5.8 according to (5.5.7) is directed in the negative z-axis direction because $F(a) > 0$. The total

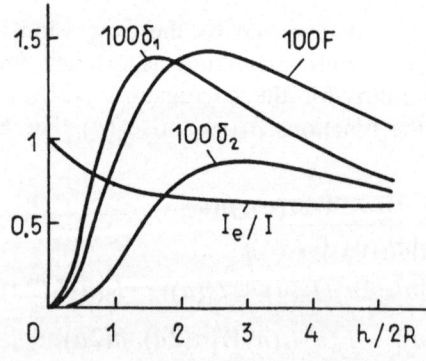

Fig. 5.10. Plot of the functions $\delta_1(a)$, $\delta_2(a)$, $F = 0.5\delta_1 + \delta_2$; $I_e/I = \sqrt{a(a + I_1(a)/I_0(a))}/(a + 2I_1(a)/I_0(a))$, $a = 2\pi R/h$.

electromagnetic force over the whole volume carrying the current is equal to zero; in the electrically conducting sections of the tube, due to the inclination of current lines, it is directed in the $(-z)$-direction, and, respectively, in the fluid in the $(+z)$-direction, i.e. oppositely to the transit flow. The account of the hydrodynamics that we have given has revealed that the tube-pump drives the fluid in the direction opposite to that obtained from the electrodynamic consideration of the device as a conduction pump.

Note also that the present direction of transit flow is opposite to the flow direction in the problem of Section 5.4. Indeed, the bending of electric current lines and the respective sense of vortex rotation are opposite for both the cases.

It is of special interest to consider EVF in isolated channels. The variation of channel's cross section leads to the non-uniformity of current and the development of electrically induced flow. For example, in liquid metal devices with a longitudinal current (such as are partly found in electrical channel furnaces) the channel walls are usually nonconducting, and the current flows only in the liquid. In a real channel, technological or functional wall irregularities are always present; they may be deliberately introduced to take advantage of the associated EVF.

In theoretical terms the problem is interesting because, even for simplest noninduction approximation, the electrodynamic part of problem is geometrically connected with the hydrodynamics. There is the single boundary between the conducting—isolating materials, the fluid—rigid wall, which by its form determine simultaneously both the electric current distribution and the flow stream lines. In addition, the fluid volume in the channel contains the whole electric current, and, respectively, the total electromagnetic force and its moment acting in the whole fluid are equal to zero. Therefore in the isolated channel (periodic or closed) the pump effect of the conduction pump type is absent. The pump effect in EVF is impossible if $\xi(z)$ is a symmetric function relative to the z-axis direction, i.e. if there exists such a z_0 that $\xi(z_0 + z) = \xi(z_0 - z)$ for any z. The possibility of developing the pump effect in an asymmetrically corrugated tube of the 'inclined sawtooth' type was examined in [14].

The electric current function $\psi_1(r, z)$ must satisfy the condition $r = R(z)$ on the surface (5.3.11), i.e.

$$2\pi\psi_1(R(z), z) = I, \tag{5.5.9}$$

where I is the total current in the channel. ψ_1 is represented in the form of expansion by powers of ε:

$$\psi_1(r, z) = \sum_n \varepsilon^n \psi_{1n}(r, z),$$

where each term can be expressed by the general solution, non-singular at $r = 0$, periodic along z with period h (see Section 5.1):

$$\psi_{1n}(r, z) = a_{n0}r^2 + \sum_{m \neq 0} a_{nm} r I_1(mkr) e^{imkz}, \qquad k = 2\pi/h.$$

At the boundary (5.3.11) $\psi_1(r, z)$ is expressed as a Taylor series by r in the vicinity of $r = R_0$. Then the boundary condition (5.5.9) yields the recursional relations to determine ψ_{1n}:

$$\psi_{10}(R_0, z) = I,$$

$$\psi_{1n}(R_0, z) = - \sum_{\ell=1}^{n} \frac{(R_0 \xi(z))^{\ell}}{\ell!} \frac{\partial^{\ell}}{\partial r^{\ell}} \psi_{1n-\ell}(r, z) \bigg|_{r=R_0}, \quad n \geq 1. \qquad (5.5.10)$$

Consecutively separating the harmonics of z, in the equations (5.5.10) we obtain the coefficients a_{nm} and the respective current distribution (and magnetic field) with the prescribed accuracy governed by the small parameter ε. The method of constructing ψ_1 corresponds completely to the method presented in Section 5.3 for the function ψ.

Consider the simplest asymmetric corrugation of the form

$$\xi(z) = c_1 \sin k(z - z_0) + c_2 \sin 2kz. \qquad (5.5.11)$$

The function for $z_0 \neq h/4$ is asymmetric relative to the z-axis direction and contains the least number of harmonics. Viewed along the positive z-axis direction, the tube (5.3.11) with the corrugation (5.5.11) consists of alternating sections of sudden expansions and more smooth channel constrictions. If $c_1 = 2c_2 = 2/\pi$ and $z_0 = 0$ are set, then (5.5.11) represents the first two terms of a Fourier expansion for the asymmetric sawtooth function $f(z) = 1 - 2z/h$, $z \in (0, h)$. The latter corresponds to the form of experimental channel (Fig. 5.11). However, even for the simplest boundary (5.5.11) the calculations necessary to reveal the pump effect, the consecutive steps of which were explained above, are quite unwieldy, and the analytic representation obtained for the quantity δ in (5.3.16) is sufficiently awkward to drop its full explicit expression here. We will only discuss some basic qualitative steps of the calculation.

The results of the calculation revealed the existence of an EVF pump effect in the tube with the corrugation (5.5.11). The non-uniformity of the dependence of $G(Q)$ (5.3.16) $\delta \neq 0$ is expressed by the terms of the order of ε^3. Hence the functions ψ_1 and ψ must be evaluated to the terms of accuracy of ε^2.

The dependence of δ on the relative step of corrugation h/R_0 for $c_1 = 2c_2 =$

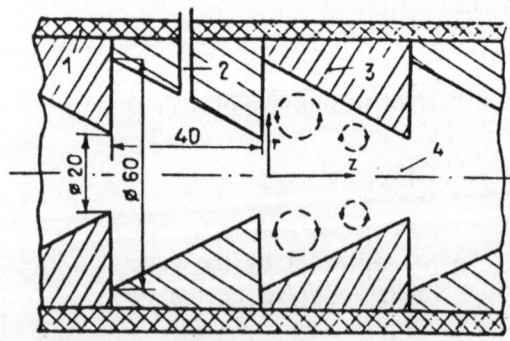

Fig. 5.11. Scheme of the channel: 1 — enclosing tube, 2 — orifice for pressure measurements, 3 — rigid nonconducting wall, 4 — electrically conducting fluid.

$2/\pi$, $z_0 = 0$ is depicted in Fig. 5.12. The quantity δ is positive, and this means the fluid is pumped in the positive z-axis direction. In Fig. 5.11 this direction is to the right, along which the constrictions of tube are more smooth than the expansions. With the increase of h/R_0 the value of δ decreases monotonously (Fig. 5.12), i.e. the pump produces the maximum flow rate and maximum pressure gradient when the period h is comparable to the amplitude of the corrugation. However, it is impossible to detect the optimal relation between the quantities d, R_0, h, because the computation employs the approximation $d \ll R_0$ and the expansion for ψ_1 satisfies the condition $j_z > 0$ only for $d \lesssim h$, where $d = c\pi R_0$ is the amplitude of corrugation.

The pressure drop along the length of a single period, $\Delta p = Gh$, in the choked regime $Q = 0$ is given by the function $\delta h/R_0$ presented in Fig. 5.12. The quantity Δp is of a maximum value for the period h comparable to the radius R_0 ($h \approx 0.8 R_0$).

The pump effect of EVF in the asymmetrically corrugated channel was also found experimentally. The experimental device was designed to investigate the pressure distribution in a liquid metal channel with dead ends and of finite length $\ell = 5.2$ m (Fig. 5.11). The channel consists of 130 separate elements 3 packed inside the stainless steel tube 1. Each element 3 is of cylindrical form with a conical central opening; the elements are made of wood impregnated with varnish. The set of elements constitute the circular channel of variable cross section with the generatrix of the asymmetric sawtooth form. The working fluid 4 is mercury.

The electric current is supplied by the copper end-electrodes. The external tube is electrically isolated from the fluid and electrodes. The form of corrugation ($R_{min} = 0.01$ m, $R_{max} = 0.03$ m, $h = 0.04$ m) is not the optimal one to obtain the maximum pump effect by EVF.

The pump effect was investigated in the choked regime $Q = 0$, i.e. the effect was realized as the nonzero mean pressure gradient along the tube's length. In the course of the experiment the pressure on the wall was measured through the

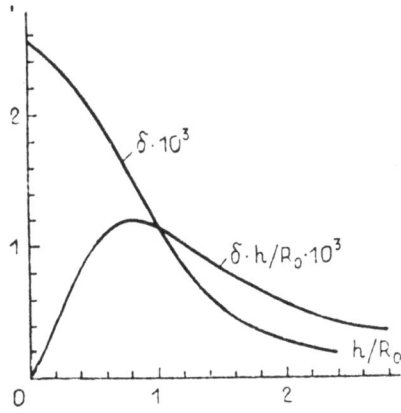

Fig. 5.12. The computed dependence of the EVF pump effect efficiency on the length of corrugation.

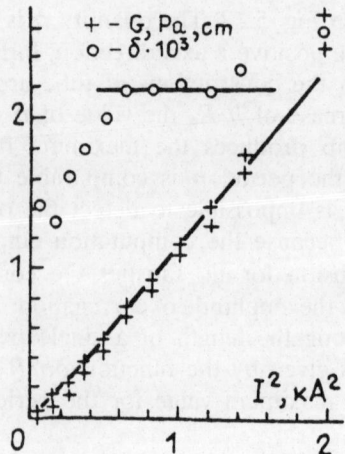

Fig. 5.13. The mean pressure gradient vs. the square of total current in the experimental tube.

radial holes 2 when the fluid flow was steady. The pressure was measured by two-fluid (mercury-paraffin) manometers in five equivalent points, relative to the corrugation, at equal intervals along the length of the channel. To avoid the end effects and other systematic errors, the piezometric curve was constructed and approximated by a straight line, and the mean pressure gradient G (Fig. 5.13) was determined from the slope line. It was found that the flow head agrees in direction with that theoretically found (from the abrupt expansion to the smooth constriction of the channel). The head grows exactly linearly along the channel length, which indicates a minority of end and other disturbing effects. The mean pressure gradient G depends linearly on the square of total current (Fig. 5.13).

To compare the experimental and theoretical results, Fig. 5.13 presents the quantity δ evaluated by the experimental data with the use of formula

$$\delta = \pi\varepsilon^3 R_0^3 G/8\mu_0 I^2,$$

where $\varepsilon = 0.5$, $R_0 = 0.02$ m. For moderate current magnitudes ($500 < I < 1500$ A) the quantity δ is constant and agrees in the order of magnitude with the theoretical value given in Fig. 5.12. A closer comparison is impossible due to the differences between the theoretical and experimental models — in the experiment the relative amplitude variation of the channel's cross section is not small.

5.6. EVF in an annular tube

The general form of the EVF in tubes considered above is the flow in an annular gap between two coaxial tubes, the walls of which can be either straight cylinders $r = R_1$ and $r = R_2$, or weakly deformed (corrugated). In addition, both the surfaces may include ring electrodes disturbing the current distribution in the annular gap, and the inner rod may carry a current producing an azimuthal magnetic field in the annular gap. The problem contains new parameters affecting the fluid flow.

One of the simplest form of electric current function in the annular channel contains a single harmonic of periodic disturbances:

$$\psi_1(r, z) = a_0 + b_0 r^2 + [a_1 I_1(kr) + b_1 K_1(kr)] r e^{ikz}. \tag{5.6.1}$$

Specifying different constants a_0, a_1, b_0, b_1, different current configurations may be obtained. Some of these configurations are shown in Fig. 5.1. In the case $a_0 \neq 0$, $b_0 \neq 0$ we have a uniform current perturbed by the terms with a_1 and b_1.

The case when the current perturbations are small compared to the uniform longitudinal current was considered in [17], and it was shown that (5.6.1) could represent the current distribution in the annular gap with the sinusoidally corrugated external and internal boundaries of the liquid conductor. The hydrodynamic part of problem was solved in [17] in an approximation neglecting the effect of wall deformation on the velocity field, i.e. actually the EVF in a straight annular tube was considered. And the development of electrically induced flow in this case might be interpreted as the result of periodic electrically conducting sections embedded either in the external tube's wall, or in the internal rod, or in both walls simultaneously.

The effect of the axial current in the internal rod on EVF depends on its direction relative to the current in the fluid. Both growth or retardation of vortices are possible, and, also, the rotation sense may change; the vortices may split both in the axial and radial directions. Yet in the case studied in [17] the effect is not selective relative to the rotation sense of the vortices, and thereby the transit flow is not generated.

In the annular tube the transit flow may be affected by the deformation of inner wall. However, due to the smaller surface area, as compared to the external wall, a substantial change of the $G(Q)$ characteristics is unlikely. The situation with a current passing through the internal rod has certain advantages. The current in the fluid can be increased up to a certain limit determined by excess heating or the pinch-effect. The limiting current in the rigid conductor is substantially higher, and it is limited by other factors.

Consider first the effect on the flow characteristics of an isolated smooth coaxial rod without current [14] in the case analogous to Section 5.4, i.e. $a_0 = b_0 = 0$ in (5.6.1). The situation shown in Fig. 5.15 yields the maximum pump effect. The method of solution was discussed above, and therefore we proceed directly to the results [13]. Fig. 5.14 demonstrates the effect of a central rod $r = R_1$ on the flow rate in an unloaded regime and the pressure drop in a choked regime. On decreasing the gap ($R_1 \rightarrow R_2$, where R_2 is the mean radius of the external corrugated wall), the flow rate Q for $\Delta p = 0$ falls monotonously down to zero; the pressure drop Δp for $Q = 0$ grows initially, reaches a maximum, and then falls to a finite value for $R_1 = R_2$.

It is evident from Fig. 5.15 that the central rod shifts the vortices to the external wall. Thereby the friction grows on this surface. In addition, the frictional force acts on the rod, too. Since in the unloaded regime the single

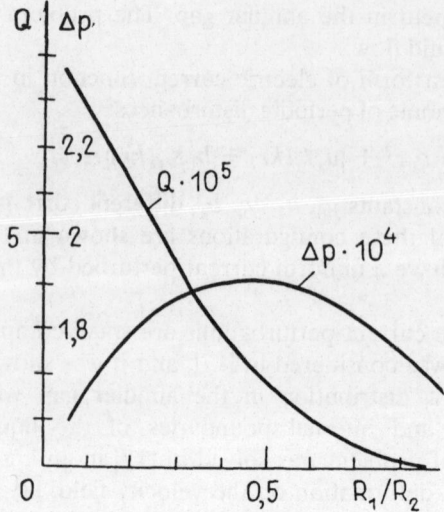

Fig. 5.14. Functions $Q(R_1/R_2)$ for $\Delta p = 0$ and $\Delta p(R_1/R_2)$ for $Q = 0$.

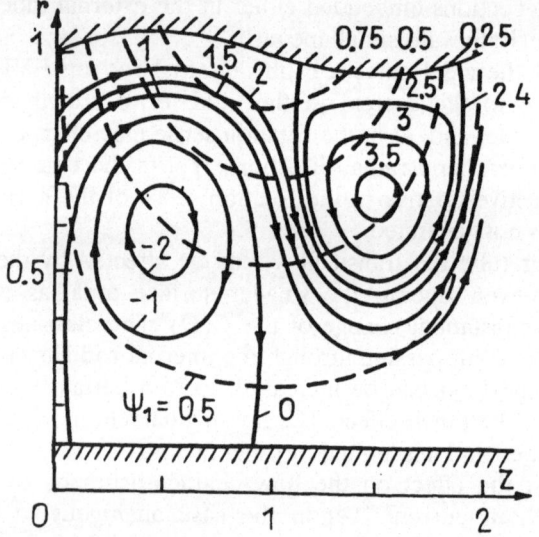

Fig. 5.15. Stream lines $10^5 \, \psi$ for zero pressure drop regime. The dashed lines show the electric current lines ($\varepsilon = 0.05$).

factor limiting the flow rate Q of the transit flow is the friction, then the growth of friction leads to the decrease of flow rate. For a higher ratio R_1/R_2 the friction (and the flow rate) decreases due to the decay of vortex intensity in view of the smaller fluid volume and the decrease of the rotational part of the electromagnetic force.

Let us now discuss the possibility of operating the electrically induced flow by passing an additional electric current in the central rod. The current distribution is determined by the expression (5.6.1), but in this case a new parameter $\kappa = I_0/I$ — being ratio of the current I_0 in the rod to the current I in the fluid

Periodic electrically induced flows 227

— enters the problem. The numerical results for the flow, similar to the flow in Fig. 5.8, but with an inner straight rod, are presented in Fig. 5.16. The dependences $Q(R_1/R_2)$ and $\Delta p(R_1/R_2)$ for $\kappa \neq 0$ are similar to the previous case, yet the maximum of pressure drop is shifted to higher values of R_1/R_2.

When a current is present in the inner rod, both the characteristics considerably increase. Thus, if the current in the rod is 5 times higher than the current in the fluid, then both the flow rate and pressure drop are by the order of magnitude higher than in the absence of current in the rod. The effect of parameter κ on the flow is demonstrated in Fig. 5.17. The inner current can increase the integral characteristics of the pump, and, by changing the direction of current in the rod, the direction of transit flow can be reversed.

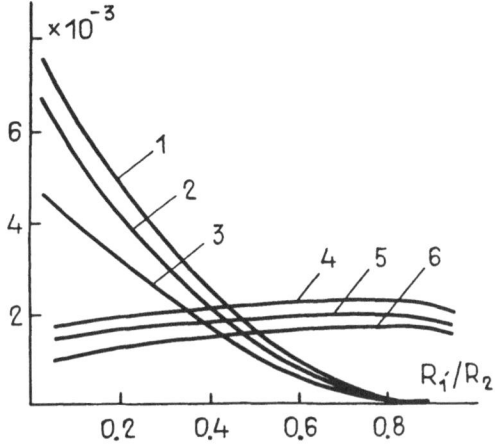

Fig. 5.16. Functions $Q(R_1/R_2)$ and $\Delta p(R_1/R_2)$ for different κ: (1) $-Q$, $\kappa = 5$; (2) Q, $\kappa = -5$; (3) $-10Q$, $\kappa = 0$; (4) $-\Delta p$, $\kappa = 5$; (5) Δp, $\kappa = -5$; (6) $-10\Delta p$, $\kappa = 0$.

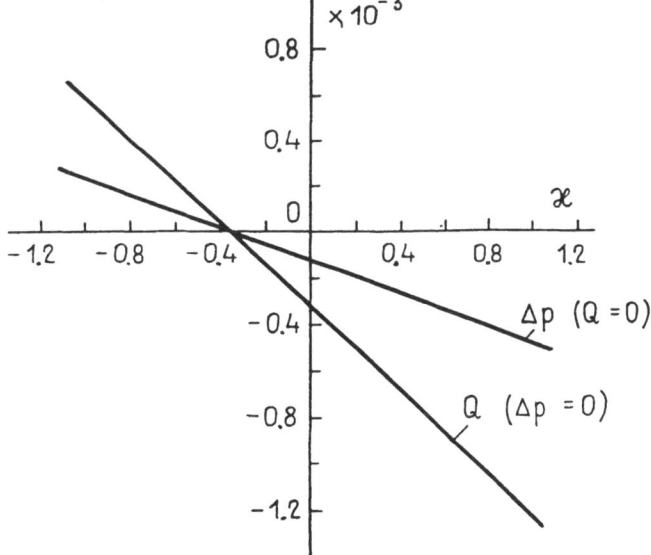

Fig. 5.17. Functions $Q(\kappa)$ for $\Delta p = 0$ and $\Delta p(\kappa)$ for $Q = 0$; $R_1/R_2 = 0.2$.

5.7. Periodic EVF in a longitudinal magnetic field

The intensity of periodic electrically induced flows and their integral characteristics depend on the configuration of the tube, the location of the electrically conducting sections of the tube, and the strength of current passed through the fluid. Only the latter factor may be considered as being suitable to operate the flow if the design of the device is fixed. Apart from this, as was shown in Section 5.6, the flow was affected by the additional current passing in the central cylindrical conductor. With the aim of finding a new possibility to operate the periodic flow, consider now the effect of an external magnetic field.

The simplest magnetic field, which does not break the axial symmetry, is the longitudinal magnetic field of a solenoid surrounding the current-carrying fluid. Consider first the simplest situation in which the current-carrying fluid column with a periodically disturbed surface is surrounded by the electrically non-conducting fluid in a rigid circular tube (Fig. 5.18). This model permits us to easily analyse the interaction mechanism of the longitudinal field with the periodic electric current and a transformation of the electrically induced flow. Apart from this, the problem, with some reservation, may be of interest in the analysis of the flows generated by a weak deformation of the current-carrying plasma column, for instance in a plasmatron or in a device of the Tokamak type, in which the plasma column is stabilized by the longitudinal magnetic field. The behaviour of the perturbed interface could be predicted if we knew the toroidal vortex flows in the plasma (the effect of other factors is not considered).

The following computational model is assumed (Fig. 5.18): the domain I bounded by the surface $r = a$ is penetrated by the electric current I, the annular domain II, $a < r < b$, is electrically nonconducting, and it is surrounded by the rigid wall $r = b$. Other physical properties of the media in the domain I and II are supposed to be equal. The interface between the domains is assumed to be perturbed periodically with a wavelength h and amplitude ε:

$$r = a(1 + \varepsilon \cos kz), \qquad k = 2\pi/h. \tag{5.7.1}$$

The periodic solution of the magnetic field equation (5.3.4), corresponding to

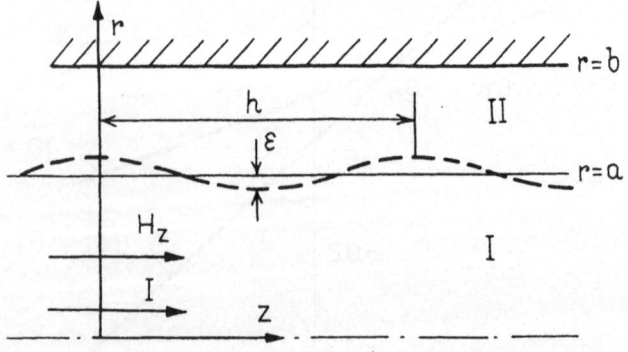

Fig. 5.18. Scheme of the problem.

Periodic electrically induced flows

the interface deformation (5.7.1) and the condition $B_\varphi|_{r=0} = 0$, follows from (5.5.1):

$$B_\varphi = c[r/2 + \tau I_1(kr)\cos kz]. \tag{5.7.2}$$

The constants c and τ are chosen in the following way. Since the current is contained in the domain I and anywhere in this domain $j_z > 0$, then from the solution

$$j_z = \frac{c}{\mu_0}[1 + \tau k I_0(kr)\cos kz] > 0$$

the constraint follows:

$$\tau < 1/kI_0(ka). \tag{5.7.3}$$

Apart from this, the electric current line $rB_\varphi = \text{const}$ must, to an accuracy ε, correspond to the surface equation (5.7.1), from which

$$\tau = -\varepsilon a/I_1(ka). \tag{5.7.4}$$

In view of (5.7.3) and (5.7.4), ε is limited by

$$\varepsilon < I_1(ka)/kaI_0(ka).$$

The constant c is determined by the total current passing across the section $r = a$, i.e. for $z = h/4$: $c = \mu_0 I/\pi a^2$.

Suppose the motion in meridional planes is much weaker than in the azimuthal direction. Then the inertial terms in the equation of motion can be neglected, apart from the centrifugal component. Then the curl φ-component and φ-component of the Navier-Stokes equation are written as

$$\rho v E^4 \psi = -\frac{1}{\mu_0}\frac{\partial}{\partial z}B_\varphi^2 + \rho\frac{\partial}{\partial z}v_\varphi^2; \tag{5.7.5}$$

$$\rho v E^2 r v_\varphi = -\frac{B_0 r}{\mu_0}\frac{\partial}{\partial z}B_\varphi, \tag{5.7.6}$$

where B_0 is a representative longitudinal magnetic field induction. The equations (5.7.5) and (5.7.6) are valid in the current-carrying domain I. The motion in the domain II is described by the same equations, dropping the electromagnetic terms.

The boundary conditions are specified by the symmetry at $r = 0$ and the no slip condition at the rigid wall:

$$v_{\varphi_I} = 0, \quad \frac{\partial \psi_I}{\partial z} = 0, \quad \frac{\partial}{\partial r}\frac{1}{r}\frac{\partial}{\partial r}\psi_I = 0 \quad \text{for} \quad r = 0;$$

$$v_{\varphi_{II}} = 0, \quad \frac{\partial \psi_{II}}{\partial z} = 0, \quad \frac{\partial \psi_{II}}{\partial r} = 0 \quad \text{for} \quad r = b. \tag{5.7.7}$$

In addition, the continuity of stream function, velocity components, and their first derivatives at the interface boundary (5.7.1) must be set:

$$v_{\varphi_I} = v_{\varphi_{II}}, \quad \psi_I = \psi_{II}, \quad \frac{\partial}{\partial z}\psi_I = \frac{\partial}{\partial z}\psi_{II}, \quad \frac{\partial}{\partial r}\psi_I = \frac{\partial}{\partial r}\psi_{II},$$

$$\frac{\partial^2}{\partial z^2}\psi_I = \frac{\partial^2}{\partial z^2}\psi_{II}, \quad \frac{\partial^2}{\partial r \partial z}\psi_I = \frac{\partial^2}{\partial r \partial z}\psi_{II}, \qquad (5.7.8)$$

$$\frac{\partial^2}{\partial r^2}\psi_I = \frac{\partial^2}{\partial r^2}\psi_{II}.$$

It is convenient to solve the problem by the use of nondimensional coordinates:

$$a_* = ka, \quad b_* = kb, \quad \psi_* = k\psi/\nu, \quad v_{\varphi_*} = v_\varphi/k\nu, \quad E_*^2 = E^2/k^2.$$

The boundary between the domains I and II is given by the nondimensional equation (the index $*$ is henceforth dropped)

$$r = a(1 + \varepsilon \cos z). \qquad (5.7.9)$$

With (5.7.2) the equation (5.7.6) takes the form:

in the domain I

$$E^2 r v_{\varphi_I} = -\varepsilon N r I_1(r) \sin z; \qquad (5.7.10)$$

in the domain II

$$E^2 r v_{\varphi_{II}} = 0, \qquad (5.7.11)$$

where $N = IB_0/\rho\nu^2 \pi k a I_1(a)$.

The solutions of (5.7.10) and (5.7.11) can be found by the method of separation of variables and then matching the solutions at the interface (5.7.9). To achieve this, all the functions of r are expanded in Taylor series at the point $r = a$, and the constants are represented in the form of an ε-power expansion retaining the first order terms. Then we have:

$$v_{\varphi_I} = 1/2 N\varepsilon[BI_1(r) - rI_0(r)] \sin z; \qquad (5.7.12)$$

$$v_{\varphi_{II}} = 1/2 N\varepsilon A[I_1(r) - \alpha K_1(r)] \sin z; \qquad (5.7.13)$$

where

$$A = a[a(I_0^2(a) - I_1^2(a)) - 2I_0(a)I_1(a)],$$
$$\alpha = I_1(b)/K_1(b);$$
$$B = A + a[2I_0(a)K_1(a) + aI_0(a)K_0(a) + aI_1(a)K_1(a)].$$

The equation for the stream function in the domain I is obtained on substituting (5.7.12) and (5.7.2) in (5.7.5):

$$E^4\psi_I = -\varepsilon Sr I_1(r) \sin z + \varepsilon^2 S \frac{I_1^2(r)}{I_1(a)} \sin 2z + \varepsilon^2 \frac{N^2}{4} [BI_1(r) -$$
$$- rI_0(r)]^2 \sin 2z, \qquad (5.7.14)$$

where $S = \mu_0 I^2/\rho v^2 \pi^2 a^3 I_1(a)$, and in the domain II we substitute (5.7.13) into (5.7.5) giving:

$$E^4\psi_{II} = \varepsilon^2 \frac{N^2}{4} A^2 [I_1(r) - dK_1(r)]^2 \sin 2z. \qquad (5.7.15)$$

Henceforth the quantity B_0 is assumed to be high enough that $\varepsilon^2 N^2/4 = O(\varepsilon)$, then denoting $M = \varepsilon N^2/4$ we obtain the electromagnetic terms of equal order ($\sim \varepsilon$).

The solutions of (5.7.14) and (5.7.15) are found by a similar method, and to the accuracy in the first order terms of ε these are of the form:

$$\psi_I = \varepsilon \sum_{n=1}^{2} [C_n r^2 I_0(nr) + D_n r I_1(nr) + \gamma_n(r)] \sin nz;$$

$$\psi_{II} = \varepsilon \sum_{n=1}^{2} [E_n r^2 I_0(nr) + L_n r^2 K_0(nr) + F_n r I_1(nr) +$$
$$+ G_n r K_1(nr) + \rho_n(r)] \sin nz.$$

The particular solutions satisfying (5.7.7) and (5.7.8) follow from Table 5.1:

$$\gamma_1(r) = -Sr^3 I_1(r)/8;$$
$$\gamma_2(r) = M\{-B^2 r^2 (I_0^2(r) + I_1^2(r))/8 - Br^2(2rI_0(r)I_1(r) - I_1^2(r))/12 +$$
$$+ [r^4(I_0^2(r) + I_1^2(r)) + 2r^2(rI_0(r)I_1(r) - 4I_1^2(r))]/120\};$$
$$\rho_1(r) = 0$$
$$\rho_2(r) = -MA^2 r^2\{(I_0(r) + dK_0(r))^2 + (I_1(r) - dK_1(r))^2\}/8.$$

The constant coefficients C_n, D_n, E_n, L_n, F_n, and G_n are functions of the nondimensional quantities a and b, and these are determined by the boundary conditions (5.7.7), (5.7.8). The terms with the index $n = 1$ describe a set of toroidal vortices induced by the electric current in the fluid and its self-magnetic field, and with $n = 2$ by the current and the external longitudinal field.

The stream lines $\psi = $ const in the absence of an external field are shown in Fig. 5.19. The electrically induced flow consists of consecutive toroidal vortices of alternating sense of rotation and of equal intensity. The flow pattern suggests a further growth of the interface perturbation amplitude. Moreover, it is evident from Fig. 5.19 that the electrically induced flow leads to mixing of the non-

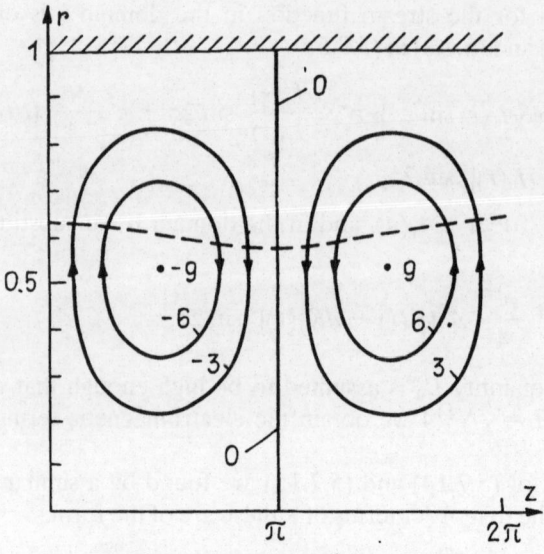

Fig. 5.19. Stream lines $10^6 \psi$ for $N = 0$.

conducting and conducting gases. This means that the plasma components are carried out of the central column and contaminations are advected from the external zone II to the core region.

In the presence of a strong longitudinal magnetic field the meridional flow takes the form of four toroidal vortices at the perturbation wavelength (Fig. 5.20). The physical mechanism generating the set of four vortices instead of two vortices could be understood by considering the vorticity of centrifugal force. Indeed, the fluid rotation is driven by the radial component of the electric current and the longitudinal magnetic field. The maximum radial current, as is

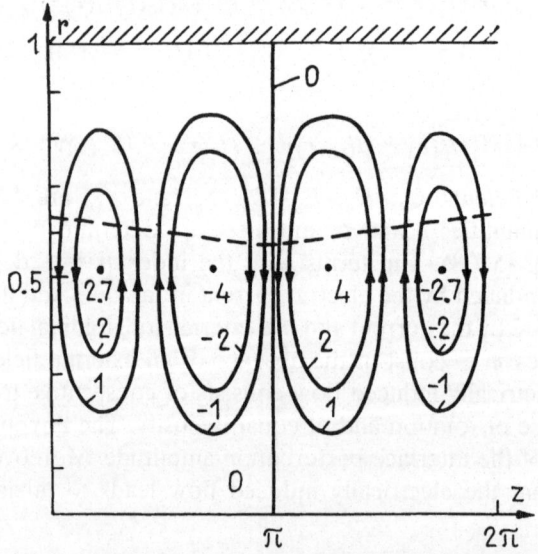

Fig. 5.20. Stream lines $10^5 \psi$ for $N = 283$.

Periodic electrically induced flows 233

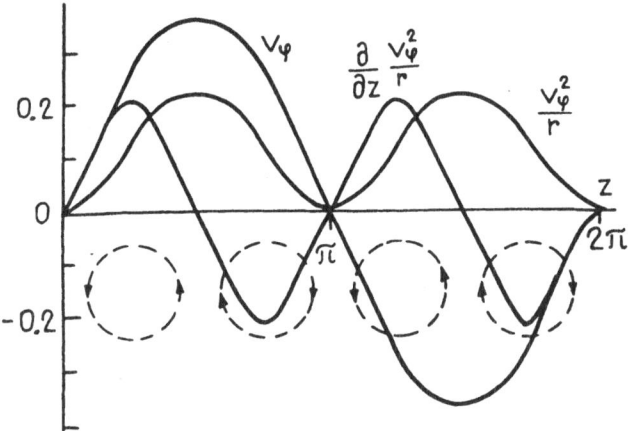

Fig. 5.21. To explain the mechanism of vortex splitting.

evident in Fig. 5.21, is in the sections $z = (2n + 1)\pi/2$ and the minimum is in the sections $z = n\pi$ ($n = 0, \pm 1, \ldots$). Consequently, the distribution of velocity component v_φ and centrifugal force v_φ^2/r along the perturbation period is of the form shown in Fig. 5.21. According to the distribution of centrifugal force, the sign of its curl, i.e. $\partial/\partial z(v_\varphi^2/r)$ changes four times within the period, and with regard to the change of sign the fluid is curled in a meridional cross section.

We may speculate on the behaviour of the interface boundary in the situation where the four vortices are present. Corresponding to the meridional flow direction at the interface, the sinusoidal boundary of period 2π will be deformed to a sinusoid of half the period. Then the number of toroidal vortices will double in the same interval $0 < z < \pi$, and this, in turn, decreases the perturbation period once again. This cascade process should end at a stable interface boundary, i.e. it leads to decay of boundary perturbations.

5.8. Longitudinal magnetic field effect on integral features of EVF

Let us now examine the effect of external field B_z on the flow rate—pressure drop characteristics for the electric current patterns described in Sections 5.4 and 5.5. Note that the main results of Section 5.7 remain valid, viz., the longitudinal magnetic field modifies the meridional flow due to the action of centrifugal force. However, the splitting of vortices within the period of electromagnetic force may not be observed. As an example consider the situation in which the electric current is supplied radially with the respective electric current function (5.4.1). Fig. 5.22 presents the distributions of the electromagnetic force curl, azimuthal velocity v_φ, centrifugal force v_φ^2/r, and its curl $\partial/\partial z(v_\varphi^2/r)$. As is evident from the figure, the signs of curls of electromagnetic and centrifugal forces driving the meridional motion coincide over the whole period of corrugation. This means that the applied magnetic field does not change the qualitative flow configuration, yet the intensity of vortices should increase with the field, thereby increasing their friction at the corrugated wall.

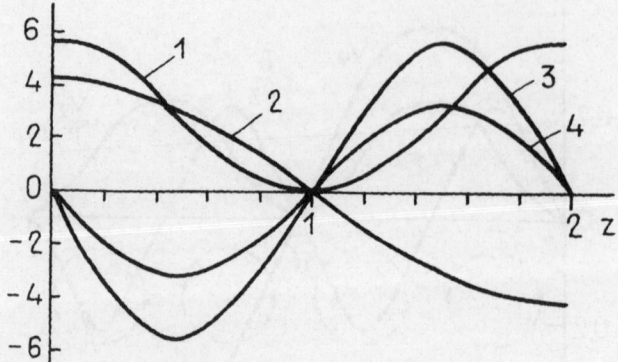

Fig. 5.22. To explain the mechanism increasing the rotation of vortex toruses: $1 - v_\varphi$, $2 - v_\varphi^2/r$, $3 - (\partial/\partial z)(v_\varphi^2/r)$, $4 - (1/\mu_0^2)(\partial/\partial z)\psi_1^2/r^3$.

The fluid motion in this case is described by the same equations, (5.7.5) and (5.7.6), and the boundary conditions are set for the specific wall configuration. The conditions are the symmetry condition on the axis (5.3.3) and the no slip condition on the rigid wall (5.3.8), or no slip in the vicinity of the central rigid core. The conditions for v_φ are analogous.

When investigating flows in tubes, the tube's radius R_0 is conveniently set as the representative length scale. Then relating the azimuthal velocity v_φ to ν/R_0 and the stream function ψ to νR_0, we obtain the equations (5.7.5) and (5.7.6) in the nondimensional form:

$$E^4\psi = -Sr^{-2}\frac{\partial}{\partial z}\psi_1^2 + N\frac{\partial}{\partial z}v_\varphi^2; \qquad E^2 rv_\varphi = -N\frac{\partial}{\partial z}\psi_1$$

where

$$S = \mu_0 I^2/\rho\nu^2; \qquad N = B_0 IR_0/\rho\nu^2.$$

The method of solution is fully described in Section 5.7, hence we proceed directly with a discussion of the results.

We represent the flow rate—pressure drop characteristics by the dependence of the nondimensional mean velocity $v_Q = Q/\pi R_0^2$ and the mean pressure gradient along z: $G = \overline{\partial p/\partial z}\,^z R_0^3/\rho\nu^2$, on a complex characterizing the relation of geometric length scales and the parameter N. In the case of radial current supply the $G(Q)$ characteristic is of the form

$$v_Q - \tfrac{1}{8}G - \varepsilon\delta\cos z_0 = 0,$$

and in the case of axial supply it is

$$v_Q - \tfrac{1}{8}G + \tfrac{1}{2}(\varepsilon_1\delta_1\cos z_1 + \varepsilon_2\delta_2\cos z_2) = 0.$$

The dependences of δ, δ_1, and δ_2 on the corrugation period h for different values of the parameter N, specifying swirl intensity, are represented in Figs. 5.23 and 5.24. When $N = 0$, these dependences correspond to those in Sections 5.4 and 5.5. The external magnetic field ($N > 0$) substantially raises

Periodic electrically induced flows 235

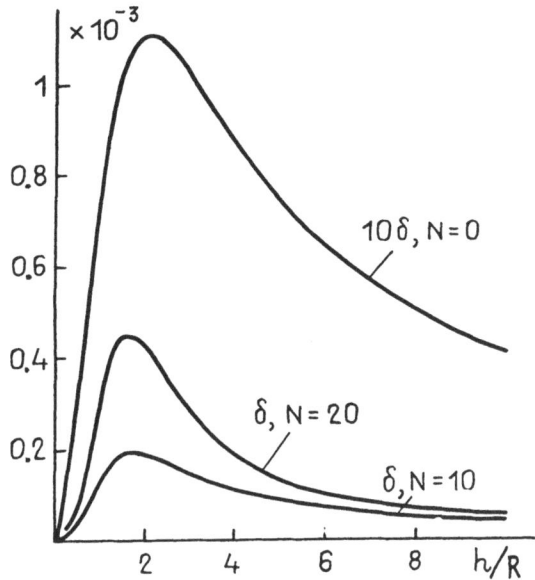

Fig. 5.23. The function $\delta(h, N)$ in the case of radial current supply.

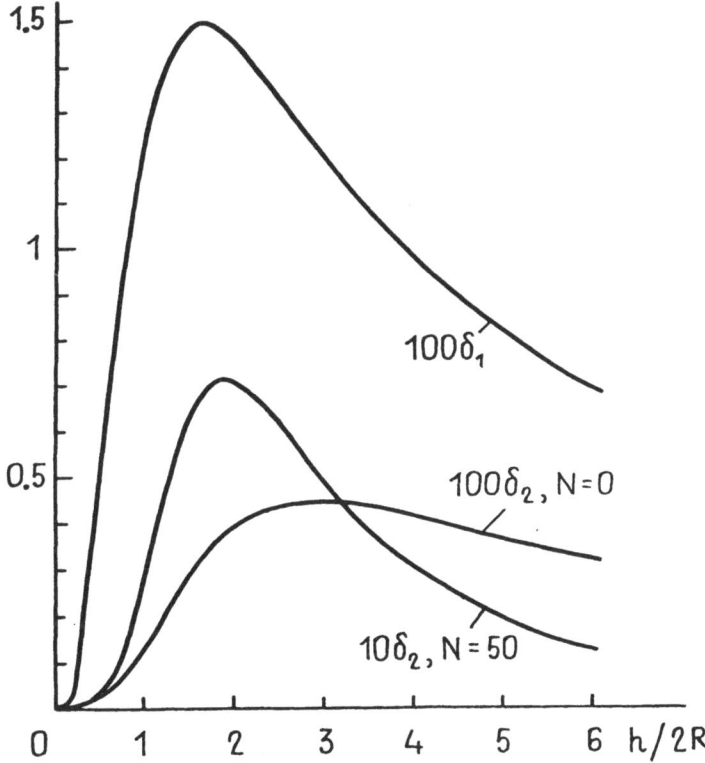

Fig. 5.24. The functions $\delta_1(h, N)$, $\delta_2(h, N)$ in the case of axial current supply.

the integral characteristics in the face of the intensified meridional flow due to the differential swirl. In this case the optimum relation between the tube's radius R_0 and the corrugation period h is also changed. If N is sufficiently high, then in the case of radial current supply the optimum ratio is $h/R_0 = 1.6$, and in the axial $h/R_0 = 3.8$.

5.9. Nonlinear interaction of periodic EVF with through-flow

The qualitative features of periodic electrically induced flows revealed here are based on the Stokes approximation of the equations of motion. The inclusion of nonlinear terms may not only modify the quantitative results, but also reveal new physical phenomena. With this aim consider the interaction of periodic EVF with the flow driven by an external motive source, for instance, by the external pressure drop at the ends of tube. In the case of weak flow ($Re \ll 1$), when the motion is determined by the linear Stokes approximation, the interaction amounts to a simple superposition of both of the flows. In fact, the interaction is absent. Superposition changes only the configuration of the stream lines, yet the integral characteristics, such as the flow rate, mean pressure gradient, and hydraulic drag, are simply additive.

Substantial interaction takes place in intense laminar flows ($Re > 1$), when the motion is determined by the full nonlinear Navier-Stokes equation. In the general case of flow in a corrugated tube the set of new factors is necessarily complex. The first effect arises due to secondary flows driven by the interaction of the external through-flow with the non-straight wall of the tube. Second, as was shown by choosing a specific phase shift of the electromagnetic force relative to the corrugation wave, the transit flow in the tube could be generated even in the absence of the external pressure drop. Thence, to reveal the interaction effect of the periodic vortices with the through-flow in a pure form, we accept the following simplified model.

Let the external pressure drop be applied at the ends of a circular straight tube; secondary flows are thus not generated. Also let an electric current induce the flow due to a periodic set of electrodes embedded in the wall and separated at a distance h from each other (Fig. 5.25a), or, alternatively, by a longitudinal current in the presence of the same ring electrodes (Fig. 5.25b). In the first case (call it the case of field $B_\varphi = B_1$) the distribution of electric and magnetic fields is given by the expression (5.4.1) and in the second by the expression (5.5.1) (the case of field $B_\varphi = B_2$).

Note that the expressions (5.4.1) and (5.5.1), strictly speaking, correspond to continuous electrodes with the specific potential distribution and thickness along the whole length of the tube's electrically conducting wall, and not to discrete electrodes. Thence the scheme shown in Fig. 5.25 is qualitative, and the ratio of isolated and conducting section lengths will not enter the solution presented below.

In the situations specified the electrically induced vortices have an opposite sense of rotation and are equivalent in intensity, hence they are not able to

Periodic electrically induced flows

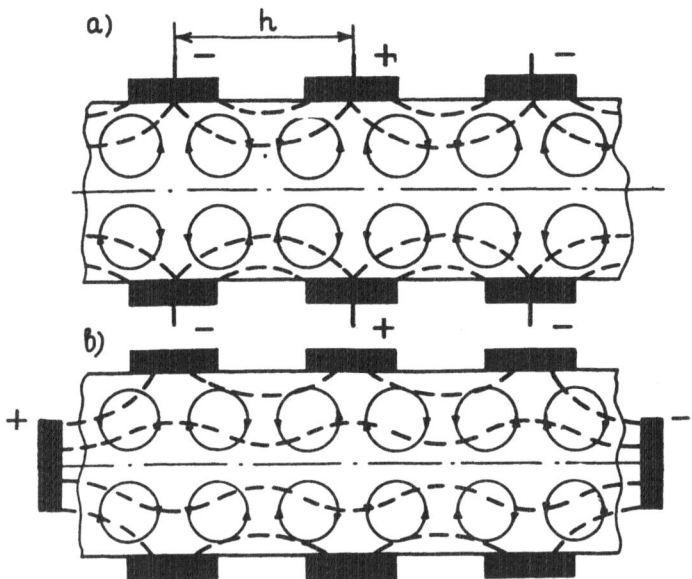

Fig. 5.25. Scheme of the model: (a) the electric current passes in the fluid from the electrodes of alternating polarity (the field B_1), (b) the longitudinal electric current is perturbed by the highly conducting ring sections (the field B_2).

generate a transit flow in the tube. Moreover the total electromagnetic force acting in the fluid volume is also zero.

A transit flow is generated by an external pressure drop, which is characterized by the constant mean pressure gradient G_0 along the tube. The stream function in the absence of EVF corresponds to Poiseuille flow $\psi = Q(2r^2 - r^4)$ with the pressure gradient, related to $\rho v^2 R_0^{-3}$, equal to $G_0 = 16Q$, where $Q = \int_0^1 v_z r \, dr = Re/2\pi$ (the velocity is related to νR_0^{-1}, R_0 is radius of the tube).

The Navier-Stokes equation for the problem was solved numerically [9]. Reliable results were obtained up to $S = 30\,000$, where the parameter S was determined by the mean-square current density as in Section 5.5. For the higher magnitudes of parameter S, in spite of the iterative procedure's convergence, the pressure drop could not be computed due to a large scatter of G values for different radial distances.

Consider the main computational results. The dependence of mean pressure gradient G on the parameter Q for several values of S and $h_0 = h/R$, i.e. the $G(Q)$-characteristic, is represented in Fig. 5.26. The straight line $G_0 = 16Q$ for $S = 0$ corresponds to Poiseuille flow. As is evident from the figure, the nonlinear interaction of transit flow ($Q \neq 0$) with the electrically induced vortices ($S \neq 0$) always leads to a substantial decrease of the pressure drop ($G < G_0$). The differences between the pressure drops initially grow in proportion to the flow rate ($G_0 - G \sim Q$), but for large rates these tend to zero. The dependence of relative pressure gradient G/G_0 on the parameter S for a constant flow rate is shown in Fig. 5.27. For small S the decrease of pressure

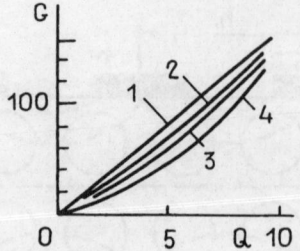

Fig. 5.26. $p(Q)$ — characteristics of the channel: (1) $S = 0$; (2) $B = B_1$, $S = 14200$, $h_0 = 2$; (3) $B = B_1$, $S = 28400$, $h_0 = 2$; (4) $B = B_2$, $S = 25000$, $h_0 = 2$.

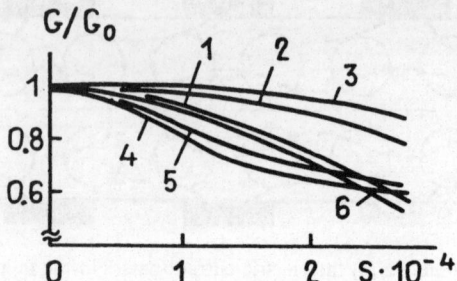

Fig. 5.27. Relative decrease of pressure drop for (1) $h_0 = 2$, $B = B_2$, $Q = 4$; for $h_0 = 1$: (2) $B = B_1$, $Q = 1$; (3) $B = B_2$, $Q = 1$; (4) $B = B_1$, $Q = 1$; (5) $B = B_2$, $Q = 2$; (6) $B = B_1$, $Q = 4$.

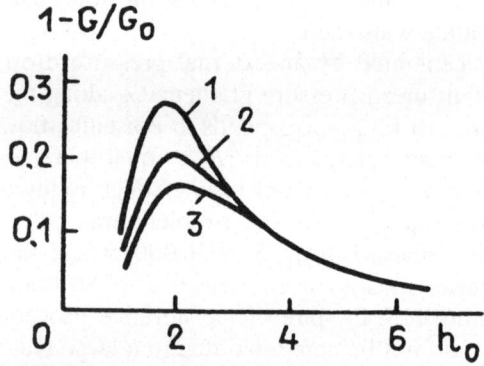

Fig. 5.28. The effect of EVF period length on the relative variation of pressure drop for $S = 14200$: (1) $B = B_1$, $Q = 1$; (2) $B = B_2$, $Q = 1$; (3) $B = B_1$, $Q = 4$.

gradient $G_0 - G$ is proportional to S^2; for high S the results suggest that $G_0 - G$ is tending monotonously to a finite value. The effect of varying h — the period of electromagnetic force and EVF — for fixed S is shown in Fig. 5.28. As is evident, the highest pressure gradient drop is observed when the period of the vortices is close to the diameter of the tube, i.e. for $h \simeq 2$. For a very small or a very large period, when the vortices are highly compressed or stretched, the pressure drop tends to the Poiseuille drop ($G \to G_0$ for $h \to 0$ and for $h \to \infty$; S, Q are constant).

To explain qualitatively the effect of pressure drop decrease due to the electrically induced vortices in the tube, consider the stream line configurations in Fig. 5.23 of the numerical nonlinear solution compared to the relevant linear solution, from which the effective action of nonlinear inertial forces is clearly evident. The inclusion of nonlinear terms, first decreases the intensity of vortices and slightly shifts them. Second, in the result of nonlinear interaction with the through-flow, the vortices become asymmetric relative to the transit flow direction — the circulation patterns are elliptic with a systematic inclination to the channel's centre. This configuration is known to maintain the momentum transfer to the channel's axis and causes the so-called negative viscosity phenomenon [12], which is observed in several natural flows [11]. The stretching of velocity profile averaged by z: $\bar{v}_z(r) = h_0^{-1} \int_0^h v_z(r, z)\,dz$ is shown in Fig. 5.29; the profile qualitatively corresponds to the experimental data [19] collected for the two-dimensional turbulent flow. The present situation is interesting from the viewpoint of the investigator studying the phenomenon of momentum transfer from vortical structures to the mean flow, although, in contrast to the set of linear vortices examined in [12] (and experimentally in [19]), we have a set of toroidal vortices. Note also that the effect of hydraulic drag reduction is found for the two different current distributions. On this basis we may suggest that the effect does not depend on a method by which the vortices are generated.

Another example of the vortex nonlinear interaction is the pump effect in a straight tube with no external through-flow. The current distribution is asymmetric relative to the direction of through-flow, and it induces a set of toroidal vortices, though the total electromagnetic force is zero. This kind of current distribution is shown in Fig. 5.1e. The asymmetric electric current lines are similar to a sawtooth form. Such a distribution may be created if each of the ring electrodes in Fig. 5.25 varies in thickness or conductivity along the tube's length. Then the electrically induced toroidal vortices are also asymmetric. The set of vortices are of alternating sense of rotation and also of alternating

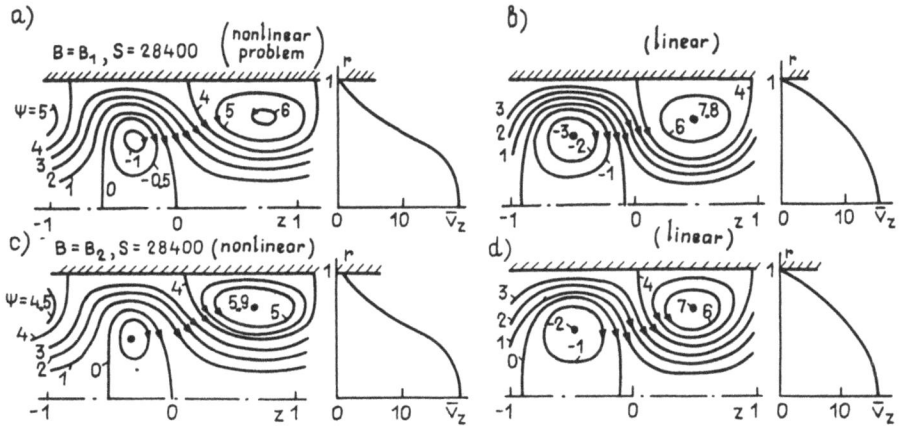

Fig. 5.29. Stream lines and the velocity profile averaged along z for the nonlinear and linear problems.

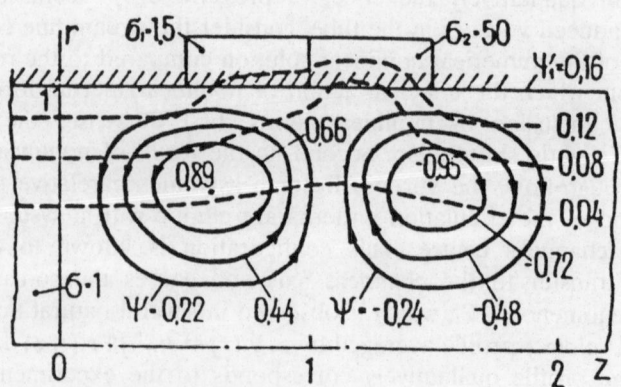

Fig. 5.30. The electric current lines and stream lines in the smooth tube with the nonsymmetric electrically conducting sections in the wall.

intensity. Physical intuition suggests that the unequal friction of the vortices at the channel's wall is able to generate a transit flow (pump effect of EVF).

It is important to emphasize, however, that, when considering the effect in the Stokes approximation there are no pump effects because of the linear superposition of the flow harmonics. The inclusion of nonlinear terms in the Navier-Stokes equations leads to multiplication of the harmonics (interaction of vortices) resulting in the appearance of new harmonics including those of zero order (transit flow). Hence the EVF pump-effect in straight tubes is found by numerical solution [10] using the same method as in the previous problem. The electric current flows along the tube, the wall of which contains a series of electrodes (Fig. 5.25); the conductivity of the first half of each electrode exceeds 1.6 times the conductivity of the fluid, and in the second half it is 50 times. In the case shown in Fig. 5.30 the nondimensional pressure drop—flow rate characteristics of the channel for the parameter S up to 10^5 is of the form

$$16Q + G(1 + 4.8 \times 10^{-12} S^2) + 6.9 \times 10^{-13} S^2 = 0. \tag{5.9.1}$$

According to (5.9.1), the channel in Fig. 5.30 has the pump-effect of EVF directed to the left and the effect of decreased hydraulic drag. Both the effects are due to the nonlinearity of Navier-Stokes equation, and the driving force arises due to the unequal friction of vortices at the channel's wall. The transit (pump) effect is proportional to I^4 for small currents.

5.10. Electrically induced flow in a loosely coiled tube

Apart from a varied cross section of the tube with straight axis, the electrically induced fluid flow may arise due to curving a constant section tube along its axis. Actually, in contrast to a straight current-carrying tube, the current density in a curved conductor is nonuniform and the self-magnetic field is not axisymmetric relative to the conductor's axis. The current density is higher for the

Periodic electrically induced flows

inner side of the bend, and lower for the external side. The magnetic lines of force are noncoaxial rings with a maximum field at the inner side of the bend. This leads, even in the absence of an external magnetic field source, to a rotational electromagnetic force in the conductor and to the development of an electrically induced flow.

The steady flow in a loosely coiled circular tube is the extensively-studied laminar flow form [1, 2, 15, 18]. In a curved tube with through-flow the secondary flow is generated due to rotational centrifugal forces. The resulting flow, viewed in the cross section, is directed along the radius of curvature in the centre, and returns along the walls to the inner side of the bend. The circulation takes the form of two vortices with opposite sense of rotation and with the axes aligned to the through-flow. Thence the external pressure drop is spent not only on maintaining the through-flow, but also on the secondary circulation. This explains the increase of hydraulic losses in a curved pipe compared to a straight tube of the same cross section. In a curved current-carrying tube the electrically induced flow interacts both with the through-flow and the secondary flow. The resulting nonlinear interaction should change the hydraulic resistance of the tube.

As a model for a loosely curved segment of the tube we choose a circular toroidal tube (Fig. 5.31). This means that we can neglect the spiral coil of a real pipe and also the entrance—exit effects. A natural way to induce the electric current in a ring shaped conductor is by induction. To simplify the problem we shall consider the steady current case, assuming this to be the first approximation for an alternating current of a relatively low frequency.

The axisymmetric current distribution in cylindrical coordinates z, r, φ (see Fig. 5.31) is obtained from the equation $\mathbf{j} = -\sigma \, \mathrm{grad} \, \Phi$, where the mode of the harmonic potential, $\nabla^2 \Phi = 0$, of the appropriate symmetry is substituted:

$$\Phi = \Phi_0 \varphi, \qquad j = j_\varphi = j_0 L/r, \qquad (5.10.1)$$

where j_0 is the magnitude of current density at distance L from the symmetry axis, L is the mean radius of the current ring (radius of the tube's axis

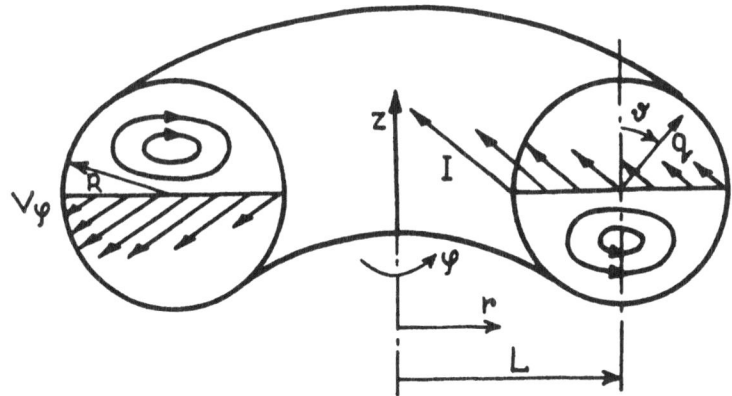

Fig. 5.31. Scheme of the model.

curvature, see Fig. 5.31). The total current in the circular conductor is specified as $I = 2\pi j_0 L^2(1 - \sqrt{1 - R_0^2/L^2})$, where R_0 is the radius of the conductor.

The magnetic field inside the conductor satisfies the equation curl $\mathbf{B} = \mu_0 \mathbf{j}$, however, in contrast to the previously considered problems of axisymmetric electrically induced flows, in the present case the symmetry of the problem does not provide enough boundary conditions to determine a unique solution inside the conductor. Hence the magnetic field is determined by the Biot-Savart law:

$$\mathbf{B}(\mathbf{r}) = \frac{\mu_0}{4\pi} \text{curl} \int \frac{\mathbf{j}(\mathbf{r}_1)}{|\mathbf{r}_1 - \mathbf{r}|} d^3 r_1 \qquad (5.10.2)$$

where $\mathbf{r}$ is the radius-vector of the observation point and $\mathbf{r}_1$ is that of the integration point. The expression (5.10.2) takes into account the fact that the field vanishes for $r \to \infty$, and integration is over the volume of a torus where the current (5.10.1) flows, because other magnetic field sources are absent.

To solve the problem for fluid flow it is necessary to express the rotational part of the electromagnetic force by its curl:

$$\text{curl } \mathbf{f}_e = \text{curl } \mathbf{j} \times \mathbf{B}. \qquad (5.10.3)$$

Substitution of (5.10.1) and (5.10.2) in (5.10.3) gives the curl of the force, which also simplifies the evaluation of the integral [8] which is singular for $\mathbf{r}_1 \to \mathbf{r}$.

Flow in the tube is determined by the Navier-Stokes equation including the external forces: a pressure gradient $G = r^{-1} \partial p / \partial \varphi$ — constant along the tube, and the electromagnetic force $\mathbf{f}_e$. The respective two nondimensional parameters governing the flow are the Reynolds number $Re = v_{\varphi_0} R_0 / \nu$ (definition of v_{φ_0}, see below), and the parameter of electrically induced flow $S = \mu_0 I^2 / \pi^2 \rho v^2$. In addition, the flow depends also on the ratio of the tube's radius R_0 to the curvature radius L. We follow the widely accepted practice in the pure hydrodynamic case [1, 15] and assume the parameter $\varepsilon = R_0/L$ to be small ($\varepsilon \ll 1$), then the solution of the Navier-Stokes equation may be sought in the form of expansion in terms of this parameter.

The expansion for the expression (5.10.3) to first order accuracy in ε yields:

$$(\text{curl } \mathbf{f}_e)_\varphi = \mu_0 j_0^2 L^{-1} q \cos \theta; \qquad (\text{curl } \mathbf{f}_e)_q = (\text{curl } \mathbf{f}_e)_\theta = 0. \qquad (5.10.4)$$

Here we have introduced the polar coordinate system q, θ in the channel's meridional cross section centred on the tube's axis (Fig. 5.29), the new coordinates are related to the z, r coordinates by the formulas: $r = L + q \sin \theta$, $z = q \cos \theta$.

Experimental studies of flows in coiled tubes, and also theoretical considerations [1], suggested a so-called Dean's approximation where for $\varepsilon \ll 1$, instead of the two nondimensional parameters ε and Re, the flow is determined by one parameter only — Dean's number:

$$K = 2Re^2 \varepsilon = \frac{G^2 R_0^7}{8\rho^2 \gamma^4 L}. \qquad (5.10.5)$$

This approximation reduces the equations of motion for a coiled tube to the equations for a straight tube with the addition of the centrifugal force $\mathbf{f}_c = \rho v_\varphi^2 r^{-1} \mathbf{i}_r$, due to the azimuthal velocity v_φ in the curved tube.

The meridional circulation is generated in view of the rotational centrifugal and electromagnetic forces oriented normally to the azimuthal velocity. The Dean number K characterizes the intensity of the secondary flow, and the electrically induced flow in this approximation is specified by the parameter $S_1 = \varepsilon S$, so that in the case of $\varepsilon = 0$ both the secondary flow and the electrically induced flow vanish. Assuming $v_\varphi = (R_0^2 - q^2)G/4\rho v$ to represent the unperturbed Poiseuille flow in a circular tube, the curl of the centrifugal force is equal to

$$(\text{curl } \mathbf{f}_c)_\varphi = -\frac{G^2}{4\rho v^2} \frac{R_0^2 - q^2}{r^2} q \cos \theta;$$

$$(\text{curl } \mathbf{f}_c)_q = (\text{curl } \mathbf{f}_e)_\theta = 0.$$

(5.10.6)

Comparing (5.10.4) and (5.10.6) we see that for any cross-sectional point of the tube $q < R_0$ the curl of the electromagnetic force is antiparallel to the curl of the centrifugal force, and both the quantities are proportional to $q \cos \theta$. For slow flow the vorticity, curl $\mathbf{v}$, is proportional to the curl of the force generating the motion, this suggests that the meridional electrically induced flow circulates in opposition to the secondary flow, and thereby compensates it to a certain extent. This compensation affects the integral characteristics of the primary flow, as was found when computing a similar flow in [9].

Denoting the rate of flow in the straight tube by $Q_0 = \pi v_{\varphi_0} R_0^2/2$, the rate of flow in the coiled tube is determined by $Q = (2Q_0/\pi) \int_0^{R_0} q \, dq \int_0^{2\pi} v_\varphi \, d\theta$. Evaluating v_φ to the accuracy of the second order terms with the small parameters K and S_1, we obtain the relative variation of the flow rate

$$\frac{Q}{Q_0} = 1 - 0.030575 \left(\frac{K}{576}\right)^2 + 0.152159 \frac{K}{576} \frac{S_1}{192} +$$
$$+ O(K + S_1)^4.$$

(5.10.7)

The second term of the expression coincides with Dean's formula [1] and expresses the decrease of flow rate in the coiled tube due to the secondary flow. The sign of the third term in (5.10.7) indicates an increase of the flow rate under the action of the electrically induced flow. Note that (5.10.7) does not include a term proportional to S_1^2 because the electrically induced flow itself does not generate a through-flow in the tube with no external pressure drop, i.e. for $K = 0$.

The drag coefficient λ of a hydraulic system is usually measured experimentally. In our case the relative variation of λ to the accuracy of (5.10.7) is expressed by

$$\frac{\lambda}{\lambda_0} = \frac{Q_0}{Q} = 1 + 0.030575 \left(\frac{K}{576}\right)^2 - 0.152159 \frac{K}{576} \frac{S_1}{192}.$$

(5.10.8)

The result (5.10.8) indicates the drag reduction for the coiled tube with the passage of electric current. This agrees with the above consideration concerning the kind of meridional electrically induced flow. The reduction of the hydraulic drag for flow in a tube was also found for the periodic electrically induced flow (see Section 5.9). In both cases the electrically induced flow is in the form of externally generated vortices, which by itself cannot create a through-flow. The drag reduction effect arises from the nonlinear interaction of the vortices and the external flow, and it does not depend on the flow direction. We may suppose that the drag reduction, resulting from nonlinear interaction between the external flow and the vortices which are symmetric relative to the flow direction, is a more general effect, and not just an exception related to the examples of electrically induced flows considered.

We estimate the application limits for the obtained results by the convergence limit of (5.10.8), i.e. by the approximate values of $K = 576$ and $S_1 = 192$, corresponding to $Re = 76$ and $I = 16$ mA in mercury at 50°C for $\varepsilon = 0.05$. These, of course, are small quantities compared to real flows. Nevertheless, the main result — drag reduction in a current-carrying coiled tube — may be found useful. In particular, a drag reduction may be expected for the curved channels in induction channel furnace (see Section 8.5).

6

Bodies in a current-carrying fluid

To explain several of the effects observed in industrial electrometallurgy and related to the passage of electric current within liquid metal or fused salts it is necessary to examine the flows arising near inhomogeneous inclusions in the current-carrying fluid. These effects include the deposition of non-conducting impurities on the walls of channels in induction smelting furnaces, an organized motion of the metal droplets during the processes of electroslag welding and remelting, and others. A knowledge of the mechanisms by which the electric current affects rigid and gaseous inclusions in the fluid may prove to be useful in the design of devices for particle separation, for composite material production, and so on. The issues concerning the behaviour of objects in the current-carrying fluid also include electrically propelled bodies containing an internal electric current source.

6.1. Effect of potential forces on a body in a current-carrying fluid

When an electric current of a uniform density passes through the fluid within a straight circular tube along its axis, a potential electromagnetic force arises, which is balanced by the relevant pressure gradient; a fluid flow is not generated in this situation. If a particle of different electrical conductivity is introduced into the fluid, the lines of electric current are deformed, and, in the general case, a rotational electromagnetic force is generated which should drive a fluid flow. We assume initially that a small enough particle does not cause a significant flow on the scale of the channel. However, even in this simplified situation, a force is acting on the particle. Thereby, in the fluid with a constant current density, a sort of buoyancy force is acting on the particle, which arises, although the current does not penetrate the particle, due to the attracting forces directed to the channel's centre on the neighbouring current-carrying fluid elements. A pressure gradient is developed along the particle's surface, and the resulting force drives the inclusion to the wall of the tube. Yet if the particle is made of a material with higher electrical conductivity than that of the fluid, it is driven to the centre of the circular channel.

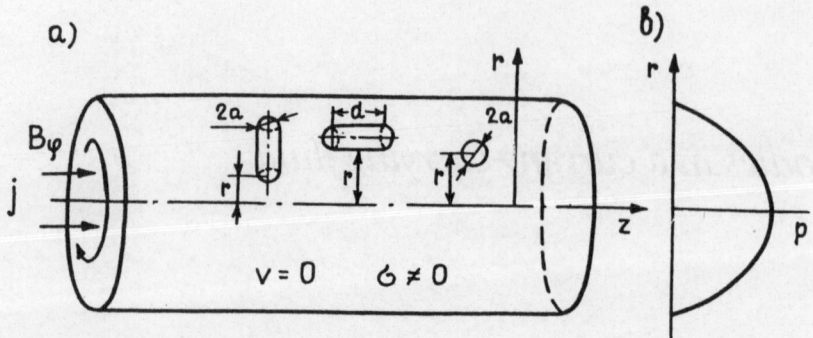

Fig. 6.1. (a) Scheme of the particle arrangement in the tube filled with fluid where the electric current is passing along the z-axis, (b) the pressure p distribution along the tube's radius.

With the aim of estimating the resulting potential force acting on the non-conducting particle consider a section of the circular channel (Fig. 6.1), through which passes an axial current of density $j_z = j = I/\pi r_0^2$, where I is the total current and r_0 is the radius of the channel [20]. The axial current generates an azimuthal magnetic field $B_\varphi = \mu_0 j r/2$. The associated electromagnetic force f_r is directed along the radius to the centre of the channel, and it is equal to $f_r = j_z B_\varphi = \mu_0 r j^2/2$. The potential force f_r is balanced by the radial pressure gradient $\partial p/\partial r = -\mu_0 r j^2/2$. The pressure in this situation does not depend on the coordinates z and φ, and it varies along the radius as $p = p_0 + \mu_0 j^2 (r_0^2 - r^2)/4$, where p_0 is the pressure at the wall (Fig. 6.1b).

The nonconducting particle in the channel is subject to the buoyant force (the rotational force due to the deformation of current lines is not taken into account) $F_r = \oint_S p \, dS$ (S is the surface area of the particle), which depends on the current magnitude ($p \sim I^2$) and on the size, configuration, and orientation of the particle relative to the channel's axis. To evaluate the force, a coordinate system fixed to the particle is introduced. For the spherical particle of radius a with the centre at a distance r from the tube's axis we obtain [20]:

$$F_r = \pi \mu_0 j^2 a^3 r = k_1 r. \tag{6.1.1}$$

To estimate the deposition time for the nonconducting particle in its way to the tube's wall, we assume the current $I = 10^4$ A, the radius of the tube $r_0 = 3 \times 10^{-2}$ m, and the particle is subject to the force (6.1.1) and the drag force determined according to Rybczynski and Hadamard [11, 13] for a slow motion as

$$F_v = 6\pi \rho_1 \nu_1 a \frac{2\rho_1 \nu_1 + 3\rho_2 \nu_2}{3\rho_1 \nu_1 + 3\rho_2 \nu_2} \frac{dr}{dt} = k_2 \frac{dr}{dt},$$

where ρ_1, ν_1 are the density and kinematic viscosity of the surrounding fluid; ρ_2, ν_2 represent the density and kinematic viscosity, respectively, for the particle's material. For a rigid particle $\nu_2 = \infty$ and the formula is reduced to the well-known Stokes formula $F_v = 6\pi \rho_1 \nu_1 a \, dr/dt$. For a spherical gas bubble ($\rho_2 \nu_2 = 0$) the expression transforms to $F_v = 4\pi \rho_1 \nu_1 a \, dr/dt$. By solving the

Bodies in a current-carrying fluid

equation of motion $k_2 \, dr/dt = k_1 r$ with the initial position of the particle $r = a$ for $t = 0$, we obtain $r = a \exp(tk_1/k_2)$. Specifying $\rho_1 = 3 \times 10^3$ kg/m^3 and $\nu_1 = 10^{-6}$ m/s^2 we find that the rigid particle of radius $a = 10$ μm $= 10^{-5}$ m will reach the wall after 32 seconds; the same size gas bubble will reach the wall after 22 seconds. A particle of greater size ($a = 10^{-4}$ m) will reach the wall as soon as in a quarter of second.

The expressions of force for particles of various forms are derived in [20]. For a cylinder of radius a, length d, and with its axis oriented normally to the axis of the tube,

$$F_r = \tfrac{1}{4}\pi\mu_0 j^2 a^2 \, dr,$$

where r is the distance from the middle of the cylinder to the axis of the tube; for a cylinder with the axis being parallel to the tube

$$F_r = \tfrac{1}{4}\pi\mu_0 j^2 \, da^2 r, \tag{6.1.1a}$$

where r is the distance between the axis of the cylinder and the axis of the tube (see Fig. 6.1).

Let us now find the force acting on an electrically conducting particle, the conductivity σ_2 of which is different from the conductivity σ_1 of the surrounding fluid (we assume perfect electrical contact at the interface). This can be done quite easily for a body of cylindrical form lying along the electric current flow direction in the circular channel (see Fig. 6.1). If the end-effects are neglected, we may assume the electric current has only a j_z-component. Then the current densities in the particle j_2 and in the fluid j_1 are related as $j_1/\sigma_1 = j_2/\sigma_2$. The magnetic field $\mathbf{B}_0$ in the cross section shown in Fig. 6.2 is determined as the superposition of the fields due to the current $j_z = j_1$ penetrating the whole circular channel: $B_\varphi = \mu_0 j_1 r/2$ (inside the channel), and the current $j_z = j_2 - j_1$ passing through only the small cylinder (particle): $B_{\varphi'} = \mu_0 (j_2 - j_1) r'/2$ (inside the particle). The system of cylindrical coordinates (r', φ', z') fixed to the small cylinder is shifted relative to the system (r, φ, z) by the distance $r = c$ along the x-axis (see Fig. 6.2). The small cylinder is subject to the force with density $\mathbf{f}_{e_2} =$

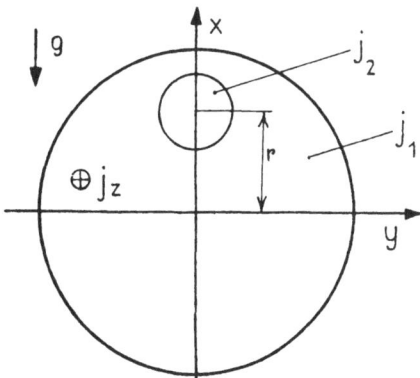

Fig. 6.2. Cross section of the circular channel and cylindrical body.

$\mathbf{j} \times \mathbf{B}_0$ having only the single component $f_{ex} = -j_z B_{0y} = -j_2 \tfrac{1}{2}\mu_0(j_2 r' \cos\varphi' + j_1 r)$. The integral electromagnetic force, acting on a length Δz of the small cylinder with the volume $V_2 = \pi r'^2 \Delta z$, is equal to

$$F_{ex} = \int f_{ex}\, dV_2 = -\frac{1}{2}\mu_0 r j_1 j_2 V_2. \tag{6.1.2}$$

If the fluid is in a hydrostatic balance

$$\nabla p = \rho_1 \mathbf{g} + \mathbf{j} \times \mathbf{B}, \tag{6.1.3}$$

where $\rho_1 \mathbf{g}$ is the gravity force density in the fluid, the stationary cylinder is subject to the total force $\mathbf{F} = \int \rho_2 \mathbf{g}\, dV_2 + \int (-p)\, d\mathbf{S}_2 + \mathbf{F}_e$ constituted by the weight of body, the pressure force on the surface, and the electromagnetic force. To express the pressure force for the situation shown in Fig. 6.2, we use (6.1.3):

$$\oint (-p)\, d\mathbf{S}_2 = -\int \nabla p\, dV_2 = -\int \rho_1 \mathbf{g}\, dV_2 - \int \mathbf{j} \times \mathbf{B}_0\, dV_2$$

$$= \mathbf{i}_x(\rho_1 g V_2 + \tfrac{1}{2}\mu_0 j_1^2 r V_2). \tag{6.1.4}$$

The expression of the total force

$$F_x = \tfrac{1}{2}\mu_0 j_1 r(j_1 - j_2) V_2 + g(\rho_1 - \rho_2) V_2 \tag{6.1.5}$$

is a generalization of Archimede's law for a current-carrying fluid. Note, the factor $\tfrac{1}{2}\mu_0 j_1 r = B_\varphi$ of the first term in (6.1.5) is the field of the uniform current of density j_1 in the channel.

If the cylinder is nonconducting, $j_2 = 0$ and the first term of (6.1.5) reduces to (6.1.1a). If the formula (6.1.5) is applied to a spherical particle, one should take into account that the ratio j_2/j_1 does not exceed 3 even for $\sigma_2 \to \infty$ (cf. (6.2.9b)). Thence for the estimate of the electromagnetic force in a highly conducting sphere the following expression may be used:

$$F_{ex} = -\mu_0 j_1^2 r V_2.$$

Assuming the drag of the sphere is given by the Stokes formula, and taking into account the acceleration of particle (including the added fluid mass $m/2$, where m-mass of the sphere), we obtain the equation of the motion

$$\frac{3}{2} m \frac{d^2 x}{dt^2} = -6\pi\rho v a \frac{dx}{dt} - \mu_0 j_1^2 V_2 x,$$

where x is the displacement of the sphere from the central axis of the tube. The solution for the initial conditions $x = x_0$, $dx/dt = 0$ ($t = 0$) is elementary. If $k_1^2 - 4k_2 > 0$, where $k_1 = 3v/a^2$, $k_2 = 2\mu_0 j_1^2/3\rho$, the sphere monotonously approaches the centre of the tube. Yet if $k_1^2 - 4k_2 < 0$ the sphere oscillates in a

Bodies in a current-carrying fluid 249

decaying motion in the vicinity of the tube's axis with the cyclic frequency $\sqrt{4k_2 - k_1^2}/2$ and the decay coefficient $k_1/2$. The oscillating motion occurs for

$$S = \frac{\mu_0 I^2}{4\pi^2 \rho v^2} > \frac{27}{32}\left(\frac{r_0}{a}\right)^4.$$

If the conditions are similar to the nonconducting particle's case, then the oscillating regime for a sphere of radius $a = 10^{-4}$ m occurs for $I > 250$ A. When the total current $I = 10^4$ A, the oscillation frequency is approximately 8 Hz, and the decay coefficient is 150.

6.2. Effect of the rotational electromagnetic forces on axisymmetric bodies

The estimates of forces (Section 6.1) are only approximate, because the deformation of electric current lines in the vicinity of the particle gives rise to rotational electromagnetic forces in the fluid; these forces cannot be balanced by a pressure gradient and drive a fluid motion. The vortical motion in the vicinity of a dopant particle may substantially change its drag, solubility, and also affect coagulation properties. If the particle is electrically conducting, a deformation of electric current lines within the particle may lead to additional components of integral electromagnetic force. To clear up the question consider the exact distribution of electric current in the fluid and the conducting body.

The magnetic Reynolds number in liquid metals is small, hence the electromagnetic forces can be determined with great accuracy in the electrodynamic approximation. We restrict ourselves to the axisymmetric situation and consider the electric current flow around a rotational body, the current at infinity being uniform and parallel to the z-axis. In this case the current distribution is determined by equation (1.6.13), which, after the substitution of the operator E^2 by its spherical expression (1.7.2) with the new variable $\mu = \cos\theta$, takes the form

$$\left(\frac{\partial^2}{\partial R^2} + \frac{1-\mu^2}{R^2}\frac{\partial^2}{\partial \mu^2}\right)\psi_1 = 0. \tag{6.2.1}$$

A solution of the equation (6.2.1) is sought for the following boundary conditions: for $R \to \infty$ the current distribution must tend to the uniform current $j_z = j_0$ with the respective

$$\psi_1 = j_0 R^2 \tfrac{1}{2}(1 - \mu^2). \tag{6.2.2}$$

On the symmetry axis $\mu = \pm 1$

$$B_\varphi = \mu_0 \frac{\psi_1}{R \sin\theta} = 0, \quad \psi_1 = 0. \tag{6.2.3}$$

For a body of arbitrary conductivity the normal component of current density is continuous when crossing its surface: $j_{1n} = j_{2n}$, but the tangential component must satisfy the condition $j_{1\tau}/\sigma_1 = j_{2\tau}/\sigma_2$, where the index 1 is related to the fluid and 2 to the body. In particular, for an ideally conducting particle ($\sigma_2 = \infty$) $j_{1\tau} = 0$. An isolated particle does not carry a current ($j_1 = 0$), and it follows from (1.10.8) and (6.2.3) on the surface of the axisymmetric body that

$$B_\varphi = \psi_1 = 0. \tag{6.2.4}$$

The solution of (6.2.1) is the infinite series (1.7.10). The condition (6.2.3) eliminates, in (1.7.10), the terms with the functions H_n, J_0, J_1, and taking into account the behaviour at $R \to \infty$ (6.2.2), the final solution is

$$\psi_1 = j_0 R^2 J_2(\mu) + \sum_{n=2}^{\infty} B_n R^{1-n} J_n(\mu). \tag{6.2.5}$$

The coefficients B_n in (6.2.5) are determined by the condition on the surface of body. If its form is close to spherical the surface equation is

$$R = a[1 + \varepsilon f(\mu)] \tag{6.2.6}$$

expressed in the coordinates with the origin in the centre of the sphere of radius a. The quantity ε is a small positive parameter, which permits to seek a solution in the form of perturbations by powers of ε. The first order approximation by ε yields on the body's surface

$$\left(\frac{R}{a}\right)^k = (1 + \varepsilon f)^k = 1 + \varepsilon k f + O(\varepsilon^2),$$

and, introducing the expansions of coefficients $B_n = B_n^0 + \varepsilon B_n^1 + \ldots$ we obtain from the condition (6.2.4) for the nonconducting body the expression to determine the coefficients:

$$j_0 a^2 (1 + 2\varepsilon f) J_2 + \sum_{n=2}^{\infty} (B_n^0 + \varepsilon B_n^1) a^{1-n} [1 + \varepsilon(1-n) f] J_n = 0. \tag{6.2.7}$$

In the zeroth approximation $B_2^0 = -j_0 a^3$, $B_n^0 = 0$ for $n \geq 3$, and the electric current function for the flow around a nonconducting sphere is

$$\psi_1 = j_0 a^2 \left[\left(\frac{R}{a}\right)^2 - \frac{a}{R} \right] J_2. \tag{6.2.8}$$

For the current flow at an arbitrary nonconducting body we obtain from (6.2.7) in the first approximation

$$3 j_0 a^2 f J_2 + \sum_{n=2}^{\infty} B_n^1 a^{1-n} J_n = 0.$$

The function $f(\mu)$ specifying the shape of the body may be expressed in the form of an expansion by the orthogonal functions J_n: $f = \Sigma_{n=0}^{\infty} b_n J_n$, however this is not necessary for the following simple examples.

The flow at a sphere shifted by the distance ε along the z-axis, $R = a(1 + \varepsilon\mu)$, is described by the equation

$$\psi_1 = j_0 a^2 \left[\left(\frac{R}{a}\right)^2 J_2 - \frac{a}{R} J_2 - 3\varepsilon \left(\frac{a}{R}\right)^2 J_3 \right]. \tag{6.2.8a}$$

The flow at an oblate spheroid can be obtained if the polar equation for a small enough ε is used:

$$R = a[1 + \varepsilon(2J_2 - 1)],$$

and for the unsymmetric body relative to the midplane $z = 0$, $R = a(1 + \varepsilon J_3)$, the electric current function is

$$\psi_1 = a^2 j_0 \left[\left(\frac{R}{a}\right)^2 J_2 - \frac{a}{R} J_2 - \frac{6}{7}\varepsilon \left(\left(\frac{a}{R}\right)^2 J_3 - \left(\frac{a}{R}\right)^4 J_5 \right) \right].$$

For the current flow at particles of arbitrary conductivity the previously set conditions for the current density components must be satisfied on the surface. Thus for a sphere of arbitrary conductivity we obtain: outside the sphere

$$\psi_1 = j_0 a^2 \left[\left(\frac{R}{a}\right)^2 - \sigma \frac{a}{R} \right] J_2; \tag{6.2.9a}$$

inside the sphere

$$\psi_1 = j_0(1 - \sigma) R^2 J_2, \tag{6.2.9b}$$

where $\sigma = 2(\sigma_1 - \sigma_2)/(2\sigma_1 + \sigma_2)$, σ_1 is the fluid conductivity, and σ_2 is the conductivity of the body. In the limiting cases for a nonconducting sphere $\sigma = 1$ and for an ideally conducting sphere $\sigma = -2$ (the electric current lines for both the cases are shown in Table 1.1: Nos. 12 and 17).

The final result of the flow description at a body is often the integral force acting on the body. The electromagnetic part of the force can be determined after the current distributions, obtained above, for bodies of different shape and conductivity according to the expression (1.6.14). Apart from this, we must take into account the fluid flow driven by the rotational electromagnetic force distribution in the fluid. Thus, for a sphere (1.6.14) gives

$$\text{curl } \mathbf{f}_e = -\mathbf{i}_\varphi 3\mu_0 j_0^2 \sigma \left[\left(\frac{a}{R}\right)^3 - \sigma \left(\frac{a}{R}\right)^6 \right] \frac{J_3}{\sqrt{1-\mu^2}} \neq 0. \tag{6.2.10}$$

The resulting fluid motion interacts with the spherical body, and the problem to determine the integral force becomes more complicated. Still more difficulties are added if it is necessary to account for a specific shape of the body. Therefore consider a method to determine the force, which is not associated with the details of the shape of the body, fluid flow, and current distribution in the vicinity of the body.

Let us choose a control fluid volume V bounded by the surface of the body and a spherical surface of large enough radius. Now recall the equation of motion in the divergent form (1.1.16). Then the integral balance of the stationary forces in the control volume is expressed by $\int_V \nabla \cdot \Pi \, dV = 0$, where Π is the tensor of momentum density flux (2.4.4). Transforming to surface integrals, the integral force on the body can be expressed by the force applied to the infinitely expanded spherical surface S_∞: $\mathbf{F} = -\int \Pi \cdot d\mathbf{S}_B = -\int \Pi \cdot d\mathbf{S}_\infty$ (the minus sign appears according to the external normal at the surface of body S_B being directed inside the control volume, but at the surface S_∞ is directed outside the control volume).

In view of the axial symmetry only the z-component of the force is different from zero. The expression for the force on the spherical surface S_∞ is of the form

$$F_z = -\int_0^\pi \int_0^{2\pi} (\Pi_{RR} \cos\theta - \Pi_{R\theta} \sin\theta) R^2 \sin\theta \, d\theta \, d\varphi. \tag{6.2.11}$$

Equal force is applied to the body. The evaluation of the force by the formula (6.2.11) is advantageous because this eliminates the integration over the surface of the body.

In many cases the electromagnetic force is dominant over other forces. If we are interested merely in the electromagnetic interaction, the expression (6.2.11) contains the components of the Maxwell stress tensor: $\sigma_{RR}^e = B_\varphi^2/2\mu_0$, $\sigma_{R\theta}^e = 0$, and reduces to

$$F_z = -\pi\mu_0 \int_{-1}^1 \psi_1^2 \frac{\mu}{1-\mu^2} \, d\mu. \tag{6.2.12}$$

Substituting (6.2.5) in (6.2.12) and passing to the limit for $R \to \infty$, we obtain

$$F_z = -\pi\mu_0 \int_{-1}^1 2j_0 J_2 B_3 J_3 \frac{\mu}{1-\mu^2} \, d\mu = -\frac{2}{15} \pi\mu_0 j_0 B_3, \tag{6.2.13}$$

since the products of even modes with J_2 does not contribute to the integral force. The final result contains only the coefficient B_3 entering the expressions of electric current functions for the nonsymmetric bodies relative to the midplane $z = 0$. The integral electromagnetic force is zero for symmetric bodies, e.g. the sphere. Asymmetric bodies are subject to a force equal to the electromagnetic force (6.2.13) at infinity. Yet this implies the paradoxical result of a finite force acting on the nonconducting sphere shifted by ε along the z-axis (cf. (6.2.8a)). The origin of the force at infinity is related to the infinite extent of the uniform current j_z with the field $B_{\varphi_1} \sim R$. In this situation the perturbation introduced by the shifted sphere with $B_{\varphi_2} \sim R^{-3}$ leads to the integral force $F \sim |\text{curl } \mathbf{B} \times \mathbf{B}| \cdot V \sim B_{\varphi_1} B_{\varphi_2} R^{-1} R^3 = \text{const}$ in the whole fluid volume, which must be compensated by an equal force (of opposite sign) applied to the infinite current-supplying electrodes. Nevertheless, the conclusion also remains valid for finite volumes with electrodes of finite width.

If the body is electrically conducting, a part of the current in the fluid penetrates the body. This current interacts with the self-magnetic field (and with the field of current in the fluid, if the body is not situated on the symmetry axis) and gives rise to an electromagnetic force applied to the elements of the body. The current-carrying body may behave differently in the fluid, depending on its shape, orientation, and conductivity.

It is instructive to consider the results of observations for the behaviour of copper objects of different shape floating in current-carrying mercury [1]. A straight copper cylinder or a sphere are attracted to the longitudinal axis of the mercury channel after the electric current is switched on, and the cylinder, in addition, is turned to assume an orientation with its longer axis parallel to the direction of current flow. If the magnitude of the current is high enough, the copper body is submerged beneath the surface of the mercury. This occurs when, according to (6.1.5), the electromagnetic force exceeds the buoyancy force.

It was also found that, in addition to the radial force, the conducting particle was subjected to an axial electromagnetic force. Consider the following experiment. In a hemicylindrical channel filled with mercury, the end walls of which are electrodes of diameter $D = 7$ cm, there floats a copper sphere of diameter $d = 3$ cm. The interelectrode distance L is variable. When the electric current is switched on, the sphere settles onto the axis of the channel. For $L = 3.5$ cm it stays stationary in the middle between the electrodes. With an increase of L the sphere moves from the centre along the axis to the electrode, where it is repelled, and it then approaches the opposite electrode, and so on: it is set into an oscillating motion.

For $L \geqslant 10$ cm the sphere does not reach the opposite electrode and oscillates along the axis relative to a point at a distance from the electrode that is approximately equal to the sphere's diameter.

To understand the strange behaviour of the copper sphere, consider the measured distribution of the axial force on the sphere along the z-axis. The distribution plotted in Fig. 6.3 reveals the positions of equilibrium ($F_z = 0$), the two points O_2 and O_3 are stable (a deviation from the position leads to a restoring force), and the third is unstable (the deviation grows under the action of axial force).

Now the behaviour of a conducting body in the current-carrying fluid can be explained. On varying the interelectrode distance L, the relative position of the equilibrium points is changed too. For small L the points O_2 and O_3 merge together. The particle remains in stable equilibrium. With an increase of L new intervals along the axis arise in which the sphere could be accelerated. If the particle deviates from the point O_1 (a small deviation may arise due to the electrically induced vortical flows at the particle – see Section 6.3), it gains speed on the interval and passes the point O_2 due to its inertia. The sphere now acquires an opposite momentum, and, if the interval O_1–O_2 is small, the sphere

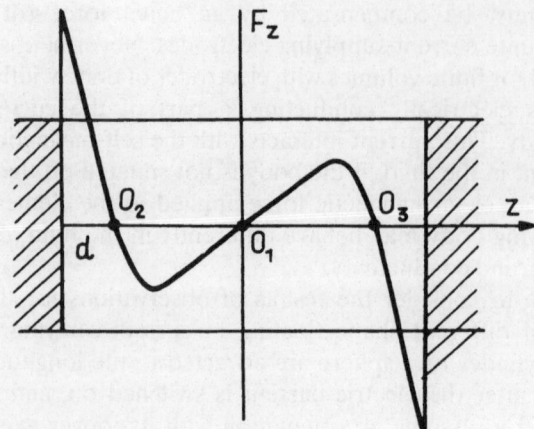

Fig. 6.3. Distribution of the axial electromagnetic force acting on a copper sphere situated on the z-axis.

may pass the interval of retardation O_2-O_1 and accelerate again in the interval O_1-O_3. Thus the sphere is oscillating from one electrode to another. When the distance L is very large, the sphere cannot pass the interval of retardation O_2-O_1, and it oscillates relative to the point O_2 (or O_3).

It remains to discover the origin of the equilibrium points and the axial force. Suppose the electrodes are ideally conducting, and the ideally conducting sphere is equal in diameter to the tube filled with a fluid of finite conductivity (Fig. 6.4a). In this situation the current is distributed uniformly in the electrodes independently of the position of the sphere, and the whole current penetrates the sphere, entering and leaving it along lines normal to the surface of the sphere. A current line consists of two curvilinear sections ℓ_1 and ℓ_3 in the fluid, and of a straight section ℓ_2 within the sphere. The problem of finding the shape

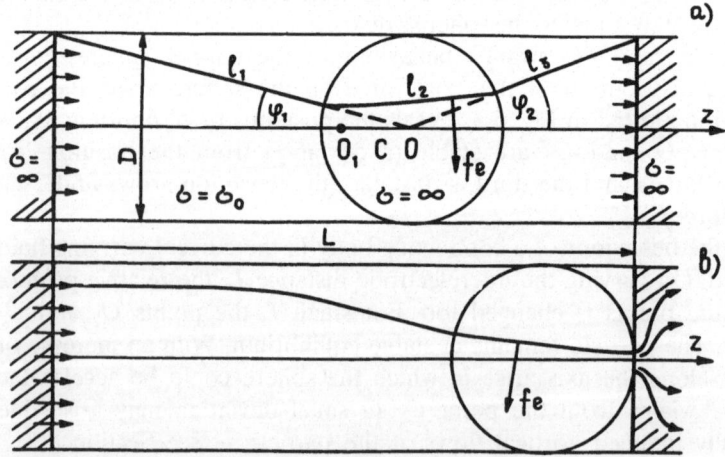

Fig. 6.4. To the origin of axial electromagnetic force in the cases: (a) a deviation from the position in the symmetry plane, (b) the position close to the electrode.

of the current line can be formulated as the problem of finding a minimum of ohmic dissipation. A simplified solution could be obtained by replacing the curvilinear sections by straight lines normal to the surface of the sphere. Then

$$\tan \varphi_1 = \frac{D}{L + 2c}, \quad \tan \varphi_2 = \frac{D}{L - 2c},$$

where c is the deviation OO_1, i.e. $\varphi_2 > \varphi_1$, and a current line in the sphere assumes the position as shown in Fig. 6.4a. In other words, there is a current concentration at the surface of the sphere facing the more distant electrode, and the current lines converge within the sphere. Since the electromagnetic force is normal to the current lines, the axial component of force f_{ez} is set up. On integrating f_{ez} over the sphere's volume we obtain the integral force, which is always directed from the centre of the channel. For a small deviation ($c/D \ll 1$) $F_{ez} \sim c/D$.

These speculations would be true if the electrodes are ideally conducting, or if the distance between the electrodes of finite conductivity is sufficiently large to retain a constant current density in the electrodes for a small deviation of the sphere. If the electrodes are of finite conductivity, the motion of the sphere leads to a redistribution of current density in the electrodes with a resultant current concentration within the sphere (Fig. 6.4b). The resulting inclination of current lines in the sphere is opposite to the previous case, and the axial electromagnetic force is directed oppositely to the z-axis. Apart from this, at the moment when the sphere touches the electrode there is a sharp increase of pressure in the liquid film of amalgam between them. The pressure increase can be estimated in a similar way as in Section 6.1, then the additional axial 'pinch' force is

$$\Delta F_z = -\frac{\mu_0 I^2}{4\pi} \cdot \frac{1}{2}$$

(the minus sign refers to the right electrode). This additional component explains the high magnitudes of force close to the electrodes in Fig. 6.3.

The effect of nonparallel current lines within a highly conducting body causes a longitudinal motion of such unsymmetric objects as a cone, a drop, a tapered cylinder, etc. In all these cases the current is concentrated at the pointed end, and its density decreases with the increase of cross-sectional area. Details of the distribution are evident from the results of numerical solution [1] for the copper cone (and copper electrodes) in mercury with the respective ratio of conductivities $\sigma_2/\sigma_1 = 50$ (Fig. 6.5). As is evident from the figure, the electric current lines within the cone are inclined to the z-axis. The electromagnetic force, normal to the current lines, has a longitudinal component, and the resulting force is directed to the base of the cone. Thus, the conical and other tapered objects are propelled along the current lines with the blunt end forward [8, 9].

It also follows from Fig. 6.5 that the current distribution in the cone, and consequently the force, depends on the location of the cone in the tube, and the highest concentration of current takes place in the situation in which the

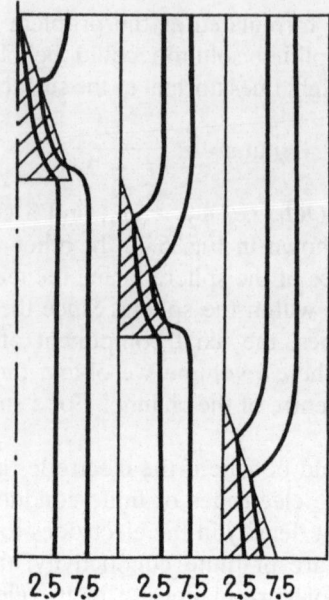

Fig. 6.5. Electric current lines ψ_1 = const in a meridional plane of the circular channel with the coaxial cone for the three positions of the cone.

pointed end touches the electrode. The integral axial force on the cone can be represented by the computed stream function ψ_1 as

$$F_{ez} = \int j_r B_\varphi \, dV = -\mu_0 I^2 2\pi \iint \frac{1}{2\bar{r}} \frac{\partial}{\partial \bar{z}} \bar\psi_1^2 \, d\bar{r} \, d\bar{z} = \mu_0 I^2 \bar{F},$$

where the integration is over the volume of the cone, and the nondimensional quantities are $\bar\psi_1 = \psi/I$, $\bar{r} = r/r_2$, $\bar{z} = z/r_2$, r_2 is the radius of the channel. The nondimensional function $\bar{F}$ varies with the cone's angle, the relative scales of the body and channel, and the position of the cone. In Fig. 6.6 the dashed lines represent the computed integral force F_{ez} vs. the square of total current $I(kA)$ for the cone with base diameter 1, height 2.5, and the diameter of the channel 2 and length 5 units. The lines 1—3 correspond to the indicated positions of the cone in the channel, the line 4 corresponds to a cone of height 1.5, shown in Fig. 6.5 in the middle of the channel. The full lines in Fig. 6.6 show the relevant experimentally measured values of force. The coincidence of the measured and simulated values is quite satisfactory; the experimental curves deviate from the straight line dependence $F_{ez} \sim I^2$ due to the growing effect of the electrically induced flow at the cone, which is not accounted for in the numerical model.

If the cone is a part of cylindrical body of length ℓ with a tapered end, the magnitude of the axial force increases. The experimental results in this case express the mean velocity of the cylinder with diameter 0.7 cm, the tapered section 1 cm long, traveling the length 26.3 cm of hemicylindrical channel with the diameter 7 cm. For the current $I = 1000$ A the velocity is 3.4, 4.4, 5.1 cm/s, respectively, for $\ell = 1.9, 2.5, 3.7$ cm. The force increases because the

Bodies in a current-carrying fluid

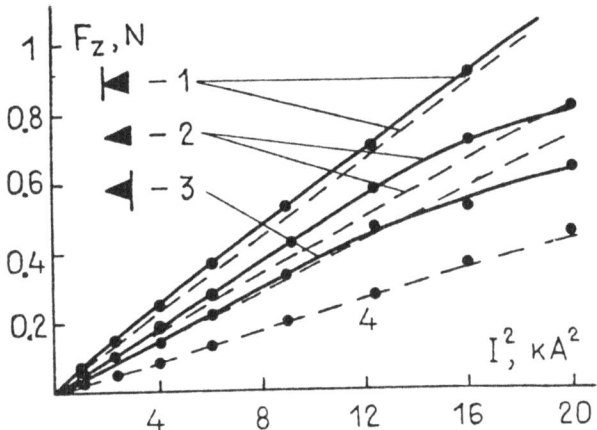

Fig. 6.6. Dependence of the axial force on the square of total current in the channel and on the position of the cone.

greater the length of the cylinder, the greater is the part of the total current that penetrates the body.

6.3. Flow at a stationary sphere

Consider a situation in which a sphere is stationary and the fluid flow is driven by rotational forces with the curl (6.2.10) due to the interaction of the electric current with the self-magnetic field. The axisymmetric velocity field is defined in spherical coordinates by the stream function ψ (3.5.3). Henceforth we will use the nondimensional ψ, obtained by dividing the dimensional stream function by $v_0 a^2$ (v_0 is a typical velocity), and the nondimensional radius $R = R^*/a$ (a is the radius of the sphere). On introducing the variable $\mu = \cos\theta$ and substituting the curl of electromagnetic force (6.2.10), we obtain the equation of motion (1.10.22) in the form:

$$(1-\mu^2)\left[\frac{\partial\psi}{\partial R}\frac{\partial}{\partial\mu}\frac{E^2\psi}{R^2(1-\mu^2)} - \frac{\partial\psi}{\partial\mu}\frac{\partial}{\partial R}\frac{E^2\psi}{R^2(1-\mu^2)}\right]$$

$$= \frac{1}{Re}E^4\psi + \frac{S}{Re^2}\sigma\left(1-\frac{\sigma}{R^3}\right)\frac{3}{R^2}J_3(\mu);$$

(6.3.1)

$$E^2 = \frac{\partial^2}{\partial R^2} + \frac{1-\mu^2}{R^2}\frac{\partial^2}{\partial\mu^2},$$

where $Re = av_0/\nu$ and the parameter $S = \mu_0 j_0^2 a^4/\nu^2\rho$ is defined by the current density at infinity.

A typical velocity is not known *a priori* in this problem, as it is, for instance, in the problem with a uniform velocity v_0 at infinity. Therefore the magnitude of v_0 is estimated by comparing the magnitudes of terms in the equation (6.3.1).

Assuming equal magnitudes of the viscous and electromagnetic forces, $Re^{-1} = SRe^{-2}$, we obtain $v_0 = Sv/a$; the inertial and electromagnetic forces $v_0 = \sqrt{S}v/a$, and equal magnitudes of the inertial and viscous forces yield $v_0 = v/a$. Thereby, all three variants lead to $v_0 = v/a$ with different normalizations by the parameter S. In the present problem we choose $Re = av_0/\nu = 1$, and the function $\psi(R, \mu, S)$ depends also on the nondimensional parameter S.

To obtain a solution of (6.3.1) we set the no slip conditions on the surface of a stationary sphere ($\mathbf{v} = 0$, $R = 1$), but the condition $\mathbf{v} \to 0$ for $R \to \infty$ does not agree, as will be shown later, with the electromagnetic term in (6.3.1). For small values of S the function $\psi(S)$ can be represented by a power series in S: $\psi = \Sigma_{n=1}^{\infty} S^n \psi_{(n)}$, and the first approximation is $\psi = S\psi_{(1)}$. We obtain for the function $\psi_{(1)}$:

$$E^4 \psi_{(1)} = \sigma \left(\frac{\sigma}{R^3} - 1 \right) \frac{3}{R^2} J_3(\mu), \tag{6.3.2}$$

the solution of which was found in [4, 7]:

$$\psi = S\sigma \frac{1}{8} \left[-R^2 - \frac{\sigma}{R} + \frac{1}{R^2} \left(\frac{\sigma}{2} - 1 \right) + \left(\frac{\sigma}{2} + 2 \right) \right] J_3. \tag{6.3.3}$$

The corresponding stream lines at the nonconducting sphere ($\sigma = 1$) are shown in Fig. 6.7. In the case of an ideally conducting sphere ($\sigma = -2$) the curl (6.2.10) is of opposite sign. The motion in this case is reversed relative to the flow in Fig. 6.7, i.e. the fluid approaches the sphere along the midplane $\mu = 0$ and flows to infinity parallel to the axis $\mu = \pm 1$.

Note that the velocity at infinity, according to (6.3.3), does not fall to zero. This follows from the particular solution $\psi_{(1)p} \sim R^2$ satisfying the right part of (6.3.2). The higher approximations in powers of S yield velocities of increasing powers of R at $R \to \infty$, for instance, $\psi_{(2)} \sim R^3$. Such a behaviour of regular perturbations (in powers of S) for small S corresponds to the well known

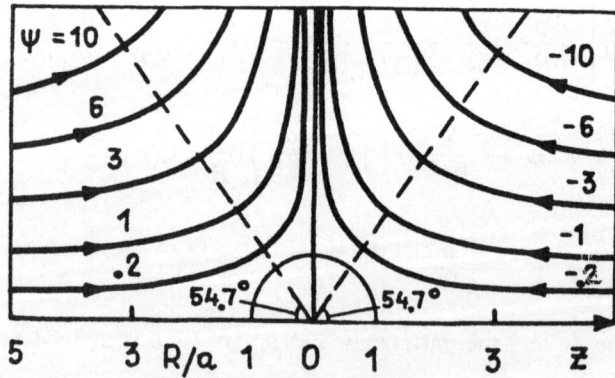

Fig. 6.7. Stream lines $\psi = \text{const}$ at the stationary nonconducting sphere. The electric current at infinity is oriented along z-axis. The dotted lines indicate the conical surface where the radial velocity is zero.

Whitehead paradox [11] for the regular perturbations $\psi = \Sigma_{n=0}^{\infty} Re^n \psi_{(n)}$ in the problem of a uniform flow around a sphere for $Re \to 0$. The first order approximation for the flow at a sphere gives $\psi \to R^2 J_2(\mu)$ for $R \to \infty$ (see Section 6.4), which is physically acceptable, because this corresponds to a uniform flow at infinity, and only the following terms of the expansion lead to growing velocities for $R \to \infty$. The Whitehead paradox is related to the fact that the first order solution does not take account of inertial forces anywhere in the flow domain, although the local Reynolds number $Re^* = aRv_0/\nu$ is not small even at a distance aR from the origin. The agreement of the function $\psi_{(0)}$ at infinity to the uniform flow with a velocity v_0 is a favourable coincidence. However, the problem with the electric current flow at a sphere leads to a first approximation which is physically unlikely.

The Whitehead paradox is resolved by applying the method of singular perturbations for $Re \to 0$ (or $S \to 0$) [25]. According to this method, the solution in the vicinity of the sphere is constructed in the form of regular perturbations (direct expansion), and far from the sphere another expansion is constructed in functions of a new variable $\lambda = ReR$; in the region $R \gg 1$ the new variable λ is assumed to be fixed when passing to the limit $Re \to 0$. Finally both the expansions are uniformly matched, giving a uniformly valid asymptotic solution in the whole flow region for $Re \to 0$.

In the case of a stationary sphere in a current-carrying fluid the matched asymptotic expansion solution is still not found, although in the case of simultaneous uniform fluid flow with velocity at infinity v_0 the solution is possible (Section 6.4).

The solution (6.3.3) remains valid in the vicinity of the sphere. With this qualification, the nonstationary solution of velocity field development after an instantaneous switch-on of the electric current in the fluid is also applicable [23]. The solution shows that the flow initially develops close to the sphere, and that it is extending with a finite velocity from the sphere; the flow field development rate is higher for the ideally conducting sphere if compared to the case of a nonconducting sphere.

The electrically induced flow at a stationary sphere for small and moderate S is studied in [16] by the use of a numerical solution of the nonlinear equations (6.3.1). The nonlinear terms are approximated by up-stream differences, and the boundary conditions at infinity are replaced by the condition $\psi/R^2 = 0$ at a finite distance R_0, which is equivalent to an impermeability condition at a spherical enclosure $R = R_0$.

The problem of electrically induced flow driven at a particle in a current-carrying fluid could be made more realistic by introducing rigid walls of a container and constraining the electric current flow by the walls. In spite of the evident complication of the geometrical conditions, this would permit us to return to the linear Stokes equations in the first approximation for small S.

6.4. Drag of a sphere in the flow of current-carrying fluid

When a sphere is immersed in a fluid that is not only carrying the electric current, but that is also flowing with a uniform velocity v_0 at infinity, then it is natural to assume the velocity v_0 as representative in the expression of Reynolds number $Re = av_0/\nu$. In the situation in which the direction of flow coincides with the direction of electric current we can use the axisymmetric function (1.5.2) to describe the fluid motion. Then, in the linear approximation, we have a solution in the form of the superposition of (6.3.3) and the classical result of Stokes [11] expressed by the nondimensional variables defined in Section 6.3:

$$\psi = \frac{S}{Re^2}\left[-R^2 - \frac{\sigma}{R} + \frac{1}{R^2}\left(\frac{\sigma}{2} - 1\right) + \left(\frac{\sigma}{2} + 2\right)\right]\frac{\sigma}{8} J_3(\mu) +$$

$$+ \left[R^2 - \frac{3}{2} R + \frac{1}{2R}\right] J_2(\mu). \tag{6.4.1}$$

According to the solution (6.4.1) the flow configuration, for example, at a nonconducting sphere ($\sigma = 1$), is changing in dependence on the current density. The electrically induced flow accelerates the fluid motion upstream from the sphere and decelerates it downstream of the sphere. In the result of interaction a closed region limited by the stream line $\psi = 0$ (Fig. 6.8), of reversed flow is formed behind the sphere. At a critical point A on the symmetry axis $\theta = 0$ the radial velocity is zero:

$$\left.\frac{v_R}{v_0}\right|_{\theta=0} = \left[1 - \frac{3}{2R} + \frac{1}{2R^3}\right] -$$

$$- \frac{S}{8Re}\left[1 - \frac{5}{2R^2} + \frac{1}{R^3} + \frac{1}{2R^4}\right] = 0.$$

With the increase of parameter S the critical point moves downstream, and for $S = 8Re$ it reaches infinity. With a further increase of S the flow configuration is transformed to the purely electrically induced flow in Fig. 6.7.

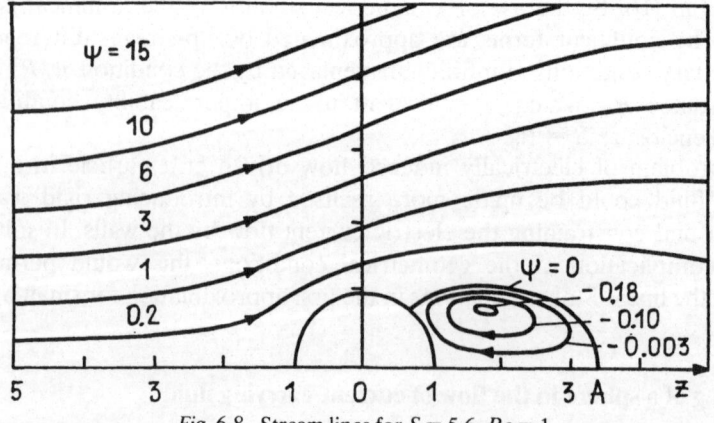

Fig. 6.8. Stream lines for $S = 5.6$, $Re = 1$.

When the conductivity of the sphere is higher than that of the fluid, the direction of electrically induced flow is reversed relative to the EVF with a nonconducting sphere, and the velocity on the symmetry axis is directed away from the sphere. Thence, the external fluid flow in the current-carrying fluid increases the velocity behind the sphere and decreases upstream, where a frontal recirculation zone is formed.

In spite of the substantial change of flow configuration, the force acting on the sphere stays the same $F_z = 6\pi\nu\rho v_0 a$ in the linear approximation, i.e. it coincides with the Stokes formula [4, 7]. This is understandable since the solution (6.4.1) is a linear superposition of the functions respectively of uniform flow at the sphere and the symmetrical electrically induced flow; the latter does not contribute to the integral force.

The force acting on a body is varied with the increase of magnitude of the electric current if the nonlinear flow interaction is taken into account. The inertial terms in the equation of motion must also be retained at large distances from the sphere, where even for an infinitesimal parameter S the electrically induced flow velocity is not zero in the linear approximation (see Section 6.3). In the situation where the sphere is surrounded by a uniform flow with velocity v_0, the stream function at infinity must be close to the function of uniform flow $\psi = R^2 J_2$. Hence, the external solution for the small numbers Re and S can be constructed in the form of uniform flow perturbations, and then it can be matched to the inner expansion valid close to the sphere.

To realize this suggested plan of solution assume that, in the limiting process for $Re \to 0$, the magnitude of the electromagnetic term coefficient $S/Re^2 = O(1)$ [6]. Then the zeroth approximation neglects the electrically induced flow. The following solution procedure is different from [6, 7] in which the solutions of simultaneous flow of fluid and electric current at spherical bodies were obtained for the first time. We proceed with the general method of matched asymptotic expansions [25], according to which in a region close to the sphere the direct expansion is valid

$$\psi = \psi_{(0)}(R, \mu) + Re\psi_{(1)}(S, R, \mu) + O(Re^2) \qquad (6.4.2)$$

for $Re \to 0$ and $S = O(Re^2)$, the independent variable R is not deformed. On substituting (6.4.2) in (6.3.1), after passing to the limit $Re \to 0$ with $S/Re^2 = O(1)$ we obtain $E^4\psi_{(0)} = 0$, the solution of which for the boundary conditions of no slip at the surface of sphere $R = 1$

$$\psi = \frac{\partial \psi}{\partial R} = 0 \qquad (6.4.3)$$

and minimum singularity for $R \to \infty$:

$$\psi_{(0)} = A_0 \left[R^2 - \frac{3}{2}R - \frac{1}{2R} \right] J_2(\mu) \qquad (6.4.4)$$

contains a temporarily unknown constant A_0, which will be determined in the matching procedure with an external expansion.

For the first approximation of the inner expansion (6.4.2) we obtain the equation

$$-9A_0^2\left(\frac{1}{R^2} - \frac{3}{2R^3} + \frac{1}{2R^5}\right)J_3 = E^4\psi_{(1)} + \frac{S}{Re^2}\frac{3\sigma}{R^2}\left(1 - \frac{\sigma}{R^3}\right)J_3,$$

the solution of which for the boundary conditions (6.4.3) is

$$\psi_{(1)} = A_1\left(R^2 - \frac{3R}{2} + \frac{1}{2R}\right)J_2 -$$

$$- \frac{3}{16}A_0^2\left(2R^2 - 3R + 1 - \frac{1}{R} + \frac{1}{R^2}\right)J_3 -$$

$$- \frac{S}{Re^2}\frac{\sigma}{8}\left[R^2 - \left(2 + \frac{\sigma}{2}\right) + \frac{\sigma}{R} + \frac{2-\sigma}{2R^2}\right]J_3.$$

Far from the sphere the external expansion is valid

$$\psi = D_0(Re)\Psi_0(\lambda, \mu) + D_1(Re)\Psi_1(S, \lambda, \mu) + o(D_1) \tag{6.4.5}$$

formulated with a new compressed variable $\lambda = \Delta(Re)R$, which is assumed to be fixed when $Re \to 0$ for large R. After substituting (6.4.5) in the equation of motion (6.3.1) rewritten for the new variable λ:

$$\Delta^5(1 - \mu^2)\left[\frac{\partial}{\partial\lambda}(D_0\Psi_0 + D_1\Psi_1)\frac{\partial}{\partial\mu}\frac{E^2(D_0\Psi_0 + D_1\Psi_1)}{\lambda^2(1 - \mu^2)} - \right.$$

$$\left. - \frac{\partial}{\partial\mu}(D_0\Psi_0 + D_1\Psi_1)\frac{\partial}{\partial\lambda}\frac{E^2(D_0\Psi_0 + D_1\Psi_1)}{\lambda^2(1 - \mu^2)}\right]$$

$$= \frac{\Delta^4}{Re}E^4(D_0\Psi_0 + D_1\Psi_1) + \frac{S}{Re^2}\sigma\Delta^2\left(1 - \frac{\sigma\Delta^3}{\lambda^3}\right)\frac{3}{\lambda^2}J_3, \tag{6.4.6}$$

$$E^2 = \frac{\partial^2}{\partial\lambda^2} + (1 - \mu^2)\lambda^{-2}\frac{\partial^2}{\partial\mu^2},$$

the coefficients Δ, D_0, D_1 can be determined by consecutively comparing the magnitudes of terms in (6.4.6).

We have assumed previously that the external flow in the zeroth approximation is the uniform flow with $D_0\Psi_0 = \Delta^{-2}\lambda^2 J_2$, whence $D_0 \sim \Delta^{-2}$. Moreover, since Ψ_0 is an eigenfunction of the operator E^2, it automatically satisfies (6.4.6) without the electromagnetic term, which is neglected in the zeroth approximation. Comparison of the inertial and viscous terms gives $\Delta^5 D_0^2 \sim \Delta^4 D_0 Re^{-1}$, whence with $D_0 = \Delta^{-2}$ we obtain

$$\Delta = Re, \qquad D_0 = Re^{-2}, \qquad \Psi_0 = \lambda^2 J_2. \tag{6.4.7}$$

Further, after substituting (6.4.7) in (6.4.6), we obtain for the first approximation $\Delta^5 D_0 D_1 \sim \Delta^4 D_1 Re^{-1} \sim \Delta^2 S Re^{-2}$ and for $S Re^{-2} = O(1)$:

$$D_1 = Re^{-1};$$

$$(1-\mu^2)\left[2\lambda J_2 \frac{\partial}{\partial \mu} \frac{E^2 \Psi_1}{\lambda(1-\mu^2)} + \mu\lambda^2 \frac{\partial}{\partial \lambda} \frac{E^2 \Psi_1}{\lambda^2(1-\mu^2)}\right]$$

$$= E^4 \Psi_1 + \frac{S}{Re^2} \frac{3\sigma}{\lambda^2} J_3.$$

The solution of equation for Ψ_1 is found in [7]:

$$\Psi_1 = \frac{\sigma}{2} \frac{S}{Re^2} \lambda J_2 + C_1(1+\mu)\left[1 - \exp\left(-\lambda \frac{1-\mu}{2}\right)\right]. \tag{6.4.8}$$

The last term of (6.4.8) is the classical Oseen solution, but with a temporarily unknown coefficient C_1.

The unknown coefficients A_0, A_1, C_1 are determined by matching the conditions of the outer and inner expansions in an intermediate domain where both the expansions are supposed to be valid. Let us apply the asymptotic matching principle stated in Section 3.4. Since in the present problem a direct expansion is constructed for the inner region at the sphere, then, according to the matching principle, the two-term formal Laurent series expansion of the inner expansion $\psi = \psi_{(0)} + Re\psi_{(1)}$ for $R \to \infty$ must be rewritten with the external variables λ and μ,

$$\psi(R \to \infty) = A_0\left[\frac{\lambda^2}{Re^2} - \frac{3}{2}\frac{\lambda}{Re}\right]J_2 + Re\left[A_1 \frac{\lambda^2}{Re^2} J_2 - \right.$$

$$\left. - A_0^2 \frac{3}{8} \frac{\lambda^2}{Re^2} J_3 - \frac{S}{Re^2} \frac{\sigma}{8} \frac{\lambda^2}{Re^2} J_3\right] + o(Re^{-1}),$$

and this is equated to the formal inner limit of the external expansion (6.4.5):

$$\psi(\lambda \to 0) = \frac{\lambda^2}{Re^2} J_2 + \frac{1}{Re}\left[C_1\left(\lambda J_2 - \frac{\lambda^2}{4}(J_2 - J_3)\right) + \right.$$

$$\left. + \frac{S}{Re^2} \frac{\sigma}{2} \lambda J_2\right] + o(Re^{-1}).$$

Equating consecutively the coefficients of $\lambda^2 Re^{-2}$, λRe^{-1}, and $\lambda^2 Re^{-1}$, we find:

$$A_0 = 1, \quad C_1 = -\left(\frac{3}{2} + \frac{\sigma}{2}\frac{S}{Re^2}\right), \quad A_1 = -\frac{1}{4}C_1.$$

When the inner expansion is found, we can evaluate the total drag coefficient for the sphere [7]:

$$c_D = \frac{2F_z}{\rho v_0^2 \pi a^2} = 12\left[\frac{1}{Re} + \frac{1}{8}\left(3 + \sigma \frac{S}{Re^2}\right)\right], \tag{6.4.9}$$

where F_z is evaluated by the formula (6.2.11). It follows from (6.4.9) that the drag of a nonconducting body ($\sigma = 1$) increases with the passage of electric current, and for a body with a conductivity higher than the fluid the drag decreases, i.e. particles of different conductivities will be separated in the flow.

The above asymptotic expansion is formally valid for $Re \ll 1$ and $S = O(Re^2)$, yet, as is shown in [6], it can be applied approximately up to $Re = 60$. For a moderate Reynolds number the solution in [7] is obtained by the numerical Galerkin method, according to which the stream function is represented in the form

$$\psi = \sum_{n=2}^{N} B_n(R) J_n(\mu),$$

and then substituted in the equation of motion (6.3.1) to determine the coefficients $B_n(R)$. The series are truncated at $N = 2$, and $B_n(R)$ are approximated by trial functions, which are required to minimize the approximation error both on the surface of the sphere and in the fluid domain.

The dependence of the computed drag coefficient on Reynolds number for different S is depicted in Fig. 6.9 [7]. It is evident from the figure that the electric current significantly decreases the drag of a body with conductivity higher than the fluid. The drag of a nonconducting body increases slightly. For higher Re the drag of an ideally conducting sphere becomes negative, which is physically impossible for the symmetric distribution of electromagnetic force. The fact of negative drag indicates an insufficient accuracy of the numerical solution, which may be improved by use of the spectral or pseudospectral method [18], and the coefficients $B_n(R)$ are then determined by solving ordinary differential equations.

In [17] a finite difference method is applied to solve the problem. The derivatives in the nonlinear terms of the equation are approximated by upstream differences, which made it possible to obtain the solution for higher Re numbers (up to $Re = 50$) than a method using only the central differences. The boundary condition at infinity ($\psi = R^2 J_2$) is replaced by the same condition imposed at a finite distance from the sphere. The stream lines constructed by this solution are shown in Fig. 6.10.

So far we have discussed the flow at rigid bodies. Consider now the

Bodies in a current-carrying fluid

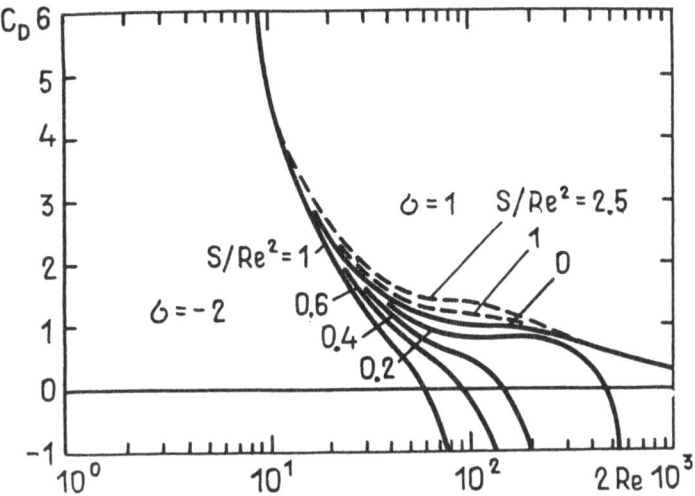

Fig. 6.9. Effect of electric current on the drag coefficient of nonconducting ($\sigma = 1$) and ideally conducting ($\sigma = -2$) spheres vs. Reynolds number.

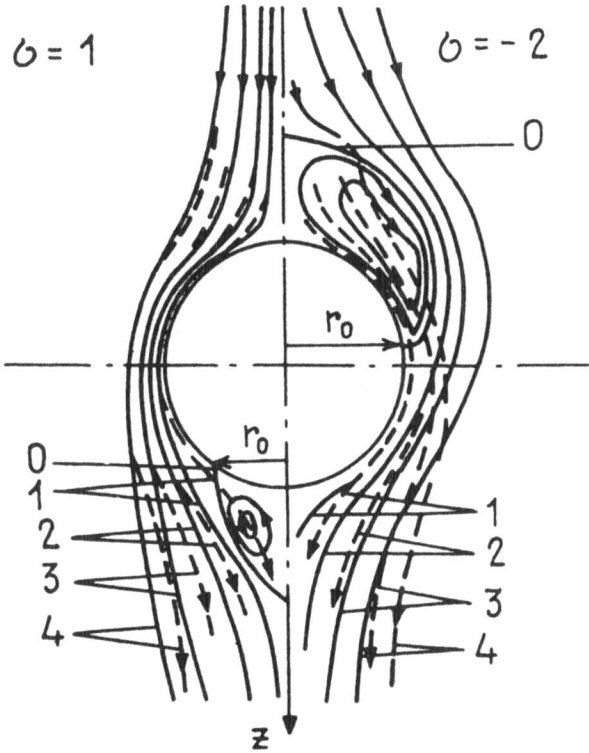

Fig. 6.10. Stream lines for $S = 0$, $Re = 10$ (dashed lines) and $S = 113$, $Re = 10$ (full lines): $0 - \psi = 0, 1 - 0.001, 2 - 0.01, 3 - 0.05, 4 - 0.1$.

behaviour of liquid drops of a different fluid in the current-carrying fluid; an internal motion in the drops could depend on the external flow. This may be of interest, for instance in applications to the electroslag remelting process, in which liquid metal drops travel through the slag bath carrying an electric current of magnitude 10^4 A. The refinement of metal in this process depends on the intensity of mass transfer at the surface of the drop and on internal mixing.

The mathematical problem of flow at a spherical undeformed drop in a current-carrying fluid differs from the previous problem of flow at a rigid sphere only by the additional motion within the drop, which is driven by the stress distribution on the surface due to the external flow. The additional condition on the surface is the continuity of the tangential velocity component v_θ. The corresponding solution is obtained in [15] in the form of a two-term Oseen expansion for $Re \to 0$ in the external fluid and a two-term Stokes expansion for the flow inside the drop. The main result of [15] is the description of mass transfer, which will be discussed in Chapter 7.

A similar problem of flow at rigid particle that is permeable for the fluid and electric current is solved in [10]. In this case the flow continuity means the continuity of the stream function ψ. The expression for the drag coefficient

$$c_D = 12\left(1 - \frac{4}{3}m_0\right)\left[\frac{1}{Re} + \frac{1}{8}\left(3 + \frac{\sigma S}{Re^2} - m_0\right)\right],$$

includes a positive constant of permeability $0 \leq m_0 \leq 1/7$, and it describes the decrease of drag for this particle.

A common feature of the problems considered above is the axial symmetry of flows with the external flow oriented along the axis of the body and the electric current oriented along the same direction. A general case with different directions of the current j_0 and velocity v_0 is quite complicated mathematically. In the case of a spherically symmetric body the flow in linear approximation can be solved by superposition of the two solutions: the first due to the external uniform flow with velocity v_0, and the second due to flow induced by the electric current j_0. When v_0 and j_0 are inclined at a prescribed angle, the resulting velocity field configuration depends substantially on the angle [5]. The solution is represented by superposition of the two terms in (6.4.1).

6.5. Flows at spheroids

Consider now the flows at spheroidal bodies in the presence of electric current penetrating the fluid and bodies. Solution of this kind of problem answers the question about the effect of a body's shape, different from the spherical, on the current-carrying fluid flow characteristics.

In the situation in which the axis of a spheroid is aligned with the uniform current, the coordinate system of a prolate spheroid can be adopted (see Section 1.4, 3). The surface of the spheroid is defined by the equation $\xi = \xi_0 =$

Bodies in a current-carrying fluid

const and the representative length scale is conveniently set as half of the focal distance (see the figure in Section 1.4, 3). Let us use the variables $\mu = \cos \eta$, $\lambda = \cosh \xi$. Then the cylindrical coordinates are expressed by the spheroidal variables as

$$z = a\lambda\mu, \qquad r = a\sqrt{(1-\mu^2)(\lambda^2-1)}.$$

Let the electric current of density j_0 at infinity be parallel to the z-axis. The electric current distribution at the spheroid is expressed by the solution of equation (1.6.13) for function ψ_1, which in the case of nonconducting spheroid was obtained in [21], and in the case of arbitrary conducting in [22]:

$$\psi_1 = j_0 a^2 \left[1 - b_0 \left(\ln \frac{\lambda+1}{\lambda-1} - \frac{2\lambda}{\lambda^2-1} \right) \right] (\lambda^2-1) J_2(\mu), \qquad (6.5.1)$$

where

$$b_0 = (\sigma_2 - \sigma_1) \bigg/ \left[\sigma_2 \left(\ln \frac{\lambda_0+1}{\lambda_0-1} - \frac{2}{\lambda_0} \right) + \sigma_1 \left(\frac{2\lambda_0}{\lambda_0^2-1} - \ln \frac{\lambda_0+1}{\lambda_0-1} \right) \right];$$

$\lambda_0 = \cosh \xi_0$;

σ_1 is the conductivity of the fluid, σ_2 of a spheroid. The equation of motion in spheroidal coordinates can be expressed from (1.10.22) by substituting the respective Lame coefficients and the function ψ_1 (6.5.1). We obtain for the nondimensional stream function ψ ($\psi^* = va\psi$):

$$D\psi(E^2\psi) = \frac{1}{\lambda_0} E^4\psi + \frac{4b_0 S}{\lambda_0^6} \frac{1}{\lambda^2-\mu^2} \left[1 - b_0 \left(\ln \frac{\lambda+1}{\lambda-1} - \frac{2\lambda}{\lambda^2-1} \right) \right] J_3(\mu);$$

$$E^2 = \frac{1}{\lambda^2-\mu^2} \left[(\lambda^2-1) \frac{\partial^2}{\partial \lambda^2} + (1-\mu^2) \frac{\partial^2}{\partial \mu^2} \right]; \qquad (6.5.2)$$

$$D\psi = -\frac{1}{\lambda^2-\mu^2} \left[\frac{\partial \psi}{\partial \mu} \left(\frac{2\lambda}{\lambda^2-1} - \frac{\partial}{\partial \lambda} \right) + \frac{\partial \psi}{\partial \lambda} \left(\frac{2\mu}{1-\mu^2} + \frac{\partial}{\partial \mu} \right) \right];$$

$$S = \frac{\mu_0 j_0^2 a^4}{\rho v^2}.$$

For small S the Stokes approximation is valid near the spheroid and the solution of (6.5.2) is of the form

$$\psi = S\psi_j = [h_0(\lambda)J_3(\mu) + h_1(\lambda)J_5(\mu)]S.$$

The corresponding boundary conditions are the no slip and impermeability of the spheroidal surface $\lambda = \lambda_0$, yet for $\lambda \to \infty$ the velocity does not fall to zero, similarly to the case with a sphere (Section 6.3).

The intensity of the electrically induced flow depends on the shape of the spheroid. In the limiting case $\lambda_0 \to 1$, when the spheroid is spindle-shaped, the electrically-induced flow vanishes, because the rotational force, proportional to the constant b_0 in (6.5.1), tends to zero if $\lambda_0 \to 1$.

When the external uniform flow of velocity v_0 is directed parallel to the z-axis, the linear Stokes solution of the flow past the spheroid is described by the stream function ψ_v^* [11]:

$$\frac{\psi_v^*}{v_0 a^2} = \psi_v = J_2(\mu)(\lambda^2 - 1) \times$$

$$\times \left[1 - \frac{(\lambda_0^2 + 1)\ln[(\lambda + 1)/(\lambda - 1)] - 2\lambda(\lambda_0^2 - 1)/(\lambda^2 - 1)}{(\lambda_0^2 + 1)\ln[(\lambda_0 + 1)/(\lambda_0 - 1)] - 2\lambda_0}\right],$$

and the superposition with the electrically induced flow is

$$\psi^* = v_0 a^2 \psi_v + vaS\psi_j.$$

The electrically induced flow in this approximation does not alter the drag coefficient of the spheroid in the uniform flow. When the inertial terms are accounted for in Oseen approximation, the drag coefficient increases for a nonconducting spheroid and decreases for the spheroid with $\sigma_2 > \sigma_1$ [22] in agreement to the spherical case.

Similarly, the flow at an oblate spheroid can be analysed in the coordinates of an oblate spheroid (see Section 1.4, 4). On introducing the variables $\mu = \cos \eta$, $\lambda = \cosh \xi$, the cylindrical coordinates are expressed by the spheroidal in the following way:

$$z = a\mu\lambda, \qquad r = a\sqrt{(1 - \mu^2)(\lambda^2 + 1)}.$$

The flow of the (uniform at infinity) electric current of density j_0 directed parallel to the z-axis past the oblate spheroid $\lambda = \lambda_0$ is expressed by the function ψ_1:

$$\psi_1 = j_0 a^2 \left[1 - b_1\left(\frac{\lambda}{\lambda^2 + 1} - \text{arccotan } \lambda\right)\right](\lambda^2 + 1)J_2(\mu), \qquad (6.5.3)$$

where

$$b_1 = (\sigma_2 - \sigma_1)/[\sigma_2(e(1-e^2)^{-1/2} - \arcsin e) + \\ + \sigma_1(\arcsin e - e(1-e^2)^{1/2})];$$

$$e = (\lambda_0^2 + 1)^{-1/2}.$$

As in the previous case, the equation of motion follows from (1.10.22), and in Stokes approximation the stream function $\psi^* = va\psi$ is of the form [21]:

$$\psi = [g_0(\lambda)J_3(\mu) + g_1(\lambda)J_5(\mu)]S.$$

The additional flow with uniform velocity v_0 at infinity is expressed by the stream function $\psi_v^* = v_0 a^2 \psi_v$ [11]:

$$\psi_v = J_2(\mu)(\lambda^2 + 1)\left[1 - \frac{\lambda(\lambda_0^2+1)/(\lambda^2+1) - (\lambda_0^2-1)\operatorname{arccotan}\lambda}{\lambda_0 - (\lambda_0^2-1)\operatorname{arccotan}\lambda_0}\right].$$

The results of [22] show that the drag coefficient in the presence of the electrically induced flow is altered only due to the inertial terms; it increases for $\sigma_2 < \sigma_1$ and decreases if the conductivity of the spheroid is higher than the fluid, $\sigma_2 > \sigma_1$, as for the bodies considered previously.

Note that the drag coefficient of an infinitely thin disc, which is the degenerate case of a spheroid when $\lambda_0 \to 0$, in the current-carrying fluid depends specifically on its conductivity. If the conductivity is small, but not zero $\sigma_2 \neq 0$, then, in view of its infinitesimal thickness, it offers zero resistance to the electric current, and the lines of uniform current penetrate the disc without a deformation: $b_1 = 0$ for $\lambda_0 = 0$, $\sigma_2 \neq 0$ (6.5.3). Thence the electrically induced flow does not arise. When $\sigma_2 = 0$, the electric current is forced to bend around the thin nonconducting obstacle and the flow is induced, which leads to an increase of the drag coefficient.

The same effect also takes place in the case of flow without axial symmetry at an infinitely thin obstacle, as was shown in [24], where the complex problem of the current flow past a spheroid in an arbitrary direction was solved. A closed form solution is obtained in [24], yet the description of the solution required that the author extended the number of labelled formulae up to a total of 121. This example shows the complexity of the electrically induced flow problems in nonaxisymmetric situations.

6.6. Discharge between electrodes of hyperboloidal form

Quite often the bodies submerged in a fluid are themselves sources of electric current, e.g. the melting electrode in the electroslag refinement process. The first step in the melting electrode problem could be the determination of electrically induced flow at an electrode that is close in shape to those found in the technology.

Consider the motion at the vertex of a slender hyperboloid of revolution, from which an electric current passes into the fluid. We use the coordinates of a prolate spheroid (ξ, η, φ) (see Section 1.4, 3), and introduce the new variables, as in Section 6.5, $\mu = \cos \eta$, $\lambda = \cosh \xi$. The approximate particular solution of the equation (1.6.13) for the electric current function is

$$\psi_1 = C_0 + D_0 \lambda. \tag{6.6.1}$$

If $D_0 = -C_0 = I/2\pi$ are set, the magnetic field has no singularities on the segment $\lambda = 1$ of the symmetry axis. The magnetic field (1.6.10) and electric current (1.6.11) are expressed as

$$B_\varphi = \frac{\mu_0 I}{2\pi a} \frac{1 - \lambda}{\sqrt{(1 - \mu^2)(\lambda^2 - 1)}};$$

$$j_\lambda = 0, \quad j_\mu = \frac{I}{2\pi a^2} \frac{1}{\sqrt{(1 - \mu^2)(\lambda^2 - \mu^2)}}. \tag{6.6.2}$$

Here the magnitude of representative current I is defined as the total current passing through the midplane $z = 0$ ($\mu = 0$) between the coordinate values $\lambda = 1$ and $\lambda = 2$ (through the circle of radius $r = a\sqrt{3}$).

The current distribution (6.6.2) shows that a hyperboloidal surface $\mu = \text{const}$ can be assumed as a surface of an infinitely conducting electrode (equipotential surface). Henceforth we shall consider a symmetric situation relative to the plane $\mu = 0$ with two electrodes $\mu = \pm\mu_1$. The respective electrically driven flow is supposed to be sufficiently slow to neglect the nonlinear terms in the equation (6.5.2) for the stream function $\psi^* = va\psi$. After replacing the expression of electromagnetic force's curl in (6.5.2) by the one appropriate to the function ψ_1 (6.6.1), the equation assumes the form [19]:

$$E^4 \psi = S \frac{\mu(1 - \lambda)}{(\lambda^2 - \mu^2)(1 - \mu^2)}; \quad S = \frac{\mu_0 I^2}{2\pi^2 \rho v^2}. \tag{6.6.3}$$

The boundary conditions are the conditions of impermeability and no slip on the surface of the electrodes,

$$\psi(\pm\mu_1) = 0, \quad \frac{\partial \psi}{\partial \mu}(\pm\mu_1) = 0, \tag{6.6.4}$$

and the absence of flow sources on the symmetry axis $\lambda = 1$

$$\psi(\lambda = 1) = 0 \tag{6.6.5}$$

The velocity field behaviour at $\lambda \to \infty$ is determined by the decay rate of the electromagnetic force.

Note, since the situation considered is symmetric relative to $\mu = 0$, then the

Bodies in a current-carrying fluid

conditions (6.6.4) on $\mu = -\mu_1$ can be equivalently replaced by the conditions of impermeability of the midplane

$$\mu = 0, \quad v_\mu \sim \frac{\partial \psi}{\partial \lambda}(\mu = 0) = 0,$$

and symmetry of the velocity profile

$$v_\lambda \sim \frac{\partial \psi}{\partial \mu}; \quad \frac{\partial^2 \psi}{\partial \mu^2} = 0,$$

relative to the midplane. The latter condition means an absence of tangential stress on the plane $\mu = 0$, and the problem can be also interpreted as a flow at the hyperboloidal electrode in presence of the free surface of fluid, where the current is withdrawn.

The solution of equation (6.6.3) can be expressed in the form [19]:

$$\psi = (1 - \lambda)M_1(\mu) + (1 - \lambda^2)M_2(\mu) + \lambda(1 - \lambda^2)M_3(\mu).$$

The final result is quite unwieldy, and we will discuss only the numerical results. The stream lines are represented in Fig. 6.11: in the case of a slender electrode $\mu_1 = 0.98$ (Fig. 6.11a) (e.g. in the electroslag welding process with a wire electrode), and in the case of a sloping electrode $\mu_1 = 0.5$ in the electroslag remelting process (Fig. 6.11b). As is evident, the electric current drives the flow directed along the electrode to the vertex, and at the plane $z = 0$ — in the form of radial jet flow. The intensity of the flow is higher for the more slender electrode (this is also evident from Figs. 6.12, 6.13 in which the radial velocity in the plane $z = 0$ and the axial velocity on the symmetry axis are depicted). This property is related to the curl of electromagnetic force which induces the flow. As follows from the right of (6.6.3), the curl increases in magnitude approaching the symmetry axis ($\mu = \pm 1$), and for the sloped, wider electrode the region of high curl is in the body of the electrode. One can also conclude that the flow close to the apex of the electrode is weak, and that it grows with distance from the apex. This kind of flow agrees with the experimental results for electrodes with nonisolated lateral surfaces (see Section 8.1).

As is evident from Fig. 6.12, the radial velocity grows at a linear rate with the distance from the symmetry axis. The second component grows similarly. Indeed, since

$$v_\lambda = -\frac{1}{a^2\sqrt{(\lambda^2 - \mu^2)(\lambda^2 - 1)}} \frac{\partial \psi}{\partial \mu}, \quad v_\mu = -\frac{1}{a^2\sqrt{(\lambda^2 - \mu^2)(1 - \mu^2)}} \frac{\partial \psi}{\partial \lambda},$$

and for large λ (or equivalently large r) $\psi_1 = O(\lambda^3)$, then in this case $v_\lambda = O(\lambda)$, $v_\mu = O(\lambda)$. The infinite velocities are related to the asymptotic behaviour for $\lambda \to \infty$, according to (6.6.2), of the driving force $f_e = j_\mu B_\varphi = O(\lambda^{-1})$. The decay rate of the force f_e is determined both by the decaying current density $j_\mu = O(\lambda^{-1})$ and the constant magnetic field at infinity due to the total current growing linearly with $\lambda \to \infty$. Of course, in a real situation the surface are of

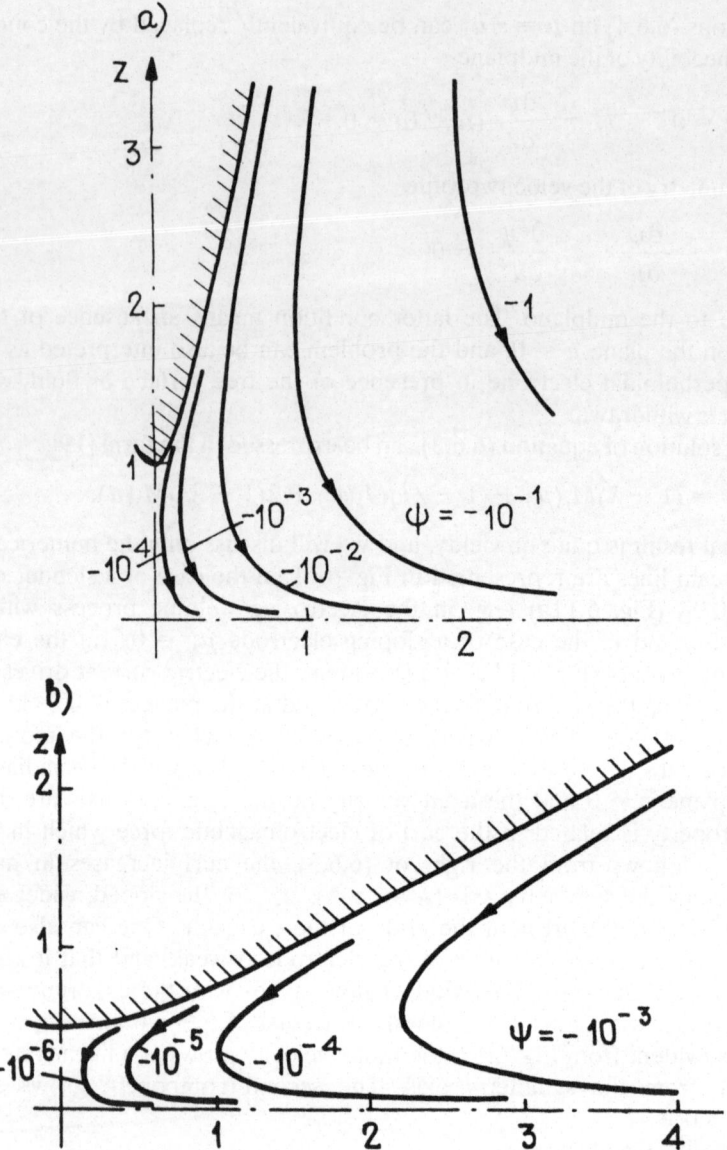

Fig. 6.11. Stream lines at the electrode of hyperboloidal form: (a) $\mu_1 = 0.98$, (b) $\mu_1 = 0.5$.

electrode is finite, hence the above solution can be applied to describe the motion at the apex of a real electrode.

Finally we need an estimate of the application bounds for the solution obtained. Assume the Stokes approximation is valid for $Re \leqslant 1$. The typical velocity is estimated on the plane $z = 0$ as

$$v_r = kSv \frac{r}{a^2},$$

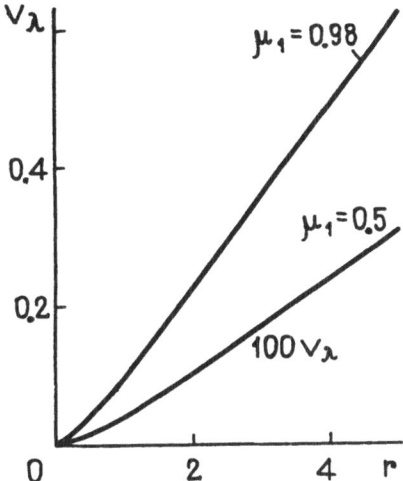

Fig. 6.12. Radial velocity distribution in the midplane $z = 0$ for various μ_1.

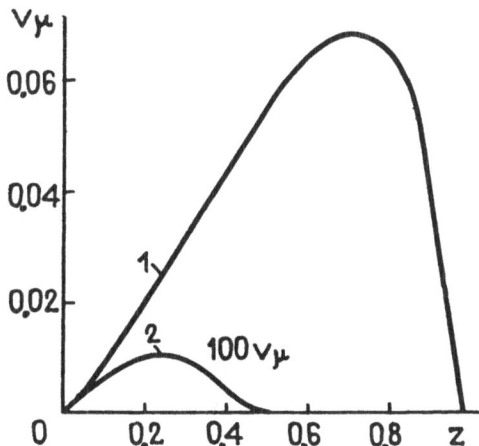

Fig. 6.13. Axial velocity distribution along the symmetry axis: (1) $\mu_1 = 0.98$, (2) $\mu_1 = 0.5$.

where, according to Fig. 6.12, $k = 0.1$ for $\mu_1 = 0.98$ and $k = 0.05$ for $\mu_1 = 0.5$. Then

$$Re = \frac{v_r r}{\nu} = kS\left(\frac{r}{a}\right)^2.$$

Thence, for $Re = 1$ the linear size of domain and the parameter S are related as $r/a = (kS)^{-1/2}$, i.e. the applicability domain of the Stokes solution decreases at a rate inversely proportional to the current growth. The domain surrounds the symmetry axis where, as noted, the flow is the most weak.

6.7. Flow at a cone with an electric current source in the apex

The issues concerning the electrically induced flows at bodies also include the situation in which an electric current supply is located within the body. In this situation the electric current enters the fluid from an electrode on the surface of the body, and the current completes the path at another electrode on the surface. We may expect an integral force acting on the body, which could be attributed, first, to the flow interaction with the body, and second, to interaction of the electric current flowing in the fluid with the closing current within the body itself (Fig. 6.14). The electric current flow, obviously, must be unsymmetric relative to the equatorial plane of the body, to ensure the imbalance of forces on the two halves of the body. Such a method of applying a force to the body in liquid media is a variant of the electromagnetic jet thruster.

Consider first the flow at a semi-infinite cone. The source of direct electric current is located within the cone with the apex angle $\pi - \theta_2$ (Fig. 6.14). We assume that the layer of a fluid or gas adjacent to the cone's surface (region 2) is electrically conducting, thereby the electric current I, supplied to the apex, passes along the layer, and then radially to the centre of electrically conducting disc A, and returns to the source. The region 1 is occupied by the same fluid, but is nonconducting. On assuming a uniform distribution of the current across the layer at a fixed radial distance, the current density is given by the expression

$$j_R = \frac{I}{2\pi R^2} \frac{1}{\mu_1 - \mu_2}, \qquad (6.7.1)$$

where $\mu = \cos \theta$. The expression (6.7.1) corresponds to (2.2.7), hence the force

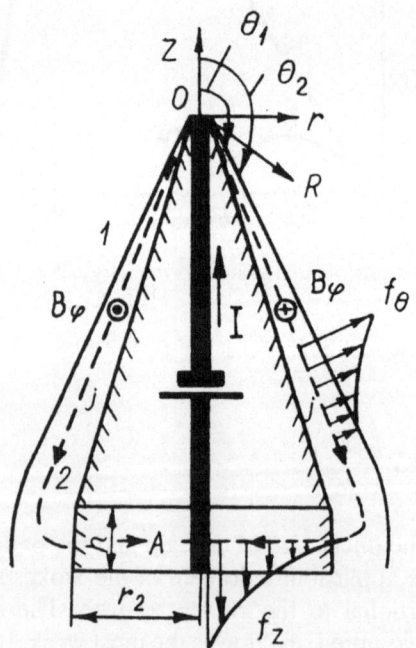

Fig. 6.14. Scheme of electric current passage at the cone.

due to the interaction with the self-magnetic field (2.2.9) drives an electrically induced flow, which can be described by the class of exact solutions with the velocity field of the form (2.5.3). The equation of motion in this case is transformed to the nonlinear ordinary differential equation of the first order (2.2.15). The boundary conditions are (2.5.6)–(2.5.8) on the symmetry axis $\mu = 1$, the no slip and impermeability conditions (2.5.5) on the rigid surface $\mu = \mu_2$, and the continuity of velocities and pressure (2.6.2) at the boundary $\mu = \mu_1$ of the conducting layer. The solution of equation (2.2.15) is discussed in Chapter 2, therefore we present here only the computed stream lines [2] (Fig. 6.15).

Let us evaluate the pressure and friction on the lateral surface of cone. Proceeding from the definition of stress tensor components:

$$\sigma_{\theta\theta} = -p + 2\nu\rho \left(\frac{1}{R} \frac{\partial v_\theta}{\partial \theta} + \frac{1}{R} v_R \right);$$

$$\sigma_{\theta R} = \nu\rho \left(\frac{1}{R} \frac{\partial v_R}{\partial \theta} + \frac{\partial v_\theta}{\partial R} - \frac{1}{R} v_\theta \right)$$

and applying the conditions on the cone's surface (2.5.5), and also the expression for the pressure $p = p_m - B_\varphi^2/2\mu_0$ (2.1.17), we obtain on the surface of the cone $\mu = \mu_2$:

$$\sigma_{\theta\theta} = -p(R, \mu_2) = \frac{\mu_0 I^2}{R^2} \frac{p_0(\mu_2)}{4\pi^2 S}$$

$$\sigma_{R\theta} = \frac{\nu^2 \rho}{R^2} \sqrt{1 - \mu_2^2}\, g''(\mu_2) = \frac{\mu_0 I^2}{R^2} \sqrt{1 - \mu_2^2}\, \frac{g''(\mu_2)}{4\pi^2 S}$$

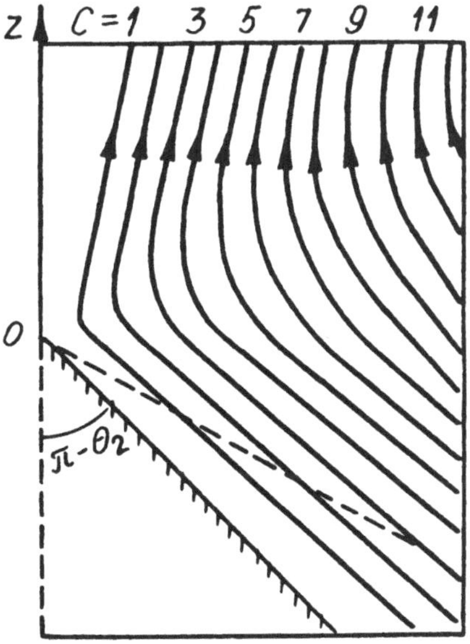

Fig. 6.15. Stream lines $\psi = C$ for $\mu_1 = -0.5$, $\mu_2 = -0.707$.

As an example we present the values of p_0 and g'' for the case $\mu_1 = -0.5$ ($\theta_1 = 120°$), $\mu_2 = 0.707$ ($\theta_2 = 135°$) represented in Fig. 6.15:

$$p_0 = 0.150S,$$
$$g'' = 0.102S. \tag{6.7.2}$$

The force by which the fluid acts on the lateral surface of cone in z-axis direction can be expressed in the following way:

$$F_1 = \int_{R_1}^{R_2} \int_0^{2\pi} (\sigma_{\theta\theta} \sin \theta_2 - \sigma_{R\theta} \cos \theta_2) R \sin \theta_2 \, dR \, d\varphi$$

$$= \mu_0 I^2 \frac{1-\mu_2^2}{2\pi S} [p_0(\mu_2) - \mu_2 g''(\mu_2)] \ln \frac{R_2}{R_1}, \tag{6.7.3}$$

where $R_2/R_1 = r_2/r_1$, r_2 is the radius of the cone's base, r_1 the radius of the electrode at the cone's apex (the value of r_1 may be specified, for instance, by the condition for a limiting current density). On substituting (6.7.2) in (6.7.3), we obtain in the above case $F_1 = \mu_0 I^2 0.0177 \ln(R_2/R_1)$.

To evaluate the total force on the conical body it is also necessary to include the interaction of currents in the fluid and in the cone. For this purpose we assume that the current at the base of the cone flows radially in a disc of height h and radius r_2 (Fig. 6.14). Assuming a current distribution that is uniform with height, i.e. $j_r = -I/(2\pi rh)$, we can find from the equation curl $\mathbf{B} = \mu_0 \mathbf{j}$ the field distribution in the disc:

$$B_\varphi = \frac{\mu_0 I^2}{2\pi rh} (h_2 + h - z), \qquad h_2 = r_2 \cotan(\pi - \theta_2),$$

where the condition $B_\varphi(z = h_2) = \mu_0 I/2\pi r$ is employed, which links the field in the disc with the field of the straight wire connecting the disc and the apex of the cone. The current circuit is completed by the conducting fluid layer.

The force in the disc is equal to

$$F_2 = 2\pi \int_{r_1}^{r_2} \int_{h_2}^{h_2+h} j_r B_\varphi r \, dr \, dz = -\frac{\mu_0 I^2}{4\pi} \ln \frac{r_2}{r_1}; \tag{6.7.4}$$

the force of equal magnitude and of opposite sign is applied to the current-carrying fluid layer:

$$F_3 = 2\pi \int_{R_1}^{R_2} \int_{\theta_1}^{\theta_2} f_\theta R^2 \sin \theta \, dR \, d\theta = \frac{\mu_0 I^2}{4\pi} \ln \frac{R_2}{R_1}.$$

Thereby the total force acting on the body

$$F = F_1 + F_2 = \mu_0 I^2 \left[\frac{1-\mu_2^2}{2\pi S} (p_0 - \mu_2 g'') - \frac{1}{4\pi} \right] \ln \frac{R_2}{R_1}, \tag{6.7.5}$$

will set it in motion. In the case of the above example $F = \mu_0 I^2 (0.0177 -$

0.0796) $\ln(R_2/R_1) < 0$, i.e. the cone will move with the base forward if it is resting before the current is switched on, or it will be decelerated by the force (6.7.5) if, prior to the current switch-on, it is moving with a constant velocity $v_{z_0} > 0$. It is instructive to compare the situation to a boat with an oarsman. In this case the role of oars belong to the magnetic field of the current within the cone, which, by interacting with the current in the fluid, generates the electromagnetic force F_3 pushing the fluid in the positive z-axis direction. The electromagnetic reaction force F_2 in the cone is opposite in direction; the part of it is lost by the drag of the body to the fluid flow.

If an external flow with the velocity $v_z = -v_0 r^m < 0$ is added, the electrically induced flow retards the fluid in the boundary layer of the oncoming flow. For a sufficiently high current magnitude this will lead to the boundary layer separation at the apex and reconnection at a distance R_0. Using the data [14], we can estimate (by the local friction behaviour) the distance from the apex of the cone to the reconnection point:

$$R_0 \sim v_0^{-\frac{1}{m+1}} I^{\frac{4}{3(m+1)}}.$$

Consequently, the electric current in the near-wall layer of fluid could be a tool to operate the flow and to generate the braking force.

6.8. Motion of a sphere with a current source

The model of a semi-infinite cone (Section 6.7) does not take into account a fluid flow at the base of the cone, which in the case of a finite body may alter substantially, for instance, the integral force acting on the body. Therefore, consider another model where the sphere with a source of electric current can move within an infinite volume of electrically conducting fluid.

The origin of the spherical coordinate system is in the centre of the sphere. Let the electric current from a source inside the sphere pass from electrodes on its surface to the surrounding fluid, moreover the electrodes are arranged in such a manner as to ensure an unsymmetric current flow relative to the equatorial plane of the sphere $z = 0$. Such the distribution of electric current can be constructed by the superposition of the dipole and quadrupole modes from the general solution of the magnetic field equation (1.7.2):

$$B_\varphi^* = b \frac{\sin \theta}{R^{*2}} + c \frac{\sin \theta \cos \theta}{R^{*3}},$$

with the corresponding current distribution:

$$j_R^* = \frac{2b}{R^{*3}} \cos \theta + \frac{c}{R^{*4}} (2 \cos^2 \theta - \sin^2 \theta);$$

$$j_\theta^* = \frac{b}{R^{*3}} \sin \theta + \frac{2c}{R^{*4}} \sin \theta \cos \theta; \qquad j_\varphi^* = 0.$$

The electric current lines in a meridional section are represented in Fig. 6.16,

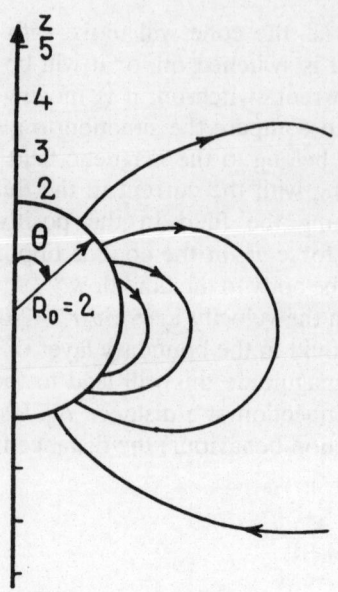

Fig. 6.16. Lines of electric current from a source within the sphere.

where the quantity $R_0 = -c/b$ is defined as a unit of length. The current of magnitude $I_0 = 0.177 b^2/c$ penetrates the upper part of the sphere $R^* = 2R_0$, intersected by the line $\theta = 102.8°$ where $j_R = 0$, and the current of the same magnitude re-enters the lower part. The current generates in the fluid ($R^* > 2R_0$) a rotational electromagnetic force $\mathbf{f} = \mathbf{j} \times \mathbf{B}$ which drives a fluid flow. Since curl $\mathbf{f}$ in this case reduces merely to a φ-component, then the curl of the Navier-Stokes equation leads to the single equation (1.10.22) to determine the stream function ψ (3.5.3). On introducing the nondimensional variables $R = R^*/R_0$, $\psi = \psi^*/\nu R_0$, $B_\varphi = B_\varphi^* R_0/I$, and $\mu = \cos\theta$, the equation of motion in linear approximation is:

$$E^4 \psi = S \left[\frac{6}{R^5} \mu(1-\mu^2) + \frac{2}{R^6}(8\mu^4 - 9\mu^2 + 1) - \right.$$

$$\left. - \frac{2}{R^7}(5\mu^5 - 6\mu^3 + \mu), \right. \tag{6.8.1}$$

where

$$S = \frac{\mu_0 b^4}{\nu^2 \rho c^2} = \frac{31.92 \mu_0 I^2}{\nu^2 \rho}.$$

For the boundary conditions of no slip and impermeability on the sphere $R = 2$

$$\psi = \frac{\partial \psi}{\partial R} = 0$$

Bodies in a current-carrying fluid

and the decay of velocity at infinity

$$\lim_{R \to \infty} = Rg(\mu) \tag{6.8.2}$$

the solution of (6.8.1) is

$$\frac{\psi}{S} = \left[A_2 R^2 + B_2 R^{-1} + D_2 R - \frac{1}{60} R^{-2} \right] J_2(\mu) +$$

$$+ \left[B_3 R^{-2} + D_3 - \frac{1}{4} R^{-1} + \frac{1}{63} R^{-3} \right] J_3(\mu) +$$

$$+ \left[B_4 R^{-3} + D_4 R^{-1} + \frac{4}{15} R^{-2} \right] J_4(\mu) +$$

$$+ \left[B_5 R^{-4} + D_5 R^{-2} - \frac{1}{14} R^{-3} \right] J_5(\mu), \tag{6.8.3}$$

where

$$B_2 = 4A_2 + \frac{1}{80}; \quad D_2 = -3A_2 - \frac{1}{8 \cdot 120};$$

$$B_3 = \frac{5}{21}; \quad D_3 = \frac{4}{63}; \quad B_4 = -\frac{4}{15}; \quad D_4 = -\frac{1}{15}; \tag{6.8.4}$$

$$B_5 = \frac{1}{14}; \quad D_5 = \frac{1}{56}.$$

The coefficient A_2 determines an external flow, uniform at infinity, directed along the z-axis. From the condition (6.8.2) it follows that $A_2 = 0$.

In the case of the sphere with the inner current source the induced velocity field falls off at $R \to \infty$ as $|\mathbf{v}| = O(R^{-1})$, and the stream function behaves as $\psi = O(R)$. The velocity decays due to the sufficiently rapid decrease of the current density $|\mathbf{j}| = O(R^{-3})$ for $R \to \infty$ and the corresponding behaviour of the force curl $|\text{curl } \mathbf{f}| = |\text{curl } \mathbf{j} \times \mathbf{B}| = O(R^{-5})$. Recall, in the problems of flows around the bodies with the uniform at infinity current density the curl of force decays as $|\text{curl } \mathbf{f}| = O(R^{-2})$, and the velocity at infinity stays finite (Section 6.3). Respectively, the Stokes approximation in that case is invalid for large distances from the body, because the inertial terms are not asymptotically small for $Re \to 0$ in the far region.

The situation for the flow at the sphere with the current source is quite different. The local Reynolds number $Re^* = |v^*| R_0 R/v = vR = \text{const} \to 0$ is small at infinity, if it is small in the vicinity of the sphere, i.e. it is uniformly small. Then the solution of the problem for small numbers Re^* can be sought in the form of regular perturbations by powers of S: $\psi = \Sigma S_n \psi_n$. This expansion will be equally valid both close to the body and far from it. The Whitehead paradox does not arise in this case. However, if the external flow with $A_2 \neq 0$ is added, one has to recover the method of matched asymptotic expansions (Section 6.4) in an attempt to account for the inertial effects.

With the aim of determining the total force on the sphere we shall evaluate, as in Section 6.7, the force F_1 by which the fluid interacts with the sphere, and the force F_2 of the electric current interaction in the sphere and in the surrounding fluid. F_1 can be evaluated by the general expression for the force acting on a finite body in an arbitrary axisymmetric flow [11]. For the solution in the first approximation (6.8.3) we obtain

$$F_1 = -\pi v^2 \rho \int_{-1}^{1} R^4 \frac{\partial}{\partial R} \left(\frac{E^2 \psi}{R^2} \right) d\mu = 4\pi v^2 \rho S \left(-\frac{1}{120} - D_2 \right).$$

An electric current circuit within the sphere is affected by the force F_2, which is equal in magnitude, but of opposite sign, to the integral electromagnetic force in the fluid:

$$F_2 = -\mu_0 I_0^2 \int_2^{\infty} \int_0^{\pi} \int_0^{2\pi} (\mathbf{j} \times \mathbf{B}) \cdot \mathbf{i}_z R^2 \sin \theta \, dR \, d\theta \, d\varphi = 4\pi v^2 \rho S \frac{1}{120}.$$

Hence the total force on the sphere is

$$F_z = F_1 + F_2 = -4\pi v^2 \rho S D_2. \tag{6.8.5}$$

Note that the expression (6.8.5) stays valid in the first approximation by S also with the addition of the external flow at infinity with $A_2 \neq 0$ if D_2 is evaluated according to (6.8.4).

In the situation where the fluid at infinity is at rest ($A_2 = 0$) the force on the sphere $F_z = 4\pi v^2 \rho S/8 \cdot 120$ is in the positive z-axis direction, what coincides with the direction of electromagnetic force. The sphere will stay stationary if a holding force $-F_z$ is applied to it. The stream lines in this situation (Fig. 6.17a) show the weaker vortex at the body, which is closed by the line $C = 0$, and the second vortex closing at infinity, which is, in effect, a transit flow. The flow rate over the annular cross section $2 < R < R_1$, $\theta = \pi/2$ grows in proportion to $\psi(R_1)$: $Q = 2\pi v R_0 \psi(R_1)$, and at a large distance $R = R_1$ it is proportional to $D_2 S R_1$, since D_2 according to (6.8.4) is of a nonzero value.

The constant D_2 will be zero if the uniform flow at infinity in the negative z-axis direction is added, formally specifying $A_2 = -1/(24 \cdot 120)$. This case is of interest because the total force on the body is zero, $F_z = 0$, i.e. the forces acting on the body are in balance. The stream lines assume the form shown in Fig. 6.17b. If an observer is moving uniformly with the flow at infinity, the sphere for him is freely drifting within the infinite fluid volume in the positive z-axis direction, driven by the electric current issuing from the source in the sphere.

The uniform flow in the opposite direction with $A_2 = 1/(24 \cdot 120)$ also leads to a closed vortex upstream of the sphere (Fig. 6.17c). In this case the drag of the sphere with the current source increases in the external flow.

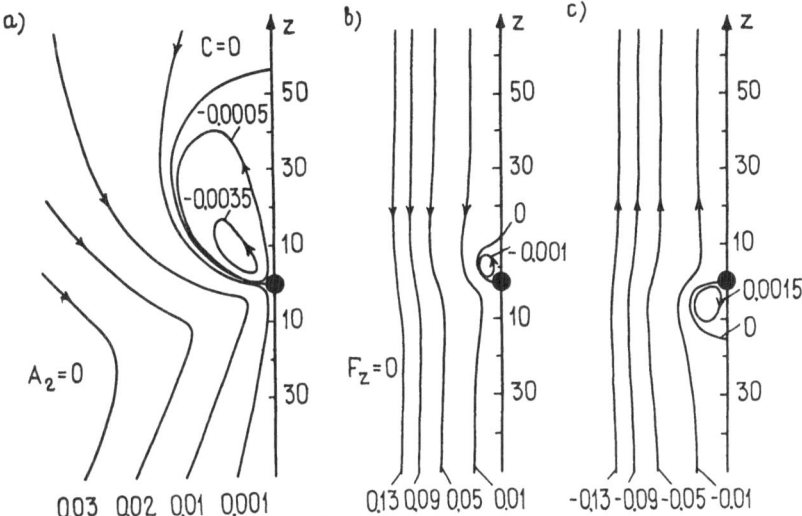

Fig. 6.17. Stream lines $\psi = -C$: (a) the electrically induced flow, (b) EVF and the uniform flow with $A_2 = -1/24 \times 120$, (c) EVF and the uniform flow with $A_2 = 1/24 \times 120$.

Thus, the model with the sphere confirms the basic result obtained for the semi-infinite cone, and it states that the integral force on the body with an unsymmetric current flow from the inner source coincides in direction with the electromagnetic force in the body. The force is able to drive the body in the surrounding liquid medium. The current passage, in addition, affects the flow configuration, thereby making it available to regulate the heat exchange between the body and the surrounding fluid.

7

Heat and mass transfer in electrically induced vortical flows

One of the most important applications of the electrically induced flow theory is the evaluation of the effect of these flows on heat- and mass-transfer processes in different technological operations, including MHD-technology. We may mention, for example, MHD refinement of liquid metals, production of composite materials, electrolytic extraction of metals, where the mass transfer is of prime importance. The integral action of heat and mass transfer finally determines the quality of welds and castings in electroslag technology, the energetic aspects of electrical arcs, productivity of smelting process in electrical furnaces, etc. We would like to emphasize here that the presence of electric current is an integral attribute of MHD-technology.

To our regret the range of works in which the heat and mass transfer in electrically induced flows is studied, is relatively small, and the spectrum of the situations considered is quite narrow. In fact the mass exchange has been studied for a spherical rigid particle and liquid drop, and heat, mass transfer in a cylindrical container with the applications to electroslag welding and remelting.

7.1. Equations of heat and mass transfer, and the nondimensional numbers

When it is necessary to include in the equation of motion the heat convection resulting from a varied density of fluid elements in a nonuniform temperature field, the Boussinesq approximation is usually applied [5]. According to this approximation the equation of state $\rho = \rho(p, T)$ for an incompressible medium is expanded in Taylor series to an accuracy of linear terms in T:

$$\rho = \rho_0(1 - \beta T),$$

where $\beta = -\rho_0^{-1}(\partial \rho/\partial T)_p$ is the thermal expansion coefficient, and the density variation is accounted for in (1.1.1) only in the buoyancy term ρg. The hydrostatic pressure $\rho_0 g$, corresponding to an equilibrium state with the mean density ρ_0, can be included in the term ∇p. Then the buoyancy term is expressed in the form $\rho_0 \beta T g \mathbf{i}_z$, where the unit vector $\mathbf{i}_z$ is directed vertically upwards.

Finally the equation of motion assumes the form

$$(\mathbf{v} \cdot \nabla)\mathbf{v} = -\frac{1}{\rho_0}\nabla p + \nu\nabla^2\mathbf{v} + \beta T g \mathbf{i}_z + \frac{1}{\rho_0}\mathbf{f}_e. \tag{7.1.1}$$

A steady-state heat transfer in a moving liquid is governed by the equation [7]

$$\rho c \mathbf{v} \cdot \nabla T = \kappa \nabla^2 T + |\mathbf{j}|^2/\sigma, \tag{7.1.2}$$

if the dissipative heat due to the internal friction is neglected and the ohmic dissipation, significant in this case, is included. In (7.1.2) c is specific heat, κ is thermal conductivity, and T is deviation of temperature from a mean constant value.

The equation of mass transport is of a simpler form, if chemical reactions are absent in the fluid, resulting in a mass exchange:

$$\mathbf{v} \cdot \nabla C = D\nabla^2 C \tag{7.1.3}$$

where C is the concentration of the transported material and D is the coefficient of diffusion. When the ohmic heat may be neglected, then the equations (7.1.2) and (7.1.3) are similar, and the results obtained for a mass transport process can be carried over to the heat transport process (of course, for identical boundary conditions).

Let us introduce the nondimensional primed quantities: $v = v_0 v'$, $j = j_0 j'$, $T - T_1 = \Delta T T'$, $C - C_1 = \Delta C C'$, $\ell = L\ell'$ where $\Delta T = T_{S_0} - T_1$, $\Delta C = C_{S_0} - C_1$, T_1 and C_1 denote respectively the mean temperature and concentration in the fluid, and T_S, C_S denote the same quantities, for instance, on a surface. Then for the nondimensional variables (primes are dropped henceforth) the equations (7.1.1)–(7.1.3) take the form:

$$(\mathbf{v} \cdot \nabla)\mathbf{v} = -P\nabla p + \frac{1}{Re}\nabla^2\mathbf{v} + A\ell \mathbf{j} \times \mathbf{B} + GrT\mathbf{i}_z \tag{7.1.4}$$

$$\mathbf{v} \cdot \nabla T = \frac{1}{Pe_T}\nabla^2 T + Qj^2; \tag{7.1.5}$$

$$\mathbf{v} \cdot \nabla C = \frac{1}{Pe_D}\nabla^2 C, \tag{7.1.6}$$

where $Re = v_0 L/\nu$ is the Reynolds number (typical ratio of inertial to viscous forces), $A\ell = \mu_0 j_0^2 L^2/\rho v_0^2$ is the Alfvèn number (ratio of electromagnetic to inertial forces), $Gr = \beta T_0 g L/v_0^2$ — Grashoff number (ratio of buoyancy to inertial forces), $Pe_T = v_0 L/\chi$ — the thermal Peclet number characterizing the relative role of convective heat transport and its transfer due to the conductivity (the molecular mechanism), $\chi = \kappa/\rho c$ — thermal diffusivity, $Q = j_0^2 L/\sigma \rho c v_0 \Delta T$ — the parameter specifying a relative power of ohmic heating, $Pe_D = v_0 L/D$ — the diffusive Peclet number characterizing the relative role of convective mass transport and molecular diffusion. Apart from these non-dimensional numbers, important parameters are also the thermal Prandtl number $Pr_T = Pe_T/Re = \nu/\chi$, characterizing the ratio of the viscous momentum transport to the heat transfer by the molecular conductivity, and the diffusive

Prandtl number $Pr_D = Pe_D/Re = \nu/D$ characterizing the ratio of the viscous momentum transport to the mass transfer by the molecular diffusion.

The nondimensional numbers listed are of the form specified above, if the respective scales of velocity v_0, temperature difference ΔT, and concentration difference Δc are known *a priori*. In most situations, typical for the electrically induced flows, the scale of velocity v_0 is not specified, since the motion of liquid medium itself is the result of current passing through the liquid. In this case the scale of v_0 must be composed of the material parameters and the quantities given in the problem, in particular the current magnitude. For example, on dimensional grounds the scale of v_0 can be represented as

$$v_0 = \nu/L \quad \text{or} \quad v_0 = \frac{I_0}{L}\sqrt{\frac{\mu_0}{\rho}} = \frac{\nu}{L}\sqrt{S}.$$

The same can be said about the temperature, which, similarly to the velocity, is determined by the electric current in the melt. A representative temperature may be assumed, for instance, in the form $\Delta T = T_0 = I_0^2/\sigma \kappa L^2$.

All this determines a variety of specific meanings of the nondimensional numbers. Particularly, for $v_0 = \nu/L$ and $\Delta T = T_0$ these are of the form: $Re = 1$, $A\ell = S = \mu_0^2 I_0^2/\rho\nu$, $Gr = \beta g L I_0^2/\sigma\kappa L^2$, $Pe_T = Pr_T$, $Pe_D = Pr_D$, $Q = Pr_T^{-1}$. For $v_0 = \sqrt{S}\nu/L$ and $\Delta T = T_0$: $Re = \sqrt{S}$, $A\ell = 1$, $Gr = \beta \rho g L/\mu_0 \sigma \kappa$, $Pe_T = Pr_T\sqrt{S}$, $Pe_D = Pr_D\sqrt{S}$, $Q = Pr_T^{-1}S^{-1/2}$. In the following we shall specify which representative scales will be assigned in any particular problem.

The final result of a heat- and mass-transfer problem is, as a rule, evaluation of a heat (mass) quantity transported from the surface of a body or to the surface. This heat quantity, on the one hand, can be expressed by the temperature gradient on the wall and the thermal conductivity κ (since, in view of the no slip constraint, the velocity on the surface is zero, and the heat is transported at the wall merely by the conductivity), on the other by the heat-transfer coefficient α for the known temperature difference in the flow and on the wall (S_0 is the surface area, t = time):

$$q = -\kappa \left.\frac{\partial T}{\partial n}\right|_{S_0} S_0 t = \alpha(T_{S_0} - T_1)S_0 t. \tag{7.1.7}$$

In the nondimensional form the relation is

$$Nu_T = \alpha L/\kappa = -\left.\frac{\partial T'}{\partial n'}\right|_{S_0}. \tag{7.1.8}$$

Obviously, the Nusselt number Nu_T is a variable along the surface, and hence it is a local number. Yet the average Nusselt number over the surface S_0

$$Nu_{av} = \frac{1}{S_0}\int_{S_0} Nu_T \, dS$$

may be considered as the ratio of the actual heat flux, determined by the intensity of convection, to the heat flux which would take place in the case of pure conduction in a layer of thickness L.

Similarly to (7.1.7) the equation of mass transport can be defined (a_D = mass transfer coefficient):

$$w = -D \left.\frac{\partial C}{\partial n}\right|_{S_0} = a_D(C_{S_0} - C_1),$$

and the local

$$Nu_D = a_D L/D = -\partial C'/\partial n'|_{S_0}, \qquad (7.1.9)$$

and the average

$$Nu_{D_{av}} = \frac{1}{S_0} \int_{S_0} Nu_D \, dS$$

diffusive Nusselt numbers introduced. The task of a heat (diffusion) computation is, therefore, to determine the dependence of the respective Nusselt number on the governing parameters (Pr, Re, and so forth).

Let us express the equations (7.1.5) and (7.1.6) in orthogonal coordinates for the axisymmetric case. In light of (1.10.1), (1.10.4), (1.10.6), and (1.10.8), we obtain:

$$\frac{\partial \psi}{\partial q_2}\frac{\partial T}{\partial q_1} - \frac{\partial \psi}{\partial q_1}\frac{\partial T}{\partial q_2}$$

$$= \frac{1}{Pe_T}\left[\frac{\partial}{\partial q_1}\frac{H_2 H_3}{H_1}\frac{\partial T}{\partial q_1} + \frac{\partial}{\partial q_2}\frac{H_1 H_3}{H_2}\frac{\partial T}{\partial q_2}\right] +$$

$$+ Q\left[\frac{H_1}{H_2 H_3}\left(\frac{\partial \psi_1}{\partial q_2}\right)^2 + \frac{H_2}{H_1 H_3}\left(\frac{\partial \psi_1}{\partial q_1}\right)^2\right]; \qquad (7.1.10)$$

$$\frac{\partial \psi}{\partial q_2}\frac{\partial C}{\partial q_1} - \frac{\partial \psi}{\partial q_1}\frac{\partial C}{\partial q_2}$$

$$= \frac{1}{Pe_D}\left[\frac{\partial}{\partial q_1}\frac{H_2 H_3}{H_1}\frac{\partial C}{\partial q_1} + \frac{\partial}{\partial q_2}\frac{H_1 H_3}{H_2}\frac{\partial T}{\partial q_2}\right]. \qquad (7.1.11)$$

In particular, for the cylindrical and spherical coordinate systems we have:

$$\frac{\partial \psi}{\partial r}\frac{\partial T}{\partial z} - \frac{\partial \psi}{\partial z}\frac{\partial T}{\partial r} = \frac{1}{Pe_T}\left[\frac{\partial}{\partial z}r\frac{\partial T}{\partial z} + \frac{\partial}{\partial r}r\frac{\partial T}{\partial r}\right] +$$

$$+ Q\frac{1}{r}\left[\left(\frac{\partial \psi_1}{\partial r}\right)^2 + \left(\frac{\partial \psi_1}{\partial z}\right)^2\right]; \qquad (7.1.12)$$

$$\frac{\partial \psi}{\partial r}\frac{\partial C}{\partial z} - \frac{\partial \psi}{\partial z}\frac{\partial C}{\partial r} = \frac{1}{Pe_D}\left[\frac{\partial}{\partial z} r \frac{\partial C}{\partial z} + \frac{\partial}{\partial r} r \frac{\partial C}{\partial r}\right]; \quad (7.1.13)$$

$$\frac{\partial \psi}{\partial \theta}\frac{\partial T}{\partial R} - \frac{\partial \psi}{\partial R}\frac{\partial T}{\partial \theta}$$

$$= \frac{1}{Pe_T}\left[\frac{\partial}{\partial R} R^2 \sin\theta \frac{\partial T}{\partial R} + \frac{\partial}{\partial \theta}\sin\theta\frac{\partial T}{\partial \theta}\right] +$$

$$+ Q\left[\frac{1}{R^2 \sin\theta}\left(\frac{\partial \psi_1}{\partial \theta}\right)^2 + \frac{1}{\sin\theta}\left(\frac{\partial \psi_1}{\partial R}\right)^2\right]; \quad (7.1.14)$$

$$\frac{\partial \psi}{\partial \theta}\frac{\partial C}{\partial R} - \frac{\partial \psi}{\partial R}\frac{\partial C}{\partial \theta}$$

$$= \frac{1}{Pe_D}\left[\frac{\partial}{\partial R} R^2 \sin\theta \frac{\partial C}{\partial R} + \frac{\partial}{\partial \theta}\sin\theta\frac{\partial C}{\partial \theta}\right]. \quad (7.1.15)$$

Table 7.1. Physical properties for some media.

Physical quantity	Liquid metal	Electrolyte
Kinematic viscosity ν, m^2/s	10^{-6}	10^{-6}
Density ρ, kg/m^3	10^3–10^4	10^3
Electrical conductivity σ, 1/Ohm m	10^6–10^7	10^1–10^2
Thermal conductivity coefficient κ, J/m K s	10^1–10^2	10^0
Specific heat c, J/kg K	10^2–10^3	3×10^3
Thermal diffusion coefficient χ, m^2/s	10^{-4}–10^{-5}	10^{-5}–10^{-6}
Thermal expansion coefficient β, K^{-1}	10^{-4}	10^{-3}
Diffusion coefficient D, m^2/s	10^{-9}	10^{-9}
Rr_T	10^{-2}–10^{-1}	10^0–10^1
Pr_D	10^3	10^3

We list for a reference in Table 7.1 the orders of magnitudes of the material parameters for liquid metals and electrolytes. By knowing the magnitudes of these quantities we can estimate the nondimensional numbers with the additional data for a typical current density, length, velocity, and external magnetic field induction.

The following three sections discuss different cases of mass transfer at spherical particles, particularly in a pure electrically induced flow, EVF with the external uniform flow $v_z = v_0$, and EVF with the external uniform longitudinal magnetic field $B_z = B_0$. In the noninduction approximation the electric current distribution, corresponding to the solution of the equation $E^2\psi_1 = 0$ (1.7.2), is given by the expression (6.2.9) (a is the radius of the sphere)

$$\psi_1 = j_0 \left(R^2 - \frac{\sigma a^3}{R} \right) \frac{\sin^2 \theta}{2}, \quad (7.1.16)$$

depending on the electrical conductivities of the fluid σ_1 and the sphere σ_2 ($\sigma = 2(\sigma_1 - \sigma_2)/(2\sigma_1 + \sigma_2)$), and the longitudinal magnetic field is given by the function (see Table 1.1 No. 5)

$$\psi_2 = \tfrac{1}{2} B_0 R^2 \sin^2 \theta. \quad (7.1.17)$$

On assuming an isothermal motion and applying the curl operator to (7.1.1), we obtain the equations of motion in the form (2.1.1), (2.1.2). Reducing these to a nondimensional form by relating ψ to $v_0 a^2$, v_φ to v_0, R to a, and applying the expression (7.1.16), (7.1.17), we obtain:

$$\sin\theta \left(\frac{\partial \psi}{\partial R} \frac{\partial}{\partial \theta} \frac{E^2\psi}{R^2 \sin^2 \theta} - \frac{\partial \psi}{\partial \theta} \frac{\partial}{\partial R} \frac{E^2\psi}{R^2 \sin\theta} \right) +$$

$$+ \frac{\sin\theta}{R} \frac{\partial v_\varphi^2}{\partial \theta} - \cos\theta \frac{\partial v_\varphi^2}{\partial R} + \frac{1}{Re} E^4\psi$$

$$= -\frac{3}{2} \frac{\sigma S}{Re^2} \left(1 - \frac{\sigma}{R^3}\right) \frac{1}{R^2} \sin^2\theta \cos\theta; \quad (7.1.18)$$

$$\frac{\partial \psi}{\partial R} \frac{\partial}{\partial \theta} R \sin\theta v_\varphi - \frac{\partial \psi}{\partial \theta} \frac{\partial}{\partial R} R \sin\theta v_\varphi + \frac{1}{Re} R^2 \sin\theta \times$$

$$\times E^2(R \sin\theta v_\varphi) = -\frac{3}{2} \frac{M\sigma}{Re^2} \sin^3\theta \cos\theta, \quad (7.1.19)$$

where

$$S = \frac{\mu_0 j_0^2 a^4}{\rho v^2}; \quad Re = \frac{v_0 a}{v}; \quad M = \frac{j_0 B_0 a^3}{\rho v^2}$$

the representative velocity v_0 depends on specific conditions of a problem.

7.2. Mass transfer from a stationary spherical particle in a current-carrying fluid

Consider the electric current passage at a rigid sphere in electrically conducting stationary fluid, when the current density j_0 is uniform far from the sphere and oriented parallel to the z-axis. If the electrical conductivities of the sphere and fluid are different, then, as was shown in Chapter 6, the flow is driven at the sphere by the rotational electromagnetic force due to the electric current line deformation. Obviously, the electrically induced flow will cause a more intense mass transfer between the fluid and particle as compared to a the quiescent fluid. The first attempt to solve the problem was made in [1]. According to this solution it is assumed that for high Pe_D numbers the variation of concentration takes place in a narrow diffusion layer at the surface of the sphere, then the approximation of diffusive boundary layer could be valid for (7.1.15):

$$\frac{\partial \psi}{\partial \theta} \frac{\partial C}{\partial R} - \frac{\partial \psi}{\partial R} \frac{\partial C}{\partial \theta} = \frac{1}{Pe_D} \frac{\partial}{\partial R} R^2 \sin\theta \frac{\partial C}{\partial R}. \qquad (7.2.1)$$

The equation (7.2.1) is solved for the boundary conditions:

$$C|_{R=1} = 1; \qquad C|_{R\to\infty} = 0; \qquad \frac{\partial C}{\partial \theta}\bigg|_{\theta=0,\,\theta=\pi} = 0, \qquad (7.2.2)$$

i.e. the maximum concentration of dissolved material is on the surface of the sphere, and far from the sphere the concentration is equal to its initial value in the fluid.

The expression for the stream function ψ/va in the case of a nonconducting sphere follows from (6.3.3):

$$\psi = -\frac{1}{16} S \left(R^2 - \frac{5}{2} + \frac{1}{R} + \frac{1}{2R^2} \right) \sin^2\theta \cos\theta, \qquad (7.2.3)$$

where S is defined in Section 7.1 and the radial coordinate is related to the radius of the sphere a.

The average Nu number expressed according to the analytical solution is equal to

$$Nu = 0.6\, S^{1/3} Pr_D^{1/3}. \qquad (7.2.4)$$

However, as has previously been noted in Section 6.3, the Stokes solution (7.2.3) does not give a decaying velocity field for $R \to \infty$. Moreover, the diffusive boundary layer, even for high Pr_D numbers, cannot be thin along the whole surface of the sphere for the electrically induced convection. Therefore a numerical solution of the equations (7.1.18) and (7.1.15) with $v_\varphi = 0$ is constructed in [10], which improves the solution (7.2.3) for the more intense electrically induced flows. In this case the velocity scale is $v_0 = j_0 a \sqrt{\mu_0/\rho}$.

The electrically induced flow at a stationary sphere is considered in Section 6.3. The mass-transfer problem is solved for the boundary conditions (7.2.2) and the additional condition of symmetry relative to the equatorial plane $z = 0$:

$$\frac{\partial C}{\partial \theta}\bigg|_{\theta=\pi/2} = 0. \qquad (7.2.5)$$

The numerical solutions are obtained for the cases of nonconducting sphere ($\sigma = 1$) and ideally conducting sphere ($\sigma = -2$); $Pr_D = 10^3$, and $S = 10^{-1}$–10^5. When comparing the flow configurations (Fig. 7.1a) with the distributions of concentration at the surface of the sphere (Fig. 7.1b), it is evident that the diffusive layer is thin in the zones of oncoming flow (for $\sigma > 0$ — in the regions $\theta = 0, \pi$; for $\sigma < 0$ — in the region $\theta = \pi/2$), and it grows in thickness and finally erupts in the zones of offcoming flow (for $\sigma > 0$ — at $\theta = \pi/2$ for $\sigma < 0$ — in the vicinity of $\theta = 0, \pi$). The distribution of local nondimensional number $Nu_\theta = \partial C/\partial R|_{R=1}$ along the surface of the sphere (Fig. 7.2) shows that the most significant mass transport takes place in the zones of oncoming flow, and the intensity of mass transfer increases with the current density. The average mass-transfer coefficient

$$Nu = -\int_0^{\pi/2} \frac{\partial C}{\partial R}\bigg|_{R=1} \sin\theta \, d\theta$$

for the nonconducting and ideally conducting spheres vs. S are shown in Fig. 7.3. As it is evident, for small S the values of Nu for both cases are indiscernible, however the quantities grow differently for $|\sigma S| > 10^2$.

If we assume that the dependence $Nu(Pr_D)$ is similar to (7.2.4), the computational data can be generalized by the formula:

$$Nu = A\ell (|\sigma|S)^k Pr_D^{1/3}, \tag{7.2.6}$$

where for $|\sigma S| < 10^2$, i.e. for weak EVF, $A\ell = 0.45$, $k = 0.305$, both for the nonconducting and ideally conducting spheres. This result differs from (7.2.4), but it could be regarded as more plausible, because it is derived under more

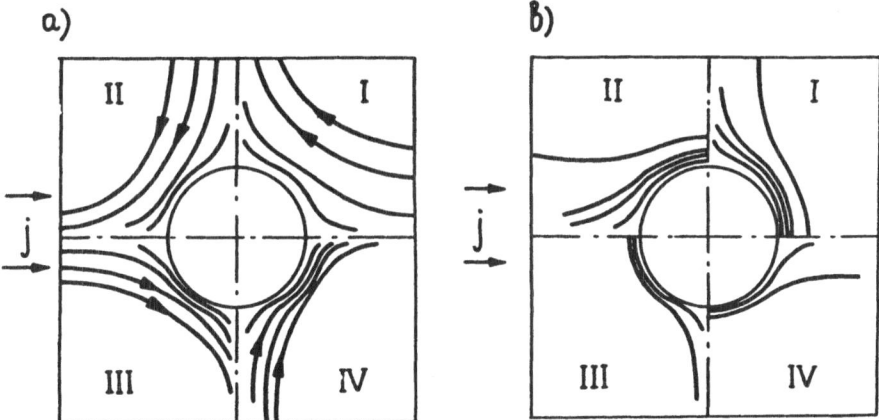

Fig. 7.1. (a) Stream lines, (b) lines of equal concentration in the case of electric current flow at a sphere. The quadrants I, III — nonconducting sphere; II, IV — ideally conducting sphere. For the quadrants I and II $S = \pi$, for III and IV — 400π. The values at the stream lines starting from the sphere are: $10^{-3}, 10^{-2}, 5 \times 10^{-2}, 10^{-1}$; the values at the concentration lines are: $C = 0.6, 0.2, 0.1, 0.001$ for $S = \pi$, and $C = 0.1, 0.001$ for $S = 400\pi$.

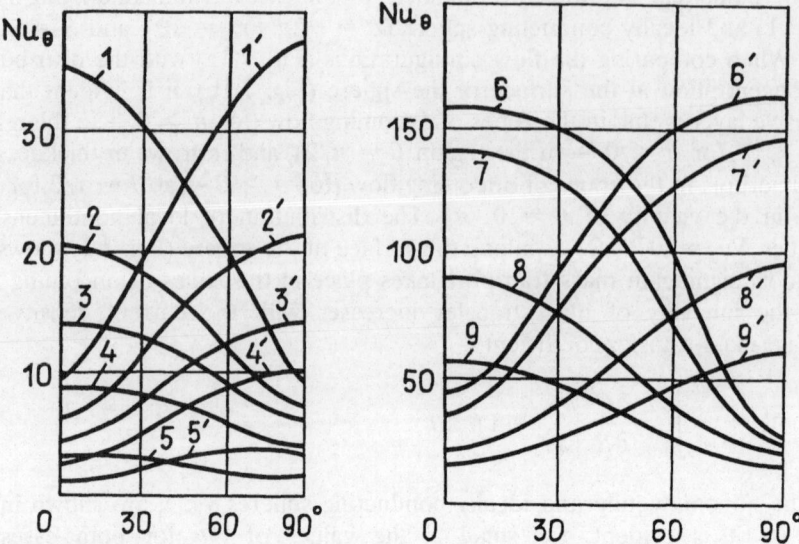

Fig. 7.2. The local Nusselt number distribution over the surface of the sphere. Numbers at the curves in the growing sequence correspond respectively to $S/\pi = 100, 16, 4, 1, 0.04, 3.2 \times 10^4, 10^4, 1.6 \times 10^3, 4 \times 10^2$. The numbers without the primes correspond to nonconducting sphere, and those with the primes to an ideally conducting sphere.

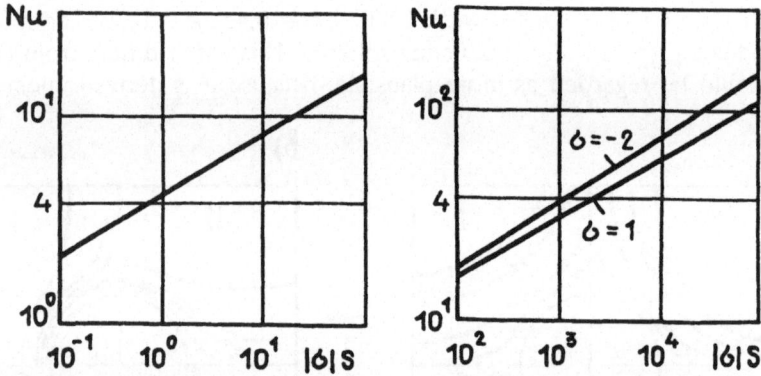

Fig. 7.3. The mean nondimensional mass-transfer coefficient of the sphere.

general assumptions. For $|\sigma S| > 10^2$, $A\ell = 0.34$, $k = 0.33$ if $\sigma = -2$, and $A\ell = 0.45$, $k = 0.275$ if $\sigma = 1$.

It is of interest to note that a single experiment [17] could be directly related to the problem considered. Although it was conducted a little earlier than the theoretical work [10], it was stimulated by the first investigation [1], which suggested the possibility of intensifying mass transfer from a particle in a current-carrying fluid.

In the experiment the dissolution of metallic spheres of diameter ranging from 9.8 to 30 mm, made of brass, steel, lead in 10% hydrochloric acid solution

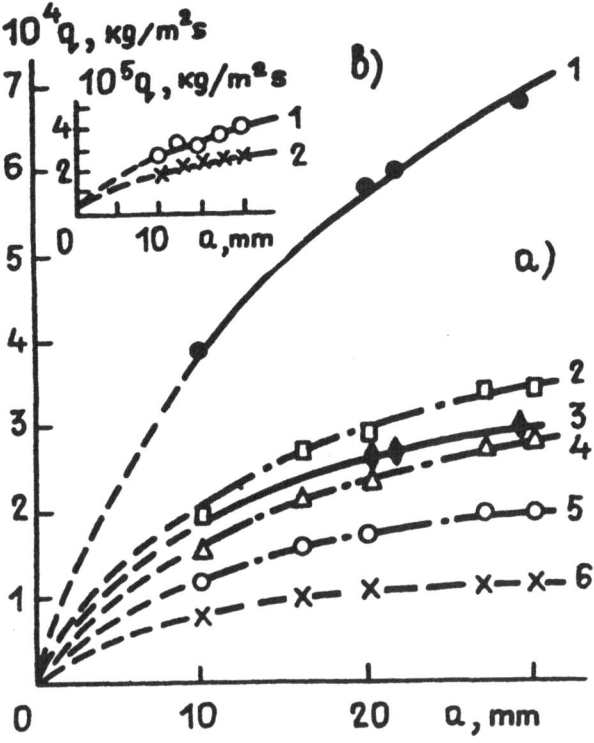

Fig. 7.4. The dependence of diffusive flux density at the sphere on its diameter: (a) 1 and 3 — steel spheres in the current of density $j = 4720$ and 2360 A/m^2; 2, 4, 5, and 6 — brass spheres $j = 4720, 3540, 2360, 1180$ A/m^2; (b) lead spheres 1 and 2 — $j = 2360$ and 1180 A/m^2.

was studied. The experimental results are represented in Fig. 7.4 in the form of diffusive flux density at the surface of spheres vs. the diameter of spheres for the different values of current density. To compare the results with the mass transfer in the absence of current, the data for the dissolution under ordinary conditions are also presented: the diffusive flux density q for the brass spheres is $(0.7-1.0) \times 10^{-6}$ kg/m^2 s in the temperature interval $T = 27-35°$C, for the steel spheres $q = (0.5-0.8) \times 10^{-6}$ kg/m^2 s, $T = 26-30°$C, and for the lead spheres $q = (0.4-0.56) \times 10^{-5}$ kg/m^2 s, $T = 30-38°$C. In spite of slight differences in the experimental conditions and the theoretical problem, the ratios of the experimental mass fluxes to the fluxes derived from the relation (7.2.6) [10] differ by less than 13 per cent for the different current densities.

7.3. Mass transfer from a translating spherical particle in a current-carrying fluid

Consider now a more general problem of flow at a sphere with unperturbed uniform velocity v_0 oriented parallel to the unperturbed electric current lines.

Then in Stokes approximation, instead of (7.2.3), we have for the nonconducting sphere (see Section 6.4)

$$\psi = \frac{1}{2} Re \left(R^2 - \frac{3}{2} R + \frac{1}{2R} \right) \sin^2 \theta -$$

$$- \frac{S}{8} \left(R^2 - \frac{5}{2} + \frac{1}{R} + \frac{1}{2R^2} \right) \sin^2 \theta \cos \theta. \quad (7.3.1)$$

The solution of the problem (7.2.1) and (7.2.2) with the stream function (7.3.1) was obtained in [1]. The main results of the solution are shown in Fig. 7.5, and for $S = 0$ the average Nusselt number

$$Nu = 0.991 \, Pe_D^{1/3}. \quad (7.3.2)$$

Consider now the nonlinear solution of mass transfer at moderate Re numbers for spheres of different conductivities [11]. The solution for the hydrodynamic part of the problem is discussed in Section 6.4. This case differs from the problem in Section 7.2 by the presence of velocity scale v_0, then (7.1.18) includes three independent parameters: $Re = v_0 a/v$, S, σ if the stream function is made nondimensional by $v_0 a^2$. The mass-transfer equation (7.1.15), in turn, contains the Peclet number in the usual form $Pe_D = v_0 a/D$. The boundary conditions are (7.2.2) (in the presence of external flow v_0 the condition (7.2.5), of course, is dropped).

The numerical solution is checked by comparing it with the experimental relation for the mass transfer from a sphere in the fluid without the current:

$$Nu = 2 + a_1 Re^{1/6} Pe_D^{1/3},$$

where Re and Pe_D are expressed by the diameter of the sphere, and the coefficient α, according to different authors [2], is equal to 0.55, 0.95, or 0.72. The numerically simulated value $\alpha = 0.815$ differs by merely 7 per cent from the recommended value $\alpha = 0.76$ in [15]. The electric current effect on the

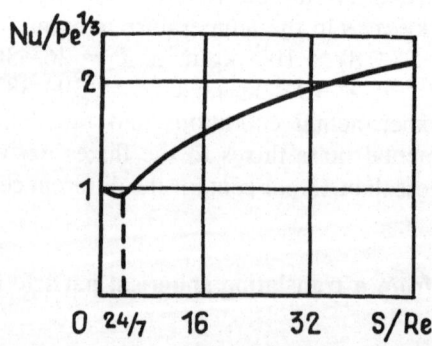

Fig. 7.5. The electrically induced flow effect on the mass transfer of nonconducting sphere in the linear approximation.

mass transfer at the nonconducting ($\sigma = 1$) and ideally conducting ($\sigma = -2$) spheres is investigated for $Re = 10$, $Pr_D = 10^3$ in the range $12.5 \leq S \leq 314$.

The concentration field (Fig. 7.6) obtained for $S = 113$ should be inspected with the respective stream line configuration (see Fig. 6.10). As is evident from the figures, in the zones of closed circulation the lines of equal concentration are pushed off the surface of the sphere, which can be attributed to the intense mixing in these zones. Yet in the zones where the electrically induced flow promotes the basic flow the concentration gradient is growing and, respectively, the thickness of diffusive layer is also increasing.

The distribution of local Nusselt number (Fig. 7.7) shows that with the increase of S the mass-transfer rate at the nonconducting sphere is increasing at the front critical point, and, when the zone of recirculating flow develops at the rear critical point, also in this zone. At the front critical point of the ideally conducting sphere the mass-transfer rate falls sharply due to the basic flow retardation by the electrically induced flow, but it increases in the zone of equatorial section and further downstream, because there the electrically induced flow promotes the basic flow. These facts may be related to the slight decrease of the average Nusselt number for the relatively small S (Table 7.2). The further increase of the current density leads to the growth of mass-transfer intensity, and the growth is higher for the ideally conducting sphere.

Table 7.2.

S/π	$Nu/Pe_D^{1/3}$	
	$\sigma = 1$	$\sigma = -2$
0	0.863	0.863
4	0.869	0.846
16	0.886	0.781
36	0.930	0.885
64	1.008	1.211
100	1.091	1.378

When two ratios are compared for the nonconducting sphere — of the mass transfer with the electric current to the mass transfer without the current, evaluated by the nonlinear solution (Table 7.2), and by the Stokes solution (Fig. 7.5) — it becomes evident that the Stokes solution overestimates the mass-transfer rate by approximately 1.5 times. Similar results also follow for the stationary sphere considered in the previous paragraph.

The solution for the mass transfer from a spherical nonconducting drop in the current carrying fluid was obtained in [9]:

$$Nu = \frac{2}{\sqrt{6\pi(k+1)}} Pe_D^{1/2} \left[1 + \frac{Re}{8} \left(\frac{3k+2}{k+1} + \frac{S}{Re^2} \right) \right]^{1/2}, \quad (7.3.3)$$

where k is the ratio of dynamic viscosities of the drop material and surrounding

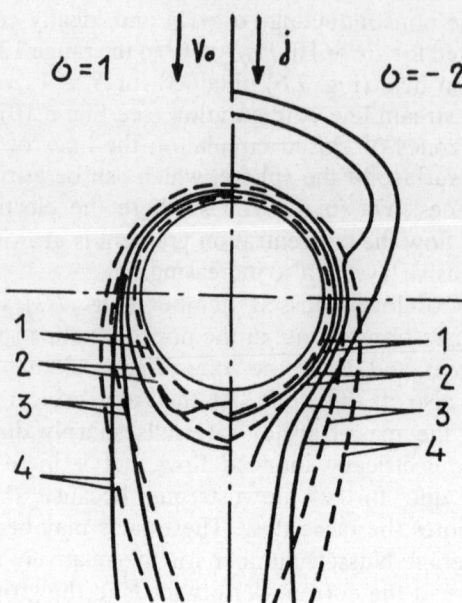

Fig. 7.6. The lines of equal concentration: (1) $C = 0.4$, (2) 0.2, (3) 0.01, (4) 0.005. The full lines — $S = 113$, the dashed — $S = 0$.

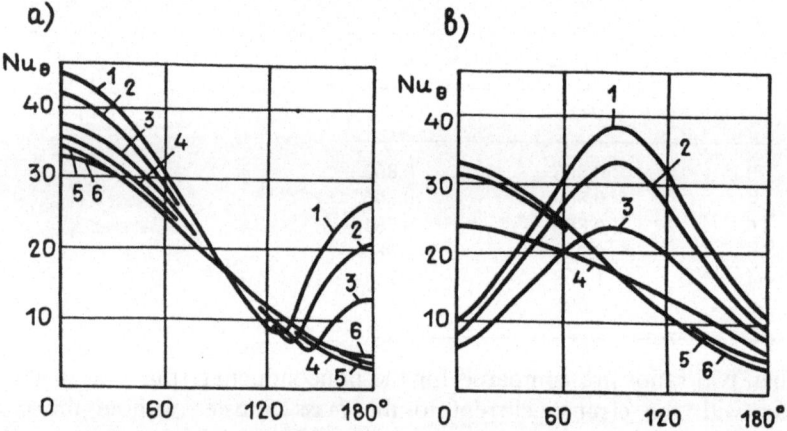

Fig. 7.7. Local Nusselt number distribution over the surface of sphere (a) $\sigma = 1$, (b) $\sigma = -2$. The numbers at the curves in growing sequence correspond respectively to $S/\pi = 100, 64, 36, 16, 4, 0$.

fluid. For $k = 0$ we obtain the solution for a gas bubble, and for $S = 0$ the solution in [6]. The formula (7.3.3) cannot be reduced to the case of a rigid sphere ($k = \infty$).

7.4. Mass transfer from a stationary sphere in a longitudinal magnetic field

Now let a rigid spherical particle be placed in the electric and magnetic fields parallel to each other and uniform far from the sphere. In this case, apart from the electrically induced meridional flow, a swirling fluid motion is driven due to the magnetic field interaction with the component of the electric current normal to it. The swirl velocity v_φ does not affect the mass-transfer process directly, because the equation (7.1.15) in the axisymmetric case does not contain v_φ. Hence, the swirling motion may influence the mass transfer merely by affecting the meridional flow, i.e. by the function ψ in (7.1.15).

The swirling motion at the sphere is differential, and the rotational centrifugal force drives the corresponding meridional flow. Indeed, consider the Stokes approximation of (7.1.19) with the typical velocity scale $v_0 = j_0 B_0 a^2/\rho v$:

$$\frac{1}{R}\frac{\partial^2}{\partial R^2} Rv_\varphi + \frac{1}{R^2}\frac{\partial^2 v_\varphi}{\partial \theta^2} + \frac{\cotan \theta}{R^2}\frac{\partial v_\varphi}{\partial \theta} - \frac{v_\varphi}{R^2 \sin^2 \theta}$$

$$= -\frac{3}{4}\sigma \frac{\sin 2\theta}{R^3}$$

the solution of which for the boundary conditions $v_\varphi|_{R=1} = 0$, $v_\varphi|_{R \to \infty} = 0$, $v_\varphi|_{\theta=0, \theta=\pi} = 0$ is

$$v_\varphi = \frac{\sigma}{8R}\left(1 - \frac{1}{R^2}\right)\sin 2\theta. \tag{7.4.1}$$

The centrifugal force

$$f_c = \frac{v_\varphi^2}{R \sin \theta} = \frac{\sigma^2}{16 R^3}\left(1 - \frac{1}{R^2}\right)^2 \sin \theta \cos^2 \theta \tag{7.4.2}$$

for a fixed R has a maximum for $\theta_0 = \arcsin 1/\sqrt{3}$, i.e. it is distributed along the surface $R = \text{const}$ as shown in Fig. 7.8a. According to this distribution, the centrifugal force drives the fluid from the regions at $\theta = 0, \pi$ and $\theta = \pi/2$ to an intermediate zone. The actual meridional flow computed in [12] is shown in Fig. 7.8b. The flow coming from the zone $\theta = \pi/2$ is weaker than the flow from

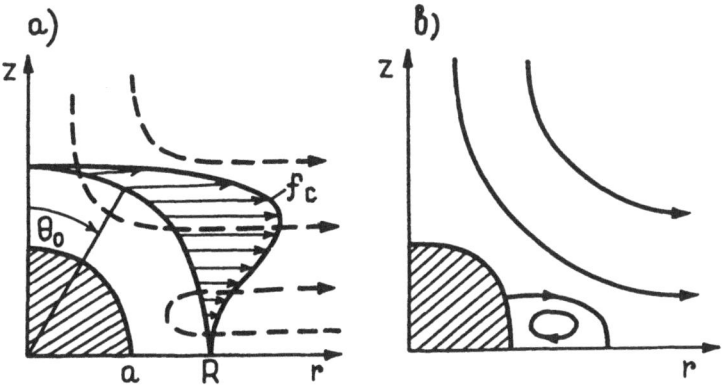

Fig. 7.8. (a) The centrifugal force distribution over the sphere of radius R, (b) the meridional flow at the sphere driven by the centrifugal force (one fourth of the flow region is shown).

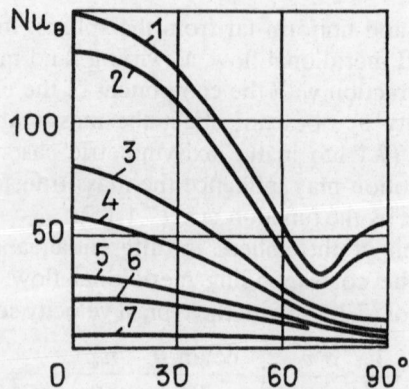

Fig. 7.9. Local Nusselt number distribution over the surface of the sphere. Numbers at the curves in the growing sequence correspond to the numbers $M = 1.5 \times 10^4$, 10^4, 2×10^3, 10^3, 2×10^2, 10^2, 10.

the polar zones $\theta = 0, \pi$, and for the weak rotation case a closed circulation zone at $\theta = \pi/2$ is organized. With the increase of swirl intensity, the size of the recirculating zone is diminished and the separation line moves to $\theta = \pi/2$.

It is of interest to note that the distribution of centrifugal force (7.4.2) does not depend on the conductivity of the sphere, yet its magnitude is varied. The same can be said about the swirl velocity v_φ (7.4.1), but the sense of swirl depends on the sign of σ (for the nonconducting sphere ($\sigma = 1$) and ideally conducting ($\sigma = -2$) the senses of swirl are opposite). This means that the meridional flow configuration will be the same for a sphere of any conductivity (excluding the trivial case $\sigma = 0$ when $v_\varphi = 0$). Consequently, the mass transfer is also similar for the spheres of different conductivity. Of course, the conclusion is valid if the meridional flow is driven only by the centrifugal force.

The latter condition is satisfied for the numerical solution of mass transfer in [12]. If $S \ll M$ is assumed, the term in the right-hand side of (7.1.18) can be neglected relative to the curl of centrifugal force. The numerical solution is obtained for $Pr_D = 10^3$ and $10 \leq |\sigma|M \leq 1.5 \times 10^4$. The computed values of local mass-transfer coefficient are represented in Fig. 7.9. In agreement with the meridional flow configuration the most significant growth of mass transfer occurs in the zones of oncoming fluid flow at $\theta = 0, \pi$. When the zone of closed circulation vanishes, the mass-transfer rate increases also in the vicinity of circle $\theta = \pi/2$. The integral mass-transfer coefficient can be satisfactorily approximated by the expression

$$Nu = 0.108 \, Pr_D^{0.33} (|\sigma|M)^{0.45}.$$

7.5. Heat and mass transfer in a cylindrical container

The hydrodynamic part of the solution is the most complicated (see Section 4.6). Yet if the velocity field has been found, the solution of linear equations for the temperature (7.1.12) and concentration (7.1.13) fields may be found

Heat and mass transfer in electrically induced vortical flows 297

relatively simply. Consider an example of the computation in a cylindrical domain with $h/R = 2$, $r_0/R = 0.2$.

In [18] the following boundary conditions are set for temperature:

$$\left.\frac{\partial T}{\partial r}\right|_{\substack{r=0 \\ r=1}} = 0, \qquad \left.\frac{\partial T}{\partial z}\right|_{z=2} = 0, \qquad T|_{z=0} = 0,$$

meaning that the upper and lateral surfaces of the container are thermally insulated, and the heat is conducted merely at the bottom of the container. The aim of these computations is to reveal the effect of electromagnetic convection on the temperature field in comparison with the conductive heat transfer.

From an inspection of Fig. 7.10a it becomes evident that, for the first, the electrically induced flow promotes an equalization of temperature in the container, sharply reducing local overheating of the melt. Further, according to the electrically induced flow configuration, the axial zone is the most heated; the heat is transported by convection from the small electrode along the axis to the bottom of the bath where the isotherm is of the form typical for the front of crystallization in, for example, electroslag welding. The temperature distribution along the z-axis is uniform, except for a narrow thermal boundary layer at the bottom where the temperature abruptly falls to the bottom wall temperature (the curve 1 in Fig. 7.11).

The maximum temperature occurs beneath the small electrode in the centre of both ($r = 0$). The maximum is significantly lower in the presence of convection (see Fig. 7.12 for $N = 0$). The decrease of nondimensional value T_{max} with the growth of S (with the intensity of convection), which is evident in the figure, is related to the nondimensional form

$$T = (T^* - T_1^*) \frac{\sigma \kappa L^2}{I_0^2}.$$

Nevertheless, the computational results indicate a decrease of the dimensional

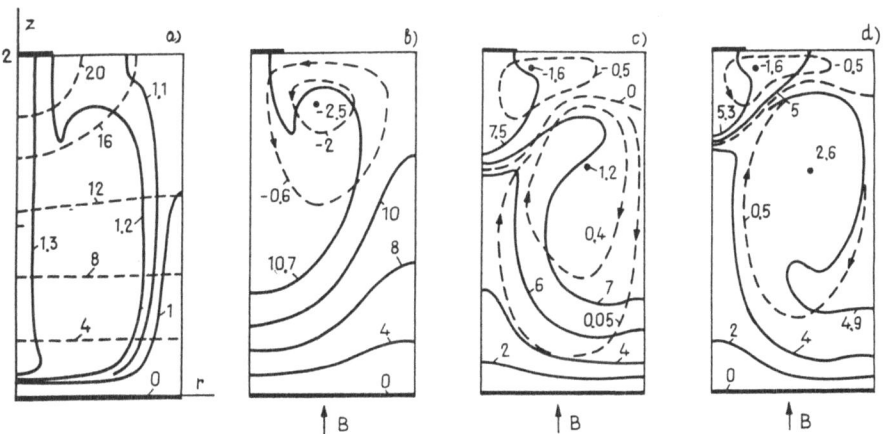

Fig. 7.10. Isotherms (full lines) in the case $S = 10^3$, $Pr = 20$, and (a) $N = 0$, (b) 0.5×10^3, (c) 10^3, (d) 1.5×10^3. The dashed lines show: (a) isotherms in the absence of convection, (b), (c), (d) stream lines.

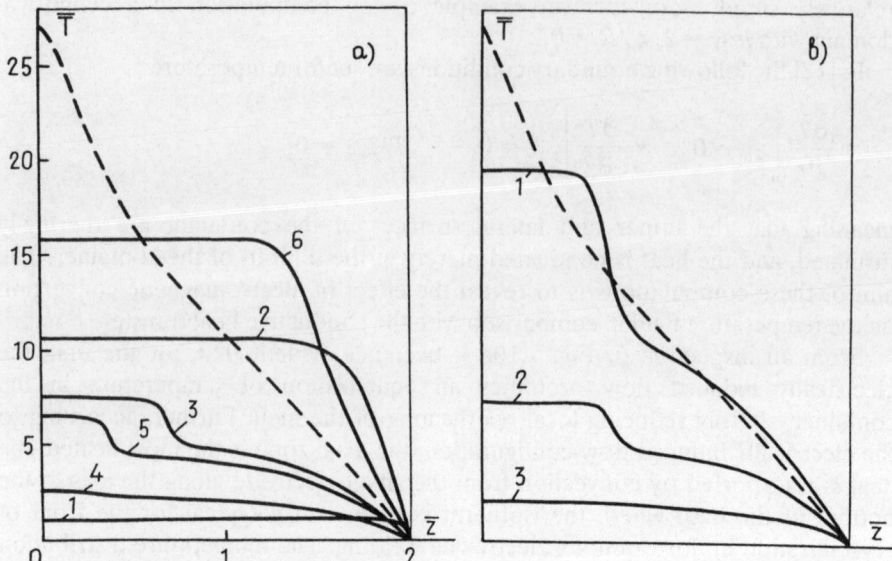

Fig. 7.11. Temperature distribution along the axis of the container: (a) $Pr = 20$ for the following S and N: $1 - 10^3, 0$; $2 - 10^3, 0.5 \times 10^3$; $3 - 10^3, 10^3$; $4 - 0, 10^3$, $5 - 0.5 \times 10^3$, $6 - 1.5 \times 10^3$; (b) $S = N = 10^3$ for the following Pr numbers: $1 - 2, 2 - 20, 3 - 200$ (the dashed line corresponds to the case of absent convection.

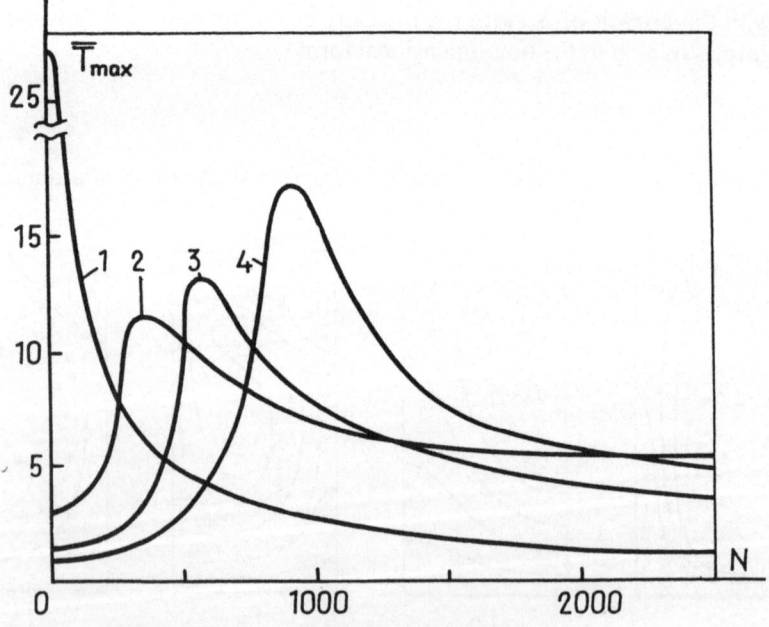

Fig. 7.12. The maximum excess temperature in the fluid as dependent on N for $Pr = 20$ and (1) $S = 0$, (2) 0.5×10^3, (3) 10^3, (4) 1.5×10^3.

temperature T^*, although at a lower rate than T. Thus, the increase of S from 500 up to 1500, i.e. three times, leads to 20 per cent temperature drop beneath the electrode.

The bottom thermal boundary layer is not of uniform thickness (Fig. 7.13 curve 1). It is lowest in the centre and grows laterally. This results in the nonuniform heat transfer at the bottom (Fig. 7.14a), which is expressed by the local Nusselt number, related to the maximum temperature in the bath,

$$Nu_T = \frac{1}{\overline{T}_{max}} \left(\frac{\partial \overline{T}}{\partial \bar{z}} \right)_{\bar{z}=0}.$$

Note that, with the growing intensity of convection, the heat transfer growth is limited. Thus, for $S = 10^3$ the heat transfer reaches a saturation, as is evident from the dependences of the local and average (by radius) Nusselt numbers $Nu_T(S)$ in Fig. 7.14b.

Similarly a concentration field may be computed for the known velocity field. For the previously discussed situation the boundary conditions for the concentration C are [19]: $C = C^*/\rho = 1$ for $z = 2$, $C = 0$ for $z = 0$, $\partial C/\partial r = 0$ for $r = 0$ and $r = 1$.

These conditions can be interpreted as a constant dissolution at the top surface of the melt and an absorption at the bottom. The side walls of the container are assumed unpermeable for the dopant material.

In the absence of convection the lines of equal concentration are horizontal, and the magnitude of concentration decreases downwards at a linear rate. The electrically induced flow leads to an increased mass transport along the symmetry axis (Fig. 7.15a), and the concentration in the central zone falls

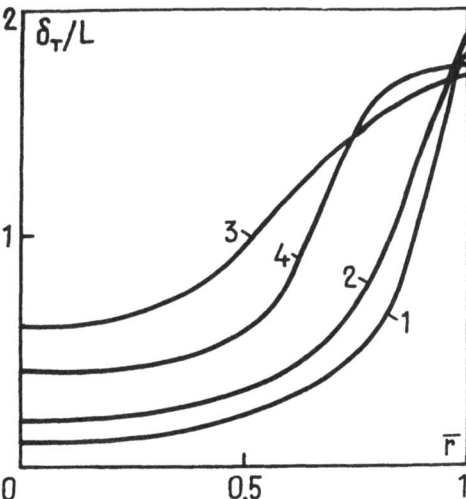

Fig. 7.13. The thermal boundary layer along the radius of cold bottom of the container $z = 2$ in the case of $Pr = 20$ and the following S, N: (1) 10^3, 0; (2) 10^3, 0.25×10^3; (3) 10^3, 5×10^3; (4) 1.5×10^3, 0.75×10^3.

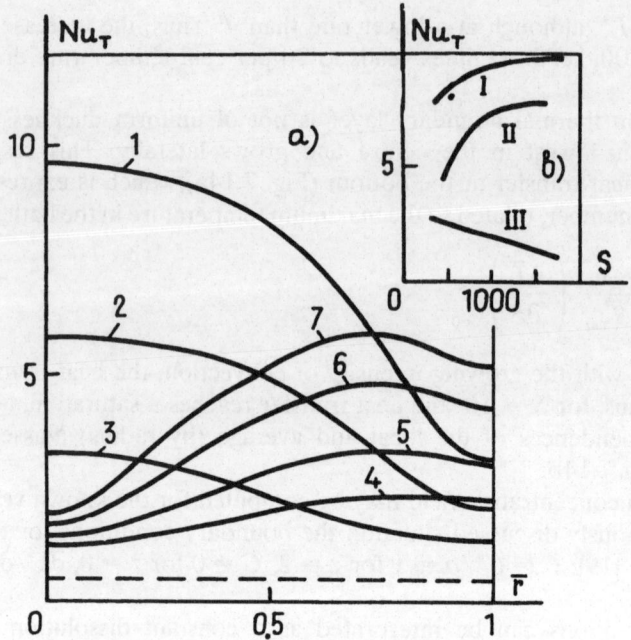

Fig. 7.14. Local heat transfer at the cold surface $z = 2$. (a) In the case of $S = 10^3$, $Pr = 20$ and the following N: $1 - 10$, $2 - 0.4 \times 10^3$, $3 - 0.5 \times 10^3$, $4 - 0.75 \times 10^3$, $5 - 10^3$, $6 - 1.5 \times 10^3$, $7 - 2 \times 10^3$ (the dashed line corresponds to the absence of convection). (b) I — heat transfer at the axis for $N = 0$; II — the average along the radius heat transfer for $N = 0$; III — the average along the radius heat transfer for $N/S = 0.6$.

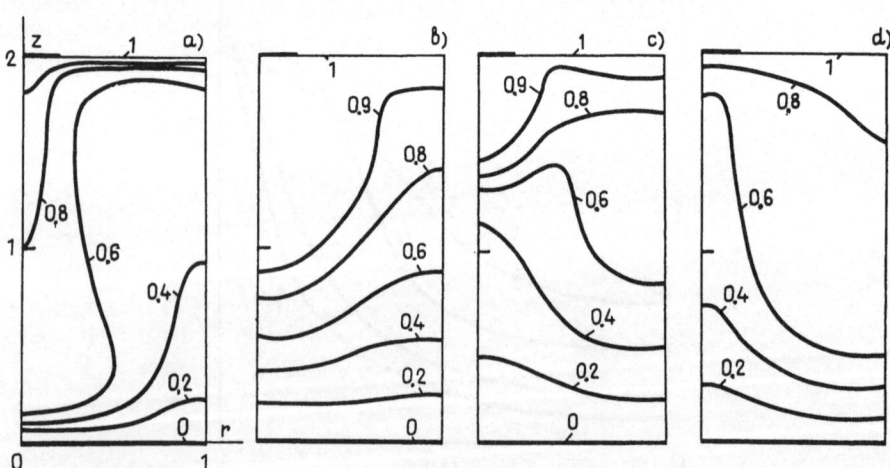

Fig. 7.15. The lines of equal concentration $C = $ const in the cases of $Pr_D = 2$ and the following S, N: (a) 10^3, 0; (b) 10^3, 0.5×10^3; (c) 10^3, 10^3; (d) 0, 10^3.

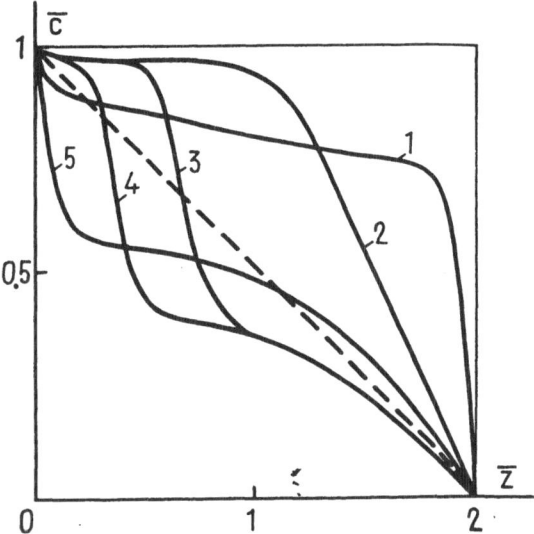

Fig. 7.16. The distribution of concentration along axis of the container for $Pr_D = 2$ and the following S, N: (1) $10^3, 0$; (2) $10^3, 0.5 \times 10^3$; (3) $10^3, 10^3$; (4) $0.5 \times 10^3, 10^3$; (5) $0, 10^3$.

weakly along the height except the narrow diffusive layer at the bottom where the concentration drops abruptly (Fig. 7.16 curve 1). As the result of axial mass transport the material is accumulated in the central part of the bottom. In accordance to the behaviour of the local diffusive Nusselt number

$$Nu_g = \frac{1}{\bar{C}_{max}} \left.\frac{\partial C}{\partial \bar{z}}\right|_{\bar{z}=0},$$

where $C_{max} = C(z=2) - C(z=0) = 1$ is the concentration difference at the top and bottom (Fig. 7.17b curve 1), the most intense mass transfer occurs in the centre of the bottom, which is undesirable for the purpose of a uniform dopant distribution in the ingot. At the top surface the highest mass transfer takes place in the middle of radius zone (Fig. 7.17a, curve 1), i.e. in the location of the most intense electrically induced flow.

7.5.1. Transient mass transfer

The mathematical model considered permits us to follow a transient process of mass transfer, e.g., after the addition of a dopant, up to the steady regime of mass transfer, and reversely, the transition from the steady regime to the total consumption after the feeding ceases. The problem is studied in [20] by solving the equation (7.1.13) with the term $r\,\partial C/\partial t$ in the left and the nondimensional time $t = t^*/t_0$, $t_0 = L^2/\nu$. The new boundary conditions for $z = 2$ are

$$C = 0 \quad \text{if} \quad t < t_1,$$
$$C = 1 \quad \text{if} \quad t \geq t_1,$$

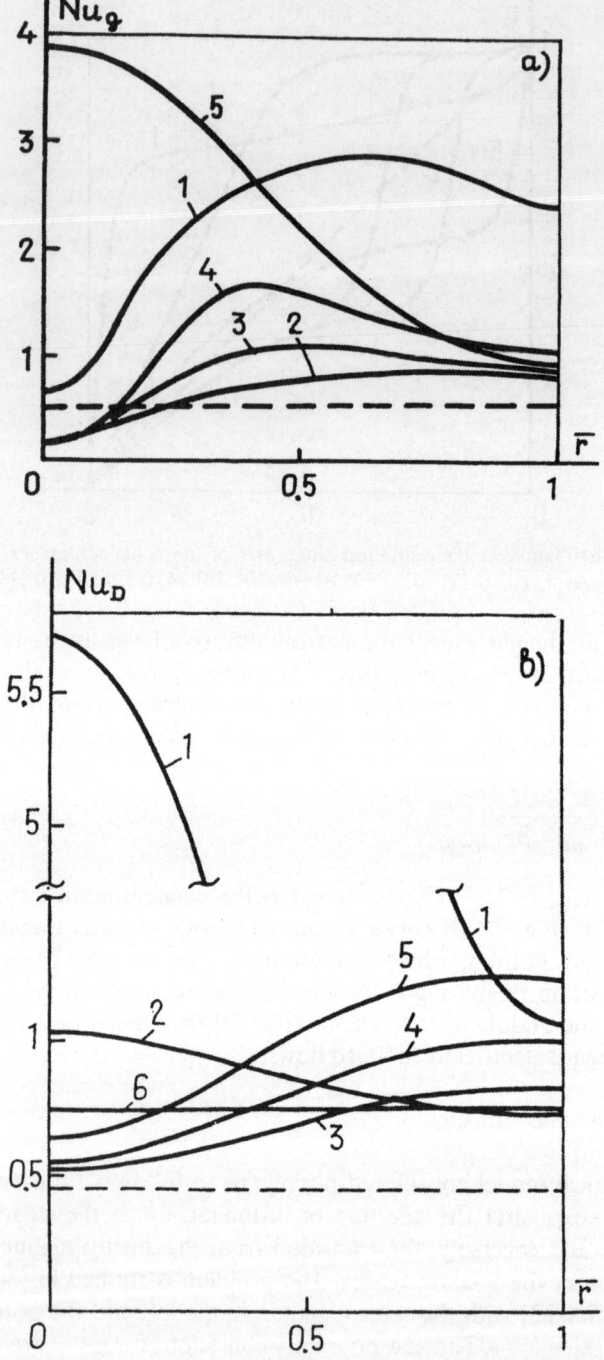

Fig. 7.17. The steady local mass transfer for $Pr_D = 2$ and the following S, N: (1) 10^3, 0; (2) 10^3, 0.5×10^3; (3) 10^3, 10^3; (4) 0.5×10^3, 10^3; (5) 0, 10^3; (6) 1.5×10^3, 10^3. (The dashed line corresponds to molecular diffusion) (a) from the surface $z = 2$, (b) from $z = 0$.

and the reversed conditions, when the feeding at $z = 2$ ceases are

$$C = 1 \quad \text{if} \quad t < t_2,$$
$$C = 0 \quad \text{if} \quad t \geq t_2.$$

The fluid flow is assumed steady, and the variation in concentration does not affect the flow.

The mass-transfer process is characterized by the integral concentration difference $\Delta C = \int_0^2 [C(r = 0) - C(r = 1)] \, dz$ [13] and the ratio Nu_D/Nu_{D_0} at the point $r = 0$, $z = 0$. Here Nu_D is the local (in time) diffusive Nusselt number and Nu_{D_0} is the respective steady-state quantity.

The numerical solution shows that, after the addition of a dopant, even in the initial stage, it is advected mainly along the axis to the bottom of the bath (Fig. 7.18a). Respectively, the concentration difference ΔC increases in a short time interval (Fig. 7.19a curve 1). The transition process ends approximately at $t_c = 0.2$. At approximately the same time a steady mass transfer is attained in the bottom centre of the bath (Fig. 7.20 curves 1 and 2), and, comparing the curves 1 and 2, it follows that the transit time decreases with the growth of electrically induced flow intensity (with S increasing). For $\nu = 10^{-4}$ m²/s, $L = 5 \times 10^{-2}$ m the value of $t_c = 0.2$ corresponds to the real time $t^* = 5$ s. The transit time for other points at the bottom of the bath is longer, for instance, at the distance $r = 0.8$ the ratio $Nu_D/Nu_{D_0} = 0.5$ is reached in the initial time interval four times

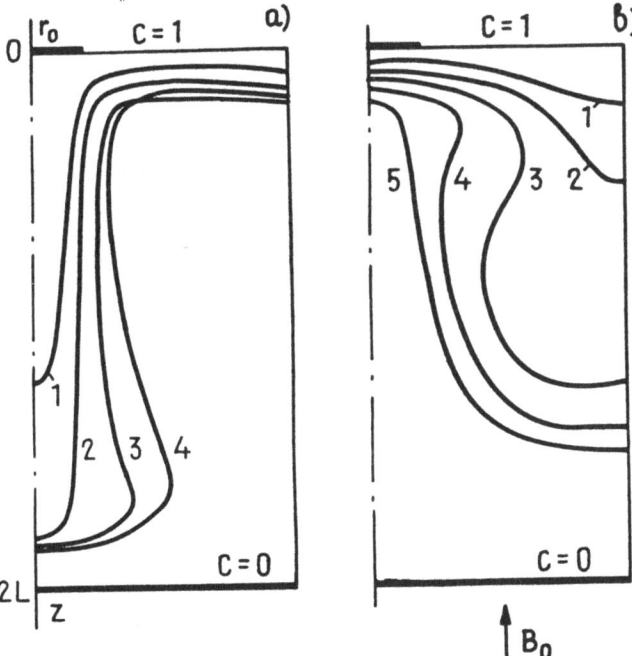

Fig. 7.18. Lines of equal concentration $C = 0.6$ at different stages of the mass transfer development starting from the introduction of additive for $Pr_D = 2$: (a) $S = 10^3$, $N = 0$, and the following t: $1 - 0.03$, $2 - 0.09$, $3 - 0.15$, $4 - t_{st}$; (b) $S = 0$, $N = 10^3$, and the following t: $1 - 0.045$, $2 - 0.135$, $3 - 0.225$, $4 - 0.315$, $5 - t_{st}$.

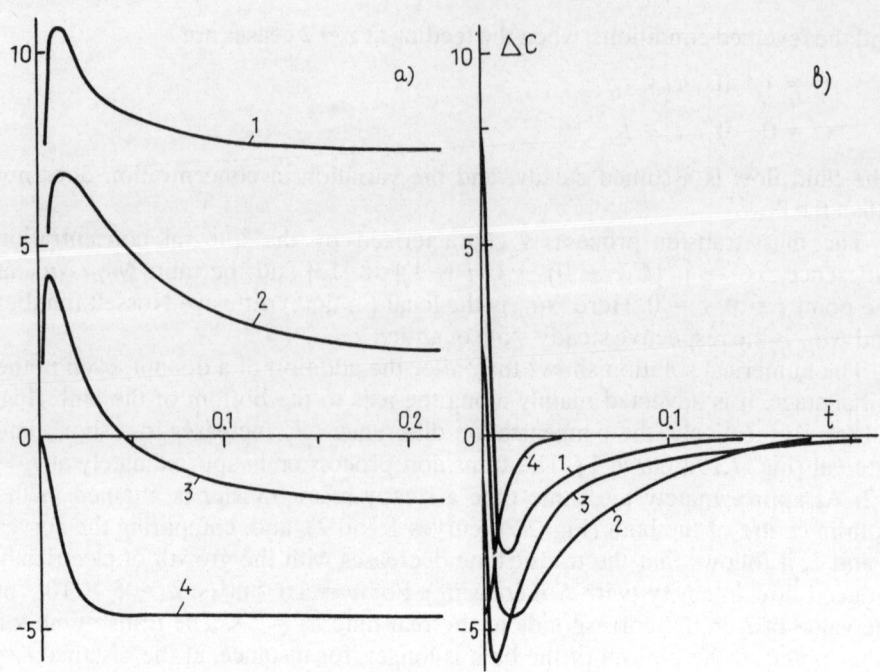

Fig. 7.19. The variation in time of ΔC for $Pr_D = 2$ and the following S, N: $1 - 10^3$, 0; $2 - 10^3$, 0.6×10^3; $3 - 10^3$, 10^3; $4 - 0$, 10^3. (a) Starting to introduce the additive, (b) ceasing the supply of additive.

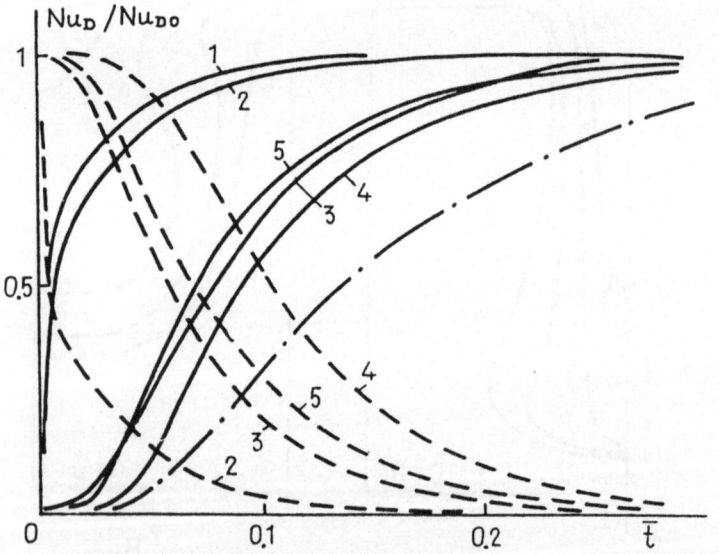

Fig. 7.20. Nonsteady local mass transfer in the centre of the container's bottom in the case of steady velocity field for $Pr_D = 2$ and the following S, N: $1 - 1.5 \times 10^3$, 0; $2 - 10^3$, 0; $3 - 10^3$, 0.6×10^3; $4 - 10^3$, 10^3; $5 - 0$, 10^3. Full lines — after starting the additive, dashed lines — after ceasing the additive; dot-dashed line corresponds to the molecular diffusion.

longer than for $r = 0$, yet after the time interval approximately equal to $0.65\ t_c$ the ratio is close to unity.

When the feeding stops, the concentration in the axial zone rapidly falls and becomes lower than at the lateral wall (Fig. 7.19b curve 1), and the mass-transfer intensity in the centre of the bottom decreases (Fig. 7.20 dashed curve 2).

7.5.2. Heat and mass transfer in an axial magnetic field

The data concerning the effect of axial field on the heat and mass transfer was obtained in [18, 19] for the same geometry of the bath and boundary conditions. The temperature and concentration fields were found to depend substantially on the ratio of the parameters N/S (see definitions in Section 4.6).

For $N/S < 0.6$ the effect of the secondary flow driven by the differential rotation results in a suppression of the electrically induced convection at the bottom of the container and in a displacement of the vortex to the melting electrode (Fig. 7.10b). The resulting drop in the total intensity of convection sharply increases an overheating of the melt (the maximum excess temperature is nearly an order of magnitude higher than in the absence of the magnetic field). At the same time isotherms at the bottom are flattening.

For $N/S > 0.6$ the convection driven by the differential rotation increases significantly. The consequence of this is a gradual decrease of overheating with the growth of field. Moreover, the secondary flow transfers the heat to the side wall of the container; the isotherms at the bottom give evidence (Fig. 7.10d) of the heat-transfer increase at the periphery of the surface. This feature could be advantageous to the quality of the ingot.

The temperature distribution along the symmetry axis of the bath shows a sudden change in the temperature from a higher temperature in the electrically induced flow region to a lower one in the secondary flow region (Fig. 7.11a curves 3 and 5). The temperature step moves to the upper electrode with the increase of N/S.

The highest overheating of the melt T_{max} is observed at the axis $r = 0$ close to the electrode. The T_{max} dependence on the material properties of the melt can be estimated by the formula $T_{max} = 1/kPr^{0.6}$. For $S = N = 1000$ (Fig. 7.11b) the coefficient $k = 0.03$. The maximum overheating dependence on the parameters S and N is more complex (Fig. 7.12). If S is fixed, then T_{max} grows with the magnetic field induction approximately up to the moment when N/S reaches 0.6. For a higher N the increased secondary flow diminishes T_{max}.

The decrease of the convection intensity at the bottom of the bath in the range $0 < N/S < 0.6$ leads to a growing thickness of thermal boundary layer δ_T (Fig. 7.13), and, thereby, to the growth of thermal resistance δ_T/λ (λ = thermal conductivity coefficient). In particular, the thickness of the thermal layer grows as $\delta_T/R \sim S^{-1}(N/S)^2$ at the symmetry axis $r = 0$ in the interval considered. The growth of thermal resistance explains the increased maximum overheating of the melt and the decreased heat transfer at the bottom of the bath for $N/S \rightarrow 0.6$ (Fig. 7.14a). For $N/S > 0.6$ the heat exchange is improved, and at an increasing rate with the distance from the centre of the bath. For $N/S = 0.6$ the local Nusselt number is practically constant along the bath radius.

The mass transfer with the axial magnetic field is quite similar to the heat transfer. The position of equal concentration lines (Fig. 7.15) shows that in the case of pure electrically induced flow the additive is transported mainly in the axial region (Fig. 7.15a); in the case of strong magnetic field the additive is transported along the lateral wall of the container (Fig. 7.15d). This is obviously related to the change of the hydrodynamic configuration in the bath. With the growth of ratio N/S from 0 to 0.6, the thickness of diffusive boundary layer grows (Fig. 7.16 curves 1 and 2), which leads to worsened conditions for the additive transport to the bottom of the bath. For $N/S > 0.6$ at the boundary, separating the electrically induced and secondary flow vortices, the change in concentration is quite abrupt. With the growth of N/S the step moves to the upper electrode.

The distribution of local diffusive Nusselt number at the surface $z = 2$ also depends on the decrease of electrically induced flow intensity in the upper section of the bath with the growth of N. The mass-transfer intensity through the surface for the refined value of $N/S = 0.66$ is close to the molecular diffusion, yet the mass transfer improves with the further growth of N/S (Fig. 7.17a). The same behaviour is observed at the bottom of the bath (Fig. 7.17b). Here the noticeable condition is the uniform concentration of additive along the bottom for $N/S = 0.66$.

Finally some words should be said about the nonstationary mass transfer in the axial field. If the field is strong enough, the electrically induced flow is not felt in the background of the field driven secondary flow ($S = 0, N \neq 0$), and a dopant introduced at the free surface of the slag bath is transported relatively slowly, yet this is not the case in the centre of the bath (Fig. 7.18a), but at the perimeter (Fig. 7.18b).

In the absence of the axial magnetic field the stabilized (for $t \to \infty$) value of the integral nondimensional concentration difference

$$\Delta C = \int_0^2 [C(r=0) - C(r=1)] \, dz$$

is positive (the curve 1 in Fig. 7.19a). This means the concentration of additive is higher at the axis of the bath than at the perimeter. When the magnetic field induction is gradually increased, the magnitude of the difference decreases (Fig. 7.19a curve 2), and for $N/S = 0.66$, $\Delta C \to 0$ at $t \to \infty$, i.e. the additive is uniformly distributed over the volume of the bath. Yet if $N/S > 0.66$, $\Delta C < 0$, i.e. the concentration of additive is higher at the perimeter of the bath than in the centre.

The development in time of the mass-transfer process in the centre of the bath bottom ($r = 0$) is shown in Fig. 7.20. As is evident, the process is retarded with the field growth, approaching pure molecular diffusion (the dashed line in Fig. 7.20). The development of the process is even slower in other points of the bath bottom if $t < 0.65 \, t_s$ (t_s is the time interval necessary to reach the steady state).

In the regime when the additive supply ceases the rate of concentration equalization is slowed down with the growth of the axial field (curves 2 and 3 in Fig. 7.19b) in comparison with the case when the field is absent (curve 1), although in the very strong field the equalization is reached in a shorter time (curve 4). The local Nu_D number in the centre of the bath $r = 0$ also behaves similarly: the time to reach the steady state increases with the field growth (Fig. 7.20).

7.6. Thermal convection in electrically induced flows

If in the problem of mass transfer we can neglect the back-effect of the concentration nonuniformity on the fluid motion, assuming that the density variation due to the mass transfer is insignificant, the effect of thermal convection due to the ohmic heating by the current driving the electrically induced flow could be significant. Unfortunately, there are few studies of simultaneous electrically induced flow and thermal convection and almost all of them are related specifically to electroslag technology (see Section 8.2) [4, 8, 14, 16]. Therefore, we are not in a position to analyse the whole variety of interactions between the electrically induced vortical flows and the associated thermal convection.

A common feature of the flows driven by the thermal convection in the axisymmetric situation is the meridional motion if the symmetry axis is aligned and antiparallel to gravity $\mathbf{g}$. Indeed, the buoyancy force is $\rho\mathbf{g} = -\rho g\mathbf{i}_z = -\rho_0 g(1 - \beta T)\mathbf{i}_z$ (in the Boussinesq approximation). From the curl expression of force in the cylindrical coordinates with the axial symmetry it follows that

$$\operatorname{curl} \rho\mathbf{g} = g \frac{\partial \rho}{\partial r} \mathbf{i}_\varphi = - \rho_0 \beta g \frac{\partial T}{\partial r} \mathbf{i}_\varphi$$

i.e. for the first, the flow is driven in meridional planes and for the second, when the density (or temperature) variation is in the radial direction, the direction of circulation depends on the decrease or increase of temperature with the distance from the symmetry axis ($\operatorname{curl} \rho\mathbf{g} > 0$ if $\partial T/\partial r < 0$, and $\operatorname{curl} \rho\mathbf{g} < 0$ if $\partial T/\partial r > 0$).

We will consider the simultaneous action of the electromagnetic force and the thermal convection due to the ohmic heating in the example based on the numerical solution for the mathematical model of electroslag remelting by Szekely and Dilavari [4, 16]. The set of equations (7.1.4) and (7.1.5) was solved for the slag and liquid metal baths for the specific material parameters of the slag and liquid metal, and the flow region shown in the following figures. The most significant feature of these computations is the use of real current magnitudes (of order 20–45 kA) typical of electroslag technology.

The use of real current magnitudes is made possible owing to the assumption of turbulent flow. The experience of numerical simulations shows that for a sufficiently high effective (turbulent) viscosity the instabilities of numerical procedure, caused by the nonlinear terms, are suppressed [3]. Hence, the solution for turbulent flows at high Reynolds numbers is relatively simple in view of the stabilizing effect of the additional viscosity on finite difference equations if compared to the laminar flow solution where the relatively smaller molecular viscosity leads to the numerical instability. For the nondimensional equation of motion the Reynolds number evaluated by the turbulent viscosity is substantially lower than by the molecular viscosity; respectively, the coefficient at the viscous term (i.e. at the higher derivatives), proportional to Re^{-1} is not small. A substantial stabilization of the numerical procedure is also achieved by the use of up-stream differences in the finite-difference approximation of nonlinear terms, which introduces an additional, uncontrollable, artificial numerical viscosity. Apart from this, it was assumed in the computations that the submerged electrode is of a sufficiently large size, which in effect decreases the difference in current densities in the melt and, consequently, the driving rotational force.

In the case of electrically induced flows in electroslag remelting (I = 20 kA) the computation of turbulent viscosity was made in [8] employing the two-parameter $K-\varepsilon$ Spolding model (Fig. 7.21). As is evident from the computed data, the ratio of the turbulent viscosity to the molecular reaches 20. The governing nondimensional parameter for electrically induced flows is $S = \mu_0 I^2/\rho \nu^2$ instead of Reynolds number. Obviously, if we replace the molecular viscosity ν by the mean turbulent viscosity ν_T the same value of parameter S would correspond to a ν_T/ν times higher current. Nevertheless, it should be noted that the application of improved numerical methods permits us also to compute the laminar flows for high S [14].

Now consider the results of the study in [16]. The stream lines for isothermal conditions (EVF without thermal convection) are presented in Fig. 7.22 for the effective magnitude of alternating current I = 30 kA, and with the thermal convection in Fig. 7.23. To explain the difference in the flow configurations represented in the figures we present also the isotherms in the slag and liquid metal. As is evident in Fig. 7.24, the hottest slag is located in the central region

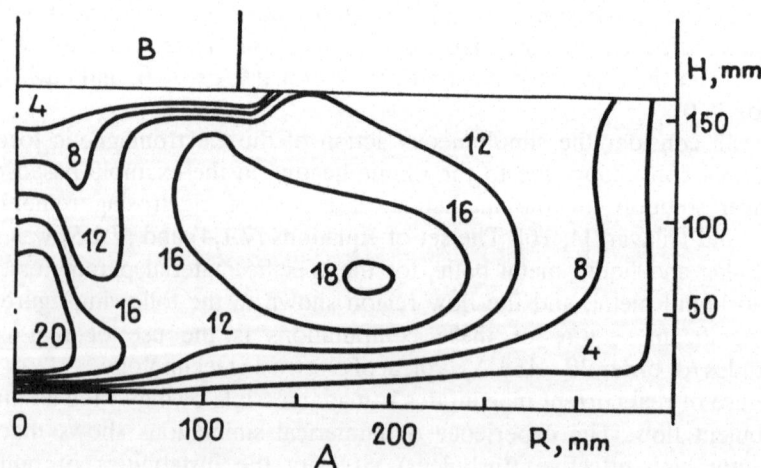

Fig. 7.21. Lines of equal ratios of the turbulent viscosity to the molecular for I = 20 kA.

Heat and mass transfer in electrically induced vortical flows 309

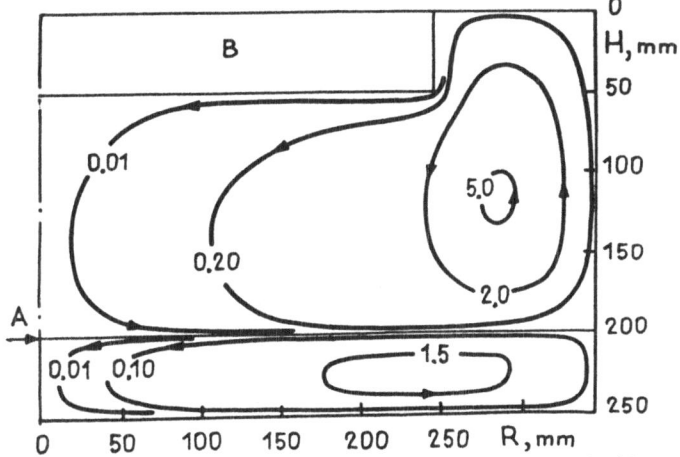

Fig. 7.22. Stream lines in isothermal conditions. A — the interface slag-liquid metal, B — the electrode.

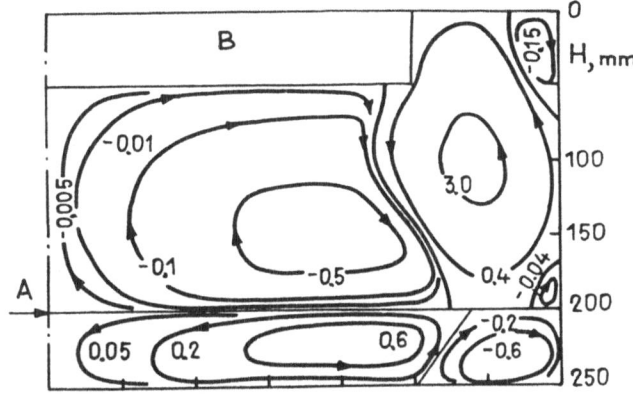

Fig. 7.23. Stream lines with the thermal convection.

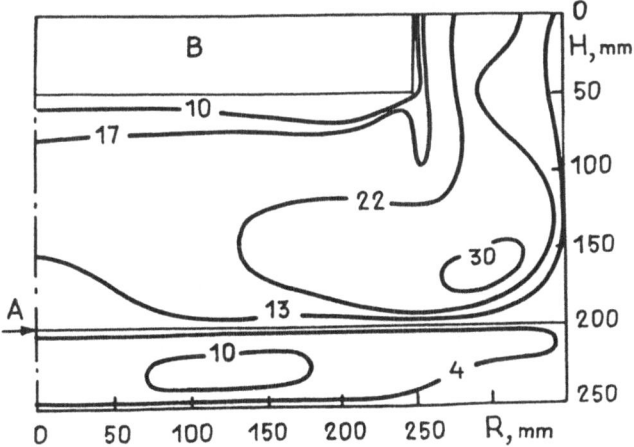

Fig. 7.24. Isotherms in the slag and liquid metal bath for $I = 36$ kA.

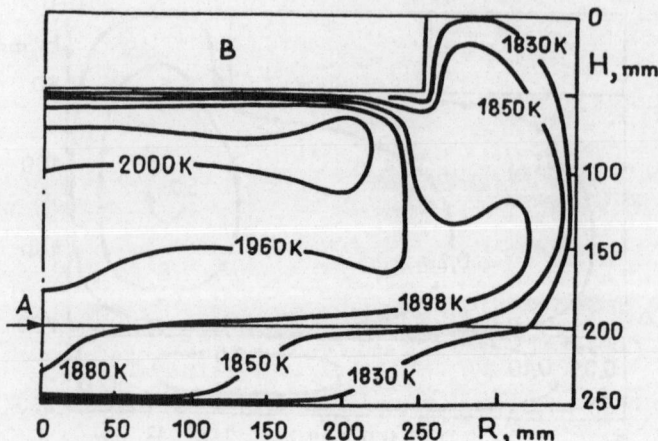

Fig. 7.25. The lines of equal ratios of the effective to the molecular conductivities.

beneath the electrode; when approaching the side wall (wall of the crystallizer) the temperature decreases. The curl of buoyancy force is positive in accordance with the previous analysis, and, thereby, the slag circulation in a meridional plane driven by the thermal convection is reverse to the electrically induced flow. As a result, the single-loop circulation for the isothermal conditions is replaced by the two-loop circulation for the nonisothermal conditions. However, the velocity of circulation driven by the thermal convection (the left loop in Fig. 7.23) is susbtantially lower than in the right loop of the electromagnetic origin. Thence, we can conclude that the electromagnetic forces play a greater role in driving the slag than the buoyancy force (the motion driven in the liquid metal bath, shown in Fig. 7.22, will be explained in Section 8.2).

Finally consider the data for the effective thermal conductivity coefficient κ_T from [16]. As is evident from Fig. 7.25, the effective thermal conductivity in the turbulent flow increases the molecular conductivity many times, which has to be accounted for in the solution of heat-transfer equation (7.1.2) and in other similar situations.

8

Experimental investigations of EVF and applications

We shall consider experiments related to the electrically induced flows jointly with the corresponding specific applications, since the experimental set-up is substantially stimulated by practical demands. Several of these actually are dictated by practical problems, because experimental methods are found to be the only way to solve the problems. However, most of the experiments are devised to demonstrate physical phenomena, including also phenomena still unpredicted by the present theories. The aims of the experiments are, first, to attract attention to the existence of certain phenomena and to motivate the necessity to include these in a mathematical model of a particular device or a process, and, second, to suggest a rational application of specific properties of EVF to solve different practical problems.

All the experiments described, except those in Section 8.7, were conducted at the Institute of Physics, Latvian SSR Academy of Sciences.

8.1. Electroslag welding

An electroslag technology, a typical example of which is electroslag welding, originated in the 50s [25]. The basic directions of investigations in welding technology were aimed at an optimum productivity and quality of weld for different welding regimes. However, only recently was it realised that the search had to be coupled with a deep understanding of magnetohydrodynamic effects in the welding bath.

The electroslag welding in schematic representation is shown in Fig. 8.1. The melted electrode 1 (electrode wire) is continuously fed in the slag bath 3 where the temperature reaches 2000°C. The slag (or flux) serves a double role. The first is related to the high electrical resistance of the slag, which leads to most of the ohmic heating that is released there, the high temperature of the slag, and melting of the metal. Second, the electrode wire is melted out of contact with the atmosphere, thus the metal does not burn-out and stays pure. Apart from this, chemically active fluxes have been widely applied, recently, with the advantage of an additional degassing and refining of the liquid metal droplets dripping off the electrode and traveling through the flux.

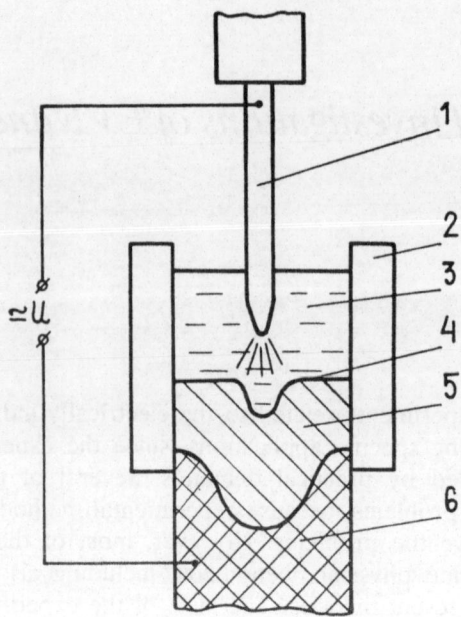

Fig. 8.1. Schematic representation of electroslag welding process.

Between the weld 6 and the slag bath there is a transition liquid metal zone, the processes in which significantly affect the quality of the weld. The slag seam is formed by the crystallizer 2, which can be translated when metal has filled the weld volume. The welding current magnitude is of the order of 1—2 kA.

It is well known that the interface separating the liquid metal zone and the weld assumes the form of a depression on the weld axis. Special-purpose investigations [38] have shown that the depth of depression depends on the feed rate of the electrode, with an increase of which the welding current grows. When the feed rate is speeded-up, which is advantageous to the productivity of the process, the depression acquires an elongated form. For the existing heat withdrawal conditions this leads to a radial growth of grains in the axial zone of the weld. In this zone different impurities and nonmetallic inclusions are accumulated, and the weld metal in this zone is of lower plasticity and porosity. Moreover, in several instances the slag inclusions are observed in this zone, which gives evidence to a mutual dependence of the depression's depth and the shape of the slag—liquid metal interface.

In electroslag welding the surface area of the melting electrode's cross section is approximately one hundredth of the weld's cross-sectional area, which is, in effect, the second electrode. Thereby, the electric current density is highly non-uniform in the slag bath, and all the conditions for driving a flow are met. To estimate the effect of the flow on the welding process, consider the experimental data collected with direct current experimental models.

We shall begin with the results of flow observation in the model consisting of a hemicylindrical container of diameter 150 mm, the end walls of which serve

Experimental investigations of EVF and applications 313

as electrodes [12]. The container of total length 300 mm is partitioned by a nonconducting plate parallel to the end walls. A copper cylinder with an insulated lateral surface is inserted in the semicircular orifice (of diameter 10 mm) at the upper part of the plate. The container is filled with mercury to a level covering half of the copper cylinder, which can now serve as an electrode (Fig. 8.2).

With certain qualification the surface of mercury could represent the section of the welding bath coinciding with the symmetry plane. To visualize the flow, a thin layer of sulfuric acid mixed with graphite powder is put on the surface of the mercury (the same technique is used in other similar experiments). The paths of graphite particles, carried along with the moving mercury, visualize the fluid flow.

As is evident from the photograph in Fig. 8.2, the flow has all the attributes of a jet spreading from the end of the cylindrical electrode, then closing in a toroidal vortex. The jet can penetrate to a great depth, and in the case of

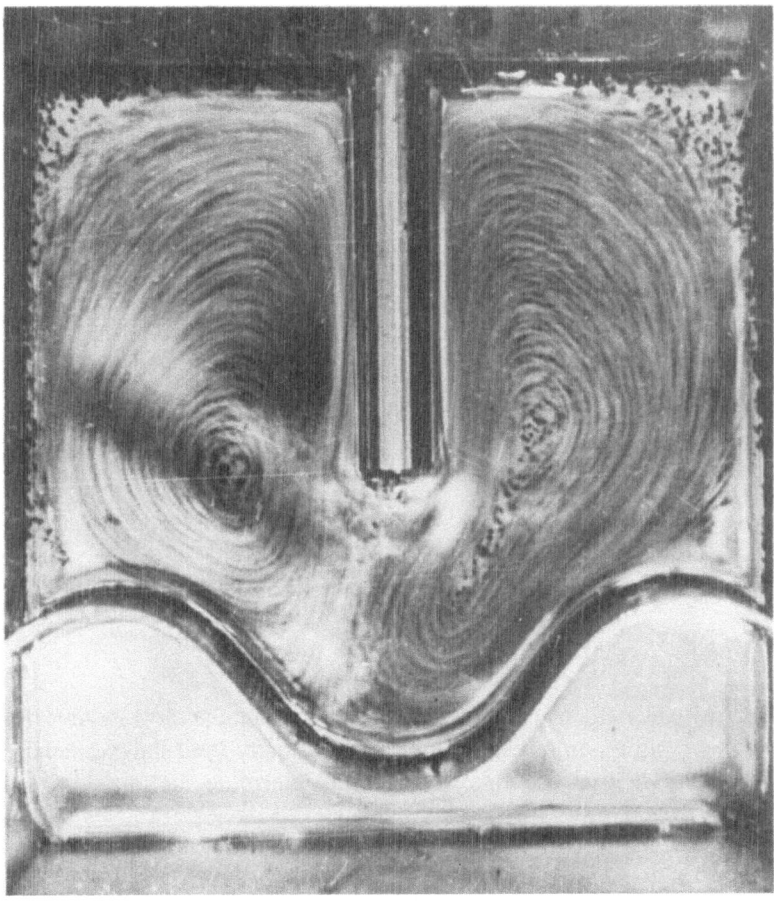

Fig. 8.2. Visualization of the motion at the free surface of mercury.

electroslag welding it can deform the interface between the slag and liquid metal.

At the present time there is a single experiment in which the velocity field of an electrically induced vortical flow has been measured, i.e. a mercury model of the slag bath [71]; the measurements have been made by the use of a fibre-optical velocity probe [69].

The experiment was conducted in a container of radius $R = 30$ mm, depth $L = R$, and for currents up to 1500 A. The radius of a smaller copper electrode (embedded in the top cover) was $r_0 = 6$ mm, the copper bottom of the bath served as the second electrode. The v_z component of velocity was measured as a function of the nondimensional coordinates $r = r^*/L$, $z = z^*/L$ (the coordinate z was measured from the small electrode) and the parameter $S = \mu_0 I^2 / 4\pi^2 \rho v^2$:

$$v_z = \alpha \frac{v}{L} \varphi(S, r, z), \qquad (8.1.1)$$

where the nondimensional coefficient α is a function of the geometric simplexes r_0/L, R/L and the ratio σ_1/σ_2 of the liquid medium conductivity (σ_1) and the electrode material (σ_2).

The measured velocity on the symmetry axis $v_z(r = 0) = v_{z_0}$ (Fig. 8.3a) reveals that the distribution of $v_{z_0}(z)$ is changing approximately up to $I = 200$ A, and the location of maximum velocity v_{z_0} moves to the centre of the container $z = 0.5$. With a further current increase, the shape of the distribution stays constant, and the location of the velocity maximum remains at $z = 0.5$; the magnitude of velocity grows at a linear rate with the current (Fig. 8.3b). Moreover, for $I > 200$ A the values of v_z normalized by I are invariant over the whole bath, i.e. the velocity field becomes self-similar. These observations permit to represent (8.1.1) in the form

$$v_z = \alpha \frac{v}{L} \sqrt{S} f(r, z) = \alpha \frac{I}{2\pi L} \sqrt{\frac{\mu_0}{\rho}} f(r, z), \qquad (8.1.2)$$

where the coefficient at $f(r, z)$ may be identified with the maximum velocity magnitude

$$v_{max} = \alpha \frac{I}{2\pi L} \sqrt{\frac{\mu_0}{\rho}}, \qquad (8.1.3)$$

and the form of $v_z(S)$ in (8.1.2) indicates the nonlinear flow regime for $I > 200$ A. This regime is interesting due to the velocity field independence of the melt viscosity, which is the most difficult to determine the quantity in natural electrometallurgical processes.

The parameter $S = 0.7 \times 10^7$ for the current $I = 200$ A in mercury ($\rho = 1.36 \times 10^4$ kg/m^3, $v = 1.15 \times 10^{-7}$ m^2/s), which is close to the lower limit of the nonlinear regime found in the numerical solution (cf. Section 4.7).

Fig. 8.3c demonstrates the self-similarity of the axial velocity normalized by

Experimental investigations of EVF and applications 315

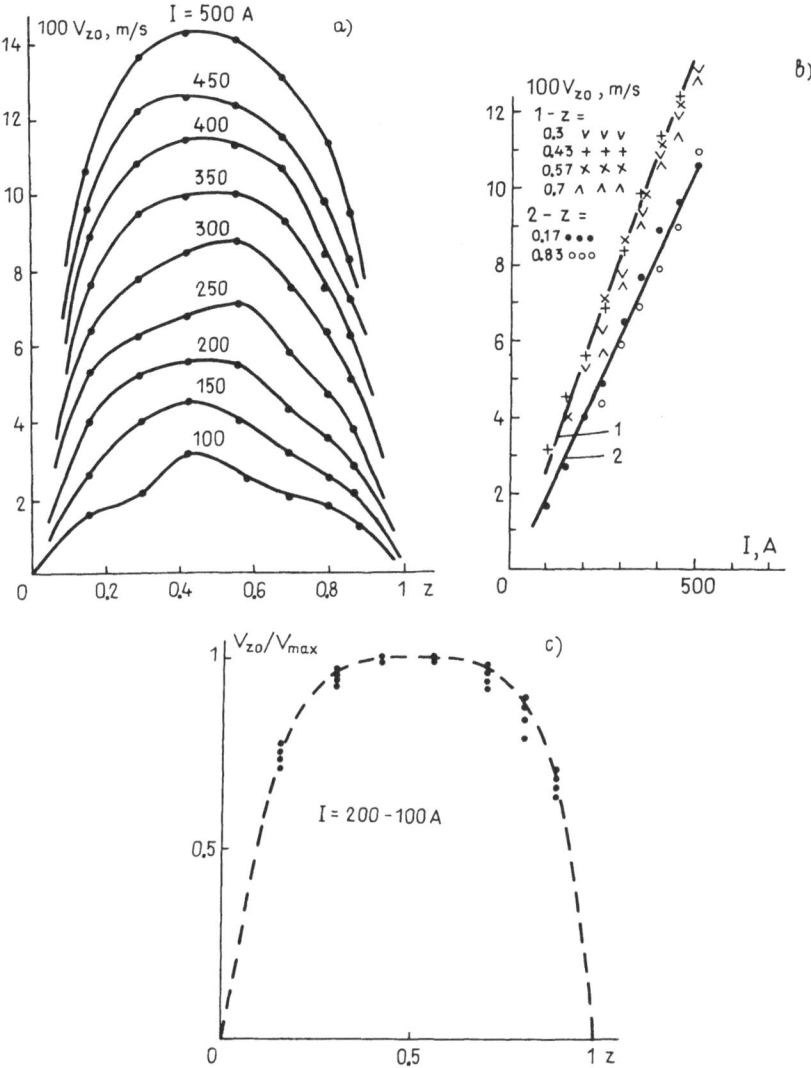

Fig. 8.3. Dependences of axial velocity: (a) $v_{z_0} = v_z\ (r = 0)$ along the z-coordinate for different magnitudes of total current I, (b) v_{z_0} vs. I for different z, (c) v_{z_0}/v_{max} vs. z for different I $(v_{max} = v_{z_0}\ (z = 0.5))$.

v_{max}. The velocity distribution is satisfactorily approximated by a parabola of the fourth order:

$$\frac{v_{z_0}}{v_{max}} = 1 - (2z - 1)^4 = f(0, z). \tag{8.1.4}$$

The value of coefficient α in (8.1.2), (8.1.3) according to the experimental data is 5.2 for $80 < I < 200$ A and 5.6 for $I > 200$ A; these are respectively

12 and 21% higher than the coefficient $\sqrt[3]{10^{3-5r_0/R}} = 4.64$ in the expression found numerically in Section 4.7 for the laminar flow with $r_0/R = 6/30 = 0.2$. There is also a correspondence in the position of the toroidal vortex centre. Thus, according to the numerical solution for $S = 10^7$ it is located in the point $z = 0.4$, $r = 0.625$, and by the experimental data in $z = 0.5$, $r = 0.66$.

Consider now the distribution of velocity v_z along the radius of the bath.

Usually experimental data for jet flows can be successfully represented in the coordinates $v_z(r)/v_{z_{max}} - r/r_{1/2}$, where $r_{1/2}$ is the coordinate of a point in a horizontal plane $z = $ const where the velocity magnitude is half of the maximum velocity magnitude in the same plane: $v_z(r_{1/2}) = \frac{1}{2} v_{z_{max}}$, $r_{1/2}$ being a function of z. This technique often succeeds in a single, universal velocity profile for different sections $z = $ const. In fact, the technique is applied to unlimited jets. In our case the jet penetrates the bounded volume with a dead end — the bottom of the bath; nevertheless, the technique is found to be useful.

The measured velocity is represented in Fig. 8.4 in the new coordinates. As is evident, the velocity profile is indeed universal in the central part of the bath where $v_z < 0$ (for small velocity magnitudes it was impossible to obtain reliable results because of insufficient sensitivity of the probe; for the same reason it was also impossible to measure the positive values of velocity at the side wall, since the velocity magnitudes for the specific flow configuration there were substantially lower than at the axis). With a good accuracy the profile is approximated by the expression ($\beta = $ const):

$$\frac{v_z}{v_{z_0}} = [1 + \beta(r/r_{1/2})^2]^{-2},$$

or, in light of (8.1.4),

$$v_z = v_{max} f(0, z)[1 + \beta(r/r_{1/2})^2]^{-2}. \qquad (8.1.5)$$

Note that the velocity profile (8.1.5) coincides with the velocity profile of axisymmetric turbulent jet in an unbounded space [58]. The only difference is for the behaviour of $v_{z_0}(z)$: in the jet $v_z \sim z^{-1}$, yet in the electrically induced flow it is determined by the formula (8.1.4). It is interesting that the experimental curve, expressing the position of the jet's half-width $r_{1/2} = r^*_{1/2}/L$ can be approximated by an expression similar to (8.1.4):

$$r_{1/2} = \gamma f(0, z) = \gamma[1 - (2z - 1)^4],$$
$$\gamma = \tfrac{1}{3}. \qquad (8.1.6)$$

This fact enables us to evaluate the velocity field in the axial region. Indeed, on substituting (8.1.6) in (8.1.5), we obtain

$$v_z = v_{max} f(0, z) \left[1 + k\left(\frac{r}{f(0, z)}\right)^2\right]^{-2}, \qquad (8.1.7)$$

where $k = \beta/\gamma^2 \approx 4$. Applying the relations (1.5.2) between the velocity and stream function, and integrating the first of these along r from 0 to 1, we obtain

Experimental investigations of EVF and applications

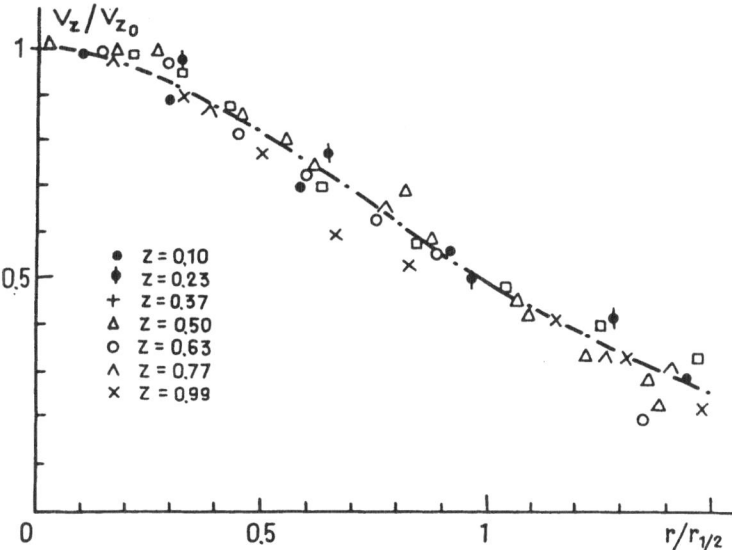

Fig. 8.4. Dependence of the axial velocity normalized by v_{z_0}, at the nondimensional radius $r/r_{1/2}$ within the different sections for $I = 10^3$ A.

an approximate expression for the dimensional stream function in the region of downflowing jet in the form:

$$\psi = v_{max} \frac{1}{2k} \frac{f^3(0, z)t}{1+t}, \qquad (8.1.8)$$

where $t = k(r/f(0, z))^2$. The second relation (1.5.2) gives the v_r component of velocity:

$$v_r = -\frac{1}{2} v_{max} f(0, z) f'(0, z) \sqrt{\frac{t}{k}} (1 + 3t)/(1 + t)^2. \qquad (8.1.9)$$

The expressions (8.1.7) and (8.1.9) describe the velocity field in the axial region.

As was mentioned above, reliable data were not obtained for the region of reversed up-flow. We could approximate the velocity field in the region in a following way. Suppose, in the region 3 of the reversed flow the v_z component of velocity is described by the second order parabola

$$\frac{v_{z_3}}{v_{max}} = Ar^2 + Br + C. \qquad (8.1.10)$$

One of the conditions for determining the constants A, B, C is the no slip condition at the side wall $v_{z_3}(r = 1) = 0$, then

$$A + B + C = 0. \qquad (8.1.11)$$

From the condition $v_{z_3} = 0$ we may find the respective radial distance $r_1 = C/A$. With the aim of evaluating the distance r_1, construct a tangent to the curve (8.1.7) (Fig. 8.4) for $r/r_{1/2} = 1$ (the latter coordinate is arbitrary; we assumed the value because this corresponds to the furthest reliable measurements). Then $r/f(0, z) = \frac{1}{3}$ in (8.1.7) and $v_{z_1}/v_{\max} = (9/13)^2 f$. By evaluating the derivative of (8.1.7) and matching the curve (8.1.7) and the parabola (8.1.10) with a straight line, we obtain a velocity profile in the matching region 2:

$$\frac{v_{z_2}}{v_{\max}} = -18 \left(\frac{6}{13}\right)^3 \left(r - \frac{f}{3}\right) + \left(\frac{9}{13}\right)^2 f, \qquad (8.1.12)$$

whence

$$r_1 = \frac{C}{A} = \frac{29}{48} f. \qquad (8.1.13)$$

The third condition to determine the constants A, B, C is the zero integral flow rate across the plane $z = $ const:

$$\int_0^{f/3} v_{z_1} r \, dr + \int_{f/3}^{r_1} v_{z_2} r \, dr + \int_{r_1}^{1} v_{z_3} r \, dr = 0,$$

where v_{z_1} is determined by the equation (8.1.7). On evaluating the integrals, we have:

$$A \frac{1 - r_1^4}{4} + B \frac{1 - r_1^3}{3} + C \frac{1 - r_1^2}{2} = -\frac{1}{15} f^3. \qquad (8.1.14)$$

The solution of the set (8.1.11), (8.1.13), (8.1.14) yields the reversed flow velocity profile

$$\frac{v_{z_3}}{v_{\max}} = \frac{4f^3}{5} \frac{r^2 - (1 + r_1)r + r_1}{(1 - r_1)^3 (1 + r_1)}. \qquad (8.1.15)$$

The expression agrees satisfactorily with the numerical solution in Section 4.7. Thus, the coordinate $r_1 = 0.6f$ coincides in both cases, and also the position of the maximum reversed velocity is the same, $r = 0.5 + 0.3f$, yet the maximum magnitudes are $|v_{z_3}|_{\max} = 0.24 |v_{\max}|$ according to the numerical solution, and the estimated $|v_{z_3}|_{\max} = 0.31 |v_{\max}|$.

The stream function and velocity component v_r can be constructed from the expressions (8.1.12) and (8.1.15), i.e. to obtain the formulae corresponding to (8.1.8) and (8.1.9).

The semi-analytical velocity field description may be applied to predict, for instance, the temperature and concentration fields. Of course, this is valid only for the specified bath geometry and position of the melting electrode.

The jet flow, originating at the melted electrode, can affect significantly the shape of the slag—liquid metal interface. To estimate the effect, consider the measured pressure distribution at the bottom of the container with $2R = 51.6$ mm, $2r_0 = 5$ mm, the height of liquid metal is 64 mm. In these experiments the interelectrode distance is varied, and the pressure is measured along the radius of the bottom electrode for different current magnitudes up to 1200 A [12].

Fig. 8.5 presents a typical pressure distribution for the fixed current $I = 1200$ A and various interelectrode distances. The maximum pressure on the axis of the jet is proportional to the square of current, and the pressure related to I^2 depends only on the interelectrode distance z (Fig. 8.5a).

Moreover, the dimensional pressure is found to depend only on the current and the typical size of the container, and it does not depend on the viscosity and density of the melt.

Indeed, the dimensional pressure may be represented in the form $p^* = p_0 p(S, z, r)$, where p_0 is the typical scale equal to $p_0 = \rho v_0^2 = \rho v^2/R^2$ if, as previously, the velocity scale is $v_0 = v/R$; p is a function determined experimentally, and the coordinates are related to the bath radius R. The experiment shows the function p is proportional to S. Whence

$$p \sim \frac{\rho v^2}{R^2} \frac{\mu_0 I^2}{4\pi^2 \rho v^2} \sim \frac{\mu_0 I^2}{R^2}.$$

In other words, the quantity $pR^2/\mu_0 I^2$ must depend only on the bath's geometric ratios and the coordinates. The conclusion agrees with the results of experiments conducted with the two media (mercury and gallium) and in geometrically similar baths, yet of different scales (Fig. 8.6) [49].

Aiming at estimation of the interface shape, we shall derive the formulae approximating the experimental pressure dependence on z, r, and I. Fig. 8.7

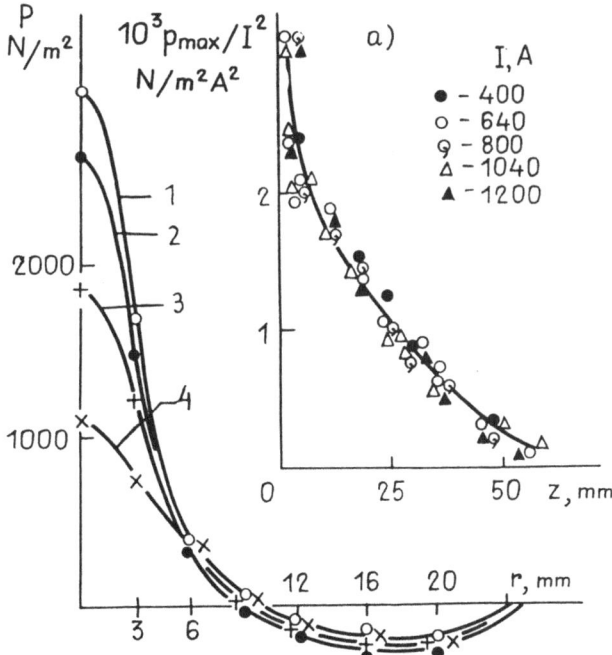

Fig. 8.5. Distribution of pressure along the radius of the container for the total current $I = 1200$ A. The distance between the electrodes z in mm is: (1) 5, (2) 9, (3) 18, (4) 24. (a) The pressure on the axis related to the square of current vs. the distance between the electrodes.

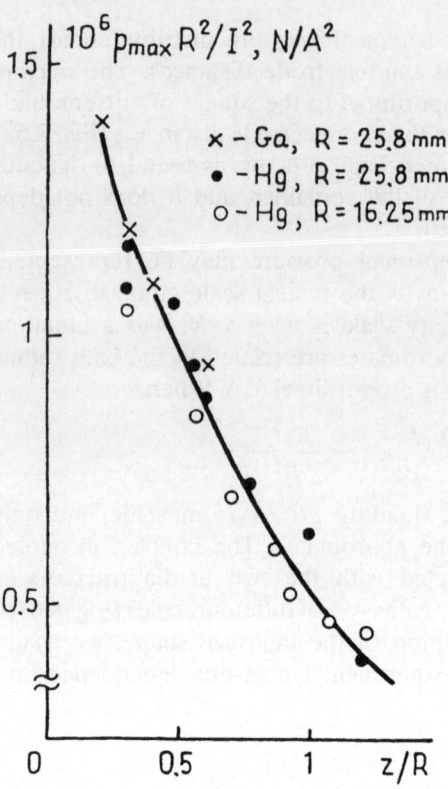

Fig. 8.6. The dependence of pressure, normalized by I^2/R^2, at the symmetry axis on the interelectrode distance in the experiments with different media and dimensions of the container.

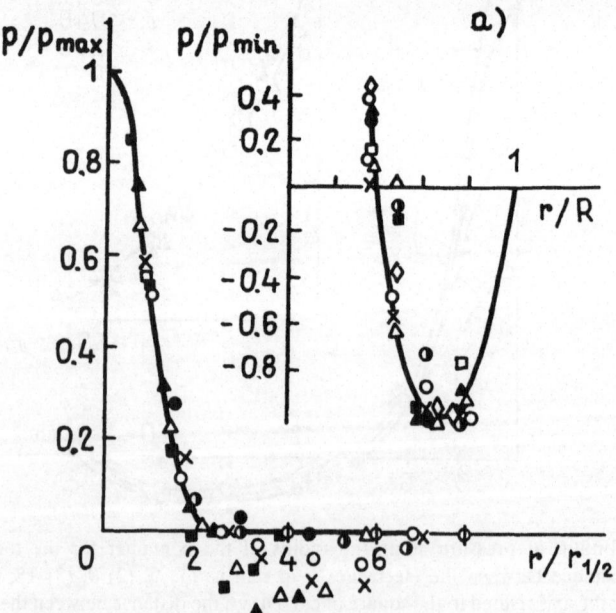

Fig. 8.7. The nondimensional pressure distribution for the total current I = 1200 A. (The symbols are defined in Fig. 8.5). (a) The zone of negative pressure.

Experimental investigations of EVF and applications 321

presents the pressure distribution, related to the axial magnitude, along the radius of the container. As is evident, in the zone of positive values the pressure distribution is close to exponential with the exponent $-(r/r_{1/2})^2/\sqrt{2}$, where $r_{1/2}$ is the distance from the axis to a point where the pressure is half the maximum one on the axis (this characteristic is related to the jet width). According to the measurements, it can be approximated by $r_{1/2} = 0.2\sqrt[4]{z}$ [12].

A general dependence in the zone of positive pressure values ($r \leqslant 0.4$) can be represented in the form

$$p = p_{\max} \exp\left[-\frac{1}{\sqrt{2}}\left(\frac{r}{r_{1/2}}\right)^2\right], \tag{8.1.16}$$

where

$$p_{\max} = \frac{10}{4\pi}\mu_0 I^2 R^{-2} z^{-1/5} \left(\frac{N}{m^2}\right) \quad \text{for} \quad 0 \leqslant z \leqslant 0.55;$$

$$p_{\max} = \frac{6}{4\pi}\mu_0 I^2 R^{-2} z^{-1} \left(\frac{N}{m^2}\right) \quad \text{for} \quad 0.55 \leqslant z \leqslant 1.25;$$

$$p_{\max} = \frac{9.3}{4\pi}\mu_0 I^2 R^{-2} z^{-3} \left(\frac{N}{m^2}\right) \quad \text{for} \quad 1.25 \leqslant z.$$

In the zone of negative pressure values ($r \geqslant 0.4$) the distribution is approximated by the formula:

$$p = -p_{\min}\left[1 - \left(\frac{r - 0.7}{0.3}\right)^2\right]$$

$$= \frac{2.6}{4\pi}\mu_0 I^2 R^{-2} \frac{\sqrt{z}}{z^3 + 1}\left[1 - \left(\frac{r - 0.7}{0.3}\right)^2\right]. \tag{8.1.17}$$

The formulae (8.1.16) and (8.1.17) may be applied to the estimation of the deformation of the slag—liquid metal interface under the action of a slag jet. The estimated interface profiles for titanium ($\rho_{Ti} = 3.8 \times 10^3$ kg/m^3) welded beneath the flux of density $\rho_f = 2.2 \times 10^3$ kg/m^3 are presented in Fig. 8.8. These profiles are calculated by the expression

$$\rho_{Ti} g(z - z_0) = p,$$

where z_0 is an initial distance from the electrode to the unperturbed interface, p is the pressure in the flux determined by the expressions (8.1.16) and (8.1.17).

Consequently, we are able to explain the effects accompanying a speeded-up welding regime: on increasing the rate of electrode advancement, which is equivalent to an increase of welding current and a decrease of the initial interelectrode distance, the depression of the interface grows, the slag of high temperature from the vicinity of melting electrode is convected by the electri-

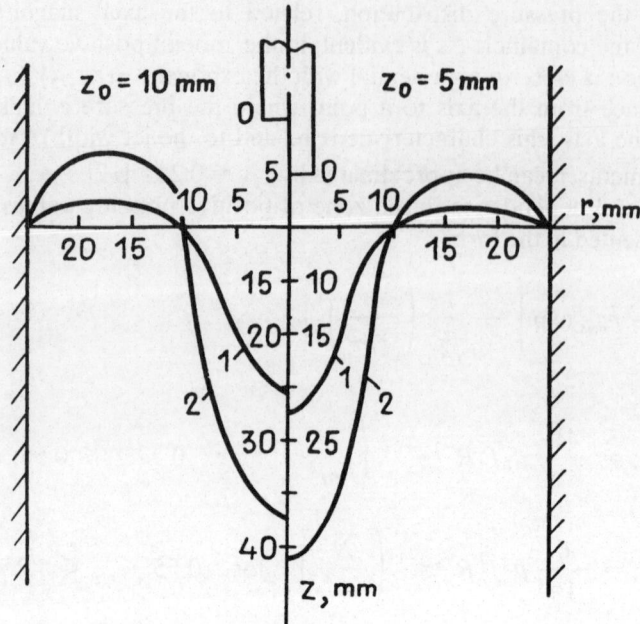

Fig. 8.8. The estimated profiles of the interface flux-liquid metal for (1) $I = 840$ A, (2) 1600 A.

cally induced flow to the depth of slag, causing a deep axial melting within the weld. The resulting crystallization front approaches the axis radially, even enclosing the slag crater, if it is sufficiently deep.

The electroslag welding conditions are quite diverse. Thus, the welding current may be direct or alternating, one or two wire electrodes, or a bar electrode may be used. The weld shapes are also divergent, e.g. cylindrical, rectangular, etc. Apart from these conditions, the electric current may pass partly from the lateral surface of electrode, and it may penetrate the crystallizer wall.

The extensive experimental data including flow visualization and pressure distributions at the interface are discussed in [11, 20, 22, 39, 59, 60]. Some of the data, corresponding to the case of multiple-electrode current supply, will be discussed in Section 8.4. Here it is appropriate to emphasize that the choice of current supply type is a tool for controlling the flow in the molten pool and, thereby, the heat and mass transfer and the crystallization of the weld.

An important method of active control uses an external magnetic field, particularly an axial one. The axial magnetic field, by means of the differential rotation mechanism, decreases the intensity of the jet propagating from the small electrode, and for a sufficiently high magnitude of the magnetic field the flow direction in the container is reversed (see Section 4.6). The respective pressure change at the bottom of the container is shown in Fig. 8.9 [39]. Experimental welds confirm that the change in the pressure distribution affects the shape of metal bath (Fig. 8.10) [39]. The field of a solenoid can be adjusted to minimize the deformation of the metal bath surface (Fig. 8.10c). Yet if the

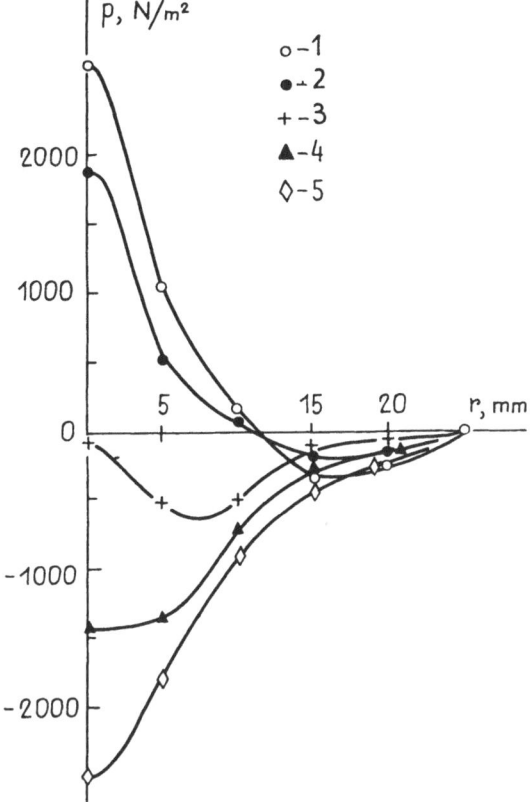

Fig. 8.9. Distribution of pressure along the radius for $z = 5$ mm, $I = 1200$ A for different magnitudes of the axial magnetic field B_z: (1) 0, (2) 6.4, (3) 9.5, (4) 15.8, (5) 19 mT.

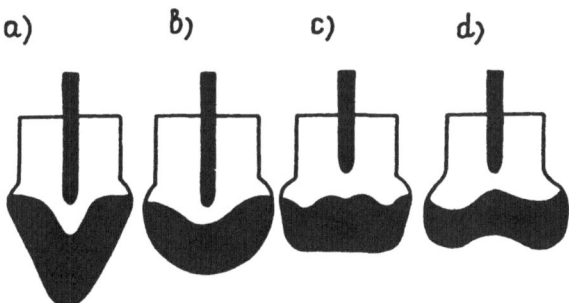

Fig. 8.10. The shape of metal pool in the axial-radial magnetic field for the welding current 800 A and the currents in the solenoid: (a) 0, (b) 80, (c) 120, (d) 160 A.

field magnitude is too high, an upward expansion of the interface (Fig. 8.10d) may replace the depression, and the process may be interrupted due to a short circuit. Note that a rotation in the weld pool may also be generated by the magnetic field of the current-supplying wires, ferromagnetic masses close to the welding zone, and even the terrestrial magnetic field, as was shown in Section 3.6, 3.7, can induce the growth of rotation.

8.2. Electroslag remelting

The electroslag remelting process is basically similar to the electroslag welding — only the purpose of the process is different, viz., electroslag remelting serves to produce a high quality metal ingot, also of a quite complex form. The term 'high quality' means either the quality of metal and that of its surface, which enables us to minimize the necessity for a following finish-machining. However the two technologies differ by the scale factor, which substantially affects the hydrodynamics of the melt; the cross-sectional size of a weld is usually 50–100 mm, and the size of an ingot produced by electroslag remelting may be up to 2 m. Equally a typical current magnitude in the remelting process is a few tens of kA. Consider now some consequences of the scale factor on the melt hydrodynamics.

First of all, the filling coefficient of crystallizer $k = r_0/R$ (r_0 = radius of melting electrode, R = radius of ingot) is usually close to unity in the remelting process, i.e., in contrast to the electroslag welding process, the melting electrode takes up a significant part of the slag pool's cross section. Obviously, the degree of electric current nonuniformity in the slag is lower, the greater is the relative diameter of the electrode. Also, the slower the electrically induced flow, then the less is its effect on the shape of the slag–liquid metal interface.

With the aim of estimating the melt velocities and pressure for electrodes of different radius, we shall apply the following method. A scalar product of the steady equation (1.1.1) by dl is integrated along the closed contour ℓ of a meridional section, e.g., the section shown in Fig. 4.14. The result of integration can be represented in the form:

$$\frac{v}{R} v_0 a = \frac{\mu_0 I^2}{\rho R^2} M, \qquad (8.2.1)$$

where v_0 is a velocity scale, a = nondimensional integral friction along the walls, $M = \oint_\ell \mathbf{f}_e \cdot \mathrm{d}\mathbf{l} = \int_S \mathrm{curl}\, \mathbf{f}_e \cdot \mathrm{d}\mathbf{S}$ is the normal component of the nondimensional integral curl of electromagnetic force in the meridional section S. Assuming a uniform current density on the surfaces of electrodes, it follows that $M = (1 - k^2)/2\pi^2 k^2$. Equation (8.2.1) can also be rewritten in the form

$$v_0 a = \frac{v}{R} SM,$$

showing that the flow intensity and the respective friction losses are determined by the integral curl of electromagnetic force, which, in turn, depends on the square of total current and the expansion of electric current ($k = r_0/R$).

Experimental investigations of EVF and applications 325

In the nonlinear regime $v_0 \sim \sqrt{S\nu/R}$, and from the above estimates

$$v_0 \sim \frac{\nu}{R}\sqrt{SM} = \frac{I}{2\pi R}\sqrt{\frac{\mu_0}{\rho}}\sqrt{\frac{1-k^2}{2\pi^2 k^2}}. \tag{8.2.2}$$

From this it follows, first, that the nondimensional friction α grows as $\sqrt{SM}$, which is equivalent to a decrease of average boundary layer thickness $\delta \sim (SM)^{-1/2}$. Second, from $p_0 = \rho v_0^2$ and (8.2.2) we could obtain an estimate of pressure:

$$p_0 \sim \frac{\rho v^2}{R^2} SM = \frac{\mu_0 I^2}{R^2} \frac{1-k^2}{2\pi^2 k^2}, \tag{8.2.3}$$

which is more elaborate than the expression in Section 8.1.

The measurements of pressure distribution for different electrode radii, described in the previous section, generally agree with the dependences (8.1.16) and (8.1.17); however, the maximum pressure is found to depend substantially on the parameter k in agreement with (8.2.3) (Fig. 8.11). The experimental data

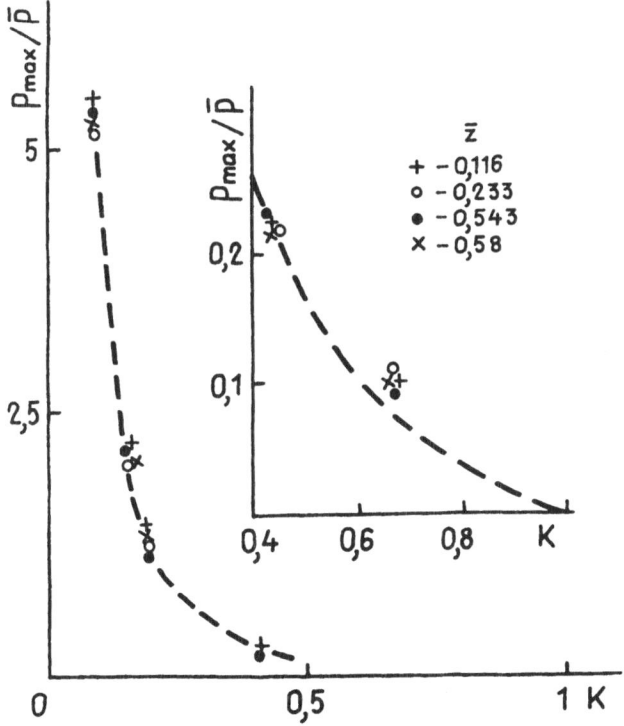

Fig. 8.11. Dependence of the normalized pressure at the symmetry axis on $k = r_0/R$. The dependence $(1 - k^2)/2\pi^2 k^2$ is shown by the dashed line.

could be approximated by the expression (8.1.16) where the value of p_{max} have to be multiplied by $0.185(1 - k^2)/2\pi^2 k^2$, for example,

$$p_{max} = \frac{10}{4\pi} \mu_0 I^2 R^{-2} z^{-1/5} 0.185 \frac{1-k^2}{2\pi^2 k^2} = p \frac{1-k^2}{2\pi^2 k^2}$$

for $0 \leq z \leq 0.55$, etc. The measured values of p_{max}/p vs. k are presented in Fig. 8.11.

The melt flow velocities for different radii of the melting electrode may be estimated by the use of the computational results presented in Section 4.7. According to these results, and also to the experimental data in Section 8.1, the formula (8.1.3) can be rewritten as

$$v_{max} = 1.2 \sqrt[3]{10^{3-5r_0/R}} \frac{I}{2\pi L} \sqrt{\frac{\mu_0}{\rho}}.$$

However, the velocity field changes significantly with the growth of the electrode radius, as is evident in Figs. 4.22 and 4.23. Hence, the results of Section 8.1 concerning the velocity field are applicable only for $r_0/R = 0.2$.

For a sufficiently prolonged remelting process the melting electrode may assume a variety of forms, most often a conical one, as is shown in Fig. 8.12. An experimental model for the flow visualization (the maximum axial velocity was also measured) was realized in a hemicylindrical container of diameter 140 mm with a conical electrode of base diameter either 108 mm, or 77 mm [54]. The second electrode was the opposite end wall of the container.

A special-purpose numerical solution of the velocity field was constructed for these geometrical conditions by Choudhary and Szekely [44]; the aim of the solution was a comparison with the experimental data, and, thus, confirmation of the validity of the computational method. A characteristic feature of the computational method was the turbulent flow equations with the effective viscosity computed according to the $K-\varepsilon$ model [41]:

$$v_{ef} = v + v_T = v + C \frac{K^2}{\varepsilon}.$$

Here K is the specific turbulent kinetic energy, ε = dissipation rate of the energy, C = an empirical constant. The quantities K and ε were determined from the corresponding transport equations [41], yet some information might be obtained in the following way. As was shown in [21], a representative scale of K was determined by $K_0 = (\ell/L)^2 v_0^2$, where ℓ was the macroscale of turbulence characterizing the scale of energy containing vortices. Since in the nonlinear flow regime $v_0 \sim \sqrt{\mu_0/\rho}(I/L)$, then $K_0 \sim I^2$. And if taking into account the fact that $\ell = cK^{3/2}/\varepsilon$, then $\varepsilon_0 \sim I^3$.

The numerical simulation confirms the quadratic dependence for K and the cubic one for ε on the current (Figs. 8.13, 8.14). In turn, these dependences determine a linear growth of the turbulent viscosity $v_T = CK^2/\varepsilon \sim I$. For high currents v_T may increase quite substantially. Thus, for $I = 10^3$ A $v_T \approx 175$ m²/s (Fig. 8.15).

The comparison of the experimental and numerical results for the maximum

Experimental investigations of EVF and applications 327

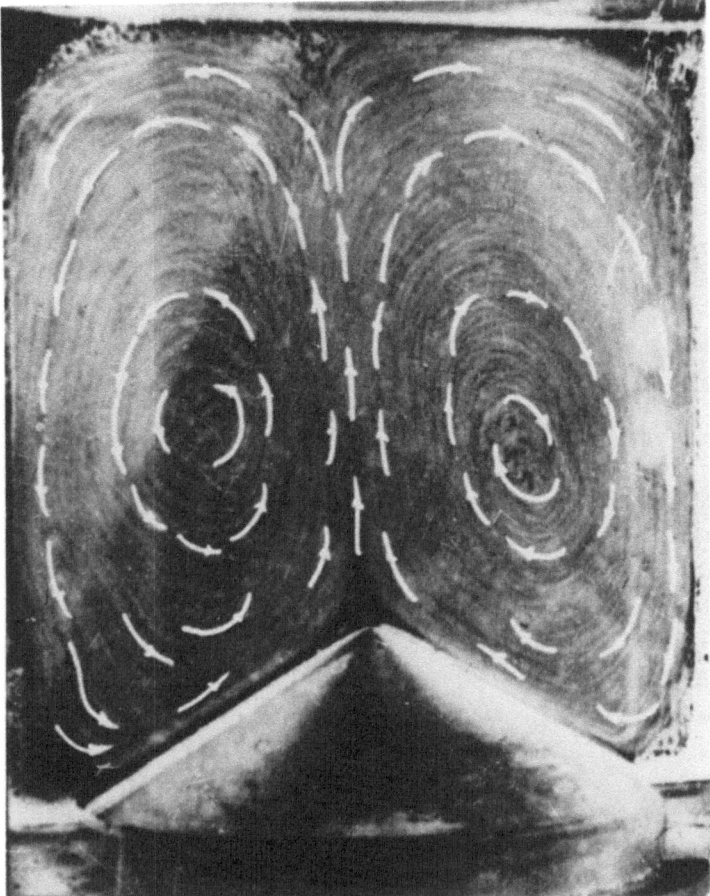

Fig. 8.12. A photograph of the flow visible on the surface of mercury in a container with a conical electrode.

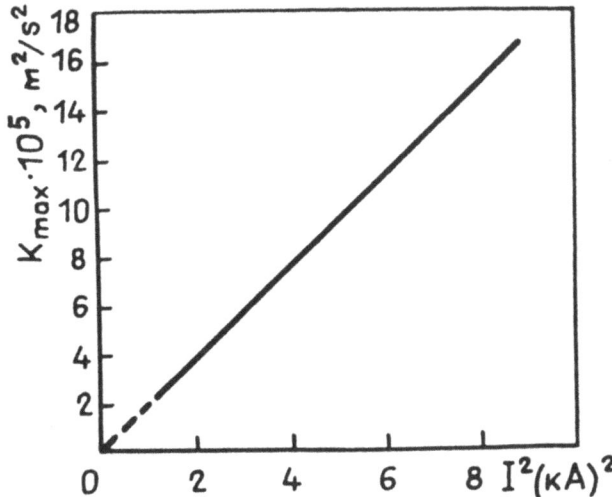

Fig. 8.13. The computed specific turbulent kinetic energy vs. square of the electric current I.

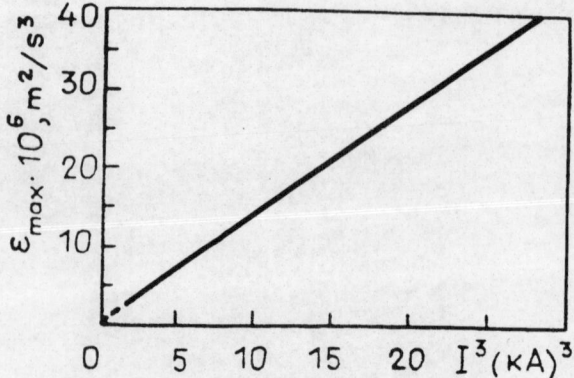

Fig. 8.14. Rate of turbulent energy dissipation vs. I^3.

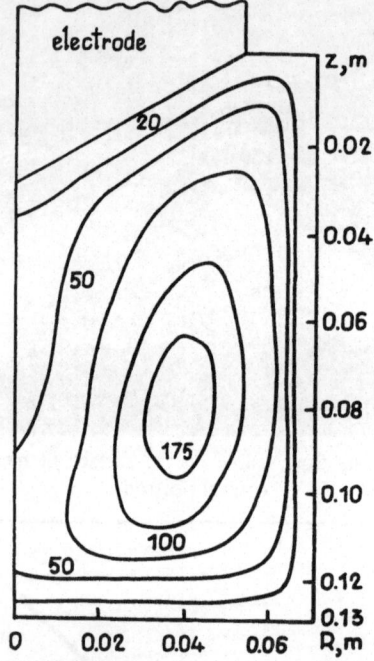

Fig. 8.15. The distribution of turbulent viscosity v_T/v over the meridional section of the container for $I = 10^3$ A.

axial velocity (Fig. 8.16) shows the discrepancy not exceeding 25 per cent, while the laminar solution in Section 8.1 leads to a 150 per cent error. However, the difference is possibly related to the measurement error, since the measurements [54] were made, in fact, of an average velocity of graphite particles traveling over a 10 mm interval on the surface of mercury. Hence, the agreement with the mathematical model is quite satisfactory.

The second consequence of the scale factor is a dependence of the melt hydrodynamics on the kind of current. In some cases of electroslag remelting a

Experimental investigations of EVF and applications 329

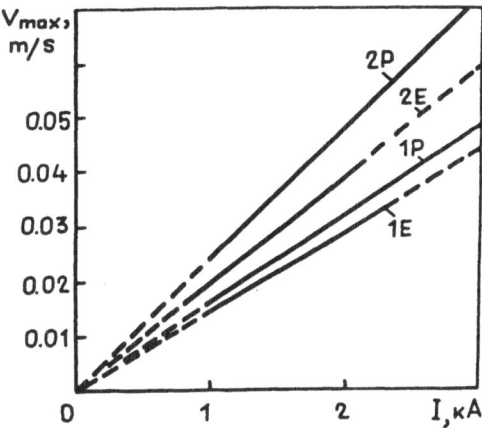

Fig. 8.16. The dependences of maximum axial velocity vs. the total current magnitude: P — obtained by the mathematical modeling, E — experimental. The diameters of electrodes are 108 mm (1) and 77 mm (2).

direct current is used [24], yet more often an alternating current is applied in the process. Thence, in a large scale current-carrying volume a skin-effect is significant even at industrial frequencies. The skin-effect contributes to a non-uniformity of the current distribution, and the hydrodynamics are mostly affected in the region where the electrical conductivity or the cross section of fluid volume change abruptly.

Consider an example in which an infinitely long cylindrical conductor has a step-like change in conductivity at $z = 0$ (Fig. 8.17), i.e. for $z > 0$ $\sigma = \sigma_1$ (slag), for $z < 0$ $\sigma = \sigma_2$ (metal) and $\sigma_1 < \sigma_2$. Far from the conductivity step the current density is uniform along z and nonuniform along r with a skin-layer of depth $\delta = \sqrt{2/w\mu_0\sigma}$ depending on the conductivity and frequency w. At the step of conductivity the skin-layer is transformed to a different thickness, i.e. in the vicinity of the step an additional nonuniformity of the electric current occurs, and the respective electromagnetic force is induced. An analysis shows that the integral curl of electromagnetic force $M = \int_V (\text{curl } \mathbf{f}_e)_\varphi \, dV$ is substantially higher in magnitude in the region of higher conductivity ($M_2 > M_1$ if $\sigma_2 > \sigma_1$) and depends on the ratio of conductivities (Fig. 8.18). After the ratio exceeds $\sigma_2/\sigma_1 = 100$, the integral curl grows weakly with σ_2/σ_1.

For the electroslag remelting process this results in an electrically induced flow driven in the liquid metal close to the interface (the flow direction is indicated in Fig. 8.17); the flow is generated due to a difference in conductivities of slag and liquid metal ($\sigma_2/\sigma_1 = 10^2 - 10^4$).

The interface in the electroslag processes is not a plane. As was noted previously, a jet emanating at the melting electrode causes a depression of the interface at the symmetry axis. This condition could affect the flow at the step of conductivity.

The effect of interface curvature can easily be analysed assuming the currents to be direct. As is shown in [29], a typical curvature of the electric current lines

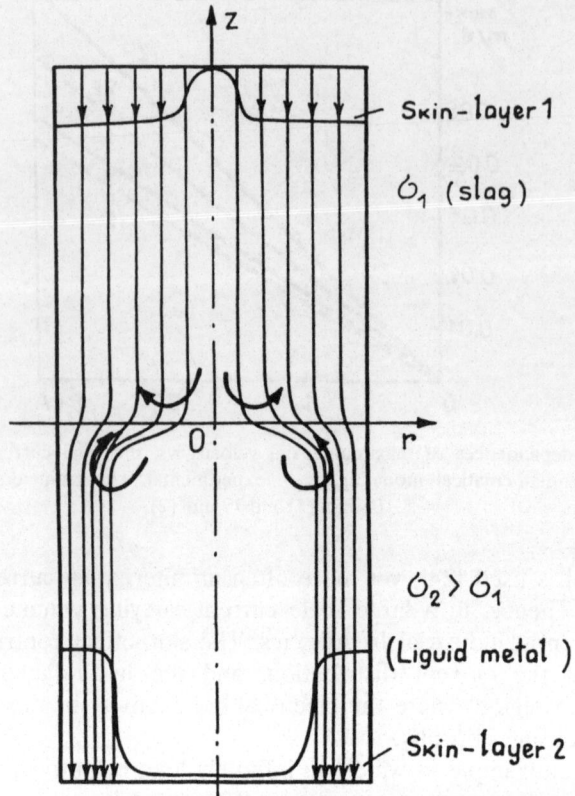

Fig. 8.17. Schematic illustration of the flow generation in the fluid with an abrupt change in the electrical conductivity.

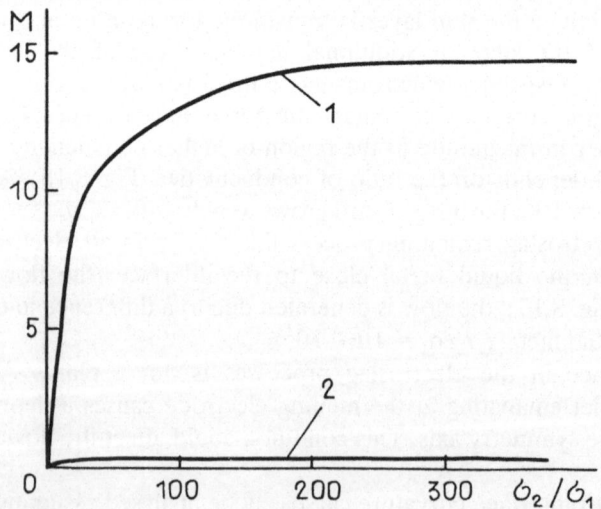

Fig. 8.18. The integral curl of electromagnetic force vs. σ_2/σ_1: (1) $M = M_2$, (2) $M = M_1$.

Experimental investigations of EVF and applications 331

and the corresponding electrically induced flow at the deformed interface can be predicted applying the boundary conditions for the current passing across the interface of two media with the different conductivities. Omitting details of the analysis, the final results are: in the case of a concave interface the electric current lines are diverging in the domain occupied by the fluid of lower conductivity (Fig. 8.19a), and the lines are converging at the convex side of interface (Fig. 8.19b). The respective electrically induced flows should take the form indicated in Figs. 8.19a, b (the dashed lines). The flow visualization confirms the predicted configurations. Note that the electrically induced flows in both cases promote the deformation of the moving interface.

It is easy to notice also a similarity of the current deformation at the concave interface and for the alternating current in the medium of lower conductivity σ_1 at the plane interface with the medium of higher conductivity $\sigma_2 > \sigma_1$. Thereby, the alternating current could intensify the electrically induced flow in the slag bath at the concave interface. In a hypothetical case of a convex interface both the effects should act in opposition.

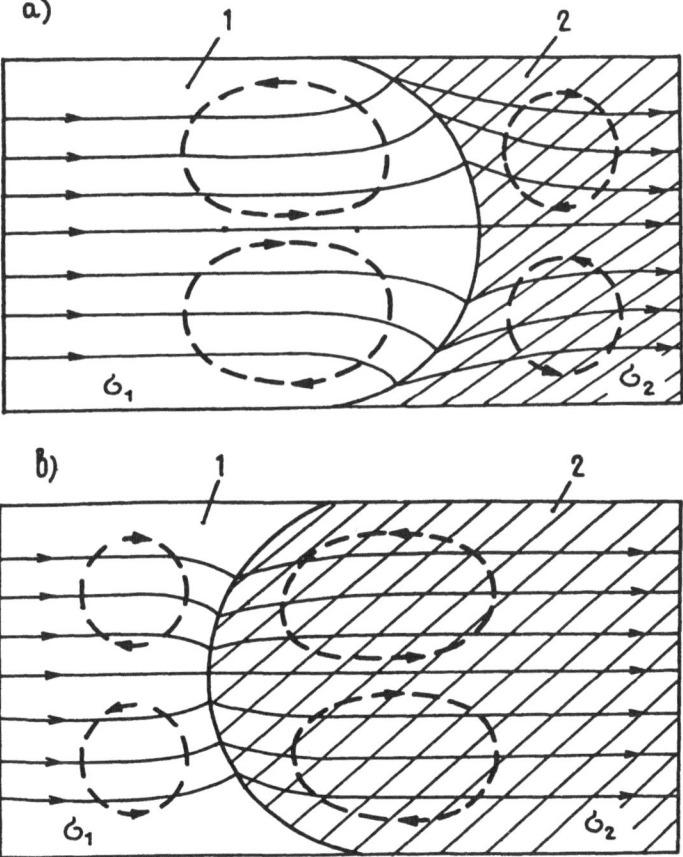

Fig. 8.19. Scheme of EVF generation; (a) in the case of concave electrode surface, (b) in the case of convex surface.

The skin-effect, as a consequence of the scale factor, could lead to local electrically induced vortices at the melting electrode. Consider several examples based on the computed alternating current distributions and the expected stream lines. For a small immersion depth of the electrode (Figs. 8.20a, b) (the depth determines energetic efficiency of the process) a thin skin-layer in the electrode is transformed to a thick layer at the walls of the slag bath. In the case of a small filling coefficient (see Fig. 8.20a) the resulting slag jet is directed to the crystallizer wall, and the closing flow will melt the sharp edge of the electrode. In the case of a large filling coefficient (see Fig. 8.20b) the slag jet from the edge of the electrode is directed to the centre of the slag bath, and the closing flow melts the central part of the electrode, shaping it into a convex form (the form is well known [24]). Finally, in the case of a large depth immersion (Fig. 8.20c) the main electrically induced flow is driven at the lateral wall of the slag bath. Note, also, that a similar electric current distribution is generated in the case when the current leaks to the crystallizer due to a melting of the insulating slag rim.

Thus, the variety of operating conditions in electroslag remelting leads to a variety of electrically induced flows. Hence, a flexible universal computational model is necessary in order to predict the hydrodynamics in the slag and liquid

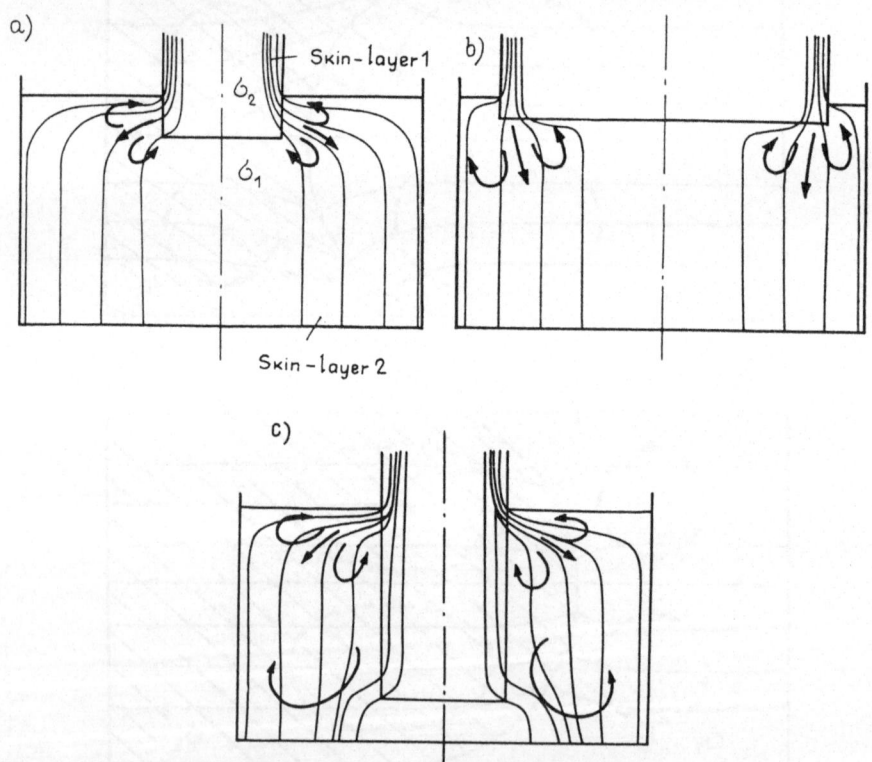

Fig. 8.20. Schemes for the flow generation: (a) the electroslag welding with alternating current; (b) electroslag remelting, (c) with an immersed electrode.

metal baths for different conditions and regimes, which would permit us to reduce the volume of laborious experimental investigations and would make these more purposeful. Certain steps in this direction have already been made (see this above, and Sections 7.5, 7.6).

8.3. Electric-arc furnaces

Electric-arc furnaces may be classified, according to their applications, into two types. The first is intended for industrial scrap remelting. The electrode, between which and the melt surface the arc burns, in this type of a furnace is usually unconsumable (e.g., made of graphite). Furnaces of the second type are designed for a metallic electrode remelting with the purpose of obtaining a high quality ingot. Here the arc is burning either in vacuum, or within an inert gas environment [26].

A common property of the arc furnaces is thermal energy release within the arc, which is transferred to the melt across the liquid metal surface. The central region of the metal surface, due to high temperature of the arc ($\sim 2500°C$), is heated to a much higher temperature than the periphery. This means for the melt hydrodynamics, that the significant radial temperature gradient could activate the thermal convection mechanism (Section 7.6), which drives the metal on the surface from the arc region to the periphery. If the length of the arc is sufficient, its shape is expanding at the melt's surface. This results in an electrically induced flow within the arc itself, and this is directed from the electrode to the melt. A consequence of this is the arc jet momentum transfer at the bath surface [27], which is manifested by a deformation of the liquid metal surface in the form of a central depression and by the presence of a radial gas flow above the metal surface, entraining the melt. The momentum transfer from the gas flow to the liquid metal drives a circulation in the melt in the same sense as the thermal convection. As is shown by the numerical solution simulating the first-type arc furnace with a relatively shallow bath [65], the inclusion of the arc momentum transfer increases the melt velocity at the surface by 3–4 times as compared to the pure thermal convection.

A circulation of the opposite sense may be driven due to the electric current's divergence from the arc spot over the melt volume. In addition, an electrically induced vortex is driven in the vicinity of the bottom electrode; the intensity of the vortex depends on the relative size of the electrode. If we note another mechanism driving the circulation due to gas bubbles in the melt, the complex nature of the problem determining the integral melt circulation becomes evident, and the various designs of furnace do not offer much chance to develop a unifying hydrodynamic model of the melt motion in the bath of a furnace.

In a vacuum-arc remelting process the melt is contained, instead of in a bath with lined walls, between the walls of a water-cooled metallic crystallizer. In this case, in view of the high arc temperature, the depth of the melting metal, i.e. the depth of the liquid metal pool, may be substantially higher than in the electroslag remelting process. These specific conditions raise new problems concerning the electrically induced flows in a container with a great depth relative to its diameter. An insight into the nature of the flow may be made in the following qualitative experiment (Fig. 8.21).

In a hemicylindrical vessel the position of a solid electrode B may be varied relative to the electrode A with a current-carrying section of diameter 10 mm. When the electrode B is closer to A than the position I–I, the usual electrically induced flow is observed, which takes the form of two vortex rings on the mercury surface shown by arrows in Fig. 8.21. When the electrode B is in the position II–II, the flow is divided into two regions: up to the section I–I the vortex flow remains intense and almost symmetric, yet in the region between the sections I–I and II–II a single weak vortex is observed, which arises consecutively at each side of the region to the right or left of the symmetry axis. Evidently, this could be related to an oscillation of the jet's axis at a sufficiently high distance from its origin (from the electrode A). What is most noticeable in this experiment is the limited distance of the jet penetration, which is obviously related to its generation conditions: the jet is driven by the electromagnetic force, the maximum curl of which is close to the small electrode where the electric current lines are diverging. The vortical force drives the fluid from the electrode, yet, in the face of continuity, the fluid is forced to return as soon as possible along the container's walls to the flow origin, i.e. the conditions force the jet to turn back along the walls, thus limiting the penetration depth. This is a principal difference between a jet issuing from the orifice in the wall and the electrically induced jet flow. When the electrode B is placed in the position III–III, an additional region arises with a weak flow similar to the flow in the first region.

Another important aspect of the vacuum-arc remelting technique is the effect of a longitudinal magnetic field, which additionally drives the metal in the bath in a swirling motion, which is often applied in practice with the aim of controlling the process. However, the application of a longitudinal magnetic field is found to be unjustified in many instances. Thus, when a liquation-sensitive steel is remelted (e.g. a carbonaceous steel), the centrifugal separation of impurities

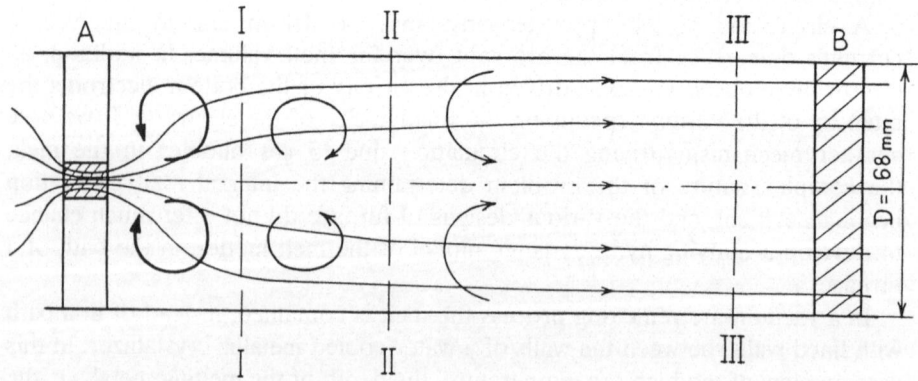

Fig. 8.21. Schematic representation of the experimental model.

Experimental investigations of EVF and applications

leads to a defect of spot-like liquation — the spots are formed in the section of the ingot with an increased concentration of carbon, sulphur, silicon, etc. [50]. This is also promoted by the secondary flow due to the differential rotation and opposite to the electrically induced flow; the metal heated in the surface layer of the bath is transported to the peripheral crystallization zone, washing out nonmetallic inclusions in the crystallization front and carrying these to the centre of the bath [61]. In this case the swirling motion of metal is unfavourable.

When steels that are insensitive to liquation are melted, it may be thought that the application of a longitudinal magnetic field is more favourable, since the rotation of metal must promote a breaking of dendrites thereby refining the structure of the ingot. However, the rotation leads to a reversed effect — a washing out of nonmetallic inclusions at the surface leads to a growth of dendrites along the bath perimeter in a direction opposite to the rotation sense. This disadvantage could probably be avoided if the sense of rotation is altered periodically.

More information concerning the experimental modelling of vacuum-arc remelting processes can be found in [1, 40].

8.4. Hydrodynamics of furnaces with multiple electrodes

Two, three, or more electrodes are used to convey the current in electroslag welding and remelting processes, in metal smelting furnaces and flux producing furnaces. The motivations for the use of multiple electrodes are different: a need to weld details of a great length, a need to uniformly heat the melt in a large-sized vessel, and a requirement of uniform power load, etc.

The complexity of hydrodynamic configurations has already been demonstrated for a single electrode current supply with the axially symmetric conditions, (Sections 8.1—8.3). Obviously, the situation with multiple electrodes is a more complex one, even at the first stage of a theoretical consideration — evaluating the three-dimensional electric and magnetic fields.

It is instructive to examine two different hydrodynamic configurations in the photographs showing the free surface of a mercury bath with a two-electrode current supply. Fig. 8.22 represents a flow configuration in which the current is supplied by two rod electrodes of equal polarity (the second electrode is the opposite wall), and in Fig. 8.23 it is supplied by two rod electrodes of opposite polarity. In both the cases jet of mercury is developed at the ends of the electrodes. Yet in the first case the two jets are 'attracted' to each other (similarly to conductors carrying like currents), and a stagnant fluid zone is formed between the electrodes; in the second case these jets fall at an angle to the side walls of the container, and a closing flow returns along the central axis, penetrating deep into the zone between the electrodes.

With the aim of clarifying the mechanism driving the flow, consider, as a somewhat simpler model, the situation with two electrodes of opposite polarity. We shall consider two limiting cases: a small and a large length of the electrodes.

336   Chapter 8

The main mechanism in the case of short electrodes, when the electric current passes along an approximately Π-shaped path between the ends of the electrodes (in the plane containing the axes of the electrodes), could be demonstrated by the following experiment [10]. Two rectangular channels 1 are filled with mercury, and they are situated at a distance L between their axes parallel to each other (Fig. 8.24). Two plastic floats 2 are placed in the mercury, and a copper wire frame 3 is attached to the floats, thus electrically connecting the two mercury channels. When an electric current source is connected to the electrodes 4, the frame is set in motion by the force F shown in Fig. 8.24. This experiment may be considered as a schematic representation of the electromagnetic rail launcher [31]. The distribution and direction of the electromagnetic force, acting on elements of the curved conductor due to its interaction with the remaining current elements, are shown in the same figure. The direction of the force can easily be determined according to the law of the force acting between arbitrary oriented line current elements. In the present situation,

Fig. 8.22. Flow in the model-pool with the two electrodes of like polarity.

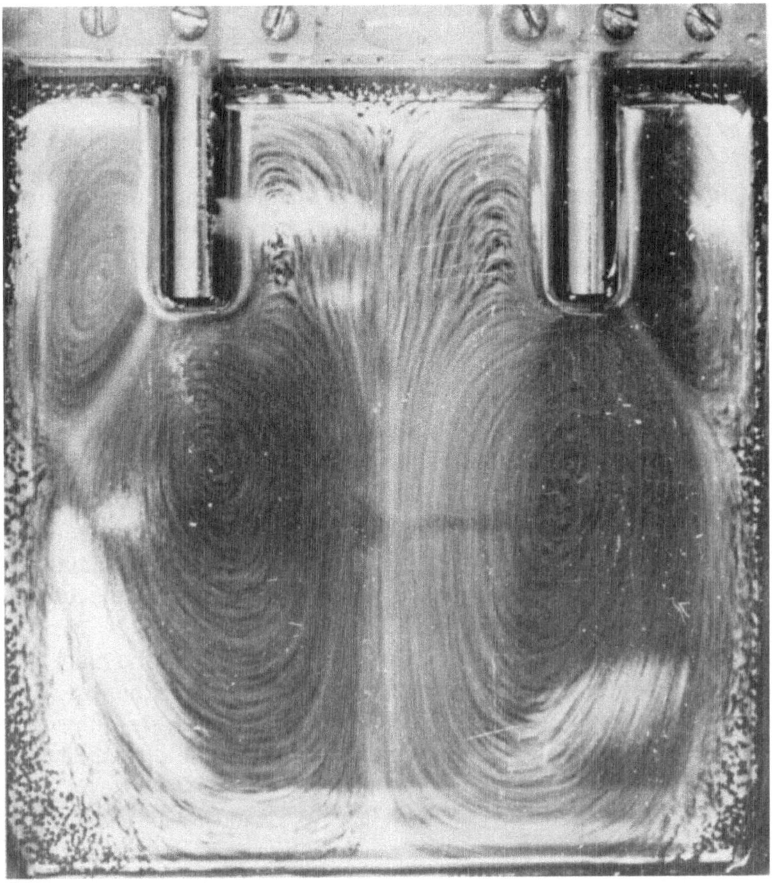

Experimental investigations of EVF and applications 337

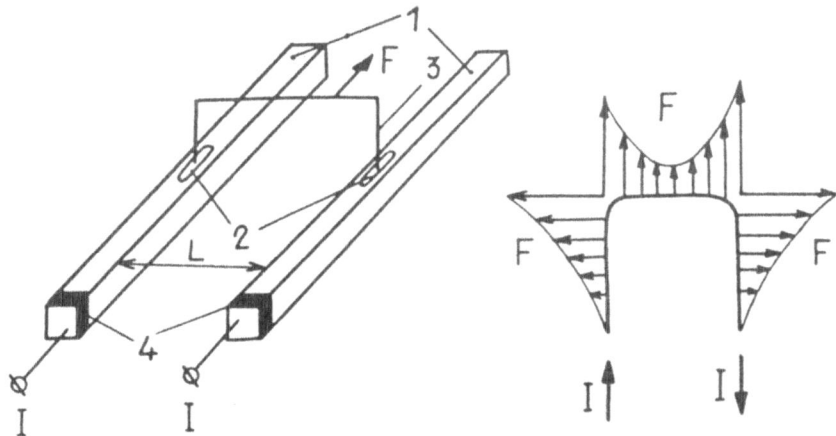

Fig. 8.23. Flow in the model-pool with two electrodes of different polarity.

Fig. 8.24. Scheme of the experiment with a floating electrically conducting bridge and the electromagnetic force distribution along the curved current-carrying conductor.

the integral force on the wire frame results from the interaction of the current in the mercury channel with the currents in the vertical and horizontal sections of the wire frame.

The experiment showed that the force depended neither on the shape of frame, nor on its size, nor on the diameter of the wire. At the same time, the force increased at a rate proportional to the square of the current, it also grew with the distance between the channels, and it depended on the dimensions (absolute and relative) of the mercury channel's cross section. The experimental data (Fig. 8.25) were satisfactorily approximated by the expression

$$F = \frac{\mu_0 I^2}{4\pi} \ln \frac{L}{R_0}, \tag{8.4.1}$$

where R_0 is the hydraulic radius of the channel, equal to the ratio of the cross-sectional area S to the perimeter P, $R_0 = S/P$.

As is evident in Fig. 8.24, the electromagnetic force is concentrated at the turn of electric current lines and the integral force is directed at the angle $3\pi/4$ measured from the straight current-carrying channel. This orientation of the force could explain the divergence of fluid jets driven at the ends of two electrodes of opposite polarity in Fig. 8.23.

Apart from this, the electric current is also distributed in the plane normal to the axes of electrodes, say $z = $ const. To analyse the consequences of this fact, we assume a simple situation in which the two electrodes are immersed to the whole depth ℓ of the liquid current-conducting layer, the distance between the electrodes being $2a$ (Fig. 8.26).

In a cylindrical coordinate system (z', r', φ') fixed to the electrode, for the condition of a uniform current density distribution over the electrode's surface, appropriate solutions for the current density and magnetic field are of the form:

$$j_{r'} = \pm \frac{I}{2\pi\ell} \frac{1}{r'}, \quad H_{\varphi'} = \pm \frac{I}{2\pi\ell} \frac{z'}{r'}$$

(the sign (+) is associated with the incoming current and (−) with the outcoming current). The superposition of the currents, expressed in Cartesian coordinates, shown in Fig. 8.26, yields the solution

$$j_x = \frac{I}{2\pi\ell} \left[\frac{x-a}{(x-a)^2 + y^2} - \frac{x+a}{(x+a)^2 + y^2} \right],$$

$$j_y = \frac{I}{2\pi\ell} \left[\frac{y}{(x-a)^2 + y^2} - \frac{y}{(x+a)^2 + y^2} \right];$$

$$B_x = \frac{\mu_0 I}{2\pi\ell} z \left[\frac{y}{(x-a)^2 + y^2} - \frac{y}{(x+a)^2 + y^2} \right], \tag{8.4.2}$$

$$B_y = \frac{\mu_0 I}{2\pi\ell} z \left[\frac{x+a}{(x+a)^2 + y^2} - \frac{x-a}{(x-a)^2 + y^2} \right].$$

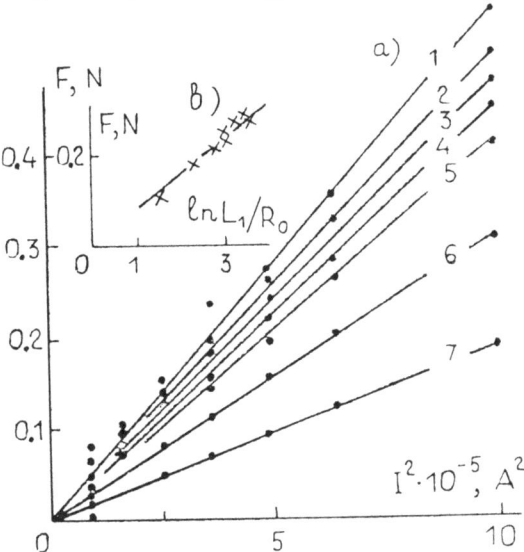

Fig. 8.25. (a) Dependence of the integral force on the hydraulic radius of current-carrying channels and the distance between the axes of the channels: $L = 150$ mm, $1 - R_0 = 4$ mm, $2 - 5$ mm, $3 - 6$ mm; $R_0 = 6.82$ mm, $4 - L = 150$ mm, $5 - 120$ mm, $6 - 70$ mm, $7 - 35$ mm. (b) The integral force vs. $\ln L/R_0$.

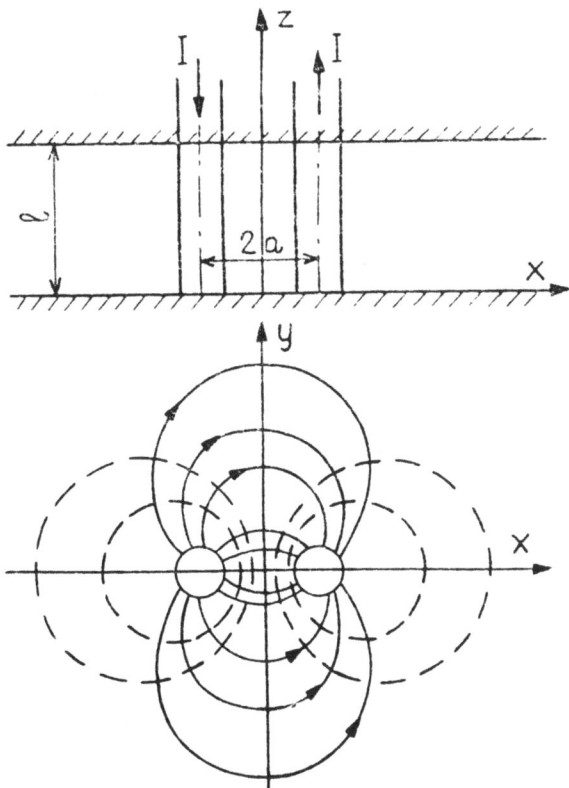

Fig. 8.26. Scheme of the electric current supply to a layer of finite thickness and the current passage in the planes $z = $ const.

The corresponding electromagnetic force has only the z-component $f_{ez} = j_x B_y - j_y B_x$, and its curl has x- and y-components. Let us consider possible fluid motions driven by the force distribution in two planes $y = 0$ and $x = 0$.

In the plane $y = 0$ $f_{ez} = -\mu_0(I/2\pi\ell)^2 4a^2 z/(x^2 - a^2)^2$ (curl $\mathbf{f}_e)_y = \mu_0(I/2\pi\ell)^2 16 a^2 xz/(a^2 - x^2)^3$, i.e. the electromagnetic force grows approaching the electrodes, and the force curls the fluid, forcing it to move downwards along the surface of the electrodes in Fig. 8.26, and the fluid could return upflowing along z-axis. In the orthogonal plane $x = 0$ a maximum of the force $f_{ez} = -\mu_0(I/2\pi\ell)^2 4a^2 z/(a^2 + y^2)^2$ is located on the z-axis, and (curl $\mathbf{f}_e)_x|_{x=0} = \mu_0(I/2\pi\ell)^2 16 a^2 yz/(a^2 + y^2)^3$ suggests a fluid downflow along the z-axis with a returning flow in a region remote from the plane $y = 0$.

Thus, two conflicting tendencies could be expected at the z-axis. If the model (Fig. 8.26) is rotated to point z-axis downwards, and the plane $z = 0$ is made a free surface, then the flow visualization on this surface shows (Fig. 8.27) that the second tendency is governing the flow in the z-axis region. I.e. similarly to the region close to the lateral surface of the electrodes, the motion is directed to the

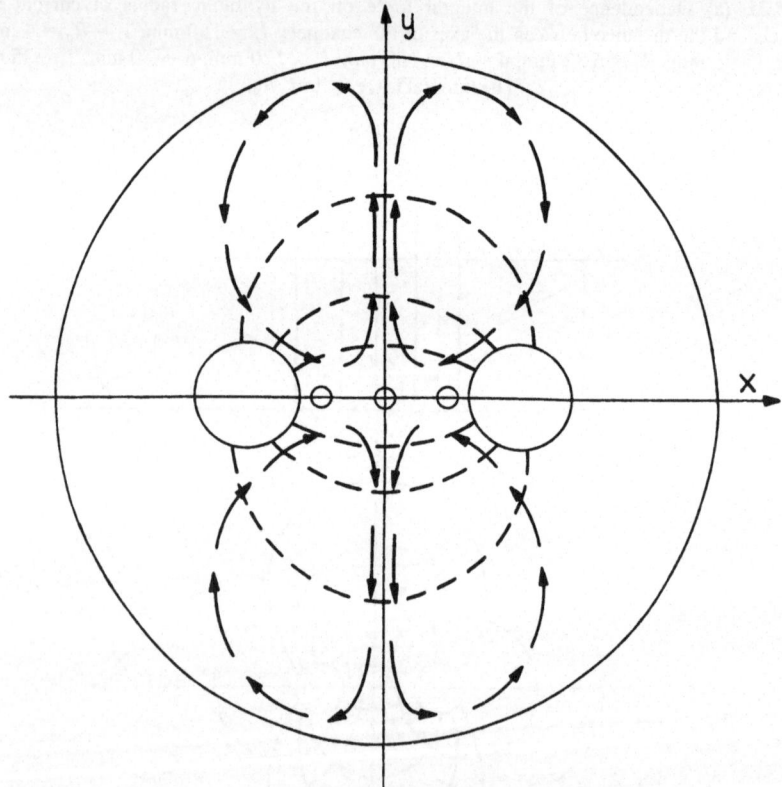

Fig. 8.27. Motion (shown by the arrows) on the free surface of mercury in the container of diameter 130 mm, the electrodes of diameter 22 mm are situated at the distance 52 mm between them, and the depth of the container, equal to the electrode immersion depth, is 8 mm. Dashed lines show the electric current passage.

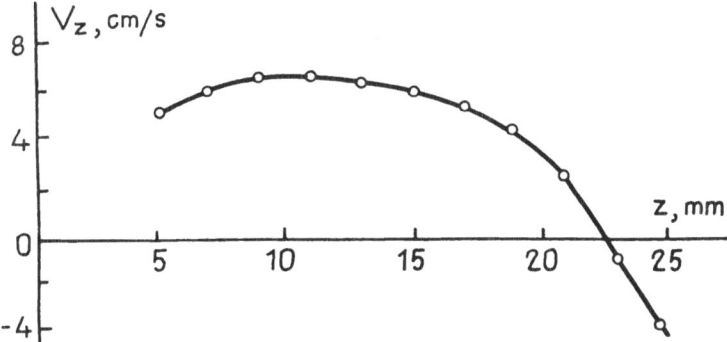

Fig. 8.28. The dependence of axial velocity v_z vs. the coordinate z for $I = 10^3$ A.

free surface (the dotted circles in Fig. 8.27). In the remaining part of the free surface the motion corresponds to the distribution of electromagnetic force. Direct measurements of the velocity distribution $v_z(z)$ along the z-axis give evidence (Fig. 8.28) that the motion to the free surface takes place only in a part of the axial region; in the remaining part the motion is oppositely directed, i.e. it corresponds to $(\text{curl } \mathbf{f}_e)_y|_{y=0}$.

The velocity field is even more complex if the electrodes are only partly submerged and in the case of a three-electrode current supply [22]; nevertheless, these flows can be explained by the mechanisms of flow generation considered.

8.5. Electrical jet thrusters

A new perspective for space-rocket engines has been opened up with the advent of electrical jet thrusters (EJT), the basic advantage of which is a capacity to accelerate a working material up to a velocity of 10^5 m/s, while in chemical engines the velocity does not exceed 10^3 m/s. By the same token, the fuel consumption is tens of times less for the EJT comparing to the chemical rockets [3, 48]. However, the efficiency of EJT is lower when a high thrust is necessary, e.g. when the space-rocket is launched. The main advantage of EJT — a possibility of working continuously during a long time period (up to several years) — could determine its use for long space flights.

An electrical arc is common to all variants of EJT. The first type of thruster employs the arc to heat a working material, and the other uses it to ionize a working material. In the electromagnetic EJT the arc takes up the thrust arising from the interaction of arc current and its self-magnetic field. Hence the electrically induced flows are essential in the operation of the electromagnetic thrusters. A typical representative of these is the coaxial-end-thruster (Fig. 8.29). With the aim of determining the thrust of an engine, consider a schematic representation in Fig. 8.30. Let the cathode be of radius b; the electric current exits from the control volume either through the end section of radius a, or

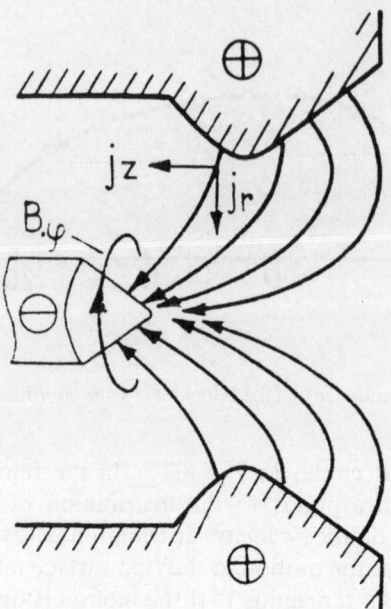

Fig. 8.29. Schematic representation of the coaxial plasma thruster.

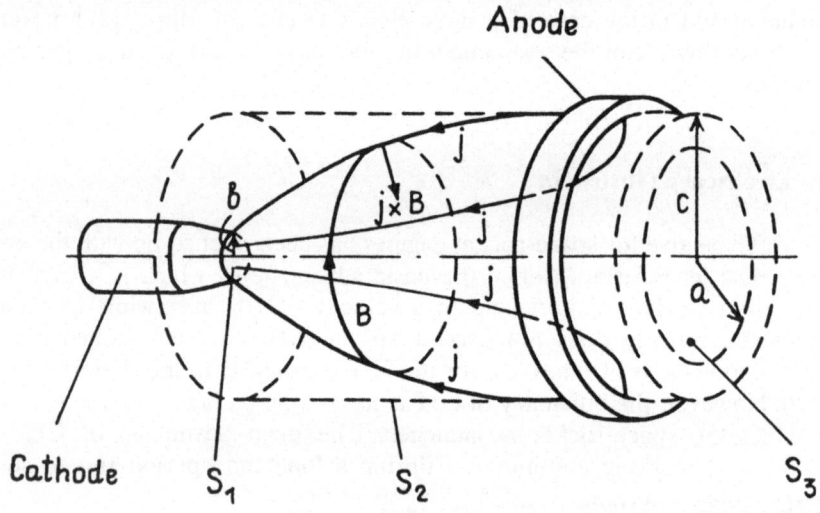

Fig. 8.30. The control volume for evaluation of integral self-magnetic force.

through the anode of radius c, situated at the lateral surface of the control volume. We shall apply the result (5.2.9), in which the zero contribution in the z-component of force $\mathbf{F}_e$ is accounted for by the electromagnetic momentum flux through the lateral surface S_2 related to the component $\sigma_{rr} = (B_r^2 - B_z^2 - B_\varphi^2)/2\mu_0 = -B_\varphi^2/2\mu_0$ of the electromagnetic stress tensor. The first integral in (5.2.9) over the surface S_1 yields $\mu_0 I^2(\frac{1}{4} + \ln c/b)/4\pi$, the second integral over the surface S_3 gives $-\mu_0 I^2(\frac{1}{4} + \ln c/a)/4\pi$ if the whole current passes the surface, and it is equal to zero if the current passes the lateral surface S_2 (in view of the fact that $B_\varphi = 0$ on the surface S_3 in this case).

Hence, if the current passes the section S_3, the composite expression is the familiar result (5.2.10):

$$F_{e_z} = \frac{\mu_0 I^2}{4\pi} \ln \frac{a}{b}, \tag{8.5.1}$$

however, if the current is absent in the section S_3, then

$$F_{e_z} = \frac{\mu_0 I^2}{4\pi} \left(\frac{1}{4} + \ln \frac{c}{b} \right). \tag{8.5.2}$$

The formulae (8.5.1) and (8.5.2) express the integral electromagnetic force in the gas volume enclosed by the surface $S_1 + S_2 + S_3$. An additional repulsive force arises between the cathode and gas volume due to a pressure increase in the region during the electrical current passage (the existence of the force acting on the cathode is confirmed experimentally). This force can be evaluated by applying the pressure distribution from Section 6.1:

$$p = p_0 + \frac{\mu_0 I^2}{4\pi^2 b^4} (b^2 - r^2). \tag{8.5.3}$$

On integrating (8.5.3) over the end surface of the cathode, we obtain the additional force $\Delta F_z = \mu_0 I^2 / 8\pi$, thence, finally, the propulsion force in the coaxial thruster is

$$F_z = \frac{\mu_0 I^2}{4\pi} \left(\frac{3}{4} + \ln \frac{a}{b} \right). \tag{8.5.4}$$

The expression (8.5.4) was derived in a different way by Maecker [43] and other authors [32, 33]. As may be seen from the expression, the higher the divergence of the jet (i.e. the higher the ratio a/b), the higher is the plasma acceleration in the self-magnetic field. Therefore the design of thrusters often abandon the usual contracting-expanding nozzle contours [23, 33, 55] in favour of contours of a greater expansion.

However, the mere presence of the force (8.5.4) is not a sufficient condition for propulsion. The effects of electrically induced flow may produce the propulsion in the opposite direction to the electromagnetic force (see Chapter 5). Therefore, the operation of EJT must be analysed in combination with the hydrodynamics in a particular device.

Several consequences of the electrically induced flows will be demonstrated in the following experiment [17]. An electric current passes through the tube of varied cross-section and of length 144 mm; the tube's ends serve as electrodes. The radius of the narrow tube section $a = 6$ mm, and of the wide section $b = 27$ mm (Fig. 8.31a). Six holes for the pressure measurements are made in the lateral wall of the tube. With the aim of damping the electrically induced flow due to the nonuniform electric current density, three stainless steel mesh

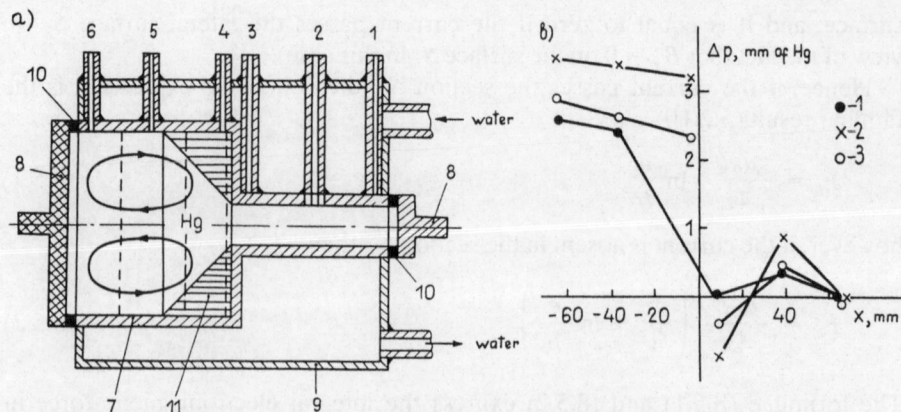

Fig. 8.31.. The experiment with a variable cross section liquid current conductor: (a) scheme of the model (1—6 — holes for pressure measurement, 7 — casing, 8 — electrodes, 9 — casing for water cooling, 10 — insulation, 11 — the plug-in plastic cone in the case of continuous expansion (with the angle 45°)); (b) the pressure distribution along the channel for $I = 1200$ A: (1) with mesh screens, (2) without screens, (3) with the insulating cone.

screens are placed in the wide section of the tube. The electrical conductivities of mercury and stainless steel are approximately equal, thereby the electric current distribution is not disturbed. Fig. 8.31b presents the pressure distribution along the wall of the tube for $I = 1200$ A, and Fig. 8.32 shows the pressure drop between the points 6—1 vs. the square of current for the experiment with meshes (curve 1) and without them (curve 2). The pressure drop is found to be proportional to the square of the current, and in the case of active EVF (without meshes) the drop is 25 per cent higher than in the case of damped EVF.

The experiment reveals a noticeable dependence of the pressure drop on the current divergence angle. If, at the step of sudden expansion, a nonconducting cone of apex angle 45° is situated, the pressure drop is almost the same as in the experiment with the meshes (curve 3 in Fig. 8.32). As it follows from Fig. 2.7, the intensity of electrically induced flow in a cone decays with decrease of the current divergence angle. These facts confirm the substantial effect of electrically induced flow on the magnitude of the axial force.

Let us now discuss the mechanism by which the electrically induced flow gives rise to the additional thrust force. As we have shown previously, the electrically induced flow in a current-carrying tube of variable cross section takes the form of a toroidal vortex with the velocity along the axis of the torus directed to the current density decrease. The flow direction is opposite at the side walls. A frictional force acts at these walls, and the fluid pushes the wall by the friction force F_f in the x-axis direction (see Fig. 8.31a). If the wall is fixed, the fluid is thrust in the negative direction of the axis. In addition, the vortical flow increases the pressure at the expansion step, 'pushing' it in the same direction as the friction force. Both these forces of hydrodynamic origin determine the additional thrust to the electromagnetic force.

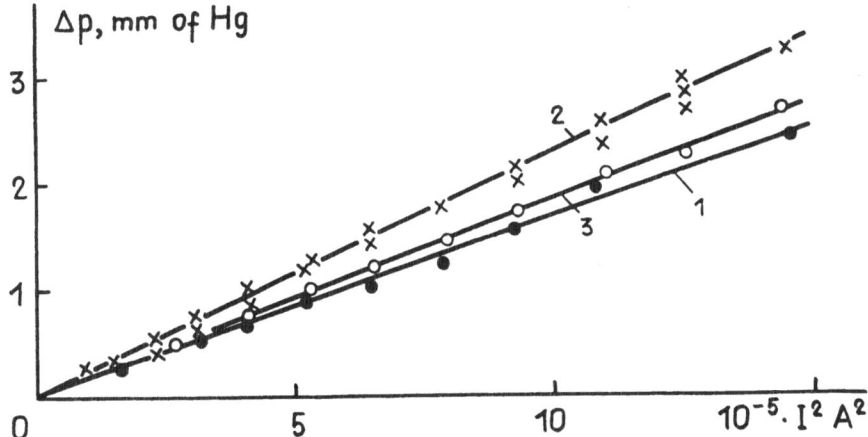

Fig. 8.32. The pressure drop between the points 1—6 vs. the square of current in the experiments: (1) with the mesh screens, (2) without screens, (3) with the insulating cone.

The thrust force depends on the arc current. Another technique to control the thrust is the application of an external magnetic field (e.g., the field of a solenoid or a current loop).

Experiments [67, 68] have shown that the axial velocity can reach high magnitudes close to the cathode, yet it falls rapidly with distance from the cathode (Fig. 8.33). When the arc is placed in the field of a short solenoid, the velocity and length of jet decrease, and in a sufficiently high field they even change their direction [67], which could be explained by the differential rotation mechanism. This observation permits not only control of the flow in the arc, but also a reversal of the thruster. This technique may expand the possibilities to orientate, manoeuvre, and brake objects in space.

Electrical arc plasma jets are also being applied in a variety of technological processes: application of different coatings and films, cutting, welding and

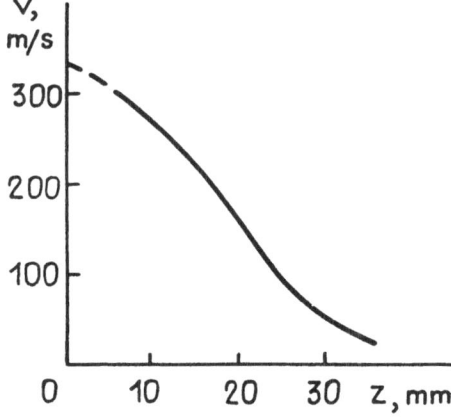

Fig. 8.33. Dependence of the axial velocity in a carbon arc on the distance from the cathode [68].

remelting of metals, alloying and carburizing, nitriding, development of atomic pure surfaces, growth of different crystals, and many others.

8.6. Induction channel furnaces

Induction channel furnaces (ICF) are a relatively widely used electromagnetic equipment for melting non-ferrous metals. It may be sufficient to mention that approximately 2/3 of the zinc and copper alloys produced are smelted in this kind of furnaces. ICF are the most effective equipment to transform electrical energy into thermal (the electrical performance coefficient amounts to 95—97 per cent [37], and the thermal one reaches 96 per cent). Moreover, these devices are characterized by low metal losses, simplicity and reliability of construction, and the possibility of easily replacing the heating element—induction unit (IU). Applications to ferrous metal melting are relatively few. The reason is the low power of IU (500—700 kW) as compared to other kinds of electrothermal equipment (coreless-type induction furnaces, arc furnaces) and, coupled with that, a limited output. This is related to the low heat exchange rate from the channel of the IU, where most of the ohmic heating occurs, to the main melt pool, in consequence of which the metal is overheated in the channels of high power the IU. The overheating for the previously specified power amounts to 100—200°. Under such circumstances, the requirements on the quality of the lining material increase sharply, the probability of failure due to a metal leakage grows, heat losses grow, etc.

A principal step in increasing the power of ICF is the transformation of the IU channels into a through-flow heater by generating an intense transit flow of metal in the heat-producing elements of the IU. One of the methods to generate the transit flow is an application of additional electromagnetic devices [52, 53, 57]. There are two kinds of such devices — those with an external magnetic field interacting with the current in the channel [53, 57], and those with an application of a travelling magnetic field (induction pump) [52]. These methods have not found application in ICF for melting metals, because they require a substantial decrease in the lining thickness; these techniques are widely exploited, however, for instance as pumps to discharge the metal, etc. Another method is based on a rational use of electromagnetic forces generated by the existing current in ICF.

Before proceeding with the discussion of feasible solutions, it seems reasonable to briefly consider physical effects in the induction channel furnaces. An ICF consists of a single, double, or triple inductor, each of which is enclosed by a loop of liquid metal (Fig. 8.34) made by the IU channel. As an electrical machine ICF is equivalent to a transformer of current, in which the secondary coil is formed from the short-circuited liquid metal loop. The current magnitude here reaches tens of kA, and in the primary coil of the transformer it is hundreds of A. A part of the loop is formed by the comparatively large volume of metal in the pool of the furnace. The current density and ohmic heating there are insignificant. The main part of the loop is the channel of the IU carrying

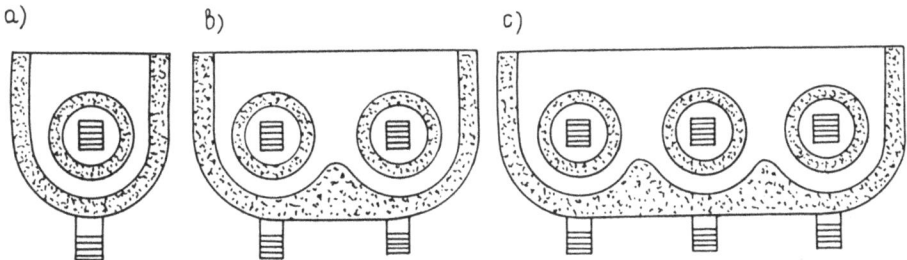

Fig. 8.34. Induction channel units: (a) single, (b) two-phase, (c) three-phase.

the high density current, where a substantial part of the electrical energy is transformed into heat. The problem, therefore, is to find an effective way to transport the large excess heat produced in the channel to the bulk of the metal in the pool of the furnace and to drive an intense mixing of the metal in the pool.

The excess temperature of metal in the channels relative to the bulk mass of metal in the pool naturally draws attention to thermal convection as a possible mechanism capable of generating the transit flow. Usually the IUs are connected to the bath of the furnace in such a way that the loop plane is oriented vertically or slightly inclined to the vertical. Since the zone of maximum heating is located in a lower section of the channel, the necessary conditions for driving a thermal convection current are met. Estimates of transit flow velocity and the resulting excess heating with the convection [17] show that with an increase of the furnace power the overheating grows at a higher rate than the mean flow rate. Thereby, thermal convection is not in a position to solve the problem of increasing power. Moreover, the thermal mechanism may come into action only after an occasional break of symmetry in the temperature distribution over the channels, hence the flow direction is unpredictable.

The second mechanism (in numerical order, not importance) is electromagnetic. The electric current is already present in an ICF due to the melting technology itself. The magnetic field, interacting with the current in the liquid metal, consists of the inductor's field and the self-magnetic field of the electric current in the fluid. However, it is easy to show that the force generated by the current and the inductor's field is alternating in time, its average in time component along the channel is zero, and hence the force cannot be used to drive the transit flow, although it drives a metal circulation in the normal cross-section of the channel [45, 47]. At the same time, the electromagnetic force due to the interaction of the current and its self-magnetic field is of a constant sign and thence may be considered for the purpose. Henceforth we shall analyse the electromagnetic force $\mathbf{f}_e$ of this nature.

One of the possibilities of driving a transit flow is related to the longitudinal component of $\mathbf{f}_e$ along the IU channel axis when the liquid conductor is of a variable cross-sectional area. Such sections in an IU are the channel openings, where the force tends to push the metal out of the channels. However, a particular feature of the force is its dependence only on the ratio of the wide and narrow section widths, and it does not depend on the shape of the expanding section (see Section 8.5). Hence, it can be exploited, as will be shown later, only when certain conditions are satisfied.

The second electromagnetic mechanism driving the transit flow is related to the turning of electric current lines at the openings of channels. The mechanism was discussed in Section 8.4, and the estimated force according to (8.4.1) is of the same order of magnitude as the force (8.5.1) arising in the expansion of the channel.

The second electromagnetic mechanism driving the transit flow is related to the turning of electric current lines at the openings of channels. The mechanism was discussed in Section 8.4, and the estimated force according to (8.4.1) is of the same order of magnitude as the force (8.5.1) arising in the expansion of the channel.

Finally another mechanism driving the transit flow is also related to the variation of channel's cross-sectional area. This mechanism is the electrically induced vortical flow, which in the case of ICF is similar to the periodic EVF considered in Chapter 5: the expansions of the channel induce vortices in the liquid metal; in the presence of rigid walls pressure and friction forces at the walls should be compensated by an increased momentum in the fluid moving along the channel, i.e. a transit flow is driven in a closed hydraulic system or a pressure drop is set up in an unclosed system. The most significant feature of this mechanism is the dependence of vortex intensity on the divergence angle, magnitude, and frequency of current, increasing with these parameters. This means that the mechanism could be exploited to control the flow by changes in the geometric current passage conditions, and its efficiency grows with the power of the IU.

Below we shall discuss several designs of different types of IU where the above principles of the transit flow generation are applied.

8.6.1. Single induction units

In the single (single-phase, single-inductor) IU the inductor is enclosed by a single liquid metal loop. The design of IU channels prior to modernization has been characterized by exact symmetry (Fig. 8.34a), hence the thermal convection mechanism comes into action only accidentally. Obviously, the electromagnetic force could drive a transit flow if the openings of the channel are made unsymmetric. However, in this case the first electromagnetic mechanism is not activated, because it does not depend on the shape of channel openings. The second mechanism could drive the flow, since the force (8.4.1) at the narrow opening with a smaller hydraulic radius is higher than the force at the wide opening. The transit flow is also driven by the electrically induced vortex mechanism. The optimum shape of the channel openings is depicted in Fig. 8.35. This shape has been found experimentally [4] and supports an intense transit flow. The tighter turn of the current in the left opening is responsible for the more intense vortex and the intense exhaust of liquid metal from the left channel opening. The resulting motion in the channel is clockwise. The over-

Experimental investigations of EVF and applications 349

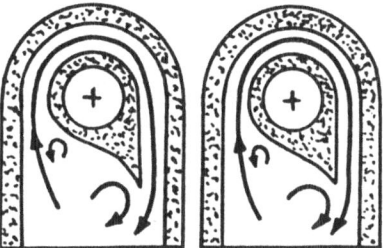

Fig. 8.35. Optimum shape channel openings in a single induction unit.

heating is reduced by a factor two compared to the symmetric form. Similar designs, including more efficient constructions with the current turning in two planes, are proposed in [37].

8.6.2. Two-phase IU

This type of IU offers significantly wider opportunities to control the intensity of transit flow. Let us discuss some of them.

8.6.2.1. The effect of phase shift. When the magnetic fluxes in the inductor are shifted in phase by $\alpha = 180°$, the electric current is circulating in two loops, doubling in the central channel. As a matter of fact, we have in this situation the two single units considered previously. Hence, the transit flow is weak and its direction is unpredictable (Fig. 8.36). The situation of coupled single units could be changed if the current in the central channel is eliminated. For this it is sufficient to set the phase shift of magnetic fluxes $\alpha = 0°$ [13, 14], which is

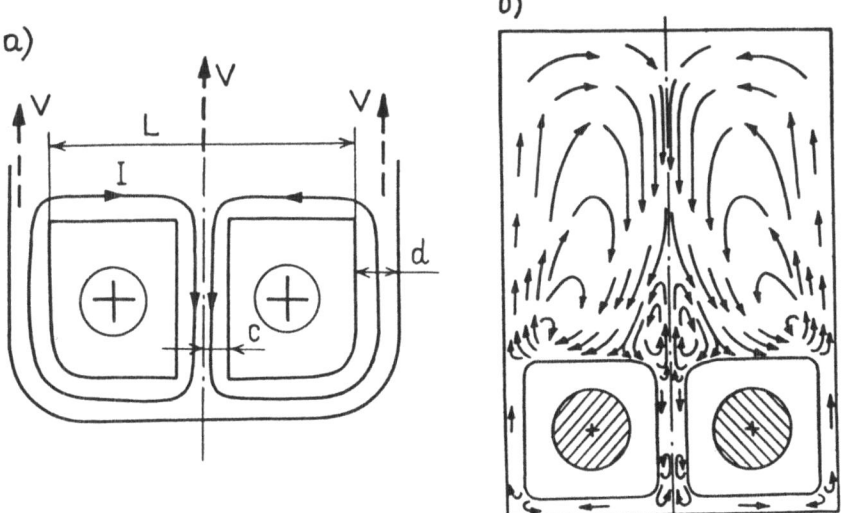

Fig. 8.36. (a) Scheme of electric current configuration, (b) schematic representation of metal flow in the model of two-phase *IU* for the phase shift $\alpha = 180°$.

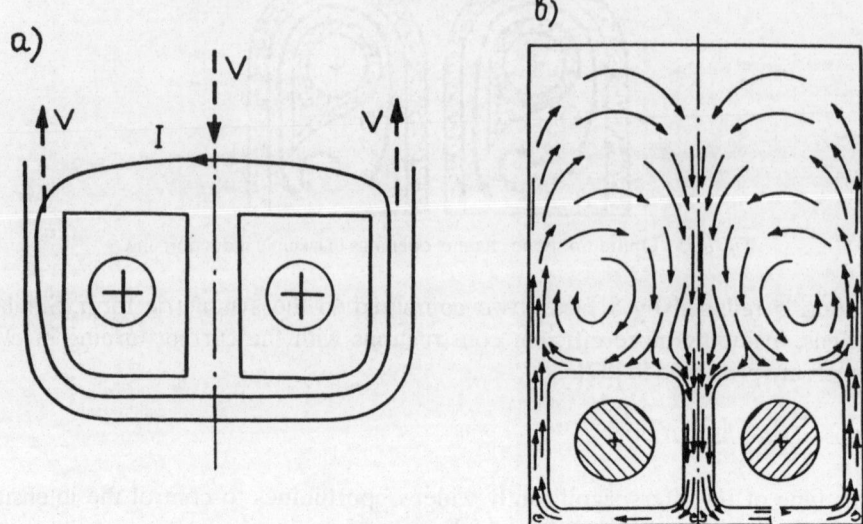

Fig. 8.37. (a) Scheme of electric current configuration, (b) schematic representation of metal flow in the model of two-phase IU for the phase shift $\alpha = 0$.

feasible if we use two separate ferromagnetic cores. In this case there is only one current loop passing the side channels and the bath of the furnace (Fig. 8.37). Apart from the electrically induced vortex mechanism, the axial components of $\mathbf{f}_e$ are activated in the side channels, since these are not compensated by the force in the central channel. The flow in the latter case is the following: the colder metal from the bath enters the central channel, then in its course it is heated in the side channels, and exits to the bath from the openings of the side channels. The intense jets from the side channels propagate deeply within the bath, maintaining there a good mixing of metal. The overheating of metal in the channels drops by 3—4 times, depending on the ratio of the central and side channel dimensions. The experimental data for the mean flow rate velocity v in the channels, depending on the frequency w and magnitude of current in the channel, and geometrical dimensions, are satisfactorily approximated by the formula:

$$Re = vL/\nu = \left(0.4 + 5 \times 10^{-3}\frac{d}{L}\bar{w}\right)\sqrt{S},$$

where $\bar{w} = \mu_0 \sigma L^2 w$, $S = \mu_0 I^2/\rho v^2$ [18]. Note that the dimensional velocity does not depend on the viscosity of the metal.

8.6.2.2. The extension of side channels [16]. The velocity of transit flow can be significantly increased (2—3 times comparing to the previous case) if extensions, made of nonconducting material, are fixed at the openings of lateral channels (Fig. 8.38). In this situation the intensity of vortices grows due to an increased

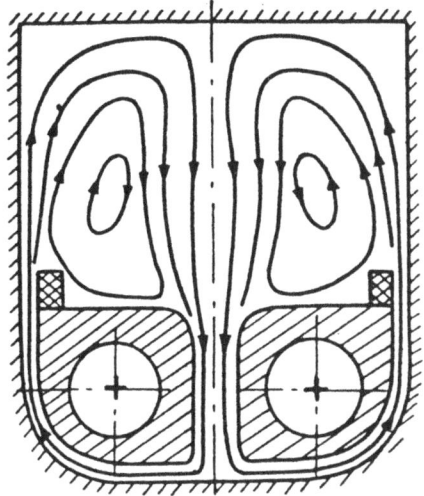

Fig. 8.38. Metal flow in the model of two-phase *IU* with the extended side channels.

angle of the current turn and the axial component of f_e grows (due to the joint interaction of the opposite currents at the edge of extension on the current turn elements). The extension of side channels is advantageous also for the electrical parameters of the IU: the pure resistance of the channels, respectively $\cos \varphi$, increases, and the thermal power losses in the central channel are compensated.

The velocity of transit flow is also affected by the material parameters of metal, the ratio of channel dimensions and their shape, the frequency of current, etc. At the present time the integrated experimental and theoretical studies have made available nondimensional equations relating the parameters governing the transit flow velocity, and a convenient design technique for the two-phase IU has been developed, in which the requested parameters include the power of the furnace and also the limiting overheating of the metal [4].

8.6.3. Three-phase IU

The methods of transit flow intensification in the three-phase IU are generally similar to the methods discussed above (the use of new channel shapes and new schemes of power connection). Up to the present time several variants of the three-phase IU have been proposed and investigated in the Institute of Electrodynamics (Ukrainian SSR Academy of Sciences) [37], and jointly in VNIIETO and the Institute of Physics (Latvian SSR Academy of Sciences) [19].

It can be summarized that, at the present time, the problem of transit flow intensification in the IU and of limiting the overheating of metal is practically solved. Two tendencies have emerged in the direction of introducing the results into practice: the modernization of existing ICFs, and the development of new generation furnaces of increased power and new technological functions.

The modernization of furnaces, as a rule, does not require additional capital investment. In practice only the profile of the lining or the electrical connection of the inductors for two- and three-phase IU must be changed. The resulting productivity of the furnaces, according to [37], is increased by 1.3—1.4 times, and in particular cases it is doubled. The durability of the lining, in view of the decreased overheating, increased twice, thus saving the cost of replacement or reconditioning. Moreover, if the electrical equipment is available, the same furnace after the modernization can be brought up to a higher power and therefore to higher productivity.

The new generation of ICF, first of all, is expected to increase productivity 2—5 times as compared to the current models at the expense of higher furnace power. At the present time ICF for non-ferrous metal smelting of power 1000—1500 kW and for ferrous metal of 2500—5000 kW have been designed and brought into commercial practice. The new furnaces in metallurgical practice have been designed for an additional heating of steel and iron in the continuous process. Alternative technological applications are also sought for the induction channel furnaces of increased power.

8.7. Electrically induced flows in a flat layer between ferromagnetic masses

Most of the results presented in this book are associated with axisymmetric electrically induced flows. At the same time, a new direction of investigation has been developed — EVF in flat current-carrying layers. A mathematical investigation of these flows is more involved than that of axisymmetric flows. This, and the few practical applications at the present time (special purpose pumps [35, 72] with additional heating of metal), could explain the small number of publications in this field.

Consider now the characteristic features necessary to drive sufficiently intense electrically induced flows in a flat layer.

When an electric current passes in a layer of infinite thickness (e.g. in the xy plane), only the z-component of the magnetic field is generated, the electromagnetic force is irrotational, and the fluid motion is not induced. Now let the current pass in a plane layer of thickness d (Fig. 8.39a). Obviously, in this situation all three components of magnetic field are generated and, generally, all three components of curl $\mathbf{f}_e$ are nonzero.

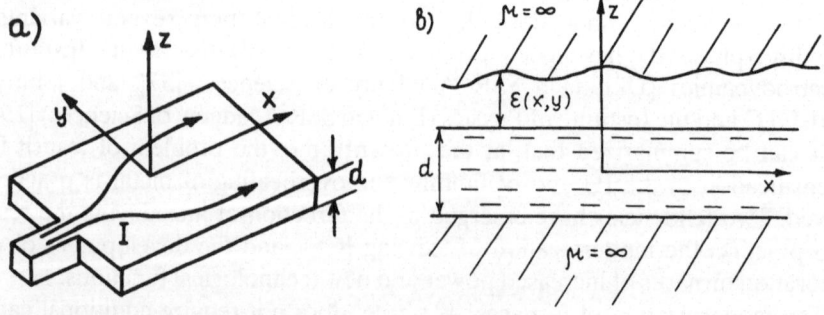

Fig. 8.39. A flat layer in a gap between ferromagnetic masses: (a) electric current configuration, (b) arrangement of the masses.

Experimental investigations of EVF and applications 353

If the thickness of layer d is small compared to the layer dimensions in the plane, the development of electrically induced flows in the planes yz and zx, respectively, due to the components (curl $\mathbf{f}_e)_x$ and (curl $\mathbf{f}_e)_y$, would be inhibited by the walls $z = \pm d/2$. Hence, sufficiently intense flows can be induced only in xy plane, i.e. these are driven by the z-component of curl $\mathbf{f}_e$. If $\mathbf{j} = (j_x, j_y, 0)$, then, in face of div $\mathbf{j} = 0$, it follows that

$$(\text{curl } \mathbf{f}_e)_z = -\left(\frac{\partial}{\partial x} j_x B_z + \frac{\partial}{\partial y} j_y B_z\right) = -\left(j_x \frac{\partial B_z}{\partial x} + j_y \frac{\partial B_z}{\partial y}\right), \quad (8.7.1)$$

i.e. the value of (curl $\mathbf{f}_e)_z$ depends on the current density and on the rate of change along the layer of the normal magnetic field component B_z.

The magnetic field component B_z can be increased if the current-carrying layer is placed in a gap $d + \varepsilon$ between ferromagnetic masses (Fig. 8.39b). Then the B_z component is increased at the expense of the B_x, B_y components. However, if $\varepsilon =$ const and the masses are sufficiently extended in the plane, we will have within the current-carrying layer only a B_z component ($B_x = B_y = 0$). Thence from $\mathbf{j} = \text{curl } \mathbf{B}/\mu_0$ it follows that

$$j_x = \frac{1}{\mu_0} \frac{\partial B_z}{\partial y}, \quad j_y = -\frac{1}{\mu_0} \frac{\partial B_z}{\partial x}, \quad (8.7.2)$$

and substituting (8.7.2) in (8.7.1) we obtain curl $\mathbf{f}_e \equiv 0$, i.e. the electromagnetic force is irrotational. Hence, the current layer of finite thickness in the constant gap between the ferromagnetic masses is equivalent to the current layer of infinite thickness.

If the thickness of the gap is variable ($\varepsilon = \varepsilon(x, y)$), other components of $\mathbf{B}$ appear. Instead of (8.7.2), we have

$$j_x = \frac{1}{\mu_0}\left(\frac{\partial B_z}{\partial y} - \frac{\partial B_y}{\partial z}\right), \quad j_y = \frac{1}{\mu_0}\left(\frac{\partial B_x}{\partial z} - \frac{\partial B_z}{\partial x}\right);$$

$$(\text{curl } \mathbf{j} \times \mathbf{B})_z = \frac{\partial B_x}{\partial z}\frac{\partial B_z}{\partial y} - \frac{\partial B_y}{\partial z}\frac{\partial B_z}{\partial x} \neq 0.$$

Consequently, at least one component B_x or B_y is necessary for the existence of a rotational electromagnetic force, which may be achieved between non-parallel surfaces of the ferromagnetic masses.

Consider a simple example of EVF driven in a flat layer. Let a uniform current pass along the x-axis in a square in a plane fluid layer, and let the gap between the ferromagnetic masses increases along the x-axis (Fig. 8.40a). In this case the electric current and the associated magnetic field $B_y(x, y, z)$, $B_z(x, y, z)$ give rise to the electromagnetic force and the curl (8.7.1):

$$(\mathbf{f}_e)_x = 0, \quad (\mathbf{f}_e)_y = -j_x B_z; \quad (\text{curl } \mathbf{f}_e)_z = -j_x \frac{\partial B_z}{\partial x}. \quad (8.7.3)$$

The magnetic field component $B_z > 0$ for $y > 0$, and $B_z < 0$ for $y < 0$. On

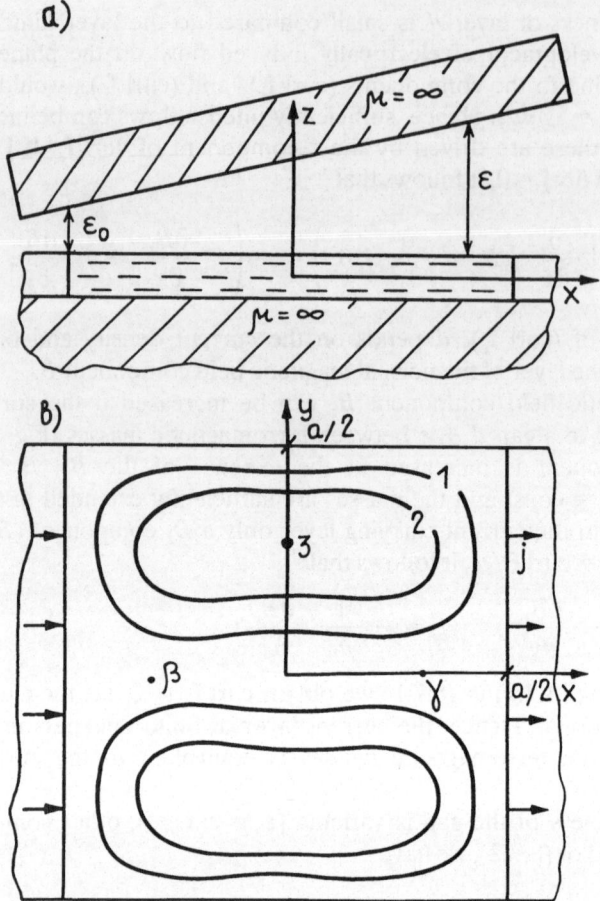

Fig. 8.40. (a) A flat square in plane cavern in the expanding gap between the ferromagnetic masses. (b) The predicted flow configuration in the cavern in the case of direct current, $S = 10^6$, $m = 0.1$. The values of ψ are indicated at the curves.

increasing the coordinate x, the magnitude of B_z decreases due to the increasing gap ε. Hence $\partial B/\partial z < 0$ for $y > 0$, and the curl of force is of positive sign; for $y < 0$, $\partial B/\partial z > 0$ and $(\text{curl } \mathbf{f}_e)_z < 0$. Corresponding to the sign of $(\text{curl } \mathbf{f}_e)_z$, two vortices could be driven in the flat cavity with the direction of circulation being such that the fluid moves along the x-axis in the centre of the cavity.

The approximate solution in [73] confirms the conclusion (see Fig. 8.40b). The flow intensity depends on the current magnitude and the angle of the plate's inclination. The experiment [7] demonstrates a pressure drop between the points β and γ (see Fig. 8.40). If the flow intensity is implicitly measured by the pressure drop, then the intensity grows initially with the angle of inclination, due to an increased nonuniformity of B_z along the x-axis, and later slowly falls due to the decrease of $|B_z|$ in the increased gap (Fig. 8.41).

Note that the non-parallel surfaces of ferromagnetics are not a sufficient condition for the rotational electromagnetic force. Thus, if in the situation depicted in Fig 8.40 $\mathbf{j} = (0, -j_y, 0)$, $\mathbf{B} = (B_x, 0, B_z)$ and B_x, B_z depend merely on the two coordinates x, z. According to (8.7.1), $(\operatorname{curl} \mathbf{f}_e)_z \equiv 0$ and a fluid motion should not be expected. Nevertheless, the authors of [8] predicted mathematically and confirmed experimentally a fluid flow, which could be generated due to the instability of current-carrying fluid in the flat cavity.

The intensity of electrically induced flow can be substantially increased for an optimum dependence $\varepsilon(x)$. Consider the first example. If the fluid cell is covered only halfway along the x-axis (Fig. 8.42), the nonuniformity of B_z is greatly increased. In this case the flow intensity grows by two orders of magnitude for the same current passing in the cell (Fig. 8.40b), and the flow configuration is a set of four vortices: the two more intense vortices are generated at the abrupt end of the ferromagnetic plate, and the other two are, evidently, secondary vortices. The computed pressure drop between the points β and γ in the case of continuous gap expansion ($\varepsilon = mx$)

$$\Delta p = 0.0188 Sm \frac{1}{(1 + \varepsilon + 0.5m)^2} ; \qquad (8.7.4)$$

in the case of abrupt expansion

$$\Delta p = 0.0321 S \frac{1}{1 + \varepsilon} \qquad (8.7.5)$$

(on multiplying Δp by $\rho v^2/d^2$ we can obtain the dimensional pressure drop).

Fig. 8.43 presents a comparison of the experimental data and the numerical ones (8.7.4), (8.7.5) [7]. Evidently, the pressure drop in the case of abrupt expansion is higher than in the case of continuous expansion. The deviation of the theoretical curve could be related to the unaccounted effect of an induced electric current. According to Fig. 8.43 the approximate limit of the non-

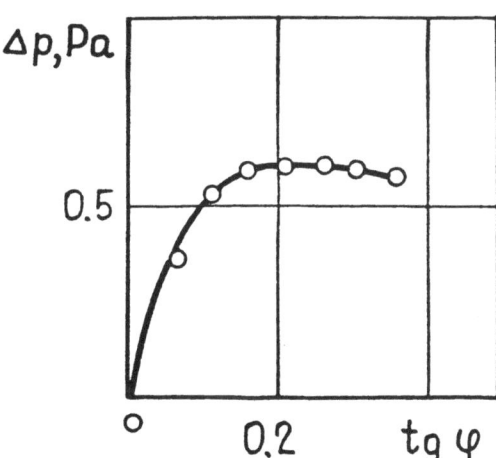

Fig. 8.41. Pressure drop between the points β and γ (see Fig. 8.40) vs. tangent of the plate's inclination angle.

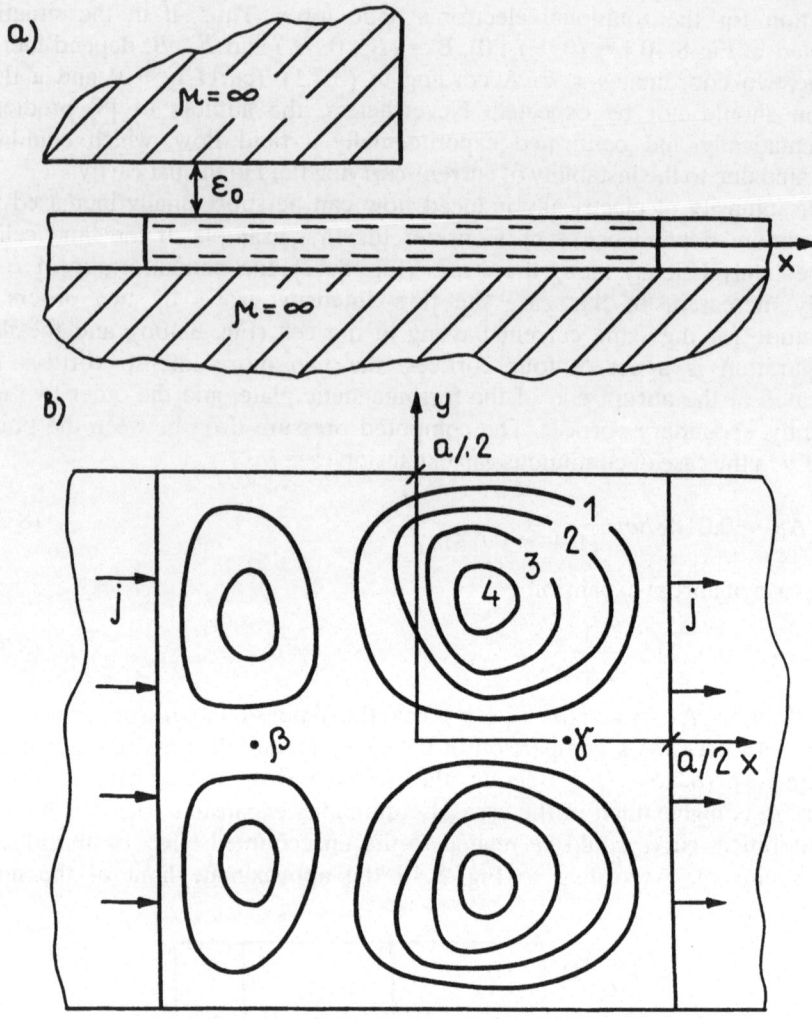

Fig. 8.42. (a) The flat square cavern in a parallel gap between ferromagnetic masses covering half its length. (b) The flow configuration for a uniform current density: $S = 10^6$, (1) $\psi = 300$, (2) 600, (3) 900, (4) 1200.

induction approximation may be estimated as $S < 10^7$ [36], which coincides with the estimate (2.9.2) derived in [62].

The set of vortices in Fig. 8.42 could drive a transit flow in a channel along the positive x-axis direction. The mechanism driving the transit flow is similar to some periodic EVF (see Chapter 5); the flow is driven due to the uncompensated friction of vortical motion at the walls $y = \pm a/2$. The electromagnetic force contains only a y-component (8.7.3), and it cannot directly drive the transit flow along the x-axis. Thereby the flat layer device is unlikely to be an efficient engineering application (e.g. a pump). Obviously, in an efficient engineering device the electromagnetic force must drive the flow directly. For

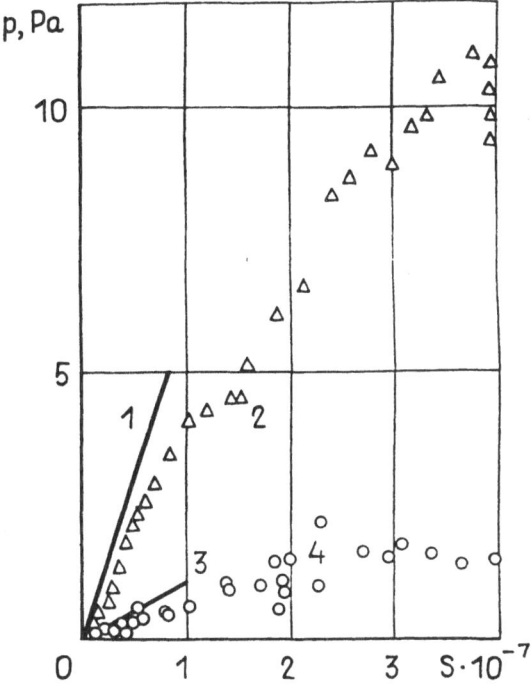

Fig. 8.43. Pressure drop between the points β and γ: (1), (3) the numerical dependences for $S < 10^7$, (2) experimental — in the case of abrupt gap expansion (Fig. 8.42), (4) in the case of continuous gap expansion (Fig. 8.40) for $m = 0.107$.

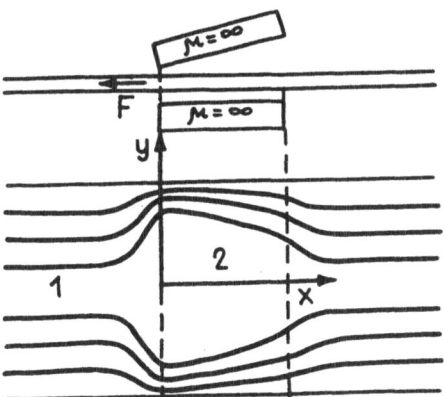

Fig. 8.44. Configuration of alternating electric current passing in a layer in the gap between ferromagnetic cores.

this purpose the electric current lines and transit flow stream lines should not coincide [6] (then the electromagnetic force, normal to the electric current lines, will have the component along the fluid motion).

In practice the electric current lines are made different to the stream lines by

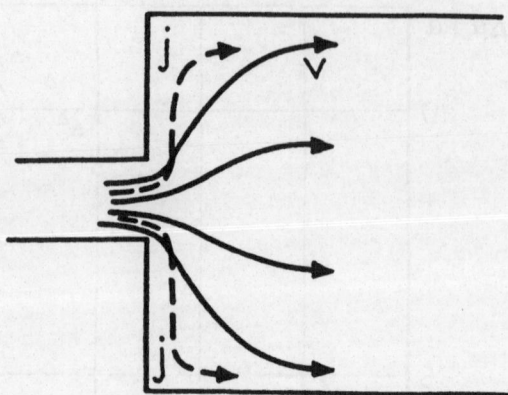

Fig. 8.45. Scheme of stream lines and alternating electric current lines in a channel with an abrupt expansion.

applying various physical techniques, e.g. introducing conducting sections within the channel [35], supplying the current through the walls of the channel, or embedding sections of higher conductivity in the wall of the channel. Yet if the electric current in the MHD-channel is alternating, then additional electrodynamic techniques are available [6]. For example, if the channel contains two fluids of different electrical conductivity, the current at the interface is highly nonuniform in regard to the different skin-layer depths; placing a section of the channel in a variable gap between the ferromagnetic masses (Fig. 8.44) the nonuniformity is achieved due to the displacement of alternating current lines to the lateral walls of the channel; in a sudden expansion of the channel the nonuniformity takes place due to a more abrupt divergence of the alternating current 'lines' than of the direct one (Fig. 8.45).

The flows induced in a flat channel by an alternating current were studied in [34, 72]. In the situations considered the ferromagnetic plates cover the channel only in the zone of sudden expansion, thereby increasing the rotational component of electromagnetic force. The experiment with direct current in the same channel led to a much lower magnitude of pressure drop, which had an opposite sign than that obtained with alternating current.

8.8. Electrolytic aluminium production

In the classical Hall-Herault process aluminium is produced by electrolysis reducing alumina Al_2O_3, which is dissolved in liquid cryolite $NaF-AlF_3$. The electric current passes from carbon anodes through the layers of electrolyte and liquid aluminium in the rectangular cell cavity, and exits through collector bars embedded in the carbon lining (Fig. 8.46). The development of industrial electrolysis cells is associated with increasing magnitudes of the total current passing through the cell. Thus, two decades ago the typical current magnitudes were 50—100 kA, a decade ago 130—150 kA, and at the present time the new

Experimental investigations of EVF and applications   359

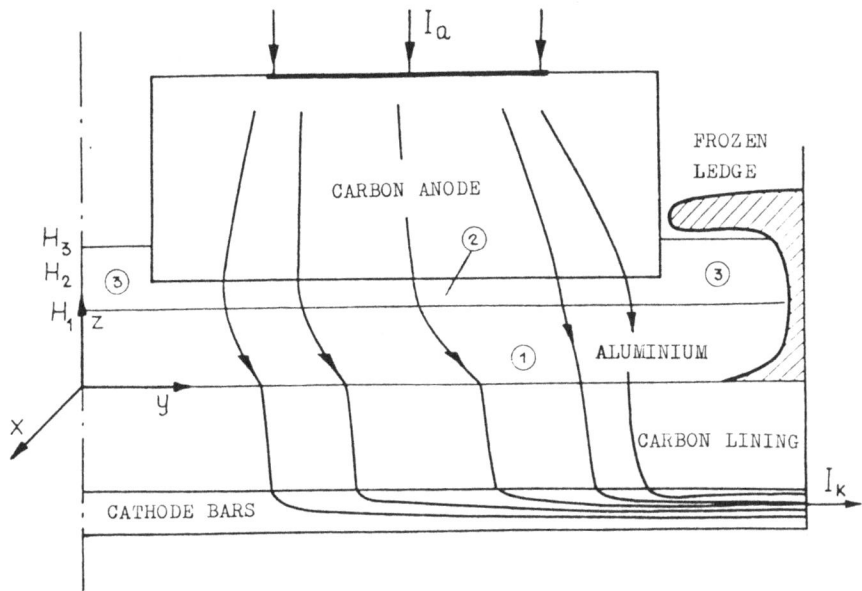

Fig. 8.46. Schematic representation of the electrolysis cell in a half of cross section $x = $ const. The division in the regions is shown: (1) liquid aluminium, (2) liquid electrolyte beneath the anode blocks, (3) liquid electrolyte in the channels.

cells are operating at 180—280 kA. At the same time, the role of electromagnetic forces have increased, affecting the hydrodynamics of liquid electrolyte and metal, and, consequently, the physicochemical processes [30]. The electromagnetic forces have the effect, first, of driving large scale flows in the liquid metal and electrolyte; knowledge of these circulation contours is necessary to predict the temperature distribution in the cell, transport of alumina and other dopants in the electrolyte, and to reveal a location of possible wall erosion. Apart from this, these forces are responsible for the interface deformation between the metal-electrolyte layers, which may lead to a series of problems, e.g. a local decrease of interpolar distance leading in extreme cases to an electrical shorting of anodes to the cathode metal, dynamic gravity waves on the surface of metal in view of the small density difference between the metal ($\rho_1 = 2.3$ kg/m^3) and electrolyte ($\rho_2 = 2.1$ kg/m^3). The necessity for optimizing the magnetic fields of different current-carrying parts of the cell was realized quite early [5], however the magnetohydrodynamic studies with the aim of controlling the process were initiated relatively recently [28, 66].

Measurements in operating industrial cells are impeded due to the high temperature (960°C) and hostility of the cryolite melt, also a re-equipment of the cell with the aim of optimizing the process is expensive. Therefore mathematical and physical models could be an acceptable alternative to study the MHD processes in the electrolysis cell. However, even the development of models runs into significant difficulties, because of the three-dimensional turbulent nature of the flow driven by the electromagnetic forces, by the gas escaping as a result of electrochemical reactions, and also by the nonuniform temperature distribution. These difficulties are aggravated by a complex, not always exactly known, shape of the boundaries determined by the frozen crust at the cell walls, sedimentations at the bottom, nonuniform burn-out of the anodes, the channels between the anode blocks, and between the anodes and the walls of cell. The necessity of rational approximations is obvious, hence the proposed models have no pretensions to describe exactly the process in the electrolysis cell, rather permitting us to analyse quantitatively a variation of flow characteristics with changes in the design and parameters.

Several mathematical models have been developed to simulate the flows and interface deformation in the cells of prescribed design. The first and most popular model [28, 42, 66] is based on the two-dimensional Navier-Stokes equations in the horizontal plane, in which the turbulent viscosity is evaluated by the well-known $K-\varepsilon$ turbulence model [41]. These equations do not account for the friction of layers at the bottom of the bath, at the lower surface of anodes, and between themselves. The order of magnitude estimates in [9, 46] demonstrate the leading role played by friction at the horizontal surfaces in the balance of forces in the equation of motion. Moreover, the model with a two-dimensional horizontal motion could not account for the effect of channels between the anode blocks and walls, which may substantially change the flow configuration and, as a result of pressure redistribution, also change the deformation of interface.

The most essential feature of the proposed mathematical model is a correct evaluation of the electromagnetic force distribution in the bath. The magnetic field $\mathbf{B}$ in the bath is a superposition of the fields: $\mathbf{B}_I$ (from the current within the bath), $\mathbf{B}_0$ (from the currents in the external circuit: leads, bus bars, several near-by cells, the other row of cells), and $\mathbf{B}_M$ (from the induced magnetization of ferromagnetic parts). The external field $\mathbf{B}_0$ is computed by integrating the Biot-Savart law over the length of approximately 250 linear current elements constituting the circuit of a single cell, then the circuit configuration is periodically translated to the nearby cells and to the other row of cells. In practice the integration is limited by five cells of the main row and five of the other.

It is necessary to evaluate the distribution of electric current density $\mathbf{j} = -\sigma \nabla \varphi$ within the bath by solving the equation $\nabla^2 \varphi = 0$ in consecutive layers of different electrical conductivity σ (Fig. 8.46). Knowing the electric current $\mathbf{j}$, the self-magnetic field is found by integrating Biot-Savart law over the whole volume occupied by the current $\mathbf{j}$:

$$\mathbf{B}_I(\mathbf{r}) = \frac{\mu_0}{4\pi} \int \frac{\mathbf{j}(\mathbf{r}') \times \mathbf{R}}{|\mathbf{R}|^3} \, dV', \quad \mathbf{R} = \mathbf{r} - \mathbf{r}'.$$

The integral magnetic field $\mathbf{B}$ is computed in the two layers of liquid aluminium and electrolyte, thereby a spatial distribution of electromagnetic force $\mathbf{f} = \mathbf{j} \times \mathbf{B}$ is known, which is the basis of the simulation of the hydrodynamics of the bath.

In the following exposition of the hydrodynamic model we shall employ some aspects of the 'shallow water' theory. This analogy could be motivated by the small depth of the liquid layers relative to the horizontal dimensions. Indeed, on introducing a typical horizontal length scale of the electrolysis bath $L = 5$ m and vertical $H = 0.25$ m (the depth of aluminium layer), the ratio $\varepsilon = H/L = 5 \times 10^{-2}$ ($\varepsilon = 10^{-2}$ for the electrolyte) is a small parameter of the problem.

Before proceeding to an analysis of the equations, consider the following three important conditions. First, the flow is turbulent, since the Reynolds number is high, e.g., in the liquid aluminium $Re = UL/\nu = 10^6$ for the typical velocity $U = 0.1$ m/s and viscosity $\nu = 5 \times 10^{-7}$ m²/s. This means that a turbulent effective viscosity ν_T should be used in the equations of motion, which, according to the data of passive additive transport in an operating electrolysis cell [30], could be specified as $\nu_T = 10^4 \nu$. The second important feature is the relation between the typical vertical W and horizontal U velocity scales: $W = \varepsilon U$, which is the consequence of the continuity equation div $\mathbf{v} = 0$. The additional property which is important for estimates of the terms in the equation of motion, is the relation between the magnitudes of vertical and horizontal derivatives, e.g., $|\partial u/\partial x|/|\partial u/\partial z| \sim H/L = \varepsilon$.

In view of the above, the steady-state Navier-Stokes equation and typical scales of the main terms are of the following form:

$$\rho(\mathbf{v}\nabla)\mathbf{v} = -\nabla p + \nabla \cdot (\rho \nu_T \nabla \mathbf{v}) + \mathbf{f} - \rho g \mathbf{e}_z,$$
$$\rho U^2/L \sim -p/L + \rho \nu_T U/L^2 + \rho \nu_T U/H^2 + \mu_0 I^2/L^3, \qquad (8.8.1)$$

where the scale of electromagnetic force $\mathbf{f} = \mathbf{j} \times \mathbf{B}$ is constructed from the typical current density $j = I/L^2$ and magnetic field $B = \mu_0 I/L$ ($\mu_0 = 4\pi \times 10^{-7}$ H/m). We can compose two independent nondimensional groups as the ratios of representative volume forces in (8.8.1):

$$\frac{\text{electromagnetic}}{\text{turbulent friction}} = S_\varepsilon = \frac{\mu_0 I^2 \varepsilon^2}{\nu_T \rho UL},$$

$$\frac{\text{inertial}}{\text{turbulent friction}} = R_\varepsilon = \frac{UL\varepsilon^2}{\nu_T}.$$

For example, in the liquid aluminium ($\rho_1 = 2.3 \times 10^3$ kg/m³) for the current $I = 175$ kA the nondimensional numbers are $S_\varepsilon = 16.7$, $R_\varepsilon = 0.25$. Since the magnitude of R_ε is almost two orders lower than S_ε, the first approximation equation of motion could neglect the inertial term, and we obtain the following three-dimensional equations of motion (for details see [9]):

$$0 = -\nabla p + \frac{\partial}{\partial z}\left(\rho \nu_T \frac{\partial}{\partial z} \mathbf{v}_h\right) + \mathbf{f} - \rho g \mathbf{e}_z; \qquad \text{div } \mathbf{v} = 0, \qquad (8.8.2)$$

where a horizontal velocity is introduced $\mathbf{v}_h = (u, v, 0)$.

The bath volume is divided in three regions (Fig. 8.46): 1 — liquid aluminium, 2 — electrolyte beneath the anode blocks, 3 — electrolyte in the channels along the bath perimeter and in the middle between the anode blocks. In the regions 1 and 2 the solution is sought as an averaged by depth of each layer velocity and pressure field, and in the region 3 — an average (by depth and width) of each channel. The turbulent friction term is substituted by either a linear or a nonlinear term given by the velocity's empirical dependence [9, 46], and the resulting equations are solved numerically (or analytically if the electromagnetic force is represented in a simplified closed form). When the averaged pressure field is known, the z-component of equation (8.8.2) can be integrated yielding the three-dimensional pressure distributions p_1 and p_2 in the respective layers. Then the shape of the deformed interface between the aluminium and electrolyte, $z_1 = H_1(x, y)$, follows from the condition $p_1 = p_2$ at the interface.

Prior to examining an example of the solution, note some features of the hydrodynamic situation. The shape of the interface is closely related to the velocity fields in the fluid layers. These flows are mainly governed by the electromagnetic force distribution within the fluid, which also depends on the design of the cell, as well as on the distribution of current loads between the individual anodes and cathode bars. If the current loads differ between the individual elements, a horizontal current ΔI along the x-axis is induced in the liquid metal. In this case the horizontal current density $(\Delta I/LH)$ is significantly higher than the vertical current $(\Delta I/L^2)$, and the density of electromagnetic force increases in proportion to $\varepsilon^{-1} = L/H$. Hence the optimum working conditions in the cell would be with uniform current loads.

This situation is also a convenient one for mathematical modelling, i.e., the current within the bath is almost invariable along the x-axis. Hence the computation may be restricted to the two-dimensional potential $\varphi(y, z)$, similarly to [28, 42], leading to the two-dimensional current distribution: $j_x = 0$, $j_y(y, z)$, $j_z(y, z)$. However, the magnetic field distribution is three-dimensional both for the external field $\mathbf{B}_0$, and the self-field $\mathbf{B}_I$ ($\mathbf{B}_M = 0$ is assumed in this example). The electric current configuration also depends on a relative bottom area covered by the frozen crust. The situation with the insulating crust covering the bottom to the half of channel width is represented in Fig. 8.46 where the computed electric current lines in a cross section $x = $ const are shown.

The flow fields and interface shape have been computed for the specified electric current passage conditions within the 175 kA cell positioned lengthwise along the line of pots. The flow configuration in the liquid aluminium (Fig. 8.47) is quite complicated and differ significantly from the configuration in the electrolyte (Fig. 8.48), where the flow is substantially affected by the presence of channels and the relative magnitude of hydraulic resistance. The respective interface deformation is depicted in Fig. 8.49.

When the operating conditions of the cell are disturbed, the MHD-processes become unsteady, i.e. the interface shape changes in time. Then the electric current, following the path of least resistance, is redistributed in such a way that a significant fraction of it passes through the convex part of the interface. The situation could become unstable, leading to a serious disruption of the technological process. The linear stability analysis for an idealized situation of infinite width fluid layers is presented in [64], however, the real situation is more

Experimental investigations of EVF and applications 363

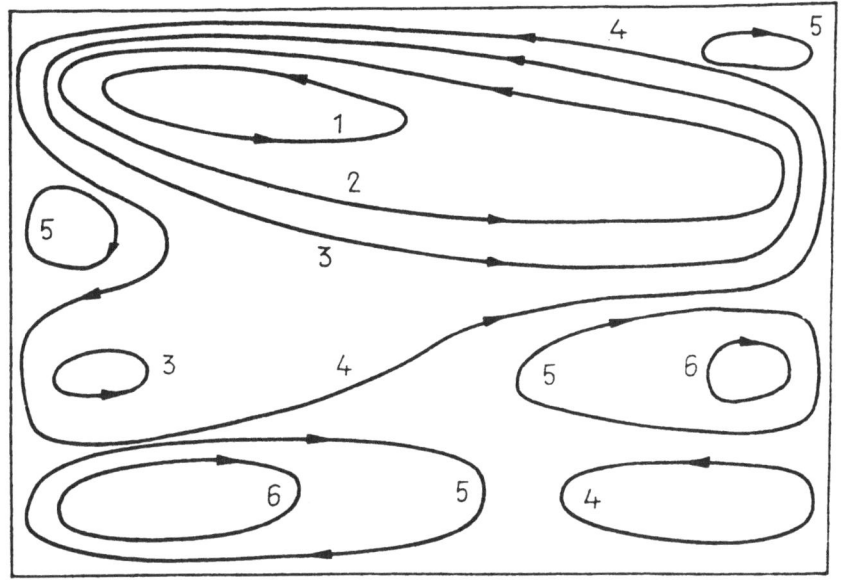

Fig. 8.47. Horizontal stream lines in the liquid aluminium layer $\psi_1 = C$. (1) $C = -0.33$, (2) -0.23, (3) -0.13, (4) -0.02, (5) 0.03, (6) 0.13.

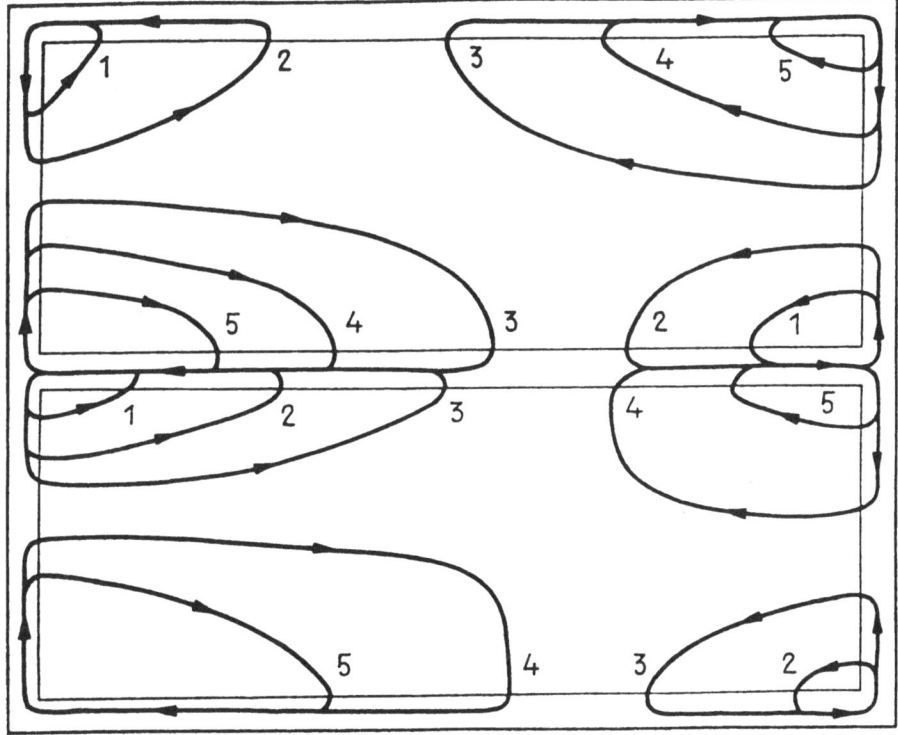

Fig. 8.48. Stream lines in the electrolyte $\psi_{2,3} = C$. (1) $C = -1.7$, (2) -0.9, (3) -0.1, (4) 0.6, (5) 1.4.

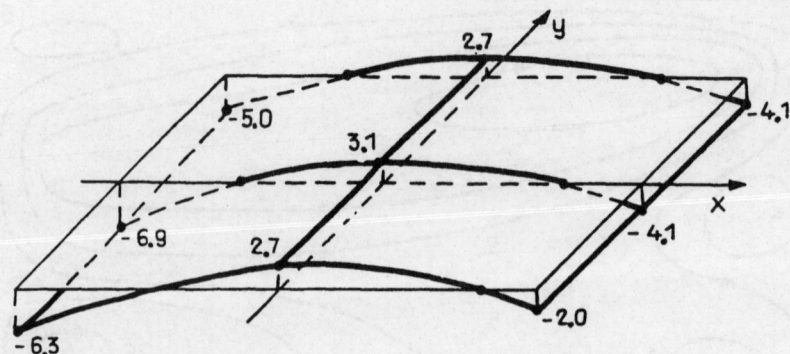

Fig. 8.49. The shape of interface between the aluminium and electrolyte (the numbers indicate the height deviation in centimeters from the initial plane).

complex, including the effect of interdependent external and internal magnetic fields, and of a bounded flow region.

References

Introduction

1. Chow C.-Y.: Flow around a nonconducting sphere in a current-carrying fluid. *Phys. Fluids* (1966), **9**(5), pp. 933–936.
2. Crane J. S. and Pohl H. A.: A study of living and dead yeast cells using dielectrophoresis. *J. Electrochem. Soc.* (1968), No. 6, pp. 584–586.
3. Dijkhuis G. C.: Threshold current for fireball generation. *J. Appl. Phys.* (1982), **53**(5), pp. 3516–3519.
4. Electromagnetic guns and launchers. *Phys. Today* (1980), No. 12, pp. 19–21.
5. Finkelnburg W. and Maecker H.: Electrische Bögen und Thermisches Plasma. *Handbuch der Physik*, Bd. XXII, S. 254–444, 1956.
6. Gelfgat Yu. M., Lielausis O. A., and Shcherbinin E. V.: *Liquid Metal under the Action of Electromagnetic Forces*. Riga: Zinatne, 1976 (In Russian).
7. Gibson E. G.: *The Quiet Sun*. Scientific and Technical Information Office National Aeronautics and Space Administration. Washington, 1973.
8. Grozdowskij G. L. et al.: Axisymmetric meridional flow of conducting fluid. *Izvestiya Akademii Nauk SSSR. Mekhanika i Mashinostroenie* (1960), No. 1, pp. 41–46.
9. Hagyard H., Low B. C., and Tandberg-Hanssen E.: On the presence of electric currents in the Solar atmosphere. *Solar Phys.* (1981), **73**, pp. 257–268.
10. Jones G. R. and Fang M. T. C.: The physics of high-power arcs. *Rep. Progr. Phys.* (1980), **43**, pp. 1415–1465.
11. Kotov V. A.: Rotation of sunspot material. *Izvestiya Krymskoj Astrofizicheskoj Observatorii* (1976), **34**, pp. 184–200.
12. Lamb H.: *Hydrodynamics*. Cambridge University Press, Cambridge, 1957.
13. Lundquist S.: On the hydromagnetic viscous flow generated by a diverging electric current. *Ark. Fys.* (1969), **40**, No. 5, pp. 89–95.
14. Maecker H.: Plasmaströmungen in Lichtbögen infolde eigenmagnetischer Kompression. *Ztsch. Phys.* (1955), Bd. 141, S. 198–216.
15. Milne-Thompson L. M.: *Theoretical Hydrodynamics*, 4th ed. Macmillan, New York, 1960.
16. Moffatt H. K.: *Magnetic Field Generation in Electrically Conducting Fluids*. Cambridge University Press, Cambridge, 1978.
17. Morozov A. I.: *Physical Principles of Electrical Space Propulsion Engines*. Moscow: Atomizdat, 1978 (In Russian).
18. Northrup E. F.: Some newly observed manifestations of forces in the interior of an electric conductor. *Phys. Rev.* (1907), **24**, pp. 474–497.
19. Potemra T. A.: Current systems in the Earth's magnetosphere. *Rev. Geophys. Space Phys.* (1979), **17**(4), pp. 640–656.
20. Ryan R. T. and Vonnegut B.: Miniature whirlwinds produced in the laboratory by high-voltage electrical discharges. *Science* (1970), **168**, pp. 67–69.

21. Serdyuk G. B.: Magnetohydrodynamic effects in an electrical arc. *Magnitnaya Gidrodinamika* (1966), No. 4, pp. 136—146.
22. Shercliff J. A.: Fluid motion due to an electric current source. *J. Fluid Mech.* (1970), **40**(2), pp. 241—250.
23. Shilov V. N. and Estrela-Leopis V. R.: Theory of suspension spherical particles motion in a nonuniform electrical field. In: *Contact Forces in Thin Films and Dispersions.* Moscow: Nauka, 1972 (In Russian).
24. Shilova E. I.: Possible manifestations of electrically induced vortical flow in dielectrophoresis. 9th Riga Conference on MHD, vol. 1. Salaspils, 1978, pp. 158—159 (In Russian).
25. Shilova E. I. and Shcherbinin Eh. V.: MHD vortex flow in a cone. *Magnitnaya Gidrodinamika* (1971), No. 2, pp. 33—38.
26. Sozou C.: On fluid motions induced by an electric current source. *J. Fluid Mech.* (1971), **46**(1), pp. 25—32.
27. Tidman D. A. and Goldstein S. A.: Acceleration of projectiles to hypervelocity using a series of imploded annular plasma discharges. *J. Appl. Phys.* (1980), **51**(4), pp. 1975—1983.
28. Uberoi M. S.: Magnetohydrodynamics at small magnetic Reynolds numbers. *Phys. Fluids* (1962), **5**(4), pp. 401—406.
29. Wienecke K.: Uber das Geschwindigkeitsfeld der Hochstromkohlebogensäule. *Ztschr. Phys.* (1955), **143**(1), S. 129—140.
30. Yantovsky Ye. I. and Apfelbaum M. S.: About the force from needle electrode in a slightly conducting liquid dielectric and simulated fluid flows. *Magnitnaya Gidrodinamika* (1977), No. 4, pp. 73—80.
31. Zhigulev V. N.: On the ejection effect due to an electrical discharge. *Doklady Akademii Nauk SSSR* (1960), **130**(2), pp. 280—283.

Chapter 1

1. Germain P.: *Cours de Mecanique des Milieux Continus.* Paris: Masson et Cie, 1973.
2. Happel J. and Brenner H.: *Low Reynolds Number Hydrodynamics.* Prentice-Hall, 1965.
3. Kalihman L. E.: *Elements of Magnetohydrodynamics.* Moscow: Atomizdat, 1964 (In Russian).
4. Kamke E.: *Differential Gleichungen, Lösungsmethoden und Lösungen.* Leipzig, 1959.
5. Kirko I. M.: *Liquid Metal in Electromagnetic Field.* Moscow: Energiya, 1964 (In Russian).
6. Landau L. D. and Lifshits E. M.: *Hydrodynamics.* Moscow: Nauka, 1986 (In Russian).
7. Loitsyanskij L. G.: *Fluid and Gas Mechanics.* Moscow: Nauka, 1973 (In Russian).
8. Milne-Thompson L. M.: *Theoretical Hydrodynamics.* New York: Macmillan, 1960.
9. Shercliff J. A.: *A Textbook of Magnetohydrodynamics.* Pergamon Press, 1965.
10. Sutton G. W. and Sherman A.: *Engineering Magnetohydrodynamics.* McGraw-Hill Book Company, 1965.
11. Vatazhin A. B., Lyubimov G. A., and Regirer S. A.: *Magnetohydrodynamic Flows in Channels.* Moscow: Nauka, 1970 (In Russian).

Chapter 2

1. Ajayi O. O.: The transient Stokes flow induced by a point source of electric current. *Meccanica* (1985), **20**, pp. 12—17.
2. Ajayi O. O., Sozou C., and Pickering W. M.: Nonlinear fluid motions in a container due to the discharge of an electric current. *J. Fluid Mech.* (1984), **148**, pp. 285—300.
3. Andrews J. G. and Craine R. E.: Fluid flow in a hemisphere induced by a distributed source of current. *J. Fluid Mech.* (1978), **84**(2), pp. 281—290.
4. Atthey D. R.: A mathematical model for fluid flow in a weld pool at high currents. *J. Fluid Mech.* (1980), **98**(4), pp. 787—801.
5. Batchelor G. K. and Gill A. E.: Analysis of the stability of axisymmetric jets. *J. Fluid Mech.* (1962), **14**(4), pp. 529—551.

References

6. Batchelor G. K.: *An Introduction to Fluid Dynamics.* Cambridge at the University Press, 1970.
7. Bojarevičs V. V.: On the application limits of an exact solution of MHD equations. *Magnitnaya Gidrodinamika* (1976), No. 4, pp. 140–141.
8. Bojarevičs V. V.: Electrovortex flow at a hemispherical electrode. *Magnitnaya Gidrodinamika* (1978), No. 4, pp. 77–81.
9. Bojarevičs V. V. and Millere R.: Amplification of rotation of the meridional electrovortex flow in a hemisphere. *Magnitnaya Gidrodinamika* (1982), No. 4, pp. 51–56.
10. Bojarevičs V. V. and Shilova E. I.: Landau-Squire jet in radially diverging electric current. *Magnitnaya Gidrodinamika* (1977), No. 3, pp. 89–94.
11. Cantwell B. J.: Transition in the axisymmetric jet. *J. Fluid Mech.* (1981), **104**, pp. 369–386.
12. Clayton B. R. and Massey B. S.: Exact similar solution for an axisymmetric laminar boundary layer on a circular cone. *AIAA J.* (1979), **17**(7), pp. 785–786.
13. Cooke J. C.: On Pholhausen's method with application to a swirl problem of Taylor. *J. Aeronautical Sci.* (1952), **19**(7), pp. 486–490.
14. Cowley M. D.: On electrode jets. In: Proc. 11th Intern. conf. phenomena ionized gases. Prague, 1973, p. 249.
15. Dangey J. W.: *Cosmic Electrodynamics.* Cambridge at the University Press, 1958.
16. Germain P.: *Cours de Mechanique des Milieux Continus.* Paris: Massan et Cie, 1973.
17. Goldshtik M. A.: A paradoxical solution of Navier-Stokes equation. *Prikladnaya Mekhanika i Matematika* (1960), **24**(4), pp. 610–621.
18. Goldshtik M. A.: *Vortex Flows.* Novosibirsk: Nauka, 1981 (In Russian).
19. Goldshtik M. A. and Silant'ev A. B.: To the theory of laminar submerged jets. *Prikladnaya Matematika i Teoreticheskaya Fizika* (1965), No. 5, pp. 149–152.
20. Golubinskij A. A. and Sichev V. V.: About a similarity solution of Navier-Stokes equations. *Uchenye Zapiski TSAGI* (1976), **7**(6), pp. 11–17.
21. Guilloud J. C. and Arnault J.: Sur une nouvelle classe de solutions exactes des equations de Navier-Stokes. *C. R. Acad. Sci., ser. A* (1971), **273**, pp. 517–520.
22. Hamel G.: Spiralformige Bewegung zäher Flüssigkeiten. *Iber. Dt. Math.-Ver.* (1916), **25**, S. 34–60.
23. Jeffery G. B.: Steady motion of a viscous fluid. *Phil. Mag., ser. 6* (1915), **29**, p. 455.
24. Joukov M. F., Koroteev A. S., and Uryukov B. A.: *Applied Dynamics of Thermal Plasma.* Novosibirsk: Nauka, 1975 (In Russian).
25. Kashkarov V. P.: Some exact solutions in the theory of incompressible fluid jets. *Investigations in physical basis of working process in stoves and furnaces.* Ed. by Vulis L. A. Alma-Ata: Kazakhian Academy of Sciences (1957), pp. 54–63 (In Russian).
26. Kidd G. J. and Farris G. J.: Potential Vortex Flow Adjacent to a Stationary Surface. *Trans. of the ASME. Ser. E. J. Appl. Mech.* (1968), **35**, No. 2.
27. Landau L. D.: New exact solution of Navier-Stokes equations. *Doklady Akademii Nauk SSSR* (1944), **43**(7), pp. 299–301.
28. Landau L. D. and Lifshits E. M.: *Hydrodynamics.* Moscow: Nauka, 1986 (In Russian).
29. Leibowich S.: The structure of vortex breakdown. *Ann. Rev. Fluid Mech.* (1978), **10**, pp. 221–246.
30. Levakov V. S. and Lyubavskij K. V.: The effect of longitudinal magnetic field on electric arc with nonmelting tungsten electrode. *Svarochnoe Proizvodstvo* (1965), No. 10, pp. 9–12.
31. Loitsyanskij L. G.: Propagation of swirling jet in unbounded space filled with the same liquid. *Prikladnaya Matematika i Mekhanika* (1953), **17**(1), pp. 3–16.
32. Loitsyanskij L. G.: Radial jet in space filled by the same liquid. *Trudy Leningradskogo Politekhnicheskogo Instituta* (1953), No. 5, pp. 5–14.
33. Loitsyanskij L. G.: *Fluid and Gas Mechanics.* Moscow: Nauka, 1973 (In Russian).
34. Lundquist S.: On the hydromagnetic viscous flow generated by a diverging electric current. *Ark. Fys.* (1969), **40**(5), pp. 89–95.
35. Moffatt H. K.: Some problems in the magnetohydrodynamics of liquid metals. *Ztschr. Angew. Math. Mech.* (1978), **58**, S. T65–T71.
36. Morgan A. J. A.: On a class of laminar viscous flows within one or two bounding cones. *Aeronautical Quart.* (1956), **7**, pp. 225–239.

37. Narain J. P. and Uberoi M. S.: Magnetohydrodynamics of conical flows. *Phys. Fluids* (1971), **14**(12), pp. 2687—2692.
38. Narain J. P. and Uberoi M. S.: Fluid motion caused by conical currents. *Phys. Fluids* (1973), **16**(6), pp. 940—942.
39. Pao H. P. and Long R. R.: Magnetohydrodynamic jet-vortex in a viscous conducting fluid. *Quart. J. Mech. Appl. Math.* (1966), **19**(1), pp. 1—26.
40. Potsch K.: Schwache Auftriebseffekte in laminaren, vertikalen, runden Freistrahlen. *Acta Mech.* **36**(1/2), pp. 1—14.
41. Ranger K. B.: Hydromagnetic momentum source. *Phys. Fluids* (1965), **8**(9), pp. 1747—1748.
42. Rumer Yu. B.: Submerged jet problem. *Prikladnaya Matematika i Mekhanika* (1952), **16**(2), pp. 255—256.
43. Rumer Yu. B.: Convective diffusion in submerged jet. *Prikladnaya Matematika i Mekhanika* (1953), **17**(6), pp. 743—744.
44. Samerville G. M.: *Electric Arc.* London: Methuen and Co., 1959.
45. Slezkin N. A.: Viscous fluid motion within two cones. *Uchenye Zapiski MGU* (1934), No. 2, pp. 83—87.
46. Schlichting H.: *Boundary Layer Theory.* New York: McGraw-Hill, 1960.
47. Schneider W.: Flow induced by jets and plumes. *J. Fluid Mech.* (1981), **108**, pp. 55—65.
48. Schwiderski E. W.: On the axisymmetric vortex flow over a flat surface. *Trans. of the ASME. Ser. E, J. Appl. Mech.* (1969), **36**(3), pp. 614—619.
49. Shcherbinin E. V.: On a kind of exact solutions in MHD. *Magnitnaya Gidrodinamika* (1969), No. 4, pp. 46—58.
50. Shcherbinin E. V.: *Viscous Fluid Jet Flows in Magnetic Field.* Riga: Zinatne, 1973 (In Russian).
51. Shcherbinin E. V.: Jet flows in an electric arc. *Magnitnaya Gidrodinamika* (1973), No. 4, pp. 66—72.
52. Shcherbinin E. V. and Yakovleva E. E.: An electrovortex flow in a spheroidal container. *Magnitnaya Gidrodinamika* (1986), No. 4, pp. 64—69.
53. Shercliff J. A.: Fluid motions due to an electric current source. *J. Fluid Mech.* (1970), **40**(2), pp. 241—250.
54. Shilova E. I. and Shcherbinin E. V.: Some exact solutions of the equations of motion in MHD. *Magnitnaya Gidrodinamika* (1969), No. 4, pp. 59—64.
55. Shilova E. I. and Shcherbinin E. V.: Some aspects of theoretical study of a three-dimensional MHD flow in a diffuser. *Magnitnaya Gidrodinamika* (1971), No. 1, pp. 11—17.
56. Shilova E. I. and Shcherbinin E. V.: MHD vortex flow in a cone. *Magnitnaya Gidrodinamika* (1971), No. 2, pp. 33—38.
57. Sozou C.: On fluid motions induced by an electric current source. *J. Fluid Mech.* (1971), **46**(1), pp. 25—32.
58. Sozou C.: On some similarity solutions in magnetohydrodynamics. *J. Plasma Phys.* (1971), **6**(2), pp. 331—341.
59. Sozou C.: On magnetohydrodynamic flows generated by an electric current discharge. *J. Fluid Mech.* (1974), **63**(4), pp. 665—671.
60. Sozou C.: Development of the flow field of a point force in an infinite fluid. *J. Fluid Mech.* (1979), **91**(3), pp. 541—546.
61. Sozou C. and English H.: Fluid motions induced by an electric current discharge. *Proc. Roy. Soc. London* (1972), **A329**, pp. 71—81.
62. Sozou C. and Pickering W. M.: The development of magnetohydrodynamic flow due to an electric current discharge. *J. Fluid Mech.* (1975), **70**(3), pp. 509—517.
63. Sozou C. and Pickering W. M.: Magnetohydrodynamic flow due to the discharge of an electric current in a hemispherical container. *J. Fluid Mech.* (1976), **73**(4), pp. 641—650.
64. Sozou C. and Pickering W. M.: The round laminar jet: the development of the flow field. *J. Fluid Mech.* (1977), **80**(4), pp. 673—683.
65. Sozou C. and Pickering W. M.: Magnetohydrodynamic flows in a container due to the discharge of an electric current from a finite size electrode. *Proc. Roy. Soc. London* (1978), **A362**, pp. 509—523.

66. Squire H. B.: The round laminar jet. *Quart. J. Mech. Appl. Math.* (1951), **4**(3), pp. 321–329.
67. Squire H. B.: Jet emerging from a hole in a plane wall. *Phil. Mag.*, ser. 7 (1952), **43**(343), pp. 942–945.
68. Squire H. B.: Radial jets. In: *50 Jahre Grenaschichforschung.* Braunschweig, 1954, S. 47–54.
69. Tsuker M. S.: Swirling jet propagating in space filled by the same liquid. *Prikladnaya Matematika i Mekhanika* (1955), **19**(4), pp. 500–503.
70. Vulis L. A. and Kashkarov V. P.: *Theory of Viscous Fluid Jets.* Moscow: Nauka, 1965 (In Russian).
71. Weber H. E.: The boundary layer inside a conical surface due to swirl. *J. Fluid Mech.* (1956), **23**(4), pp. 587–592.
72. Wu Ch.-Sh.: A class of exact solutions of the magnetohydrodynamic Navier-Stokes equations. *Quart. J. Mech. Appl. Math.* (1961), **14**(1), pp. 1–19.
73. Yatseyev V. I.: About a class of exact solutions of viscous fluid equations of motion. *Zhurnal Tekhnicheskoj Fiziki* (1950), **20**(11), pp. 1031–1034.
74. Yih C.-S., Wu F., Garg A. K., and Leibovich S.: Conical vortices: a class of exact solutions of the Navier-Stokes equations. *Phys. Fluids* (1982), **25**(12), pp. 2147–2158.

Chapter 3

1. Atthey D. R.: A mathematical model for fluid flow in a weld pool at high currents. *J. Fluid Mech.* (1980), **98**, pp. 787–801.
2. Batchelor G. K.: *An Introduction to Fluid Dynamics.* Cambridge at the University Press, 1970.
3. Bershadskij A. G.: Stability of axisymmetric meridional flows to azimuthal rotations. *Magnitnaya Gidrodinamika* (1985), No. 1, pp. 49–54.
4. Bojarevičs V. V.: MHD flows at an electric current point source. Part I. *Magnitnaya Gidrodinamika* (1981), No. 1, pp. 21–28.
5. Bojarevičs V. V.: MHD flows at an electric current point source. Part II. *Magnitnaya Gidrodinamika* (1981), No. 2, pp. 41–44.
6. Bojarevičs V. V. and Millere R.: Amplification of rotation of meridional electrovortex flow in a hemisphere. *Magnitnaya Gidrodinamika* (1982), No. 4, pp. 51–56.
7. Bojarevičs V. V., Sharamkin V. I., and Shcherbinin E. V.: Effect of longitudinal magnetic field on the medium motion in electrical arc welding. *Magnitnaya Gidrodinamika* (1977), No. 1, pp. 115–120.
8. Bojarevičs V. and Shcherbinin E. V.: Azimuthal rotation in the axisymmetric meridional flow due to an electric current source. *J. Fluid Mech.* (1983), **126**, pp. 413–430.
9. Burgers J. M.: A mathematical model illustrating the theory of turbulence. *Adv. Appl. Mech.* (1948), **1**, pp. 197–198.
10. Chow Ch. Y. and Uberoi M. S.: Stability of an electrical discharge surrounded by a free vortex. *Phys. Fluids* (1972), **15**(12), pp. 2187–2192.
11. Craine R. E. and Weatherill N. P.: Fluid flow in a hemispherical container induced by a distributed source of current and superimposed uniform magnetic field. *J. Fluid Mech.* (1980), **99**, pp. 1–12.
12. Dudko D. A. and Rublevskij I. N.: Electromagnetic mixing of slag and metal bath in electroslag process. *Avtomaticheskaya Svarka* (1960), No. 9, pp. 12–16.
13. Gagen Yu. G. and Taran V. D.: *Welding by the Magnetically Controlled Arc.* Moscow: Mashinostroenie, 1970 (In Russian).
14. Howells P. and Smith R. K.: Numerical simulations of tornado-like vortices. *Geophys. Astrophys. Fluid Dynamics* (1983), **27**, pp. 253–284.
15. Kawakubo T., Tsutchiya Y., Sugaya M., and Matsumura K.: Formation of a vortex around a sink. *Phys. Letters* (1978), **A68**(1), pp. 65–66.
16. Kawakubo T., Shingubara S., and Tsutchiya Y.: Coherent structure formation of vortex flow around a sink. *J. Phys. Soc. Jpn.* (1983), **52** (Suppl.), pp. 143–146.

17. Lavrent'ev M. A. and Shabat B. V.: *Problems in Hydrodynamics and Mathematical Models.* Moscow: Nauka, 1977 (In Russian).
18. Levakov V. S. and Lyubavskij K. V.: Effect of longitudinal magnetic field on electric arc with nonmelting tungsten cathode. *Svarochnoe Proizvodstvo* (1965), No. 10, pp. 9—12.
19. Millere R. P., Sharamkin V. I., and Shcherbinin E. V.: Effect of longitudinal magnetic field on electrovortex flow in the cylindrical volume. *Magnitnaya Gidrodinamika* (1980), No. 1, pp. 81—85.
20. Moffatt H. K.: Some problems in the magnetohydrodynamics of liquid metals. *Ztschr. Angew. Math. Mech.* (1978), **58**, T65—T71.
21. Morton B. R.: The strength of vortex and swirling core flows. *J. Fluid Mech.* (1969), **38**, pp. 315—333.
22. Okorokov N. V.: *Electromagnetic Mixing of Metal.* Moscow: Metallurgizdat, 1961 (In Russian).
23. Orszag S.: Numerical simulation of incompressible flows within simple boundaries. 1. Galerkin (spectral) representations. *Studies Appl. Math.* (1971), **50**, pp. 293—327.
24. Rotunno R.: Vorticity dynamics of a convective swirling boundary layer. *J. Fluid Mech.* (1980), **97**(3), pp. 623—640.
25. Ryan R. T. and Vonnegut B.: Miniature whirlwinds produced in the laboratory by high-voltage electrical discharges. *Science* (1970), **168**, pp. 1349—1351.
26. Shcherbinin E. V.: *Viscous Fluid Jet Flows in Magnetic Field.* Riga: Zinatne, 1973 (In Russian).
27. Shercliff J. A.: Fluid motion due to an electric current source. *J. Fluid Mech.* (1970), **40**(2), pp. 241—250.
28. Shercliff J. A.: The dynamics of conducting fluids under rotational magnetic forces. *Sci. Progress* (1979), **66**, pp. 151—170.
29. Shivamoggi B. K.: Method of matched asymptotic expansions — asymptotic matching principle for high approximations. *Ztschr. Angew. Math. Mech.* (1978), **58**(8), S. 354—356.
30. Smyslov Yu. N. and Shcherbinin E. V.: Nonlinear magnetohydrodynamic model of tornado. *Problems of Mathematical Physics.* Ed. by Tuchkevich V. M. Leningrad: Nauka, 1976, pp. 271—282 (In Russian).
31. Sozou C. and English H.: Fluid motion induced by an electric current discharge. *Proc. Roy. Soc. London* (1972), **A329**, pp. 71—81.
32. Sozou C. and Pickering W. M.: Magnetohydrodynamic flow due to the discharge of an electric current in a hemispherical container. *J. Fluid Mech.* (1976), **73**(4), pp. 641—650.
33. Torrance K.: Natural convection in the thermally stratified enclosures with localized heating from below. *J. Fluid Mech.* (1979), **95**, pp. 474—495.
34. Van Dyke M. D.: *Perturbation Methods in Fluid Mechanics.* New York: Academic Press, 1964.
35. Weir A. D.: Axisymmetric convection in a rotating sphere. 1. *J. Fluid Mech.* (1976), **75**(1), pp. 49—79.
36. Wilson T. and Rotunno R.: Numerical simulation of a laminar endwall vortex and boundary layer. *Phys. Fluids* (1986), **29**(12), pp. 3993—4005.

Chapter 4

1. Abdullah A. J.: Some aspects of the dynamics of tornadoes. *Month. Weather Rev.* (1955), No. 83, pp. 83—88.
2. Axford W. I.: Axisymmetric stagnation point flow in magnetohydrodynamics. *Appl. Sci. Research* (1961), **9**, sect. B, No. 1, pp. 213—229.
3. Bacon F.: Natural history of winds — Extraordinary winds and sudden blasts. In: 1622. *The Works of Francis Bacon*, Carey and Hart. Philadelphia, 1844, vol. 3, p. 449.
4. Batchelor G. K.: *An Introduction to Fluid Dynamics.* Cambridge at the University Press, 1970.

5. Bojarevičs V. V. and Millere R. P.: Electrovortex flow in Karman's class. 10th Riga Conference on MHD. Salaspils, 1981, vol. 1, pp. 157−158 (In Russian).
6. Bojarevičs V. V. and Sharamkin V. I.: MHD flow due to the discharge of an electric current in an axially symmetric layer of finite thickness. *Magnitnaya Gidrodinamika* (1977), No. 2, pp. 55−60.
7. Bojarevičs V. V., Freibergs J. Ž., Shilova E. I., and Shcherbinin E. V.: *Electro-Vortex Flows*. Riga: Zinatne, 1985 (In Russian).
8. Brady J. F. and Acrivos A.: Steady flow in a channel or tube with an accelerating surface velocity. *J. Fluid Mech.* (1981), **112**, pp. 127−150.
9. Brady J. F.: Flow development in a porous channel and tube. *Phys. Fluids* (1984), **27**(5), pp. 1061−1067.
10. Braham R. R.: The water and energy budgets of the thunderstorm and their relation to thunderstorm development. *J. Meteorol.* (1952), No. 9, pp. 237−243.
11. Brooks E. M.: The tornado-cyclone. *Weatherwise* (1949), **2**, pp. 32−33.
12. Bucenieks I. E., Petersons D. E., Sharamkin V. I., and Shcherbinin E. V.: MHD flows caused by diverging currents passing through closed volumes of liquid. *Magnitnaya Gidrodinamika* (1976), No. 1, pp. 92−97.
13. Cham T. S.: The laminar boundary layer of a source and vortex flow. *Aeronautical Quart.* (1971), **22**(2), pp. 196−206.
14. Glazov O. A.: Rotation of conducting liquid under the stationary disc in the presence of magnetic field. *Magnitnaya Gidrodinamika* (1967), No. 2, pp. 75−80.
15. Goldshtik M. A.: A class of exact solutions of Navier-Stokes equations. *Zhurnal Prikladnoj Mekhaniki i Tekhnicheskoj Fiziki* (1966), No. 2, pp. 106−109.
16. Goldshtik M. A.: *Vortex Flows*. Novosibirsk: Nauka, 1981 (In Russian).
17. Golubinskij A. A. and Sychev V. V.: About a similarity solution of Navier-Stokes equations. *Uchenye Zapiski TSAGI* (1976), **7**(6), pp. 11−17.
18. Greenspan H. P.: *The Theory of Rotating Fluids*. Cambridge at the University Press, 1968.
19. Gribben R. J.: Magnetohydrodynamic stagnation-point flow. *Quart. J. Mech. Appl. Math.* (1965), **18**(3), pp. 357−384.
20. Gutman L. N.: Theoretical Model of Tornado. *Izvestiya Akademii Nauk SSSR, Ser. Geophys.* (1957), N 1, pp. 79−93.
21. Gutman L. N.: *Introduction to the Nonlinear Theory of Mesometeorological Processes*. Leningrad: Gidrometeoizdat, 1969 (In Russian).
22. Hare R.: On the causes of the tornado or waterspout. *Amer. J. Sci. Arts* (1837), **32**, pp. 153−158.
23. *Intense Atmospheric Vortices*. Ed. by Bengtsson L., Lighthill J. Berlin-Heidelberg: Springer Verlag, 1982.
24. Kakutani T.: Axially symmetric stagnation-point flow of an electrically conducting fluid under transverse magnetic field. *J. Phys. Soc. Jpn.* (1960), **15**(4), pp. 688−695.
25. Karman Th.: Über laminare und turbulente Reibung. *ZAMM* (1921), **1**, S. 233−251.
26. Kislykh V. I. and Smulskij I. I.: To the hydrodynamics of vortex chamber. *Inzhenerno-Fizicheskij Zhurnal* (1978), **35**(3), pp. 543−544.
27. Lentini M. and Keller H. B.: Computation of Karman swirling flows. *Lecture Notes in Computer Sci.* Ed. by Goos G., Hartmanis J., 1979, vol. 76, pp. 89−100.
28. Lewellen W. S. and King W. S.: Boundary-layer similarity solution for rotating flows with and without magnetic interaction. *Phys. Fluids* (1964), **7**(10), pp. 1674−1680.
29. Lin C. C.: Note on a class of exact solutions in magnetohydrodynamics. *Arch. Rational Mech. Anal.* (1958), **1**(4), pp. 391−395.
30. Loitsyanskij L. G.: *Laminar Boundary Layer*. Moscow: GIFML, 1962 (In Russian).
31. Lucretius: *On Nature of Objects*. Part I. Moscow: Akademiya Nauk SSSR, 1946 (Russian translation).
32. Malbakhov V. M.: Investigation of tornado structure. *Izvestiya Akademii Nauk SSSR. Fizika Atmosfery i Okeana* (1972), **8**(1), pp. 17−28.
33. Malbakhov V. M. and Gutman L. N.: Nonstationary problem of mesoscale atmospheric vortices with vertical axis. *Idem* (1968), **4**(6), pp. 586−598.

34. Meksyn D.: Integration of the boundary-layer equations. *Proc. Roy. Soc. London*, ser. A (1956), **237**(1211), pp. 543—559.
35. Mikhailov A. O.: *About Storms*. Morskoj Sbornik, SPb., 1888, No. 3, pp. 1—37.
36. Millere R. P., Sharamkin V. I., and Shcherbinin E. V.: Effect of longitudinal magnetic field on electrovortex flow in a cylindrical volume. *Magnitnaya Gidrodinamika* (1980), No. 1, pp. 81—85.
37. Millsaps K. and Nydahl J. E.: Heat transfer in a laminar cyclone, *ZAMM* (1973), **53**, pp. 241—246.
38. Nalivkin D. V.: *Hurricanes, Storms and Tornadoes*. Leningrad: Nauka, 1963 (In Russian).
39. Nanbu K.: Vortex flow over a flat surface with suction. *AIAA J.* (1971), **9**(8), pp. 1642—1645.
40. Pao H. P.: Magnetohydrodynamic Flows over a Rotating Disc. *AIAA J.* (1968), **6**(7).
41. Peltier J. C. A.: Translation by Robert Hare. *Amer. J. Sci. Arts* (1840), **38**, p. 73.
42. Petrovskij I. G.: *Lectures in the Theory of Ordinary Differential Equations*. Moscow: Nauka, 1964.
43. Reznikov B. I.: A method to integrate asymptotically the equations of laminar boundary layer. *Aerophysical Investigations of Supersonic Flows*. Ed. by Dunaev Yu. A. Leningrad: Nauka, 1967, pp. 284—300 (In Russian).
44. Reznikov B. I. and Smyslov Yu. N.: Magnetohydrodynamic flow in the vicinity of critical point in purely azimuthal magnetic field. *Izvestiya Akademii Nauk SSSR. Mekhanika Zhidkosti i Gaza* (1967), No. 4, pp. 3—8.
45. Sandler V. Yu.: Numerical study of temperature and velocity fields in a slag pool. *Magnitnaya Gidrodinamika* (1982), No. 2, pp. 113—119.
46. Schlichting H.: Laminare Strahlungsbreitung. *ZAMM* (1933), **13**, S. 260—263.
47. Schlichting H.: *Boundary Layer Theory*. New York: McGraw-Hill, 1960.
48. Shcherbinin E. V.: *Viscous Fluid Jet Flows in Magnetic Field*. Riga: Zinatne, 1973 (In Russian).
49. Shilova E. I. and Shcherbinin E. V.: Some aspects of theoretical study of MHD flow in a diffuser. *Magnitnaya Gidrodinamika* (1971), No. 1, pp. 11—17.
50. Shilova E. I. and Shcherbinin E. V.: MHD model of weak whirlwind. *Magnitnaya Gidrodinamika* (1974), No. 2, pp. 77—86.
51. Sychev V. V.: On viscous electrically conducting fluid motion under the action of rotating disc in the presence of magnetic field. *Prikladnaya Matematika i Mekhanika* (1960), **24**, No. 5, pp. 906—908.
52. Smirnov E. M.: Similarity solutions of Navier-Stokes equations for a swirling flow of incompressible fluid in a circular tube. *Idem* (1981), **45**, No. 5, pp. 833—839.
53. Smyslov Yu. N. and Shcherbinin E. V.: Nonlinear magnetohydrodynamic model of tornado. *Problems of Mathematical Physics*. Leningrad: Nauka, 1976, pp. 271—282 (In Russian).
54. Sozou C.: Electrical discharges and intense vortices. *Proc. Roy. Soc. London* (1984), **A392**, pp. 415—426.
55. Srivastava A. C. and Sharma S. K.: The effect of a transverse magnetic field on the flow between two infinite discs — one rotating and the other at rest. *Bull. Acad. Polon. Sci., Ser. Sci. Techn.* (1961), **9**(11), pp. 639—645.
56. Uman M. A.: *Lightning*. New York: McGraw-Hill Book Co., 1969.
57. Van Dyke M.: Semi-analytical applications of the computer. *Fluid Dynamics Trans.* Ed. by W. Fiszdon *et al.*, Warszawa, 1978, vol. 9, pp. 305—320.
58. Vlasyuk V. Kh.: Effect of melting electrode radius on the electrically induced vortical flow in a cylindrical container. *Magnitnaya Gidrodinamika* (1987), No. 4, pp. 101—106.
59. Vonnegut B.: Electrical theory of tornadoes. *J. Geophys. Research* (1960), **65**(1), pp. 203—212.
60. Wegener A.: Wind und Wassershosen in Europa. In: *Die Wissenschaft*. Braunschweig, 1917, Bd. 60, S. 301.
61. Yuan S. W., Finkelstein A. B.: Laminar pipe flow with injection and suction through a porous wall. *Trans. ASME* (1956), **78**(4), pp. 719—724.

Chapter 5

1. Dean W. R.: The stream-line motion of liquid in a curved pipe. *Phys. Mag. A* (1928), ser. 7, Vol. 5, No. 30, pp. 673–695.
2. Dennis S. C. R.: Calculation of the steady flow through a curved tube using a new finite-difference method. *J. Fluid Mech.* (1980), **99**(3), pp. 449–468.
3. Freibergs J. Ž.: Transit electrovortex flow in a corrugated tube with longitudinal current. *Magnitnaya Gidrodinamika* (1978), No. 2, pp. 27–31.
4. Freibergs J. Ž.: Flow in a toroidal tube with current. *Magnitnaya Gidrodinamika* (1981), No. 4, pp. 61–66.
5. Freibergs J. Ž. and Shcherbinin E. V.: Axisymmetric electrovortex flow in the tube with periodical contractions. *Magnitnaya Gidrodinamika* (1977), No. 4, pp. 46–54.
6. Gradshteyn I. C. and Ryzhik I. M.: *Tables of Integrals, Sums, and Series*. Moscow: Publishing House of Physical and Mathematical Literature, 1963 (In Russian).
7. Happel J. and Brenner H.: *Low Reynolds Number Hydrodynamics*. Prentice-Hall, 1965.
8. Landau L. D. and Lifshits E.M.: *Electrodynamics of Continuous Media*. Moscow: Nauka, 1982 (In Russian).
9. Millere R. P. and Freibergs J. Ž.: Effect of periodic electrovortex flow on laminar flow through a tube. *Magnitnaya Gidrodinamika* (1980), No. 2, pp. 52–56.
10. Millere R. P. and Sulejmanov R. Kh.: EVF in a tube with varying electrical conductivity of wall. 12th Riga Conference on MHD. Salaspils, 1987, vol. 1, pp. 223–226 (In Russian).
11. Olson D. B. and Spence T. W.: Asymmetric disturbances in the frontal zone of a Gulf Stream. *J. Geophys. Research* (1978), **83**(C9), pp. 4691–4695.
12. Starr V. P.: *Physics of Negative Viscosity Phenomena*. McGraw Hill Book Co., 1968.
13. Sulejmanov R. Kh.: Axisymmetric electrovortex flow in a variable cross-section area tube with an internal cylinder. *Magnitnaya Gidrodinamika* (1983), No. 3, pp. 90–94.
14. Sulejmanov R. Kh. and Freibergs J. Ž.: Pump effect of electrovortex flow in a corrugated tube. *Magnitnaya Gidrodinamika* (1985), No. 4, pp. 98–104.
15. Toyakoglu H. C.: Steady laminar flows of an incompressible fluid in curved pipes. *J. Math. Mech.* (1967), **16**(12), pp. 1321–1337.
16. Uberoi M. S.: Magnetohydrodynamics at small magnetic Reynolds numbers. *Phys. Fluids* (1962), **5**(4), pp. 401–406.
17. Uberoi M. S. and Chow Ch.-Y.: Magnetohydrodynamics of flow between two coaxial tubes. *Phys. Fluids* (1966), **9**(5), pp. 927–932.
18. Van Dyke M.: Extended Stokes series: laminar flow through a loosely coiled pipe. *J. Fluid Mech.* (1978), **86**(1), pp. 129–145.
19. Votsish A. D. and Kolesnikov Yu. B.: The anomalous impulse transfer in MHD shear flow with two-dimensional turbulence. *Magnitnaya Gidrodinamika* (1976), No. 4, pp. 47–52.

Chapter 6

1. Bojarevičs V. V., Millere R. P., and Chudnovsky A. Yu.: Forces acting on bodies in a current-carrying fluid. *Magnitnaya Gidrodinamika* (1985), No. 1, pp. 67–72.
2. Bojarevičs V. V. and Shcherbinin E. V.: The electric current flow past the cone. *Magnitnaya Gidrodinamika* (1974), No. 4, pp. 38–42.
3. Bucenieks I. E., Kompan Ya. Yu., Sharamkin V. I., Shilova E. I., and Shcherbinin E. V.: Experimental study of MHD processes in electrical welding. *Magnitnaya Gidrodinamika* (1975), No. 3, pp. 143–148.
4. Chow Ch. Y.: Flow around a nonconducting sphere in a current carrying fluid. *Phys. Fluids* (1966), **9**(5), pp. S. 33–36.
5. Chow Ch. Y.: Hydromagnetic wake around a nonconducting sphere. *Phys. Fluids* (1967), **10**(1), pp. 234–236.

6. Chow Ch. Y. and Billings D. F.: Current-carrying fluid past a nonconducting sphere at low Reynolds number. *Phys. Fluids* (1967), **10**(4), pp. 871—873.
7. Chow Ch. Y. and Halat J. A.: Drag of a sphere of arbitrary conductivity in a current-carrying fluid. *Phys. Fluids* (1969), **12**(11), pp. 2317—2322.
8. Gel'fgat Yu. M., Lielausis O. A., and Shcherbinin E. V.: *Liquid Metal under the Action of Electromagnetic Forces*. Riga: Zinatne, 1976 (In Russian).
9. Graneau P.: Electromagnetic jet-propulsion in the direction of current flow. *Nature* (1982), **295**, pp. 311—312.
10. Gupta R. K.: Flow induced by the presence of a conducting porous sphere in a fluid carrying a uniform current. *ZAMM* (1976), **56**(5), S. 191—196.
11. Happel J. and Brenner H.: *Low Reynolds Number Hydrodynamics*. Prentice-Hall, 1965.
12. Kompan Ya. Yu., Hizhnyak K. K., Bucenieks I. E., Sharamkin V. I., and Shcherbinin E. V.: Modeling studies of flows in slag bath of titanium electrical welding. *Titanium Casting*. Kiev: Naukova Dumka, 1976, pp. 109—115 (In Russian).
13. Lamb H.: *Hydrodynamics*. New York, 1945.
14. Loitsyanskij L. G.: *Laminar Boundary Layer*. Moscow: GIFML, 1962 (In Russian).
15. Oreper G. M.: Effect of the electric current on the velocity field around the nonconducting drop and on its dissolution rate in a conducting liquid. *Magnitnaya Gidrodinamika* (1974), No. 3, pp. 52—56.
16. Oreper G. M.: Numerical research of the spherical hydrodynamics and mass-transfer in a current-carrying fluid. *Magnitnaya Gidrodinamika* (1979), No. 3, pp. 38—42.
17. Oreper G. M.: Influence of electrovortex flow on hydrodynamics and mass transfer of a sphere moving in current-carrying fluid. *Magnitnaya Gidrodinamika* (1980), No. 1, pp. 72—76.
18. Orszag S.: Numerical simulations of incompressible flows within simple boundaries. 1. Galerkin (spectral) representations. *Studies Appl. Math.* (1971), **50**, pp. 293—327.
19. Sharamkin V. I. and Shcherbinin E. V.: Electrovortex flow at discharge between hyperboloidal electrodes. *Magnitnaya Gidrodinamika* (1978), No. 2, pp. 32—38.
20. Shilova E. I.: On the purification of liquid metals from nonconducting particles in the magnetic field produced by the current. *Magnitnaya Gidrodinamika* (1975), No. 2, pp. 142—144.
21. Sozou C.: Flow induced by the presence of a non-conducting ellipsoid of revolution in fluid carrying a uniform current. *J. Fluid Mech.* (1970), **42**(1), pp. 129—138.
22. Sozou C.: Slow flow of a fluid carrying a uniform current past a nonconducting ellipsoid of revolution. *J. Fluid Mech.* (1970), **43**(1), pp. 121—127.
23. Sozou C.: The development of magnetohydrodynamic flow due to the passage of an electric current past a sphere immersed in a fluid. *J. Fluid Mech.* (1972), **56**(3), pp. 497—503.
24. Sozou C.: Flow induced by the presence of a spheroid in a fluid carrying a uniform electric current. *Intern. J. Eng. Sci.* (1977), **15**, pp. 345—358.
25. Van Dyke M. D.: *Perturbation Methods in Fluid Mechanics*. New York: Academic Press, 1964.

Chapter 7

1. Akselrud G. A. and Oreper G. M.: Mass-transfer between a rigid spherical body and current-carrying fluid. *Inzhenerno-Fizicheskij Zhurnal* (1974), **27**(6), pp. 1015—1018.
2. Brounshtein B. I. and Fishbein B. I.: *Hydrodynamics, mass and heat transfer in disperse systems*. Leningrad: Khimija, 1977 (In Russian).
3. Daly B. J. and Harlow F. H.: Turbulent effects in the numerical solution of gas-dynamic problems. *Proc. 2nd Int. Conf. Meth. Fluid Dyn.* Berlin: Springer, 1971.
4. Dilavari A. H. and Szekely J. A.: A mathematical model of slag and metal flow in the ESD process. *Metallurgical Trans.* (1977), **8B**, pp. 227—236.
5. Gershuni G. Z. and Zhukhovitskij E. M.: *Convective Stability of Incompressible Fluid*. Moscow: Nauka, 1972 (In Russian).

References

6. Gupalo Yu. P., Ryazantsev Yu. S., and Chalyuk A. T.: Diffusion in a drop for high Peclet numbers and finite Reynolds numbers. *Izvestiya Akademii Nauk SSSR. Mekhanika Zhidkosti i Gaza* (1972), No. 2, pp. 161—162.
7. Landau L. D. and Lifshits E. M.: *Electrodynamics of Continuous Media*. Moscow: Nauka, 1982 (In Russian).
8. Kreyenberg J. and Schwerdtfeger K.: Stirring velocities and temperature field in the slag during electroslag remelting. *Arch. Eisenhüttenwessen* (1979), **50**(1), S. 1—6.
9. Oreper J. M.: Effect of the electric current on the velocity field around the nonconducting drop and on its dissolution rate in a conducting liquid. *Magnitnaya Gidrodinamika* (1974), No. 3, pp. 52—56.
10. Oreper G. M.: Numerical research of the spherical hydrodynamics and mass-transfer in a current-carrying fluid. *Magnitnaya Gidrodinamika* (1979), No. 9, pp. 38—42.
11. Oreper G. M.: Influence of electrovortex flow on hydrodynamics and mass-transfer of a sphere moving in a current-carrying fluid. *Magnitnaya Gidrodinamika* (1980), No. 1, pp. 72—76.
12. Oreper G. M.: Effect of electromagnetic field on conducting liquid flow near a sphere and on the mass-transfer from its surface. *Magnitnaya Gidrodinamika* (1980), No. 4, pp. 69—72.
13. Polezhaev V. I. and Fedyushkin A. I.: Hydrodynamic effects of concentration stratification in closed volumes. *Mekhanika Zhidkosti i Gaza* (1980), No. 3, pp. 11—18.
14. Sandler V. Yu.: Numerical study of temperature and velocity fields in a slag pool. *Magnitnaya Gidrodinamika* (1982), No. 2, pp. 113—119.
15. Sherwood T. K.: Diffusion effects in heterogeneous catalysis. *Teoreticheskie osnovy khimicheskoj tekhnologii* (1967), **1**(1), pp. 17—30.
16. Szekely J. and Dilavari A. H.: A mathematic description of heat and mass transfer effects during the electroslag refinement. *Problems of Special Electrometallurgy*. Kiev: Naukova Dumka, 1979, pp. 45—65 (In Russian).
17. Vitkov G. A.: On the dissolution of metallic spheres in electric field. *Magnitnaya Gidrodinamika* (1976), No. 3, pp. 142—143.
18. Vlasyuk B. Kh. and Sharamkin V. I.: Numerical study of heat and mass transfer in an electrovortex flow in longitudinal magnetic field. I. *Magnitnaya Gidrodinamika* (1986), No. 3, pp. 78—84.
19. Vlasyuk B. Kh. and Sharamkin V. I.: Numerical study of heat and mass transfer in an electrovortex flow in longitudinal magnetic field. II. *Magnitnaya Gidrodinamika* (1987), No. 1, pp. 86—90.
20. Vlasyuk V. Kh. and Sharamkin V. I.: Nonstationary mass transfer in a cylindrical volume with electrovortex flow in longitudinal magnetic field. *Magnitnaya Gidrodinamika* (1987), No. 3, pp. 97—100.

Chapter 8

1. Abricka M. Yu., Mikelsons A. Eh., and Moshnyaga V. N. *et al.*: Effect of constant magnetic field on the liquid metal motion in vacuum arc melting. *Magnitnaya Gidrodinamika* (1979), No. 3, pp. 105—110.
2. Andrienko S. Yu., Chaikovskij A. I., and Chudnovskij A. Yu.: An investigation of electrically induced flows in the slag bath of electroslag remelting. 10th Riga Conference on MHD. Salaspils, 1981, vol. 3, pp. 39—40 (In Russian).
3. Artsimovich L. A.: *Elementary Plasma Physics*. Moscow: Atomizdat, 1966 (In Russian).
4. Aref'ev A. V., Bucenieks I. E., and Levina M. Ya. *et al.*: Heat and mass transfer intensification in induction channel furnaces. *Preprint*. Inst. of Physics, Latvian SSR Academy of Sciences. Salaspils, 1981 (In Russian).
5. Baimakov Yu. V. and Vetyukov M. M.: *Electrolysis of Fused Salts*. Moscow: Metallurgiya, 1966 (In Russian).
6. Barannikov V. A. and Khripchenko S. Yu.: The dynamical processes causing the transit flow in MHD channel carrying electric current. *Magnitnaya Gidrodinamika* (1981), No. 1, pp. 132—135.

7. Barannikov V. A. and Khripchenko S. Yu.: Electrovortex flow in a flat closed channel. *Magnitnaya Gidrodinamika* (1981), No. 2, pp. 137—139.
8. Barannikov V. A. and Zimin V. D.: The rest state instability of isothermal conducting fluid in a slightly diverging gap in the presence of electric current. *Magnitnaya Gidrodinamika* (1982), No. 3, p. 141.
9. Bojarevičs V.: A mathematical model of MHD-processes in an alumimium electrolysis cell. *Magnitnaya Gidrodinamika* (1987), No. 1, pp. 107—115.
10. Bojarevičs V. V., Chajkovskij A. I., Chudnovskij A. Yu., and Shcherbinin E. V.: On the force acting on a moving bridge in electromagnetic railgun launchers. *Magnitnaya Gidrodinamika* (1986), No. 2, pp. 105—115.
11. Bojarevičs V. V., Sharamkin V. I., and Shcherbinin E. V.: Effect of longitudinal magnetic field on the fluid motion in electric-arc and electroslag processes. *Magnitnaya Gidrodinamika* (1977), No. 1, pp. 115—120.
12. Bucenieks I. E., Kompan Ya. Yu., Sharamkin V. I., Shilova E. I., and Shcherbinin E. V.: Experimental study of MHD processes in electrical welding. *Magnitnaya Gidrodinamika* (1975), No. 3, pp. 143—148.
13. Bucenieks I. E., Levina M. Ya., Stolov M. Ya., Sharamkin V. I., and Shcherbinin E. V.: On liquid metal flow induced by electromagnetic forces in an induction channel furnaces. *Magnitnaya Gidrodinamika* (1977), No. 4, pp. 103—106.
14. Bucenieks I. E., Levina M. Ya., and Stolov M. Ya. *et al.*: A method of metal smelting in induction channel furnaces. U.S.A. Patent No. 4185159; W. Germany Patent No. 2655393; British Patent No. 1573433. 1978.
15. Bucenieks I. E., Levina M. Ya., and Stolov M. Ya. *et al.*: Induction channel furnace. U.S.S.R. Certificate of Authorship 664009. Otkrytiya. Izobreteniya. Promyshlennye Obraztsy. Tovarnye Znaki, 1979, No. 19, p. 149.
16. Bucenieks I. E., Levina M. Ya., and Prostyakov A. A. *et al.*: Induction channel furnace. U.S.S.R. Certificate of Authorship 723800. *Idem*, 1980, No. 11, p. 261.
17. Bucenieks I. E., Levina M. Ya., Stolov M. Ya., and Shcherbinin E. V.: Physical principles of MHD and thermal effects in induction channel furnaces. *Preprint*, Inst. of Physics, Latvian SSR Academy of Sciences. Salaspils, 1980 (In Russian).
18. Bucenieks I. E., Levina M. Ya., Stolov M. Ya., and Shcherbinin E. V.: Study of liquid metal motion in induction channel furnaces. *Magnitnaya Gidrodinamika* (1980), No. 3, pp. 123—130.
19. Bucenieks I. E., Levina M. Ya., and Stolov M. Ya. *et al.*: Three-phase induction channel furnace. U.S.S.R. Certificate Authorship 890561. Otkrytiya. Izobreteniya. Promyshlennye Obraztsy. Tovarnye Znaki, 1981, No. 46, p. 287.
20. Bucenieks I. E., Sharamkin V. I., and Shcherbinin E. V. *et al.*: Pressure modeling in electroslag welding metal bath. Avtomaticheskaya Svarka, 1977, No. 7, pp. 26—28.
21. Choudhary M.: Sc.D. Thesis. Massachusetts Inst. Technol. Cambridge, Mass., 1980.
22. Dement'ev S. B., Zhilin V. G., and Ivochkin V. P. *et al.*: Investigation of electrically induced flow patterns in a cylindrical bath with multiple electrodes. 12th Riga Conference on MHD. Salaspils, 1987, vol. 1, pp. 207—210 (In Russian).
23. Ducati A. C., Giannini G. M., and Muehlberger E.: Experimental results in high-specific-impulse-thermo-ionic acceleration. *AIAA J.* (1964), **2**(8), pp. 116—118.
24. *Electroslag Furnaces*. Ed. by Paton B. E., Medovar B. I. Kiev: Naukova Dumka, 1976 (In Russian).
25. *Electroslag Welding*. Ed. by Paton B. E. Moscow: Mashgiz, 1959 (In Russian).
26. Erokhin A. A.: *Arc-Plasma Smelting of Metals and Alloys*. Moscow: Nauka, 1975 (In Russian).
27. Erokhin A. A., Bukarov V. A., Ishchenko Yu. S., and Kublanov V. Ya.: Momentum transfer from an impulse arc to the weld. Avtomaticheskaya Svarka, 1976, No. 5, pp. 6—7.
28. Ewans J. W., Zundelevich Y., and Sharma D.: A mathematical model for prediction of currents, magnetic fields, melt velocities, melt topography and current efficiency in Hall-Hérault cells. *Metallurg. Trans.* (1981), **12B**, pp. 353—360.
29. Gelfgat Yu. M., Lielausis O. A., and Shcherbinin Eh. V.: *Liquid Metal under the Action of Electromagnetic Forces*. Riga: Zinatne, 1976 (In Russian).

References

30. Grjotheim K., Krohn C., and Malinovsky M. et al.: *Aluminium Electrolysis, The Chemistry of Hall-Hérault Processes.* Düsseldorf, 1977.
31. Hawke R. S., Brooks A. L., Fowler C. M., and Peterson D. R.: Electromagnetic railgun launchers: direct launch feasibility. *AIAA J.* (1982), **20**(7), pp. 978–985.
32. Hügel H., Kruelle G., and Peters T.: Investigation of Plasma Thrusters with Thermal and Self-Magnetic Acceleration. *AIAA J.* (1967), **5**(3).
33. John R. R., Bennett S., and Connors J. F.: Experimental performance of a high specific impulse arc jet engine. *Astronautica Acta* (1965), **11**, pp. 97–103.
34. Khripchenko S. Yu.: Experimental research of expanding channel with axial current. *Magnitnaya Gidrodinamika* (1977), No. 2, pp. 134–144.
35. Khripchenko S. Yu.: A submerged conduction MHD-pump without special-purpose current-leads to the active zone. In: Seminar on Applied Magnetohydrodynamics. Part 2, Perm', 1978, pp. 123–125 (In Russian).
36. Khripchenko S. Yu.: The application limits of low magnetic Reynolds number approximation for electrovortex flows in a flat channel placed between ferromagnetic pole pieces. *Magnitnaya Gidrodinamika* (1981), No. 4, pp. 137–139.
37. Kolesnichenko A. F.: *MHD-Installation and Processes in Technology.* Kiev: Naukova Dumka, 1980.
38. Kompan Ya. Yu., Khizhnyak K. K., and Bucenieks I. E. et al.: Metal bath depth dependence on electroslag welding regime. *Avtomaticheskaya Svarka* (1977), No. 5, pp. 17–20.
39. Kompan Ya. Yu., Petrov A. N., Sharamkin V. I., and Shcherbinin E. V.: Some specific features of electroslag welding in the axial-radial magnetic field. *Idem* (1978), No. 9, pp. 39–43.
40. Lasis U. A., Moshnyaga V. N., and Sharamkin V. I.: Effect of current feed scheme on MHD flow in vacuum arc melting furnace liquid bath. *Magnitnaya Gidrodinamika* (1980), No. 2, pp. 127–130.
41. Launder B. E. and Spalding D. B.: Computer Methods. *Appl. Mech. and Eng.* (1974), **3**, pp. 269–289.
42. Lympany S. D. and Ewans J. W.: The Hall-Hérault cell: some design alternatives examined by a mathematical model. *Metallurg. Trans.* (1983), **14B**, pp. 63–70.
43. Maecker H.: Plasmaströmungen in Lichtbögen infolde eigenmagnetischer Kompression. *Ztschr. Phys.* (1955), **141**(1), S. 198–216.
44. Medovar Yu. G., Emel'yanenko Yu. G., Shcherbinin E. V., Choudhary M., and Szekeley J.: Comparison of the physical and mathematical modeling of velocity field in an electroslag remelting slag bath. *Problemy Spetsial'noj Elektrometallurgii* (1982), **17**, pp. 9–15.
45. Mestel A. J.: On the flow in a channel induction furnace. *J. Fluid Mech.* (1984), **147**, pp. 431–447.
46. Moreau R. and Ewans J. W.: An analysis of the hydrodynamics of aluminium reduction cells. *J. Electrochem. Soc.* (1984), **131**(10), pp. 2251–2259.
47. Moros A., Hunt J. C. R., and Lillicrap D. C.: Study of electromagnetic features in channel induction furnaces. 4th Beer-Sheva Seminar on MHD Flows and Turbulence. Beer-Sheva, 1984, pp. 706–715.
48. Morozov A. I.: *Physical Principles of Electrical Space Propulsion Engines.* Moscow: Atomizdat, 1978 (In Russian).
49. Moshnyaga V. N. and Sharamkin V. I.: Experimental study of electrovortex flow in the cylindrical volume. *Magnitnaya Gidrodinamika* (1980), No. 4, pp. 77–80.
50. Okorokov G. N., Boyarshinov V. Ya., and Shamil' Yu. P. et al.: Improvement of macrostructure of X15 stell smelted in vacuum-arc furnace. *Stal'* (1963), No. 1, pp. 30–34 (In Russian).
51. Oreper G. M. and Szekely J.: Heat and fluid-flow phenomena in weld pools. *J. Fluid Mech.* (1984), **147**, pp. 53–79.
52. Patent 3363044 (USA). Induction channel furnace. (I. Pectius, B. Fredricson). Publ. 9.01.68.
53. Patent 1281377 (England). An induction furnace. (S. Granstrom, I. Goranson). Publ. 12.07.72.
54. Paton B. E., Medovar B. I., and Emel'yanenko Yu. G. et al.: Study of magnetic hydrodynamic phenomena in slag pool during ESR. 8th Japan-USSR Joint Symposium on Physical Chemistry of Metallurgical Processes. 1981, Tokyo, Japan, pp. 241–253.

55. Patrick R. M. and Powers W. E.: Plasma flow in a magnetic arc nozzle. Proc. 3rd Symp. Adv. Propulsion Concepts. New York-London, 1963, vol. 1, pp. 115—136.
56. *Plasma Processes in Metallurgy and Technology of Inorganic Compounds*. Moscow: Nauka, 1973 (In Russian).
57. Polishchuk V. P. and Cin M. R.: Induction channel furnace. U.S.S.R. Certificate Authorship 288183. Otkrytiya. Izobreteniya. Promyshlennye Obraztsy. Tovarnye Znaki, 1970, No. 36, p. 91.
58. Schlichting H.: *Boundary Layer Theory*. New York: McGraw-Hill, 1960.
59. Sharamkin V. I.: Experimental investigation of longitudinal magnetic field influence on the pressure for some electrical welding processes. *Magnitnaya Gidrodinamika* (1977), No. 2, pp. 139—141.
60. Sharamkin V. I.: MHD flow caused by electrical current between two parallel electrodes. *Magnitnaya Gidrodinamika* (1977), No. 4, pp. 121—125.
61. Sharamkin V. I. and Chernov Yu. V.: Character of flows in a liquid metal bath of vacuum arc remelting. *Stal'* (1977), No. 8, pp. 713—714.
62. Shcherbinin E. V.: *Jet Flows of Viscous Fluid in Magnetic Field*. Riga: Zinatne, 1973 (In Russian).
63. Shcherbinin E. V.: Jet flows in an electric arc. *Magnitnaya Gidrodinamika* (1973), No. 4, pp. 66—72.
64. Sneyd A. D.: Stability of fluid layers carrying a normal electric current. *J. Fluid Mech.* (1985), **156**, pp. 223—236.
65. Szekely J., McKelliget J., and Choudhary M.: Heat-transfer fluid flow bath circulation in electric-arc furnaces and dc plasma furnaces. *Ironmaking and Steelmaking* (1983), **10**(4), pp. 169—179.
66. Tarapore E. D.: The effect of some operating variables on flow in aluminium reduction cells. *J. of Metals* (1982), No. 2, pp. 50—55.
67. Ungurs I. A. and Shilova E. I.: Electric arc in the magnetic field of a solenoid. *Magnitnaya Gidrodinamika* (1982), No. 1, pp. 141—143.
68. Wienecke R.: Über das Geschwindigkeitsfeld der Hochstromkohlebogensäule. *Ztschr. Phys.* (1955), **143**(1), S. 129—140.
69. Zhilin V. G.: *Fiber Optical Velocity and Pressure Measurement Transducer*. Moscow: Energoatomizdat, 1987 (In Russian).
70. Zhilin V. G., Ogorodnikov V. P., Osipov V. V., and Petukhov B. S.: Optical-mechanical anemometer. *Teplofizika Vysokikh Temperatur* (1976), **14**(4), pp. 834—840.
71. Zhilin V. G., Ivochkin Yu. P., and Oksman A. A. *et al.*: An experimental investigation of the axisymmetric electrovortex flow velocity field in a cylindrical container. *Magnitnaya Gidrodinamika* (1986), No. 3, pp. 110—116.
72. Zimin V. D., Troshin V. M., and Khripchenko S. Yu.: MHD-pump with flat channel and abrupt expansion. 10th Riga Conference on MHD. Riga, 1978, vol. 1, pp. 76—77.
73. Zimin V. D. and Khripchenko S. Yu.: Two-dimensional representation of MHD equations for flows in flat channels with ferromagnetic pole pieces. *Magnitnaya Gidrodinamika* (1979), No. 4, pp. 117—127.

Index

Alfvèn number 283
angular momentum 149
Archimede's law, generalization 248
asymptotic solution 128, 261
azimuthal motion 30

batchelor number 67, 73, 105
Bessel functions 40, 220
Biot-Savart law 2, 242, 360
boundary layer 78, 129, 172
Boussinesq approximation 282

circulation 72, 85, 122, 234
charge conservation 19
corrugated tube 208
curl of velocity (vorticity) 33

Dean's approximation 242
dielectrophoresis 15
differential rotation 123, 133, 188, 295
diffusive boundary layer 288
drag coefficient 243
drag reduction 240, 243, 264

earth's magnetic field 137
electric current point source (electrode) 6, 84
electric current stream function 34
electric field potential 1, 20, 41, 196
electrical arc 12, 91, 341
electrical jet thrusters 14, 341
electrodynamic approximation 59, 71
electromagnetic accelerator 14
electromagnetic rail launcher 336
energy of lightning discharge 178
energy of rotating motion 151, 189
equation of mass transport 283
exact solution 62

filling coefficient of crystallizer 324

flow detachment 172
flow rate per unit length 73, 80, 170
frequency 329, 350
full pressure function 67

Galerkin method 139, 264
Gegenbauer function 37, 108
Grashoff number 283

Hartman number 73

induced electric current 55, 104, 168, 185
induction unit 346
irrotational flow 32

Jeffrey-Hamel flow 80

Karman's class of exact solutions 160, 162

lame coefficients 22, 23, 26
Landau-Squire-Yatseyev's solution class 74, 162
Landau problem 76, 101
Legendre polynomials 38
line source 73, 169
liquation 335
low magnetic Reynolds number approximation 58, 67, 104

magnetic field induction 2, 34
magnetic stream function 55
Maxwell stress 20, 199, 200
meridional motion 30
momentum flux 72, 76, 99, 150, 199, 202, 252

natural coordinates 36, 202
negative viscosity 239
Nusselt number 284

Ohm's law 19

parameter of electrically induced flow 73
parameter of electrically induced flow, critical
  96, 101, 104, 110, 122, 173
Peclet number 283
physical properties of medium 18, 286
pinch effect 4, 8
point electrode 5, 107
Poiseuille flow 237, 243
potential force 3, 245
Prandtl number 283

Reynolds number 79, 82, 90, 242, 283, 361
ring electrode 209, 236
rotational force 3, 121, 249

saddle point 217
similarity solution 63, 155
skin-effect 329, 332
skin-layer depth 329

spin-up of converging flow 136, 148
Stokes approximation 87, 204, 240, 268, 272
Stokes stream function 31, 33
strain tensor 20
stream line 30
stream line of electric current 34
stress tensor 20

thermal convection 297, 347
thermoelectric effect 17

vector potential 33
viscous friction 121, 123, 150, 215
volume flux 31, 72
vortex line 73, 82, 85
vortex, rotating 143, 146, 148
vortex, toroidal 99, 144, 211
vorticity 121, 123, 133, 142, 149

wave number 209
Whitehead paradox 259, 280

MECHANICS OF FLUIDS AND TRANSPORT PROCESSES
Editors: R. J. Moreau and G. Æ. Oravas

1. J. Happel and H. Brenner, Low Reynolds Number Hydrodynamics. 1983.
   ISBN 90-247-2877-0
2. S. Zahorski, Mechanics of Viscoelastic Fluids. 1982.
   ISBN 90-247-2687-5
3. J. A. Sparenberg, Elements of Hydrodynamic Propulsion. 1984.
   ISBN 90-247-2871-1
4. B. K. Shivamoggi, Theoretical Fluid Dynamics. 1984.
   ISBN 90-247-2999-8
5. R. Timman, A. J. Hermans and G. C. Hsiao, Water Waves and Ship Hydrodynamics: An Introduction. 1985.
   ISBN 90-247-3218-2
6. M. Lesieur, Turbulence in Fluids. 1987.
   ISBN 90-247-3470-3
7. L. A. Lliboutry, Very Slow Flows of Solids. 1987.
   ISBN 90-247-3482-7
8. B. K. Shivamoggi, Introduction to Nonlinear Fluid-Plasma Waves. 1988.
   ISBN 90-247-3662-5